Illuminating and Advancing the Path for Mathematical Writing Research

Madelyn W. Colonnese
Reading and Elementary Education Department, Cato College of Education, University of North Carolina at Charlotte, USA

Tutita M. Casa
Department of Curriculum and Instruction, Neag School of Education, University of Connecticut, USA

Fabiana Cardetti
Department of Mathematics, College of Liberal Arts and Sciences, University of Connecticut, USA

A volume in the Advances in Library and Information Science (ALIS) Book Series

IGI Global
PUBLISHER of TIMELY KNOWLEDGE

Published in the United States of America by
IGI Global
Information Science Reference (an imprint of IGI Global)
701 E. Chocolate Avenue
Hershey PA, USA 17033
Tel: 717-533-8845
Fax: 717-533-8661
E-mail: cust@igi-global.com
Web site: http://www.igi-global.com

Library of Congress Cataloging-in-Publication Data

Names: Colonnese, Madelyn W., 1987- editor. | Casa, Tutita Maria, editor. | Cardetti, Fabiana, editor.
Title: Illuminating and advancing the path for mathematical writing research / edited by Madelyn W. Colonnese, Tutita M. Casa, and Fabiana Cardetti.
Description: Hershey, PA : Information Science Reference, [2024] | Includes bibliographical references and index. | Summary: "This book will share strategies to support practitioners and teacher educators with the implementation of mathematical writing. We envision three major sections with each chapter highlighting the existing research and identifying and prioritizing future research by: 1. Realizing the significance of mathematical writing 2. Engaging k-12 students in mathematical writing 3. Supporting educators with the implementation of mathematical writing Ultimately, k-12 students will benefit from a more informed field because they will have access to a vital mode of mathematical reasoning and communication"-- Provided by publisher.
Identifiers: LCCN 2023026429 (print) | LCCN 2023026430 (ebook) | ISBN 9781668465387 (hardcover) | ISBN 9781668465394 (paperback) | ISBN 9781668465400 (ebook)
Subjects: LCSH: Mathematics--Authorship--Study and teaching. | Communication in mathematics--Study and teaching.
Classification: LCC QA20.M38 .I44 2023 (print) | LCC QA20.M38 (ebook) | DDC 510.71--dc23/eng/20230815
LC record available at https://lccn.loc.gov/2023026429
LC ebook record available at https://lccn.loc.gov/2023026430

This book is published in the IGI Global book series Advances in Library and Information Science (ALIS) (ISSN: 2326-4136; eISSN: 2326-4144)

British Cataloguing in Publication Data
A Cataloguing in Publication record for this book is available from the British Library.

For electronic access to this publication, please contact: eresources@igi-global.com.

Advances in Library and Information Science (ALIS) Book Series

Alfonso Ippolito
Sapienza University-Rome, Italy
Carlo Inglese
Sapienza University-Rome, Italy

ISSN:2326-4136
EISSN:2326-4144

Mission

The **Advances in Library and Information Science (ALIS) Book Series** is comprised of high quality, research-oriented publications on the continuing developments and trends affecting the public, school, and academic fields, as well as specialized libraries and librarians globally. These discussions on professional and organizational considerations in library and information resource development and management assist in showcasing the latest methodologies and tools in the field.

The **ALIS Book Series** aims to expand the body of library science literature by covering a wide range of topics affecting the profession and field at large. The series also seeks to provide readers with an essential resource for uncovering the latest research in library and information science management, development, and technologies.

Coverage

- Discovery Tools
- Libraries as Community Centers
- Alternatives to Dewey Decimal System
- International Library Statistics
- Journal Collections
- Discoverability of E-Content in Libraries
- Storage Facilities
- E-Books in Libraries
- K-12 Librarianship
- Effective Weeding Practices

IGI Global is currently accepting manuscripts for publication within this series. To submit a proposal for a volume in this series, please contact our Acquisition Editors at Acquisitions@igi-global.com or visit: http://www.igi-global.com/publish/.

The Advances in Library and Information Science (ALIS) Book Series (ISSN 2326-4136) is published by IGI Global, 701 E. Chocolate Avenue, Hershey, PA 17033-1240, USA, www.igi-global.com. This series is composed of titles available for purchase individually; each title is edited to be contextually exclusive from any other title within the series. For pricing and ordering information please visit http://www.igi-global.com/book-series/advances-library-information-science/73002. Postmaster: Send all address changes to above address.

Titles in this Series

For a list of additional titles in this series, please visit: www.igi-global.com/book-series

Emerging Technology-Based Services and Systems in Libraries, Educational Institutions, and Non-Profit Organizations
Dickson K. W. Chiu (The University of Hong Kong, Hong Kong) and Kevin K. W. Ho (University of Tsukuba, Japan)
Information Science Reference • copyright 2023 • 353pp • H/C (ISBN: 9781668486719) • US $225.00 (our price)

Applying Positivist and Interpretivist Philosophies to Social Research Practices
C.C. Jayasundara (University of Kelaniya, Sri Lanka)
Information Science Reference • copyright 2023 • 300pp • H/C (ISBN: 9781668454312) • US $215.00 (our price)

Handbook of Research on Advancements of Contactless Technology and Service Innovation in Library and Information Science
Barbara Holland (Brooklyn Public Library, USA (retired))
Information Science Reference • copyright 2023 • 388pp • H/C (ISBN: 9781668476932) • US $270.00 (our price)

Perspectives on Justice, Equity, Diversity, and Inclusion in Libraries
Nandita S. Mani (University of Massachusetts, Amherst, USA) Michelle A. Cawley (University of North Carolina at Chapel Hill, USA) and Emily P. Jones (University of North Carolina at Chapel Hill, USA)
Information Science Reference • copyright 2023 • 320pp • H/C (ISBN: 9781668472552) • US $215.00 (our price)

Handbook of Research on Technological Advances of Library and Information Science in Industry 5.0
Barbara Jane Holland (Independent Researcher, USA)
Information Science Reference • copyright 2023 • 549pp • H/C (ISBN: 9781668447550) • US $270.00 (our price)

Global Perspectives on Sustainable Library Practices
Victoria Okojie (University of Abuja, Nigeria) and Magnus Osahon Igbinovia (Ambrose Alli University, Nigeria)
Information Science Reference • copyright 2023 • 376pp • H/C (ISBN: 9781668459645) • US $240.00 (our price)

Cases on Establishing Effective Collaborations in Academic Libraries
Mary E. Piorun (University of Massachusetts Chan Medical School, USA) and Regina Fisher Raboin (University of Massachusetts Chan Medical School, USA)
Information Science Reference • copyright 2023 • 358pp • H/C (ISBN: 9781668425152) • US $215.00 (our price)

Handbook of Research on Academic Libraries as Partners in Data Science Ecosystems
Nandita S. Mani (University of North Carolina at Chapel Hill, USA) and Michelle A. Cawley (University of North Carolina at Chapel Hill, USA)

IGI Global
PUBLISHER of TIMELY KNOWLEDGE

701 East Chocolate Avenue, Hershey, PA 17033, USA
Tel: 717-533-8845 x100 • Fax: 717-533-8661
E-Mail: cust@igi-global.com • www.igi-global.com

Table of Contents

Foreword xv

Preface xvii

Acknowledgment xx

Section 1
An Introduction

Chapter 1
Framing Our Understanding of Mathematical Writing 1
Tutita M. Casa, Department of Curriculum and Instruction, Neag School of Education, University of Connecticut, USA
Fabiana Cardetti, Department of Mathematics, College of Liberal Arts and Sciences, University of Connecticut, USA
Madelyn W. Colonnese, Reading and Elementary Education Department, Cato College of Education, University of North Carolina at Charlotte, USA

Section 2
Engaging Teachers in Mathematical Writing

Chapter 2
Pre-Service Teachers and Mathematics Writing 16
R. Alex Smith, University of Nevada, Las Vegas, USA
Madeline Price, University of Nevada, Las Vegas, USA
Tessa L. Arsenault, The University of Texas at Austin, USA
Sarah R. Powell, The University of Texas at Austin, USA
Erin Smith, University of Nevada, Las Vegas, USA
Michael Hebert, University California, Irvine, USA

Chapter 3
Explanatory Writing: A Window Into K-8 Prospective Teachers' Reasoning About Contextual Problems 35
Marta T. Magiera, Marquette University, USA
Mohammad Al-younes, Marquette University, USA

Chapter 4
Enhancing Preservice Teachers' Mathematical Knowledge Through Writing: From Peer Review to Inter-Institutional Conversations ... 62
Christopher Dennis Nazelli, Wayne State University, USA
S. Asli Özgün-Koca, Wayne State University, USA
Deborah Zopf, Henry Ford College, USA

Chapter 5
Dialogically Focused Secondary Mathematics Teachers' Orientations Toward Mathematical Writing: A Multiple-Case Study ... 93
Ethan P. Smith, Washington State University Tri-Cities, USA

Chapter 6
Mathematical Writing, Teacher Noticing, and Text Clubs ... 115
James P. Bywater, James Madison University, USA
Sarah Lilly, University of Virginia, USA
Jennifer L. Chiu, University of Virginia, USA

Chapter 7
Assessing Mathematical Writing: Comparative Judgment and Professional Learning ... 135
Ian Jones, Loughborough University, UK
Jodie Hunter, Massey University, New Zealand

Chapter 8
A Genre-Based Approach to Improving Students' Mathematical Writing: Practical Support for Educators ... 155
Lauren Haggis, South Australia Department for Education, Australia
Tutita M. Casa, Department of Curriculum and Instruction, Neag School of Education, University of Connecticut, USA

Section 3
Engaging Students in Mathematical Writing

Chapter 9
Supporting Equity Through Mathematical Writing ... 174
Christie Sullivan Martin, University of South Carolina, USA
Drew Polly, University of North Carolina at Charlotte, USA

Chapter 10
A Mathematics Intervention to Engage Elementary Students in Mathematical Writing ... 193
Madelyn W. Colonnese, Reading and Elementary Education Department, Cato College of Education, University of North Carolina at Charlotte, USA
Fabiana Cardetti, Department of Mathematics, College of Liberal Arts and Sciences, University of Connecticut, USA

Chapter 11
A Framework for Assessing Students' Written Explanations of Numerical Reasoning 209
Sarah Quebec Fuentes, Texas Christian University, USA
Lindy Crawford, Lee Pesky Learning Center, USA
Ashley Rowe, Lee Pesky Learning Center, USA
Jacqueline Huscroft-D'Angelo, Oregon Research Institute, USA

Chapter 12
What to Assess for Mathematical Writing: The Dimensionality of Written Expression for Mathematical Problem Solving and Reasoning 238
Joo-Young Lee, The University of Maine, USA
Elizabeth M. Hughes, The Pennsylvania State University, USA
Paul J. Riccomini, The Pennsylvania State University, USA

Section 4
Engaging All Learners in Mathematical Writing

Chapter 13
Academic Literacy in Mathematics to Frame Mathematical Writing Research and Practice: Writing and Revising for Mathematical Practices 264
Judit N. Moschkovich, University of California, Santa Cruz, USA

Chapter 14
Mathematical Writing and English Language Learners: Strengths-Based Instructional Strategies ... 291
Salvador Huitzilopochtli, University of California, Santa Cruz, USA

Chapter 15
Linking Practices and Standards to Strategies that Enhance Writing for Students With Disabilities 316
Destini N. Braxton, St. Mary's College of Maryland, USA
Karen L. Terrell, Loyola University Maryland, USA

Chapter 16
Constructing Written Arguments for Developing Mathematical Reasoning for Students With Learning Difficulties in the Elementary Grades 328
Sharlene A. Kiuhara, University of Utah, USA

Compilation of References 343

About the Contributors 381

Index 386

Detailed Table of Contents

Foreword xv

Preface xvii

Acknowledgment xx

Section 1
An Introduction

Chapter 1
Framing Our Understanding of Mathematical Writing 1
Tutita M. Casa, Department of Curriculum and Instruction, Neag School of Education, University of Connecticut, USA
Fabiana Cardetti, Department of Mathematics, College of Liberal Arts and Sciences, University of Connecticut, USA
Madelyn W. Colonnese, Reading and Elementary Education Department, Cato College of Education, University of North Carolina at Charlotte, USA

In this chapter, the authors situate mathematical writing within the context of discourse, define mathematical writing, describe the classroom environment that fosters mathematical writing, and share experiences engaging elementary students in mathematical writing. At their core, the types of mathematical writing are defined by the purposes for which students write in the mathematics classroom. The authors further describe the types of mathematical writing teachers prompt and are nurtured by the classroom environment. The insights shared draw from the authors' previous and current research endeavors, advancements in curriculum development, collaborations with mathematics coaches, and experiences teaching elementary students and supporting preservice and inservice teachers. They end the chapter by highlighting some directions for future research to expand our understanding of this crucial area in mathematics education.

Section 2
Engaging Teachers in Mathematical Writing

Chapter 2
Pre-Service Teachers and Mathematics Writing 16
R. Alex Smith, University of Nevada, Las Vegas, USA
Madeline Price, University of Nevada, Las Vegas, USA
Tessa L. Arsenault, The University of Texas at Austin, USA
Sarah R. Powell, The University of Texas at Austin, USA
Erin Smith, University of Nevada, Las Vegas, USA
Michael Hebert, University California, Irvine, USA

This chapter focuses upon research related to building pre-service teachers' (PSTs') own competencies in mathematics writing (MW) as well as their skills in teaching MW and integrating MW into their future K-12 classrooms. The authors provide a review of the research with PSTs and MW. Although there is limited research on PSTs and MW, it is likely that many PSTs enter their preparatory programs with limited MW experience and competencies despite expressing beliefs in the value of MW. Therefore, there is a responsibility of mathematics teacher educators (MTEs) to build PSTs' knowledge of the various genres of MW and their purposes, to explicitly teach PSTs to compose high-quality MW themselves, and to provide frequent models of how MW may be integrated into the mathematics classroom across a range of genres. The authors conclude with suggestions for current practice and future research.

Chapter 3
Explanatory Writing: A Window Into K-8 Prospective Teachers' Reasoning About Contextual Problems 35
Marta T. Magiera, Marquette University, USA
Mohammad Al-younes, Marquette University, USA

This chapter contributes to broader research on mathematical discourse and its multimodal forms that reveal how students interpret, make meanings in mathematics, and communicate their understandings. Central to this chapter is explanatory writing in mathematics, which targets a specified audience to explain and transmit mathematical ideas. In the context of a mathematics problem-solving course and focusing on contextual problem-solving situations, this chapter illustrates how explanatory writing in mathematics offers a window into prospective elementary school teachers' (PTs') reasoning and communication skills. The chapter highlights PTs' strategic reasoning and sense-making and describes how they communicate their problem solutions. It provides mathematics teacher educators with insights into how PTs think about contextual problem situations, offering directions for planning how to help PTs develop new ways of thinking. This chapter also includes suggestions for future research on PTs' learning to construct written mathematical explanations.

Chapter 4
Enhancing Preservice Teachers' Mathematical Knowledge Through Writing: From Peer Review to Inter-Institutional Conversations 62
Christopher Dennis Nazelli, Wayne State University, USA
S. Asli Özgün-Koca, Wayne State University, USA
Deborah Zopf, Henry Ford College, USA

In this chapter, the authors present and analyze a mathematical writing activity for preservice elementary school teachers with a novel peer review component. The products of this activity, referred to as "journals", are detailed representation-rich explanations of problems solved in small groups. These journals were exchanged across institutions, and the addition of a computer-mediated communication platform transformed the peer reviews into inter-institutional conversations. An analysis of the journals and the conversations they sparked reveal four opportunities to develop mathematical knowledge for teaching. Evidence, in the form of journal and review excerpts as well as two case studies, suggests that the process enabled preservice teachers to deepen their understanding of foundational concepts from the elementary curriculum, to communicate mathematics more effectively to others, to better make sense of the mathematical thinking and writing of others, and to incorporate their colleagues' suggestions into their future writing. Potential future research paths are discussed.

Chapter 5
Dialogically Focused Secondary Mathematics Teachers' Orientations Toward Mathematical Writing: A Multiple-Case Study 93
Ethan P. Smith, Washington State University Tri-Cities, USA

Because mathematics classrooms entail expectations for students to communicate understanding through both writing and spoken discourse, it is important to consider the place of writing in the context of such discourse. This chapter details a case study centered around three dialogically focused high school mathematics teachers and their orientations toward mathematical writing. Teachers in this study were first asked to describe their orientations toward mathematical writing. Then, they were asked to reflect on samples of student writing and recordings from their classrooms to further understand how they orient toward students' mathematical writing. Findings suggested that these teachers' orientations toward mathematical writing are embedded in broader views that they hold toward discourse, feedback, and student and teacher self-efficacy around writing. Findings also indicated the value of using the construct of teacher noticing to help investigate and strengthen teachers' orientations toward mathematical writing.

Chapter 6
Mathematical Writing, Teacher Noticing, and Text Clubs 115
James P. Bywater, James Madison University, USA
Sarah Lilly, University of Virginia, USA
Jennifer L. Chiu, University of Virginia, USA

This chapter explores how using explanatory mathematical student writing artifacts can be used to position teachers to attend to, interpret, and respond to student mathematical thinking from an iterative and engaged stance. This chapter argues that as conceptualizations of teacher noticing shift from a view that is linear and detached to a view that is iterative and engaged, a corresponding shift should occur in the learning to notice activities that pre-service teachers engage in. The chapter proposes the idea of text clubs, and explores, with two illustrative examples, how pre-service teacher discussions that center explanatory mathematical student writing artifacts offer a potential alternative to the predominant role of video clubs that center teaching video artifacts, and how this approach may support the advancement of new paths for mathematical writing research.

Chapter 7
Assessing Mathematical Writing: Comparative Judgment and Professional Learning 135
Ian Jones, Loughborough University, UK
Jodie Hunter, Massey University, New Zealand

The chapter discusses potential professional learning benefits for educators who engage in assessing students' mathematical writing. It draws on interview data from twelve mathematics educators who were experienced in assessing primary students' written responses to free response prompts covering a range of topics. The first stage of the interviews used a stimulated recall protocol that followed a comparative judgment procedure in which each participant was presented with pairs of students' written responses and asked to decide which was 'better'. The second stage was semi-structured with questions about how participants made their comparative judgment decisions, and whether doing so improved their understanding of students' thinking. The findings are that assessing mathematical writing can provide educators with insights into students' representations, underlying ideas and learning trajectories, and can also provide stimulus for changing classroom practice.

Chapter 8
A Genre-Based Approach to Improving Students' Mathematical Writing: Practical Support for Educators 155
Lauren Haggis, South Australia Department for Education, Australia
Tutita M. Casa, Department of Curriculum and Instruction, Neag School of Education, University of Connecticut, USA

This chapter introduces the resources and practical guides that the South Australian Department for Education has developed to support teachers to teach writing in mathematics classrooms and develop their understanding of the types of writing mathematics students should be proficient in. Developing this understanding has the potential to improve classroom practice and positively impact learning outcomes for students in both writing and mathematics. This chapter may offer insights for other jurisdictions seeking to support the teaching of mathematical writing.

Section 3
Engaging Students in Mathematical Writing

Chapter 9
Supporting Equity Through Mathematical Writing 174
Christie Sullivan Martin, University of South Carolina, USA
Drew Polly, University of North Carolina at Charlotte, USA

This chapter establishes the relationship between the National Council of Teachers of Mathematics NCTM principles to action, Gutiérrez equity framework, and mathematical writing. The chapter discusses the research that supports mathematical writing as an instructional practice that enhances opportunities for reasoning learning and equity. The chapter provides four cases of activities that embed mathematical writing in the classroom. Each case describes implementation, examples of students' work, discussion of their work, extensions, and outlines the activities connection to the equity framework. The cases are concluded with takeaways that reinforce the purpose of each activity and strategies for implementation. The chapter closes with implications to further support the use of mathematical writing in the classroom and future research direction.

Chapter 10
A Mathematics Intervention to Engage Elementary Students in Mathematical Writing.................... 193
Madelyn W. Colonnese, Reading and Elementary Education Department, Cato College of Education, University of North Carolina at Charlotte, USA
Fabiana Cardetti, Department of Mathematics, College of Liberal Arts and Sciences, University of Connecticut, USA

Mathematical writing is an important yet underutilized mode of discourse. Teachers play an important role in implementing and engaging their students' mathematical writing. However, research on mathematical writing does not offer suggestions for how teachers can engage students in mathematical writing. The purpose of this chapter is to contribute to this area of research by describing how teachers implemented a systematic mathematical writing intervention during a summer school program. The student writing resulting from the intervention demonstrated that students can successfully engage in mathematical writing. Further, students' written work showed key elements of mathematical writing and indicated an understanding of the subconstructs of fraction conceptualization that appeared to be absent prior to the intervention.

Chapter 11
A Framework for Assessing Students' Written Explanations of Numerical Reasoning..................... 209
Sarah Quebec Fuentes, Texas Christian University, USA
Lindy Crawford, Lee Pesky Learning Center, USA
Ashley Rowe, Lee Pesky Learning Center, USA
Jacqueline Huscroft-D'Angelo, Oregon Research Institute, USA

Disciplinary writing in mathematics supports the use of words, symbols, and visual representations, allowing one to communicate more fully in writing than through speech alone. This chapter explores the disciplinary writing of 394 fourth-grade students who shared their numerical reasoning in written explanations to seven whole-number and fraction comparison tasks. Data were collected via whole group administration procedures and students' explanations were scored using a validated framework for evaluating numerical reasoning. Results include descriptive statistics and qualitative analyses of student responses using the framework's five categories. A difference was found between the types of reasoning students shared for whole-number tasks versus those shared for fraction comparison tasks, favoring the incorporation of more conceptual reasoning for whole-number tasks. The framework is a practical and effective tool that teachers can use to examine the depth of student reasoning and disciplinary writing to document such reasoning.

Chapter 12
What to Assess for Mathematical Writing: The Dimensionality of Written Expression for Mathematical Problem Solving and Reasoning.. 238
Joo-Young Lee, The University of Maine, USA
Elizabeth M. Hughes, The Pennsylvania State University, USA
Paul J. Riccomini, The Pennsylvania State University, USA

This chapter investigates the internal structure of informative/explanatory mathematical writing (MW) as a latent variable. The study defines it as written communication in mathematics to explain and describe one's problem solving and reasoning by using mathematical language and representation. To assess the dimensions of MW, ten scales were developed based on literature and expert reviews to quantify students' MW performance. The dataset utilized for statistical analyses consisted of MW samples from 153 4th-grade students attending a public school in a Mid-Atlantic state. Through exploratory factor analysis (EFA) conducted on the three MW prompts, two factors emerged: mathematics problem solving and reasoning and writing. Subsequently, a confirmatory factor analysis (CFA) was conducted, resulting in a cross-factor loading of reasoning onto both factors. The role of representations in MW remains uncertain. This finding contributes to an understanding of the underlying dimensions, as well as offers a validated approach for assessing MW.

Section 4
Engaging All Learners in Mathematical Writing

Chapter 13
Academic Literacy in Mathematics to Frame Mathematical Writing Research and Practice: Writing and Revising for Mathematical Practices 264
Judit N. Moschkovich, University of California, Santa Cruz, USA

Oral student contributions are important for learning mathematics, but without a written record of students' reasoning, revising is difficult or impossible. While writing makes revising possible, research needs to theoretically frame what to revise, how to revise, and how to support revisions. The chapter addresses the question: How can the academic literacy in mathematics framework inform research and practice on mathematical writing? The examples illustrate how that framework provides a complex view of language and a focus on mathematical practices to frame research and practice on mathematical writing and revising. The chapter considers why student writing and revising are important and describes how both research and instruction can emphasize mathematical practices. The examples focus on explanatory and argumentative writing and several mathematical practices central to revising explanations or arguments: generating conjectures, connecting representations, justifying claims, constructing arguments, and attending to precision.

Chapter 14
Mathematical Writing and English Language Learners: Strengths-Based Instructional Strategies ... 291
Salvador Huitzilopochtli, University of California, Santa Cruz, USA

The purpose of this chapter is to examine the role of mathematical writing to enhance mathematics instruction for students bureaucratically labeled as English Language Learners (ELLs) and multilingual students. The chapter addresses the specific needs of these students and focuses on two major claims: first, mathematical writing is useful for ELLs learning mathematics, and second, mathematics instruction for ELLs should support them to generate mathematical claims and provide opportunities for them to revise their work. The claims are illustrated using data from a dissertation study that examined how multilingual students learned mathematical arguments in a classroom that used mathematical writing and conferencing. This chapter per the author frames mathematical arguments using proof schemes and Toulmin's model and takes a strengths-based approach to instruction. The final section suggests directions for future research and focuses on expanding beyond writing as an individual exercise.

Chapter 15
Linking Practices and Standards to Strategies that Enhance Writing for Students With Disabilities 316
Destini N. Braxton, St. Mary's College of Maryland, USA
Karen L. Terrell, Loyola University Maryland, USA

Because mathematics and language are interconnected, often students are asked to "explain" their answers or "justify" their choices. It is critical that students are able to verbally express their thoughts, as this is considered an "authentic practice of the discipline." The Standards for Mathematical Practice require that all students be able to make sense of problems, and also communicate both in writing and orally; skills which the communication standard has promoted for over two decades. Moreover, in the case of SWDs, those who are dually identified often show more confidence when asked to write their logic, as opposed to speaking. This chapter, therefore, explores strategies that will assist and allow students to exhibit their acquired knowledge through their writing skills in mathematics and enable the building of their confidence when explaining their logic and reasoning – especially those with diverse learning needs.

Chapter 16
Constructing Written Arguments for Developing Mathematical Reasoning for Students With Learning Difficulties in the Elementary Grades 328
Sharlene A. Kiuhara, University of Utah, USA

Using argument writing to enhance students' mathematics learning has been shown to increase students' content knowledge. Investigating argument writing has also been found to increase the content knowledge and quality of mathematical reasoning of students with learning challenges. This chapter examines the instructional challenges for implementing mathematics writing to students who experience learning difficulties, evidence-based practices to address these students' math and writing needs, and a closer look at how constructing written arguments can enhance fraction knowledge and genre knowledge for students with learning difficulties.

Compilation of References 343

About the Contributors 381

Index 386

Foreword

Writing has received growing recognition as an important tool for learning in the mathematics classroom. This is important because writing is tied to complex cognitive skills that are part of critical thinking. This thinking inherently includes analysis and problem solving, both critical in mathematical learning. The writing-to-learn movement has become an integral part of mathematics teaching and learning prompted by growing evidence that writing supports learning. Yet, mathematics classrooms have been slow to value writing activities.

My journey with writing in mathematics began in the eighties when I was teaching high school. I recognized that I needed an approach that would engage students and provide opportunities for developing mathematical thinking. There were few resources, so I tried multiple formats with my students to see what worked best. Over the year, I saw that students' writing related to their mathematics resulted in more engagement with the content, supported metacognitive thinking, and reinforced problem solving.

Writing in math engaged more students to reason mathematically. Initially, it was a challenge to get my students to write but with modeling and reporting out with student created exemplars, the expectations for the writing tasks became clearer for the students and their resistance to write became less and less. Students needed to use written communication to reason about the mathematics they were encountering. The writing served as a foundation for broader discourse around concepts and problems. As students communicated their mathematical ideas, I noticed that their conceptual understanding improved and they could reason about their approaches to problems and mathematical tasks. Writing provided a powerful tool for my students to think more deeply about the procedures and concepts they were applying. Requiring writing pushed them to be more engaged and to think critically about their own work and the work of their peers.

As my experiences with writing in mathematics became more seasoned, I expanded my ideas about the approach. I was noting a level of student reflection that I knew resulted from the writing process. After doing some reading and research on this observation, I came to realize that this was developing my students' metacognitive thinking. "Thinking about thinking" was the tag that resonated with me. That is, they were more aware of their own thinking as a result of writing. Students were more successful in their mathematics and demonstrated how their thinking influenced their strategies and approaches as they interpreted, organized, and strategized their 'doing mathematics.' I began to explore students modeling their thinking and sharing their writing with the goal of getting them to regularly rely on the monitoring of their mathematical thinking. As a teacher, I found this to be very powerful. The development of metacognition from writing in mathematics would later become the research focus of my dissertation.

As I explored the metacognition link, the relationship between writing and mathematical problem solving also became clearer. Writing in mathematics meant that students regularly shared their thinking about the hows and whys of the mathematics they were using. I no longer asked my students to "show all your work," but, instead, I prompted them to "explain your thinking" and "justify your process." Students became more adept at solving problems because of their writing about their thinking, their interpretation of the problem and the context, the strategies selected, and the steps involved in the process of finding a solution. Students took ownership of their work and wrote and talked often about 'my thinking' or 'my reasoning.' As a teacher, I was convinced—writing in the mathematics classroom was an approach to improved student learning.

My experiences as a teacher highlight how writing in mathematics classrooms provides an opportunity for creating meaning behind the mathematics and promoting new learning for students. However, the instructional environments of classrooms must become positive ecosystems to maximize the potential for writing in mathematics. My journey underscores why ***Illuminating and Advancing the Path for Mathematical Writing Research*** is so important. This book will provide research perspectives that develop the current state of practice for mathematical writing to inform educators as well as researchers. This comprehensive resource is greatly needed to provide guidance on the use of mathematical writing as part of an effective teaching and learning mathematics ecosystem.

David K. Pugalee
Center for STEM Education, University of North Carolina at Charlotte, USA

David Pugalee *is Professor of Education at the University of North Carolina at Charlotte where he serves as Director of the Center for Mathematics, Science, & Technology Education. He earned his Ph.D. in mathematics education from the University of North Carolina at Chapel Hill and taught at the elementary, middle, and secondary levels. With a decade of experience teaching mathematics and science, he has published research articles in Review of Educational Research, Educational Studies in Mathematics, and School Science and Mathematics. His works include several books and book chapters published by the National Council of Teachers of Mathematics. In addition, he has published three books relative to communication and mathematics: Writing to Develop Mathematical Understanding, Developing Mathematical and Scientific Literacy, and Lesson Imaging in Science and Mathematics. His research interest includes strategies for effective STEM teaching and learning.*

Preface

As editors of this comprehensive reference book, *Illuminating and Advancing the Path for Mathematical Writing Research*, we are honored to present an exploration into the multifaceted realm of mathematical writing. Our journey into the world of mathematical communication has been a collaborative effort, bringing together diverse expertise and experiences. It is our collective belief that mathematical writing, as a mode of communication, plays a pivotal role in the realm of mathematics education.

Mathematical writing is more than just a method of recording mathematical concepts and solutions. It is a conduit that grants teachers and researchers profound insight into their students' thought processes, and in turn, offers students an invaluable opportunity to deepen their understanding of mathematics, engage in mathematical reasoning, and embrace the fundamental skill of communicating mathematically. Unique in its nature, mathematical writing requires the harmonious integration of mathematical content knowledge, general writing skills, and specialized writing skills specific to the field of mathematics. It necessitates the ability to judiciously blend mathematical symbols, representations, and text, thereby bridging the gap between mathematical theory and its effective articulation.

We have chosen to organize the 16 chapters in this book into four sections to differentiate the perspectives helping to inform the research and implementation of mathematical writing. These sections highlight the current state of this area across its multifaceted aspects. The first section, "An Introduction," consists of a chapter where we lay the foundation for the conceptualization of mathematical writing within the broader context of mathematical discourse, as well as for key considerations for classroom implementation. Subsequent sections are composed of several chapters where contributing authors discuss their work that focuses on the engagement in mathematical writing of a specific population; namely, preservice and in-service teachers, the broad student population, and traditionally marginalized student populations.

The second section, "Engaging Teachers in Mathematical Writing," highlights efforts being made to support future and practicing teachers to implement mathematical writing across grade levels. Marta Magiera shares the outcomes of how prospective elementary school teachers taking a mathematics problem-solving course engaged in explanatory writing. Christopher Dennis Nazelli, S. Asli Özgün-Koca, and Deborah Zopf present results from inter-institutional conversations that positioned peers to review one another's written explanations from problems that were solved in small groups. A team comprised of R. Alex Smith, Madeline Price, Tessa Arsenault, Sarah R. Powell, Erin Smith, and Michael Hebert present a study where they utilized an asynchronous mathematical writing module to share an example of integrating mathematical writing within mathematics methods courses to help future elementary and/or special education teachers increase their understanding of mathematical writing. Ethan Smith describes a multiple-case study of three high school mathematics teachers oriented dialogically in relation to spoken discourse and its influence on their orientation to mathematical writing. James Bywater, Sarah Lilly, and

Jennifer Chiu detail how using explanatory writing artifacts during text clubs can facilitate preservice teachers' noticing and iterative analysis of elementary and high school students' mathematical thinking. Ian Jones and Jodie Hunter highlight an investigation of how elementary, middle, and high school teachers comparatively judged students' mathematical writing after participating in a series of workshops. Lauren Haggis and Tutita Casa share lessons learned from a years-long collaboration that informed Lauren's leadership in taking a genre-based approach to guiding educators in South Australia with the intention of allowing other stakeholders to implement mathematical writing across grades and schools.

The third section, "Engaging Students in Mathematical Writing," consists of chapters that present concrete ideas for classroom implementation as well as for the assessment of student-generated mathematical writing pieces. Christie Sullivan Martin and Drew Polly discuss activities that draw the connection between mathematical writing as a practice that enhances reasoning, learning, and equity. Madelyn Colonnese and Fabiana Cardetti present the implementation of mathematical writing intervention to shed light on what is possible with purposeful mediation. Sarah Quebec Fuentes, Lindy Crawford, Ashley Rowe, and Jacqueline Huscroft-D'Angelo, share the use of a framework for assessing students' explanatory writing including a range of student responses that illustrate the complexity and richness of this type of writing. Joo-Young Lee, Elizabeth M. Hughes, and Paul J. Riccomini focus on the internal structure of the dimensions of mathematical writing and offer a validated approach for assessing it.

Section 4, "Engaging All Learners in Mathematical Writing," includes chapters that address ways students from traditionally marginalized populations can participate in and benefit from mathematical writing. Judit Moschkovich makes a case for a complex view of mathematical language that acknowledges the strengths of students whose first language is not the language of instruction when composing and revising explanatory and argumentative pieces in relation to several mathematical practices. Salvador Huitzilopochtli follows with an examination of the role mathematical writing, particularly argumentation, can play in enhancing instruction for multilingual students bureaucratically labeled as English language learners (ELLs). Destini Braxton and Karen Terrell emphasize the interconnectedness of mathematics and language and present strategies to promote writing authored by middle school students with disabilities, classified as ELLs, and dually identified. Lastly, Sharlene Kiuhara examines instructional considerations to support students who experience learning difficulties in writing mathematically with a focus on how constructing written arguments can enhance their understanding of fractions and this type of mathematical writing.

Over the years, a growing body of research has revealed the positive impact of mathematical writing on students' mathematical learning and their metacognitive development. However, despite the mounting evidence, mathematical writing remains underutilized as a means of mathematical communication in educational settings. This incongruity has raised important questions for both researchers and practitioners: What types of writing should students engage in while studying mathematics? What considerations should be taken into account to facilitate the successful incorporation of mathematical writing for all students? How can teachers empower students to become proficient mathematical writers? And how can teacher educators support and equip teachers to embark on this transformative journey?

Our ultimate aspiration is that this book serves as a valuable resource for researchers, teachers, mathematics coaches, district leaders, graduate students, and curriculum developers. We hope that the knowledge and strategies shared within these pages will help to enrich mathematics education by infusing mathematical writing, fostering mathematical discourse, and expanding our collective comprehension of mathematical writing.

In doing so, we believe that the younger generation of K-12 students will reap the benefits of a more informed field, gaining access to a vital mode of mathematical reasoning and understanding that will empower them to navigate the world of mathematics with confidence.

Sincerely,

Madelyn W. Colonnese
Reading and Elementary Education Department, Cato College of Education, University of North Carolina at Charlotte, USA

Tutita M. Casa
Department of Curriculum and Instruction, Neag School of Education, University of Connecticut, USA

Fabiana Cardetti
Department of Mathematics, College of Liberal Arts and Sciences, University of Connecticut, USA

Acknowledgment

We are sincerely thankful to all the authors who devoted much time and effort crafting, drafting, and revising their contributions in an effort to advance mathematical writing research. We also thank the reviewers who generously provided feedback, including Mohammad Al-younes, Julie Bacak, James Bywater, Abigail Clifford, Michelle Cook, Kaitlyn Evans, Janine Firmender, Lauren Haggis, Elizabeth Hughes, Salvador Huitzilopochtli, Jodie Hunter, Ian Jones, Sharlene Kiuhara, Joo-Young Lee, Wendy Lewis, Marta Magiera, Christie Martin, Judit Moschkovich, Christopher Nazelli, Holly Pinter, Sarah Quebec Fuentes, Terry Rose, Ethan Smith, and Michael Urbanski. The unique perspectives they shared were invaluable in helping ensure the authors' important contributions are evident in the chapters. We appreciate their expertise and efforts.

We are grateful to the advisory board members for their trust in our vision for this work from the start and their willingness to devote time and effort to ensure the quality of the contributions. Jack Dieckmann, Karen Karp, Susie Katt, Karl Kosko, David Pugalee, and Jennifer Suh collectively brought important insights from their notable scholarship and practice-based experiences.

There also undoubtedly are administrators, coaches, colleagues, graduate students, specialists, students, and teachers who deserve recognition for supporting the advancement of this book's ideas, directly and indirectly. We hope, in turn, that this book contributes to their critical work.

Finally, we thank our families for their patience and support as we took time away to complete this work.

Madelyn, Tutita, and Fabiana

Editorial Advisory Board

Section 1
An Introduction

Chapter 1
Framing Our Understanding of Mathematical Writing

Tutita M. Casa
Department of Curriculum and Instruction, Neag School of Education, University of Connecticut, USA

Fabiana Cardetti
https://orcid.org/0000-0002-3178-4659
Department of Mathematics, College of Liberal Arts and Sciences, University of Connecticut, USA

Madelyn W. Colonnese
Reading and Elementary Education Department, Cato College of Education, University of North Carolina at Charlotte, USA

ABSTRACT

In this chapter, the authors situate mathematical writing within the context of discourse, define mathematical writing, describe the classroom environment that fosters mathematical writing, and share experiences engaging elementary students in mathematical writing. At their core, the types of mathematical writing are defined by the purposes for which students write in the mathematics classroom. The authors further describe the types of mathematical writing teachers prompt and are nurtured by the classroom environment. The insights shared draw from the authors' previous and current research endeavors, advancements in curriculum development, collaborations with mathematics coaches, and experiences teaching elementary students and supporting preservice and inservice teachers. They end the chapter by highlighting some directions for future research to expand our understanding of this crucial area in mathematics education.

INTRODUCTION

Our collective intrigue about mathematical writing began many years ago at the University of Connecticut in the United States when our paths crossed during several ventures. The one that most influenced our work in this area was when we collaborated on the Elementary Mathematical Writing Task Force that

DOI: 10.4018/978-1-6684-6538-7.ch001

was convened in 2016 with funding from the National Science Foundation (Grant No. 1545908, *Task Force on Conceptualizing Elementary Mathematical Writing: Implications for Mathematics Education Stakeholders*). The task force aimed to define more robust notions of mathematical writing because it was not yet fully realized in the research and practitioner literature. We invited educators with various backgrounds, experience, and expertise to help ensure the outcomes were comprehensive and would attend to all students' needs. The task force members included researchers, mathematics coaches, curriculum developers, assessment developers, and teachers with experience and expertise in mathematics and writing education; elementary, secondary, and tertiary education; and regular, special, and gifted education.

Tutita led the Elementary Mathematical Writing Task Force, Maddie participated in the discussions and gathered data as a doctoral graduate assistant, and Fabiana enhanced the task force's work by sharing her insights as a mathematician about writing within the discipline. We have since expanded on this project and continued to advance the research base that has demonstrated that engaging students in mathematical writing benefits their learning (e.g., Bangert-Drowns et al., 2004; Bicer et al., 2018; Casa et al., 2017; Cohen et al., 2015; Cross, 2009; Gavin et al., 2009; Gavin, Casa, Adelson et al., 2013; Gavin, Casa, Firmender et al., 2013; Graham et al., 2020; Kostos & Shin, 2010; Pugalee, 2004; Tan & Garces-Bascal, 2013). We are honored to have the opportunity through this book, *Illuminating and Advancing the Path for Mathematical Writing Research*, to promote other scholars' work in this area. Following Graham's (2019) lamentation that "many students do not receive the writing instruction they need or deserve" (p. 277), it is essential for mathematics educators to coalesce around the definition of mathematical writing for students to accrue the greatest benefits from research and practice.

The purposes of this chapter are to situate mathematical writing within the context of discourse, expand on the types of mathematical writing that are prompted by teachers or nurtured by the classroom environment, describe the classroom environment that supports students' engagement in all types of mathematical writing, share some experiences supporting elementary students, and suggestions for future research. The insights we share draw from our previous and current research endeavors. We too build off our efforts in developing curriculum materials that involved collaborating with mathematics coaches, guiding preservice and inservice teachers, and teaching elementary students.

BROADENING NOTIONS OF DISCOURSE

Educators may readily agree with the notion that "discourse practices warrant the attention of mathematics educators because discourse is the primary medium of education" (Wagner, Herbel-Eisenmann & Choppin, 2012, p. 1) if they envisage speaking mathematically. Researchers have long documented the benefits of such talk on student's learning of mathematics (cf., Cobb et al., 1997; Kazemi & Stipek, 2001; Stein et al., 1996; Wood et al., 2006). Subsequently, teachers have been able to capitalize on this research base by drawing from quality instructional resources (cf., Chapin et al., 2009; Kazemi & Hintz, 2014; Kersaint, 2015; Smith & Stein, 2011) to benefit their students.

The National Council of Teachers of Mathematics' (NCTM) seminal definition of discourse presented in the *Professional Standards for Teaching Mathematics* (1991) suggests that discourse comprises more than talk:

The discourse of a classroom—the ways of talking, agreeing and disagreeing—is central to what students learn about mathematics as a domain of human inquiry with characteristic ways of knowing. Discourse is both the way ideas are exchanged and what the ideas entail: Who talks? About what? In what ways? What do people write, what do they record and why? What questions are important? How do ideas change? Whose ideas and ways of thinking are valued? (p. 34).

Of significance, *discourse* involves not just oral exchanges that have long been valued but also writing as a means to express ideas. The definition also suggests that students are to engage in a process where ideas are exchanged among the community of learners.

It is essential to consider how one can express mathematical ideas in writing to ensure researchers and practitioners approach their work in a cohesive manner. Although symbolic notation and visual representations are ways to do so, we contend that students' ideas need to be expressed using text. That is, mathematical writing includes prose and may or may not be infused with symbols and/or visual representations. There are several reasons we have come to adopt this position. First, we seek to distinguish recording an answer, label, or record something to fill in the blank from more comprehensive ways of engaging in discourse. Second, readers can interpret symbols (e.g., expressions, equations) and visual representations (e.g., drawings, graphs) in ways not intended by the student who authored the piece. Third, we are committed to gaining insight into individual students' ideas, particularly those traditionally marginalized in the mathematics classroom, and desire to understand the unique perspectives they can share using their own words. Fourth, we posit that teachers and researchers can gain accurate or sufficient insight into students' computational approaches without asking students to recount their steps, which can be particularly laborious for younger students. We believe that a better use of time would be to have students advance their learning rather than repeating the explanation of their procedures in writing. Importantly, mathematical writing is distinct from other writing genres as it requires a combination of mathematics content knowledge, skills unique to the mathematics register, and general writing skills (Namkung et al., 2020).

While Morgan (1998) called for an increase in writing during math class, we agree with her and others (cf. Baxter, Woodward, & Olson, 2001; Morgan, 2001; Reys, 1994) that it is essential to consider what kinds of writing students engage in to ensure that they accrue the benefits of this practice. The Elementary Mathematical Writing Task Force's distinction between *writing about math* and *mathematical writing* (Casa et al., 2016; Casa, Evans, Firmender, & Colonnese, 2017; Firmender, Casa, & Colonnese, 2017) provided an important distinction. Reflecting *writing about math*, students may be asked to respond to writing prompts addressing affective components (cf., Jurdak & Rihab, 1998); e.g., "what frustrates me most about math is…"), metacognitive aspects (cf., Dougherty, 1996; e.g., "I study my times tables by…), experiences with math (cf., Wilcox & Monroe, 2011; e.g., "write your math autobiography"), and creative pieces (cf., Burns, 1995; e.g., "write a poem about math"). We envision these exercises as positioning students to write about a math-related topic rather than to think mathematically.

In contrast, engaging students in mathematical writing positions them "*to reason* mathematically and *to communicate ideas*" (Casa et al., 2016, p. 3, original emphasis). That is, the act of writing results from students engaging in the mathematical processes (NCTM, 2000) and practices (National Governors Association Center for Best Practices [NGA Center] and Council of Chief State School [CCSS], 2010) that collectively suggest students make sense of their ideas, conjecture, argue, debate, problem solve, and prove. Such a position reflects a writing-to-learn approach that considers "writing as a vehicle for strengthening, extending, and deepening students' knowledge" (Graham et al., 2020, p. 180). Using this definition, research shows benefits to students' problem solving (Pugalee, 2004), and meta-analyses

Table 1. This is an example of the kind of shift we seek in students' written responses that foster reasoning

<table>
<tr><td colspan="2">Mathematical Writing Prompt: "Riva drew this picture  to show the fraction ⅘. Do you agree or disagree that her drawing shows 4/5? Explain why."</td></tr>
<tr><th>Sample Student Writing Before They Learn How to Write Mathematically</th><th>Sample Student Writing After They Learn How to Write Mathematically</th></tr>
<tr><td>My ways right. Riva's drawing does not match mine She marked the wrong tickmark. She needs to add one more like me.</td><td>I donot agree with Riva She is not showing $\frac{1}{5}$ths She is thinking about 5 tickmarks. She marked the place for $\frac{4}{5}$ on line 4. Her drawing shows $\frac{3}{4}$.</td></tr>
</table>

indicate it advances students' mathematical learning (Graham et al., 2020) and achievement (Bangert-Drowns et al., 2004; Bicer, 2018).

The potential benefits of mathematical writing are not being properly sought, given that research has shown that students' oral arguments are often better than their written ones (Campbell et al., 2021; Stylianides, 2019). Unfortunately, when students do write, they lack reasoning (Evens & Houssart, 2004; Nachowitz, 2018) and students sometimes explain "that they are correct for other nonmathematical reasons, such as 'I tried it, and I was right'" (Lepak & Going, 2019, p. 300). Table 1 describes the shift we would like to see in students' mathematical writing using responses we generated based on our experiences supporting elementary students learning how to write mathematically.

TYPES OF MATHEMATICAL WRITING

Types Teachers Can Prompt

The Elementary Mathematical Writing Task Force (Casa et al., 2016) distinguished two writing genres teachers can promote in their instruction (Firmender, Casa, & Colonnese, 2017). The purpose of engaging in *informative writing* is for students to describe or explain their reasoning. They may be prompted to allow them to continue to unpack key concepts, express a generalization, and address solution strategies, all of which have the potential to reflect the depth of students' understanding. These pieces may allow mathematics educators to formatively and summatively assess individuals' and the class's progress. Such prompts can be used to promote the learning of important mathematical ideas. For instance, students could be asked to describe the role of the numerator within a fraction, compare and contrast squares and rectangles, explain the values of the digit 2 in the number 291,236, or highlight the patterns in a basic facts table.

Teachers can also promote *argumentative writing*. The purpose of this type is for students to persuade the reader of one's position and/or critique others' reasoning of a mathematical solution (Casa et al., 2016; Firmender, Casa, & Colonnese, 2017). These pieces allow mathematics educators to judge the appropriateness of a solution strategy and the level of conviction students have with respect to the content being addressed in a prompt. For example, students might write to convince others which measurement tool is appropriate to calculate the amount of grass seed needed to re-seed a lawn. When prompted to take a stand on a given solution, students may agree that although counting individual pennies is a viable strategy to determine the amount of money in a piggy bank, combining groups of 10 is a more efficient way to do so.

Types That Are Nurtured by the Classroom Environment

There are additional types of mathematical writing that come forth organically out of students' needs or desire to engage in them rather than the teacher dictating they do so (Firmender, Casa, & Colonnese, 2017; Firmender, Dilley, et al., 2017). Classroom environments that promote students' working out their solutions, including by using manipulatives and talking with peers, may motivate students to produce *exploratory writing*. Because the author and intended reader of this type of writing is the student themself, the neatness, organization, spelling, and grammar are inconsequential for others. Further, the mathematical validity of all the recorded ideas reflects the student's developing reasoning and perhaps their final conclusion. Naturally, it would be inappropriate to formally grade students' exploratory writing for writing-specific components and the validity of the solution.

A writing-rich mathematics classroom environment can also nurture *mathematically creative writing* (Firmender, Casa, & Colonnese, 2017). To be able to produce such a piece of writing, students would need to have the freedom to think creatively during mathematics class. Firmender, Dilley, and colleagues (2017) explain that

> *originality, fluency, and flexibility are inherent to creativity in mathematics (Balka, 1974; Guilford, 1957). These are demonstrated through divergent thinking and require the elaboration of ideas that are many, varied, and unique. In addition, originality is demonstrated through the generation of new and appropriate problems and solutions or approaches to problems (Plucker, Beghetto, & Dow, 2004)—whether imposed on or proposed by students. This "newness" or originality may be novel only to the student (e.g., a student discovering a well-known formula to solve a problem), or it could be new to the field of mathematics as a whole (e.g., positing a new conjecture; Kaufman & Beghetto, 2009). (p. 206, original emphasis)*

Students on their own accord may opt to write a mathematically creative piece if the class already has been introduced to this writing type and encouraged to produce it, or the teacher could uplift someone's mathematically creative ideas by suggesting they record their thoughts. Importantly, while engaging in mathematically creative writing is one way to meet the needs of students who have been identified as having talent in mathematics (Firmender, Dilley, et al., 2017), educators should seek to capture creative ideas produced by any student regardless of whether they have been identified as having gifts in this area.

Figure 1. Mathematical writing community components that were informed by Graham's (2018) writer(s)-within-community model

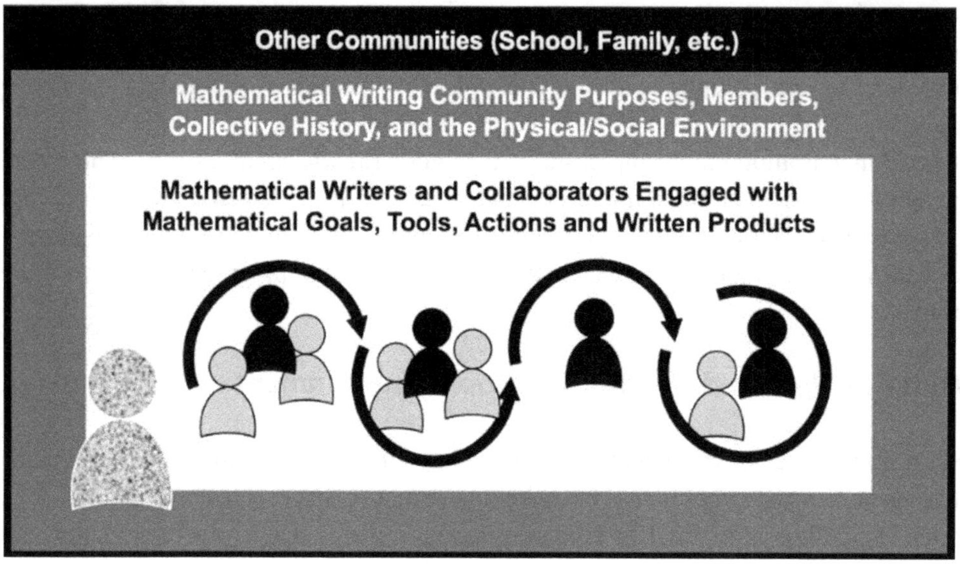

Supporting students to record such ideas has the potential of providing evidence that they may benefit from receiving enriched learning opportunities.

THE CLASSROOM ENVIRONMENT THAT NURTURES MATHEMATICAL WRITING

We have argued thus far that it is important to engage students in mathematical writing that allows them to communicate their mathematical ideas in writing. Although to be able to achieve this successfully, it is necessary to use, adapt, or design prompts that align with specific types of mathematical writing (cf., Casa et al., 2022), doing so is not sufficient. Research further recommends the development of an active writing community (Graham et al., 2012). The influence of the classroom environment in successfully implementing any activity is undeniable. In particular, understanding the significance of a mathematical writing community to create a classroom environment that nurtures the development of *mathematical* writing skills is essential.

We draw upon Graham's (2018) revised writer(s)-within-community model that originally was informed by cognitive and sociocultural views, and is a perspective reflected in Vygotsky's (1978) proposition that an individual's cognitive development is mediated by their social context. Informed by an adaptation of Graham's model, we identified the predominant components that drive a mathematical writing community. Figure 1 depicts these components that simultaneously inform and are informed by each other.

Starting with the broadest perspective, student mathematical writers are influenced by *other communities* outside their class, such as the school and family culture. The experiences created within these other communities that students bring to the mathematics classroom can profoundly influence how they think and communicate their mathematical ideas reflected in the discourse that suggests what is valued

(NCTM, 1991). These background experiences vary across many different contexts ranging from more informal ones, such as when playing games with family members and parents/guardians managing children's allowances, to more educational ones like learning opportunities experienced in prior grades.

Moving inward, the next layer of influence resides inside their class. The *mathematical writing community's purposes* drive the practices calling for students to construct their arguments, critique others' ideas, and be precise with their language when expressing themselves orally and in writing (NGA Center & CCSSO, 2010) in ways that are acceptable within the classroom (Stylianides, 2019). The class's *collective history* is influenced by their learning of mathematics and language arts, with the mathematical writing lessons guiding how they can advance these experiences focused on the community's newfound purposes. As they share common learning experiences about important mathematical ideas that elicit their written perspectives, they unpack what is necessary to include in a piece of mathematical writing so that it makes sense to them, for example, when co-developing characteristics of mathematical arguments. The encompassing *social environment* is informed by the *physical environment* which should include essential elements that facilitate and encourage engagement in mathematical writing. Including a mathematics word wall and allowing students agency about what goes on it to benefit the entire community is an example of how these aspects of the environment interweave.

The innermost layer of the mathematical writing community model comprises its *members* who shape the mathematical written pieces. All are involved in the writing process, depicted by the curved arrows in Figure 1. The writing community members include:

1. the *mathematical writers* depicted as the small and darker silhouettes in each curved arrow in Figure 1 who take full or partial responsibility for creating a particular piece of writing;
2. the *collaborators* who may include students (small lighter silhouettes in Figure 1) and teachers and/or mentors (larger patterned silhouette overseeing the students) who assist the mathematical writers in different ways, including teacher and peer feedback; and
3. the *readers*, who could be members within or be invited into the mathematical writing community that would allow writers to consider their perspectives throughout the writing process to help ensure what is being written would be understandable by them.

Ultimately, teachers' *mathematical goals* (e.g., encourage students to use more precise language), the *tools* made available to students (e.g., types of paper), *actions* (e.g., deciding whether to draw visual representations like a graph), and *written products* (e.g., arguments) further define the mathematical writing community.

FROM THEORY TO PRACTICE

Since 2016, our team has been devoted to developing ways to engage students in the types of writing defined by the task force. This included implementing lessons we developed in two Grade 3 (~ages 8-9) classes in two public urban schools in Connecticut, US. During the first of these, Tutita taught 3 weeks of a 5-week unit developing students' fluency with basic multiplication facts before stopping due to the COVID-19 pandemic school closures. The school's principal invited her to use an empty classroom to teach. The students in her class represented the school population, almost 100% of whom were Black and brown, and almost 00% who qualified for free or reduced lunches.

There were several ways in which Tutita purposefully established the mathematical writing community. She first attended to the physical space to ensure students had appropriate tools to engage in any type of mathematical writing, including various kinds of paper (e.g., scrap, graph) and writing utensils (e.g., highlighters, dry erase markers), and that these materials would be accessible to all students. These materials were kept in a class meeting area demarcated by a rug, tables with containers that contained numerous manipulatives, and a small whiteboard. When welcoming the class to the room, she had them take a tour of the space and let them know they could use the materials when needed and did not need to ask permission to use them. Furthermore, Tutita initiated the unit by asking the class if they could help their parents and guardians better understand the math they were learning since so many of her own children's friends' parents had struggled to understand what their kids were learning in school. The third graders enthusiastically agreed to do so and were excited to learn they would be hosting their families at the end of the unit to share their written ideas.

The students responded to one to two informative or argumentative prompts during the 3-week implementation, which was videotaped to allow the team to follow the students' responses to the lessons. Throughout them, we observed students' immediate enthusiasm to write, full engagement in class discussions, independent reasoning, intrigue about how their peers solved problems differently, and growth in their understanding of the mathematical concepts and characteristics of their writing. There was a time when two girls enthusiastically asked to continue to write on a whiteboard in the back of the room to make sense of their ideas rather than play games of their choosing. After Tutita invited the students to share their strategies with the class and record their final ideas on the whiteboard in the front of the room, the class relied on their writing to continue the lesson the following day. Also, students' enthusiasm for hosting their families at the end of the unit was a regular topic of discussion. On one occasion, a student asked whether to write his piece in Chinese to let his mother understand his thinking clearly. This question opened the discussion further to address how some ways of communicating mathematical ideas in writing can be understood by people outside of the United States if the symbols and visual representations are familiar to them even though the language may not be without it being translated.

Because of the initial promise of the unit's implementation (despite the truncated schedule), there was a second feasibility exploration in Spring 2021 consisting of additional lessons that were implemented by a math coach in a different school in the same district. Like the previous school, most students were from economically limited households. More than 90% of students were classified as being minorities, with the majority speaking languages other than English.

Although we were not permitted to visit the school because of pandemic-related restrictions, meetings with the math coach showcased several promising implementation features. She noted how her students were "listening to understand, not to respond," which improved their ability to explain their thinking. She realized an advancement in their understanding of math concepts and their ability to reason mathematically. This reflected an overall improvement in students' mathematical writing and in general writing skills (e.g., writing in complete sentences, understanding the writing process). The math coach also relayed how the implementation of the lessons benefited a diverse spectrum of learners, including a large group of multilingual learners. Most of them came from families who had emigrated from Brazil and spoke fluent Portuguese. The students were encouraged to speak and write in whatever language worked for them at the time and to meet their purposes. For some, this was the first time they freely expressed themselves orally and in writing when the math coach was present.

FUTURE RESEARCH DIRECTIONS

As we open our understanding of mathematical discourse in the classroom to include writing, new areas of inquiry emerge to understand better the practical implications of this broadened notion of discourse. The preliminary results of our classroom implementations are promising and indicate the need for further investigation. Additional research will help illuminate the path forward and advance our understanding of what is being hinted at in current efforts and help expand the existing literature. We see three immediate directions of future research needed to be pursued; namely, student learning, teacher education, and curriculum development.

Studies are needed that investigate student learning of mathematical writing, shedding light on how mathematical writing skills can be developed and supported within each grade. Longitudinal studies would help clarify changes in the development over time and effective ways to support students across grade levels. In addition, research studies that focus on the impact that this learning has on students' understanding of mathematical content are necessary. Attention to supporting the learning of different student populations is also clearly warranted.

Future research must carefully consider the needs of inservice and preservice teachers. It is natural to assume that most teachers have not experienced mathematical writing -as presented in this chapter- in their own mathematical learning journey in school. Therefore, studies that focus on professional development designed to support teachers' understanding of mathematical writing and its classroom implementation are required. Considering teachers' professional views regarding the challenges, limitations, and opportunities within the classrooms is also paramount. Careful attention needs to be paid to the preparation of future teachers. Questions concerning when, where, and how to incorporate the study of mathematical writing in their coursework need to be investigated.

Furthermore, research-based design and development of curriculum materials that intentionally target the facilitation of mathematical writing in the classroom is needed. Investigating the implementation and uptake of such materials will help inform revisions or modifications of new and existing materials, recommendations for practice, and implications for learning. Of significance, the implementations of these curricular resources need to be considered within the local school context that is informed by the communities in which they reside. In turn, this may open new opportunities for research into student and professional learning.

Although "each student's knowledge of mathematics is uniquely personal" (NCTM, 1991, p. 2), we have failed to support their abilities to express their individual reasoning through writing for more than 3 decades since discourse was first emphasized in the teaching and learning of mathematics. Therefore, it is important for researchers and practitioners to understand better how to promote students' engagement with mathematical writing in ways that can advance their learning of content and broaden how they can express their reasoning ideas, receive insights from their peers, and potentially co-create new mathematical knowledge.

ACKNOWLEDGMENT

We would like to acknowledge Janine Firmender's contributions to several ideas in this chapter. Most importantly, she co-led the Elementary Mathematical Writing Task Force and contributed to our curriculum development efforts.

REFERENCES

Balka, D. S. (1974). Creative ability in mathematics. *The Arithmetic Teacher, 21*(7), 633–636. doi:10.5951/AT.21.7.0633

Bangert-Drowns, R. L., Hurley, M. M., & Wilkinson, B. (2004). The effects of school-based writing-to-learn interventions on academic achievement: A meta-analysis. *Review of Educational Research, 74*(1), 29–58. doi:10.3102/00346543074001029

Baxter, J., Woodward, J., & Olson, D. (2001). Effects of reform-based mathematics instruction in five third-grade classrooms. *The Elementary School Journal, 101*(5), 529–547. doi:10.1086/499686

Bicer, A., Perihan, C., & Lee, Y. (2018). The impact of writing practices on students' mathematical attainment. *International Electronic Journal of Mathematics Education, 13*(3), 305–313. doi:10.12973/iejme/3922

Burns, M. (1995). *Writing in math class: A resource for grades 2-8.* Math Solutions.

Burton, L., & Morgan, C. (2000). Mathematicians writing. *Journal for Research in Mathematics Education, 31*(4), 429–453. doi:10.2307/749652

Campbell, T. G., King, S., & Zelkowski, J. (2021). Comparing middle grade students' oral and written arguments. *Research in Mathematics Education, 23*(1), 21–38. doi:10.1080/14794802.2020.1722960

Casa, T. M., Evans, K., Firmender, J. M., & Colonnese, M. W. (2017). Why should students write in math class? *Educational Leadership, 74*(5). ASCD. http://www.ascd.org/publications/educational-leadership/feb17/vol74/num05/Why-Should-Students-Write-in-Math-Class¢.aspx

Casa, T. M., Firmender, J. M., Cahill, J., Cardetti, F., Choppin, J. M., Cohen, J., Cole, S., Colonnese, M. W., Copley, J., DiCicco, M., Dieckmann, J., Dorl, J., Gavin, M. K., Hebert, M. A., Karp, K. S., LaBella, E., & Moschkovich, J. N., Moylan, K. Olinghouse, N. G., & Zawodniak, R. (2016). *Types of and purposes for elementary mathematical writing: Task force recommendations.* Elementary Mathematical Writing Task Force. http://mathwriting.education.uconn.edu

Casa, T. M., Firmender, J. M., Gavin, M. K., & Carroll, S. R. (2017). The influence of challenging geometry and measurement units on the mathematics achievement of kindergarteners. *Gifted Child Quarterly, 61*(1), 52–72. doi:10.1177/0016986216671806

Casa, T. M., Gilson, C., Bruce-Davis, M. N., Gubbins, E. J., Hayden, S. M., & Canavan, E. J. (2022). Maximizing the potential of mathematical writing prompts. *Mathematics Teacher: Learning and Teaching Prek-12, 115*(8), 538–550.

Casa, T. M., MacSwan, J. R., LaMonica, K. E., Colonnese, M. W., & Firmender, J. M. (2019). An analysis of the amount and characteristics of writing prompts in grade 3 mathematics student books. *School Science and Mathematics, 119*(4), 176–189. doi:10.1111sm.12333

Cobb, P., Boufi, A., McClain, K., & Whitenack, J. (1997). Reflective discourse and collective reflection. *Journal for Research in Mathematics Education, 28*(3), 258–277. doi:10.2307/749781

Cohen, J. A., Casa, T. M., Miller, H. C., & Firmender, J. M. (2015). Characteristics of second graders' mathematical writing. *School Science and Mathematics*, *115*(7), 344–355. doi:10.1111sm.12138

Colonnese, M. W. (2020). The development of instructional guidelines for elementary mathematical writing. *School Science and Mathematics*, *120*(3), 129–143. doi:10.1111sm.12391

Cross, D. I. (2009). Creating optimal mathematics learning environments: Combining argumentation and writing to enhance achievement. *International Journal of Science and Mathematics Education*, *7*(5), 905–930. doi:10.100710763-008-9144-9

Dougherty, B. J. (1996). The write way: A look at journal writing in first-year algebra. *Mathematics Teacher*, *89*(7), 556–560. doi:10.5951/MT.89.7.0556

Evens, H., & Houssart, J. (2004). Categorizing pupils' written answers to a mathematics test question: 'I know but I can't explain.'. *Educational Research*, *46*(3), 269–282. doi:10.1080/0013188042000277331

Firmender, J. M., Casa, T. M., & Colonnese, M. W. (2017). Write on: Reasoning through mathematical writing. *Teaching Children Mathematics*, *24*(2), 84–92. doi:10.5951/teacchilmath.24.2.0084

Firmender, J. N., Dilley, A., Amspaugh, C., Field, K., LeMay, S., & Casa, T. M. (2017). Beyond *doing* mathematics: Engaging talented students in mathematically creative writing. *Gifted Child Today*, *40*(4), 205–211. doi:10.1177/1076217517722180

Gavin, M. K., Casa, T. M., Adelson, J. L., Carroll, S. R., & Sheffield, L. J. (2009). The impact of advanced curriculum on the achievement of mathematically promising elementary students. *Gifted Child Quarterly*, *53*(3), 188–202. doi:10.1177/0016986209334964

Gavin, M. K., Casa, T. M., Adelson, J. L., & Firmender, J. M. (2013). The impact of advanced geometry and measurement units on the achievement of grade 2 students. *Journal for Research in Mathematics Education*, *3*(44), 478–509. doi:10.5951/jresematheduc.44.3.0478

Gavin, M. K., Casa, T. M., Firmender, J. M., & Carroll, S. R. (2013). The impact of advanced geometry and measurement curriculum units on the mathematics achievement of first-grade students. *Gifted Child Quarterly*, *57*(2), 71–84. doi:10.1177/0016986213479564

Graham, S. (2018). A revised writer(s)-within-community model. *Educational Psychologist*, *53*(4), 258–279. doi:10.1080/00461520.2018.1481406

Graham, S. (2019). Changing how writing is taught. *Review of Research in Education*, *43*(1), 277–303. doi:10.3102/0091732X18821125

Graham, S., Bollinger, A., Booth Olson, C., D'Aoust, C., MacArthur, C., McCutchen, D., & Olinghouse, N. (2012). *Teaching elementary school students to be effective writers: A practice guide* (NCEE 2012-4058). National Center for Education Evaluation and Regional Assistance, Institute of Education Sciences, U.S. Department of Education. http://ies.ed.gov/ncee/wwc/publications_reviews.aspx#pubsearch

Graham, S., Kiuhara, S. A., & MacKay, M. (2020). The effects of writing on learning in science, social studies, and mathematics: A meta-analysis. *Review of Educational Research*, *90*(2), 179–226. doi:10.3102/0034654320914744

Guilford, J. P. (1957). Creative abilities in the arts. *Psychological Review*, *64*(2), 110–118. doi:10.1037/h0048280 PMID:13420286

Jurdak, M., & Rihab, A. Z. (1998). The effect of journal writing on achievement in and attitudes toward mathematics. *School Science and Mathematics*, *98*(8), 412–419. doi:10.1111/j.1949-8594.1998.tb17433.x

Kaufman, J. C., & Beghetto, R. A. (2009). Beyond big and little: The Four C model of creativity. *Review of General Psychology*, *13*(1), 1–12. doi:10.1037/a0013688

Kazemi, E., & Stipek, D. (2001). Promoting conceptual thinking in four upper-elementary mathematics classrooms. *The Elementary School Journal*, *102*(1), 59–80. doi:10.1086/499693

Kersaint, G. (2015). *Orchestrating mathematical discourse to enhance student learning*. Ready Curriculum Associates. https://ttaconline.org/Document/zxbIhX_YCJNP0qvIYsAjT0x-qdzE3VlX/WP-Curriculum_Associates%20Orchestrating_Mathematical_Discourse.pdf0.pdf

Kostos, K., & Shin, E. K. (2010). Using math journals to enhance second graders' communication of mathematical thinking. *Early Childhood Education Journal*, *38*(3), 223–231. doi:10.100710643-010-0390-4

Lepak, J., & Going, T. (2019). Designing scaffolds for students' written arguments. *Mathematics Teaching in the Middle School*, *24*(5), 300–303. doi:10.5951/mathteacmiddscho.24.5.0300

Morgan, C. (1998). *Writing mathematically: The discourse of investigation*. Falmer Press.

Morgan, C. (2001). Mathematics and human activity: Representation in mathematical writing. *Research in Mathematics Education*, *3*(1), 169–182. doi:10.1080/14794800008520091

Nachowitz, M. (2018). Intent and engagement: Writing in mathematics for conceptual understanding. *Investigations in Mathematics Learning*, *11*(4), 245–257. doi:10.1080/19477503.2018.1461051

Namkung, J. M., Hebert, M., Powell, S. R., Hoins, M., Bricko, N., & Torchia, M. (2020). Comparing and validating four methods for scoring mathematics writing. *Reading & Writing Quarterly*, *36*(2), 157–175. doi:10.1080/10573569.2019.1700858

National Council of Teachers of Mathematics. (1991). *Professional standards for teaching mathematics*. NCTA.

National Council of Teachers of Mathematics. (1991). *Professional standards for teaching mathematics*. NCTA.

National Governors Association Center for Best Practices and Council of Chief State School Officers. (2010). *Common core state standards for mathematics*. National Governors Association Center for Best Practices and Council of Chief State School Officers. http://www.corestandards.org/Math/

Plucker, J. A., Beghetto, R. A., & Dow, G. T. (2004). Why isn't creativity more important to educational psychologists? Potentials, pitfalls, and future directions in creativity research. *Educational Psychologist*, *39*(2), 83–96. doi:10.120715326985ep3902_1

Pugalee, D. K. (2004). A comparison of verbal and written descriptions of students' problem solving processes. *Educational Studies in Mathematics*, *55*(1), 27–47. doi:10.1023/B:EDUC.0000017666.11367.c7

Renzulli, J. S., Leppien, J. H., & Hays, T. S. (2000). *The multiple menu model: A practical guide for developing differentiated curriculum*. Creative Learning Press.

Reys, B. (1994). Promoting number sense in the middle grades. *Mathematics Teaching in the Middle School*, *1*(3), 114–120. doi:10.5951/MTMS.1.2.0114

Smith, M. S., & Stein, M. K. (2011). *5 practices for orchestrating productive mathematics discussions*. National Council of Teachers of Mathematics.

Stein, M. K., Grover, B. W., & Henningsen, M. (1996). Building student capacity for mathematical thinking and reasoning: An analysis of mathematical tasks used in reform classrooms. *American Educational Research Journal*, *33*(2), 455–488. doi:10.3102/00028312033002455

Stylianides, A. J. (2019). Secondary students' proof constructions in mathematics: The role of written versus oral mode of argument representation. *Review of Education*, *7*(1), 156–182. doi:10.1002/rev3.3157

Tan, T., & Garces-Bacsal, R. M. (2013). The effect of journal writing on mathematics achievement among high-ability students in Singapore. *Gifted and Talented International*, *28*(1), 173–184. doi:10.1080/15332276.2013.11678412

Tomlinson, C. A., Kaplan, S. N., Renzulli, J. S., Purcell, J., Leppien, J., & Burns, D. (2002). *The parallel curriculum: A design to develop high potential and challenge high-ability learners*. Corwin Press.

Vygotsky, L. S. (1978). *Mind in society: The development of higher psychological processes*. Harvard University Press.

Wagner, D., Herbel-Eisenmann, B., & Choppin, J. M. (2012). Inherent connections between discourse and equity in mathematics classrooms. In B. Herbel-Eisenmann, J. M. Choppin, D. Wagner, & D. Pimm (Eds.), *Equity in Discourse for Mathematics Education: Theories, Practices and Policies* (pp. 1–16). Springer. doi:10.1007/978-94-007-2813-4_1

Wilcox, B., & Monroe, E. E. (2011). Integrating writing and mathematics. *The Reading Teacher*, *64*(7), 521–529. doi:10.1598/RT.64.7.6

Wood, T., Williams, G., & McNeal, B. (2006). Children's mathematical thinking in different classroom cultures. *Journal for Research in Mathematics Education*, *37*(3), 222–252. doi:10.2307/30035059

ADDITIONAL READING

Bossé, M. J., & Faulconer, J. (2008). Learning and assessing mathematics through reading and writing. *School Science and Mathematics*, *108*(1), 8–19. doi:10.1111/j.1949-8594.2008.tb17935.x

Casa, T. M., Choppin, J. M., & Moschkovich, J. N. (2020). *Crafting a research agenda for writing to reason mathematically*. https://drive.google.com/file/d/17DtgO9gfXwpzZdfRmwrukeijGmXnn_16/view

Dorl, J., Casa, T. M., & LaBella, E. (2022). Developing young students' expressive mathematical language. *Young Children*, *77*(3), 32–41.

Hebert, M., & Powell, S. (2016). Examining fourth-grade mathematics writing: Features of organization, mathematics vocabulary, and mathematical representations. *Reading and Writing*, *29*(7), 1511–1537. doi:10.100711145-016-9649-5

Kenney, J. M., Hancewicz, E., Heuer, L., Metsisto, D., & Tuttle, C. L. (2005). *Literacy strategies for improving mathematics instruction*. Association for Supervision and Curriculum Development.

Kosko, K. W., & Zimmerman, B. S. (2019). Emergence of argument in children's mathematical writing. *Journal of Early Childhood Literacy*, *19*(1), 82–106. doi:10.1177/1468798417712065

Powell, S., & Hebert, M. (2016). Influence of writing ability and computation skill on mathematics writing. *The Elementary School Journal*, *117*(2), 310–335. doi:10.1086/688887

Powell, S. R., Hebert, M. A., & Hughes, E. M. (2021). How educators use mathematics-writing in the classroom: A survey of mathematics educators. *Reading and Writing*, *34*(2), 417–447. doi:10.100711145-020-10076-8

KEY TERMS AND DEFINITIONS

Argumentative Writing: A type of mathematical writing to present one's own position or critique one or more solutions backed with evidence.

Explanatory Writing: A type of mathematical writing to describe or explain and the same as informative writing.

Exploratory Writing: A type of mathematical writing for the author to make sense of the problem and figure out a possible solution(s).

Informative Writing: A type of mathematical writing to describe or explain and the same as explanatory writing.

Mathematically Creative Writing: A type of mathematical writing that captures mathematical thinking beyond what has been taught.

Reasoning: Engagement in the process of determining a mathematical solution that is foundational to the mathematical writing experience.

Section 2

Engaging Teachers in Mathematical Writing

Chapter 2
Pre-Service Teachers and Mathematics Writing

R. Alex Smith
https://orcid.org/0000-0002-7143-3562
University of Nevada, Las Vegas, USA

Sarah R. Powell
https://orcid.org/0000-0002-6424-6160
The University of Texas at Austin, USA

Madeline Price
https://orcid.org/0000-0002-5194-6002
University of Nevada, Las Vegas, USA

Erin Smith
https://orcid.org/0000-0002-6726-628X
University of Nevada, Las Vegas, USA

Tessa L. Arsenault
https://orcid.org/0000-0003-0027-2236
The University of Texas at Austin, USA

Michael Hebert
https://orcid.org/0000-0003-4235-1266
University California, Irvine, USA

ABSTRACT

This chapter focuses upon research related to building pre-service teachers' (PSTs') own competencies in mathematics writing (MW) as well as their skills in teaching MW and integrating MW into their future K-12 classrooms. The authors provide a review of the research with PSTs and MW. Although there is limited research on PSTs and MW, it is likely that many PSTs enter their preparatory programs with limited MW experience and competencies despite expressing beliefs in the value of MW. Therefore, there is a responsibility of mathematics teacher educators (MTEs) to build PSTs' knowledge of the various genres of MW and their purposes, to explicitly teach PSTs to compose high-quality MW themselves, and to provide frequent models of how MW may be integrated into the mathematics classroom across a range of genres. The authors conclude with suggestions for current practice and future research.

INTRODUCTION

Communicating mathematical ideas clearly, coherently, and effectively to teachers, peers, and others, has been a distinct goal of mathematics instruction for over 20 years (National Council of Teachers of Mathematics [NCTM], 2000). Writing represents a common method of communicating mathematical

DOI: 10.4018/978-1-6684-6538-7.ch002

ideas to others, as well as one's self. Writing plays a critical role in advancing students' mathematical learning by promoting reflection and clarification of ideas through explanations, descriptions, definitions, and critiques (Freeman et al., 2016; Graham et al., 2020; Marks & Mousley, 1990; NCTM, 2000). In addition, mathematical writing (MW), which we broadly define here as any form of writing in or about mathematics, can support students' development of productive dispositions in mathematics while reducing math anxiety (Boaler, 2002; Cobb & Hodge, 2002; Ivanič, 1998; Murphy & Hall, 2008).

MW represents a meta genre (i.e., a genre with several subgenres) with distinctive features and purposes that demand specific knowledge and skill sets of teachers and students (Casa et al., 2016). Research has demonstrated that writing varies distinctly across genres, wherein skills in one genre (e.g., narrative) do not readily transfer to a different genre of writing (e.g., argumentative; Graham et al., 2016), this also applies to MW (Hughes et al., 2020). Therefore, teachers should not expect students to transfer general writing skills to MW but should instead provide MW-specific instruction to support students' MW development. While a growing body of research is expanding our understanding of what MW is (Casa et al., 2016), how MW can support K-12 students (Powell et al., 2017), and the MW practices of current teachers (Powell et al., 2021); little research has focused upon pre-service teachers (PSTs) and MW.

The purpose of this chapter is to describe why research on PSTs is critical to the MW performance of future K-12 students and summarize what we know about PSTs and MW in order to provide guidance to instructors of mathematics methods coursework. These instructors may be responsible for the mathematics education of a PST, whether that be mathematics content knowledge or mathematics pedagogical knowledge. Additionally, we use students throughout this manuscript to refer to K-12 students, including terms like 'future students' to refer to students PSTs will teach once they become in-service teachers, and 'pseudo-students' or 'real students' to distinguish between real and pseudo students when discussing the assessments and activities within our MW module. We begin the chapter with a broad overview of what MW is and what current K-12 practices are. Next, we provide a review of the MW research with PSTs, which includes PSTs' prior MW knowledge and competencies and the impact of MW embedded within mathematics methods coursework with PSTs. Given the limited research with PSTs, we describe a study conducted by the authors that uses an asynchronous MW module to provide an example of how mathematics methods instructors might begin integrating MW within their PST coursework. We conclude the chapter by providing recommendations for current instructors in regards to PSTs as well as provide directions for future research.

Four Types of MW

In MW, students use written composition with embedded visuals and symbols to communicate mathematical concepts. The purpose of MW is to assess student mathematical understanding or develop mathematical understanding (Powell et al., 2021). Students may use four types/subgenres of MW to achieve different communication purposes: exploratory, informative/explanatory, argumentative, and creative mathematics writing (see Casa et al., 2016).

In exploratory MW, students act as their own audience and use writing to make sense of a mathematics concept or problem. Students might write mathematics vocabulary definitions in a journal for their own reference (Bruun et al., 2015) Alternatively, they might engage in pre-writing before they start problem solving. For example, students may complete a written paraphrase of a word problem before solving the problem (Swanson et al., 2019). Although students may regularly practice exploratory writing in the classroom, it is rarely the focus of MW interventions (Powell et al., 2017; Powell et al., 2021).

In informative/explanatory MW, students write an explanation about a mathematics concept or procedure with a typical audience member being a teacher or another student. The explanation may describe a mathematics concept in isolation or the concepts used during problem solving (Hebert & Powell, 2016; Hughes et al., 2020). For instance, Bundock et al. (2021) included explanatory MW with problem solving instruction for students to express their understanding of their own rate of change problem solving. Explanatory MW is the most common MW type to be included on both high-stakes and researcher-created assessments (Powell & Hebert, 2022; Powell et al., 2017). Similarly, over half of educators reported that they have their students write mathematics explanations at least once a week (Powell & Hebert, 2022; Powell et al., 2017).

In argumentative MW, students communicate their reasoning to others in the form of a mathematical argument or by critiquing the reasoning of others. For example, students may justify their choices when solving fraction problems (Kiuhara et al., 2019). Although argumentative MW closely aligns with the Common Core State Standards focused on constructing viable arguments and critiquing the reasoning of others within mathematics, it occurs less frequently than explanatory MW in intervention studies and in educator reported classroom instruction (National Governors Association Center for Best Practices & Council of Chief State School Officers, 2010; Powell et al., 2017; Powell et al., 2021).

In creative MW, students communicate novel mathematical ideas to others through creative writing (e.g., creating word problems; writing mathematical stories) to demonstrate fluent and flexible mathematical thinking. For example, students may participate in Problem Posing in which they write their own word problems (Bicer et al., 2020). Teachers report infrequently using creative MW in classroom instruction (Powell et al., 2021).

Skills Needed for Effective MW

To effectively communicate through MW, students must use (a) general writing skills (e.g., spelling, grammar), (b) mathematics knowledge (e.g., computation, vocabulary), and (c) MW specific skills, such as the integration of mathematical representations within connected text. However, students sometimes have trouble with each of these components of MW (Arsenault et al. 2023; Hebert & Powell, 2016; Powell & Hebert, 2016), and may need instruction in one or more areas.

MW performance moderately correlates with general-writing performance (Graham et al, 2016). One of the important writing components in MW is organization (Hebert & Powell, 2016; Stonewater, 2002), which may change by the type of MW. For example, in informative/explanatory MW, students are often required to restate the problem, state their problem solving procedures, and include a conclusion. For argumentative MW, students need to develop their argument, list reasons to support their argument, and potentially acknowledge and refute counter arguments. These organizational structures are similar to general writing genres, and students draw on their writing knowledge for them. However, students do not always generalize their writing knowledge to MW. For example, researchers found that students will regularly include a conclusion statement with a solution, but tend to be less likely to include their problem solving procedures and restate the problem (Hughes et al. 2020).

Of course, students must also have, and demonstrate their knowledge of, mathematical knowledge, such as mathematics specific vocabulary (Hebert & Powell, 2016; Stonewater, 2002) and computational knowledge (Powell & Hebert, 2016; Hughes & Lee, 2020). Vocabulary may reflect the four types of mathematics vocabulary terms defined by Monroe and Panchyshyn (1995): technical (mathematics specific terms such as 'subtraction' or 'decimal'), subtechnical (terms that have multiple meanings including mathematics specific meanings such as 'combine' or 'change'), general terms used in mathematics that don't have a specific mathematics meaning (e.g., answer, right), and symbolic terms such as numerals (e.g., 8, 16) and symbols (e.g., =, +). Students also often have to demonstrate their knowledge of computational mathematics, such as when they may need to check the work of a pseudo student and identify the computational errors (Powell & Hebert, 2016), or complete their own computations during problem solving before responding to a MW prompt (Hughes & Lee, 2020).

Lastly, MW requires specific skills related to integrating mathematics within connected text. For example, students need to understand the conventions for how to include symbolic vocabulary (e.g., 7, 10, +, =) within connected text (Arsenault et al., 2023; Hughes et al., 2020). Students must communicate their computations and problem-solving within their MW by embedding and referring to equations, math facts, number lines, arrays, or pictorial representations. Students also sometimes need to integrate and demonstrate their problem solving within their written explanation. However, some research has shown that students infrequently attempt solving the mathematics problem separately from the content included in the written response (Arsenault et al., 2023). When students do attempt the computations, they regularly make number selection, operation, missing step errors, and computational errors (Hughes et al. 2020). Teachers need to understand the unique skills students need for MW, and the difficulties they may encounter, in order to help them communicate through MW more effectively.

Current K-12 Instructional Practices in MW

The use of MW in the classroom relates to the standards requiring students to reason and construct arguments in mathematics (National Governors Association Center for Best Practices, 2010; Powell & Hebert, 2022). Across grade levels, educators most frequently include MW in Grades 3 to 5, which aligns with when MW most regularly occurs on high-stakes assessments (Powell & Hebert, 2022; Powell et al., 2021). When teaching and assessing MW, in-service teachers most frequently focus on explanatory MW, aligning with what is most frequently included on high stakes assessments (Powell et al. 2017; Powell & Hebert, 2022). However, teachers reported infrequent use of assessment and opportunities for student revision to MW (Powell et al., 2021). When supporting MW tasks through instruction, in-service teachers reported practicing components of explicit instruction (i.e., explanations, modeling, practice, feedback) as well as providing instruction in the use of pictures and symbols within MW (Powell et al., 2021).

Despite the inclusion of MW on high-stakes tests and in NCTM standards, fewer than half of teachers reported conducting assessments or providing instruction in MW (Powell et al., 2021). Considering K-12 students' recent difficulties on high-stakes mathematics assessments that include MW (National Center for Education Statistics, 2023), MW instruction has not adequately prepared K-12 students to demonstrate their mathematics knowledge through MW. Further, given the potential benefits of writing to support mathematical learning in general (Graham et al., 2020) and to promote positive dispositions towards mathematics (Kim & Oh, 2014), there is a clear need to improve both the quantity and quality of MW instruction across a range of genres and purposes in K-12 classrooms. One route by which we can improve MW practices in future K-12 classrooms is the education and training of PSTs by instructors of mathematics methods coursework.

PRE-SERVICE TEACHERS (PSTS) AND THE FUTURE OF MW

PSTs engage in coursework and structured field-based experiences intended to provide them the requisite content-knowledge and pedagogical skills needed to sustain high-quality teaching practices in schools, as well as improve upon existing practices by implementing new or improved pedagogical approaches. Considering the lack of MW instruction and assessment in K-12 classrooms (Powell et al., 2022) and research indicating the educational benefits of MW for K-12 students (Graham et al., 2020), mathematics methods instructors can serve a critical role in bridging this research-to-practice gap by providing the learning necessary for PSTs to improve their own MW as well as the MW instruction PSTs deliver in their future K-12 classrooms. Instructors must first understand the existing competencies and prior experiences of their PSTs in order to build their content knowledge and skills in their own MW. One avenue for this may be through modules embedded within coursework, as we highlight in this chapter.

PSTs' Prior Experiences, Beliefs, and Competencies in MW

PSTs report limited experiences with MW during their K-12 education (Kenney et al., 2014; McCarthy, 2008; Smith et al., 2021) and therefore are likely to enter college preparatory programs with limited prior experience with MW and, subsequently, restricted understandings of, and competencies in, MW. Smith et al. (2021) surveyed 111 PSTs across two universities in the Southern United States and found that, when PSTs had engaged in MW in K-12, the majority of those experiences were relegated to explanatory MW activities, with few reporting explicit instruction received in *how* to compose explanatory MW. Given the lack of prior experiences and instruction in MW, it is likely that many PSTs begin their collegiate coursework with narrow conceptions of what MW is (e.g., just writing your answer in a complete sentence), and the many purposes it can serve. In short, many PSTs begin their preparatory program with limited prior experience or instructional models to draw from when generating their own MW, much less when designing instruction to teach MW skills or leverage MW to support others' mathematical learning. In addition to coming into their preparatory programs with limited prior experience with MW, PSTs are not likely to observe MW instruction and assessment during their field-based experiences, especially across a range of MW subgenres (Powell et al., 2021). As such, mathematics methods instructors will need to focus instructional time determining their PSTs' prior knowledge of MW, building background knowledge of MW, and modeling effective practice across a range of genres within their course(s).

Despite limited prior experience with MW, PSTs do indicate that they value MW and believe it is important for K-12 students' mathematics learning (Smith et al., 2021), similar to practicing teachers (Powell et al., 2021). If PSTs enter their programs already believing in the value of MW, then they may be motivated to learn more about MW and, in turn, implement more MW instruction in their future classrooms. However, PSTs' competencies in their own MW, both perceived (i.e., self-efficacy) and demonstrated, prior to engaging with math methods coursework influences the amount of content learning PSTs must do, as well as the degree to which PSTs will likely engage with said content.

In terms of MW competencies, Smith et al. (2023) examined the explanatory MW samples of 111 PSTs in response to a third-grade mathematics task during their first mathematics methods course. The PSTs demonstrated only partial understanding of the content embedded in the mathematical task. Additionally, their MW responses did not provide evidence of mastery in mathematics or writing (Smith et al., 2023). Given that the PSTs in the study had already completed collegiate coursework on the content area embedded within the task, the authors were surprised there was not a ceiling effect, at least in regard to the mathematical dimensions on the rubric employed (See Figure 1). These findings support other research indicating that PSTs' conceptual understanding is related to their MW (Magiera & Zambak, 2020), but also suggest that communicating one's mathematical thinking via writing is more complex than reporting an answer to a mathematical problem. Smith et al. (2023) determined that their findings indicate a need for additional research examining PSTs' MW mechanics and organization specifically–an area where little research exists. Further, Smith and colleagues (in progress) found that PSTs reported lower MW self-efficacy (average of a three out of five on a five-point Likert scale, neutral) at the beginning of their first math methods course than either writing self-efficacy or mathematics self-efficacy, both of which averaged a four out of five on a five-point Likert scale (i.e., somewhat positive).

Other research into PSTs' MW competencies is limited and varied in its assessment. Using a rubric which contained three "mathematical practices" elements and three "communicative" elements, Wall et al. (2016) found that PSTs scored much lower in "communication" than "math" when solving a middle school math problem. On the scale of Novice, Apprentice, Practitioner, or Expert, the majority of PSTs scored at Novice or Apprentice on the communicative elements of Completeness, Clarity, and Reflection when entering their first math methods course (Wall et al., 2016). However, the majority of PSTs scored at least Practitioner level on each of the three mathematical practice elements. PSTs' scores increased significantly across all rubric components through MW practice and a virtual field experience, with communicative elements showing the most drastic growth in competency (Wall et al., 2016).

Magiera and Zambak (2020) assessed PSTs' explanatory MW based on the inclusion of four functional elements: conclusion, supporting evidence, reasoning articulated within an explanation, and conceptual reference. Across a range of MW tasks PSTs achieved 73.9% of available points, with greatest weakness in reasoning (scoring only 66.5%). Notably, PSTs' explanatory MW competencies were significantly higher than their competencies in critiquing and providing feedback on student MW, and the two competencies were correlated within individuals (Magiera & Zambak, 2020). These results suggest not only the difficulty for PSTs to communicate their mathematical understanding in writing upon beginning math methods coursework, but also that high levels of generative MW competency are needed to enable competency in critiquing students' MW. Strong MW activities guided by mathematics methods instructors are needed to enable PSTs to communicate mathematical explanations clearly to their future students and to engage productively with future student work.

Although there is limited research on the prior experiences, beliefs, and competencies of PSTs in MW, the research reviewed here indicates that mathematics methods instructors likely need to dedicate significant instructional time explicitly building background knowledge in MW while providing multiple models of and frequent opportunities to practice MW. Although PSTs indicate that they value MW, so do practicing teachers that also report low rates of implementing MW instruction. Furthermore, despite practicing teachers reporting to value MW, they report using predominately explanatory MW and rarely provide instruction to their students on *how* to compose MW. Therefore, how teachers, and likely PSTs, value MW alone is not sufficient to promote implementation. Questions remain as to why both PSTs' and in-service teachers' values of MW do not appear to translate to MW practice. For PSTs specifically, improving their own MW competencies and providing frequent models of high-quality MW instruction and assessment during their preparatory program could promote their future teaching practices. Within these practices, PSTs are likely to benefit greatly from instruction emphasizing their argumentative MW and ability to provide productive feedback on other's MW.

Research Examining Instructors' Uses of MW in Coursework

Much research on mathematics methods instructors' practices incorporating MW into math methods coursework has focused on the use of exploratory and reflective writing. The implementation of exploratory MW in education coursework has led to growth of PSTs' conceptual understanding (Helstad et al., 2017), increased pedagogical content knowledge (Kenney et al., 2013), and positive shifts in their beliefs surrounding mathematics (Salinas, 2004) and about themselves as mathematicians (Zarestky & Bigler, 2021). Writing in reflection journals while in mathematics content or methods courses can help PSTs move from a view of mathematics that emphasizes procedural skills and memorization toward one that foregrounds active problem-solving and conceptual understanding, and can support PSTs' agency as learners and as future educators (Salinas, 2004; Zarestky & Bigler, 2021). Incorporating exploratory MW into the learning process of PSTs in coursework has also led to increasing willingness to use it in their future instruction (Kenney et al., 2014; Sezer, 2010).

MW activities in undergraduate and graduate education coursework can also support PSTs in their mathematical communication skills and thus their delivery of mathematical content to students. For example, writing, assessing, and revising definitions of high-leverage mathematical vocabulary like "solve," "evaluate," or "simplify" can draw PSTs' attention to the value of clear and precise language in the classroom (Kenny & Montan, 2015). Imprecise teacher explanations and terminology can cause student confusion, highlighting the importance of training PSTs in mathematical communication. Research also suggests the quality of PSTs' written mathematical explanations are strongly correlated with the quality of their oral explanations in math lessons, reinforcing the benefits of integrating MW in PST coursework (Abidin, 2018). There is also evidence that engaging PSTs in MW activities during their coursework improves their understanding of, and feedback on, student mathematical work (Magiera & Zambak, 2020; Wall et al., 2016).

Most research about PSTs' MW focuses on the effects of writing in math coursework specific to beliefs, pedagogical content knowledge, and future practices. Thus, there is little examination of specific strategies for improving PSTs' MW. That being said, there is some work which suggests that a mnemonic strategy (e.g., PSOLVE) can support undergraduate students' problem solving and MW (Hauk & Isom, 2009). Students assigned to sections of college algebra which included semi-regular MW using the mnemonic strategy PSOLVE significantly outperformed their peers on a routine algebra test, and provided explanations with greater clarity and internal consistency (Hauk & Isom, 2009). It is likely that instructional scaffolds like the PSOLVE mnemonic or graphic organizers can support PSTs' competencies in mathematical communication and their procedural performance.

Although the research is limited, there is evidence that when mathematics methods instructors embed a variety of MW tasks within their coursework that PSTs' own MW competencies improve, as well as their dispositions towards mathematics and abilities to communicate their mathematical thinking. These critical skills cut across grade levels. For example, while a PST training to be a secondary mathematics teacher may be able to confidently complete a mathematics task, their ability to communicate how they did so and why to their future students (i.e., teaching) is promoted by MW within their preparatory coursework. Given the limited research examining ways in which to build PSTs' own MW competencies and the general absence of assessment in MW research, our team developed an instructional module to introduce PSTs to MW and explore how instructors might train PSTs to use MW in their future classrooms. We provide an overview of the intervention and our exploratory study in the next section, along with some initial findings to illustrate how instructors might begin incorporating MW into their coursework.

AN ASYNCHRONOUS MW MODULE FOR PSTS: AN EXPLORATORY STUDY

Our team developed an asynchronous module for PSTs about MW to help increase their understanding of MW, with a specific focus upon MW assessment. Consented participants (n = 111) included undergraduate PSTs enrolled in a mathematics methods course at one of two universities in the United States taught by one of three instructors. The universities were in the Southeast (two instructors) and South (one instructor). At each university, the mathematics methods course was the first mathematics methods course for the PSTs, which PSTs took after completing at least one elementary mathematics content course. The goal of the mathematics methods course was to prepare PSTs seeking initial licensure in elementary education, special education, or elementary and special education.

The MW module was asynchronous, approximated one week of course content, and delivered via a learning management system (Canvas). The study was multifaceted and used a randomized control trial embedded within a pre-posttest design. First, we examined the overall impact of the MW module upon PSTs' values and self-efficacy in MW instruction and assessment as well as the impact of the module upon the PSTs' own MW utilizing a one-group pre/posttest design. Within that design, we experimentally examined the impact of a video lesson on assessing MW with PSTs' ability to apply a rubric to score elementary student's MW utilizing a randomized control trial design. We randomly assigned participants to either access the video lesson prior to being asked to score two samples of elementary students' MW (i.e., treatment), or after scoring the elementary students' MW (i.e., control). This allowed us to examine the impact of the video on PSTs' ability to evaluate MW. Also, participants evaluated whether or not they had received enough training to teach and assess MW prior to and after completing the module. Although analyses are ongoing, we will describe the module and report an overview of preliminary findings.

The MW Module

After accessing the MW module via Canvas, PSTs read an introductory page that included the seven steps to completing the module. The first step included reading a consent form, and an option to sign the consent form. If PSTs did not sign the consent form, they still had to complete the MW module (they were also informed that not consenting to the study would not impact their grade). If PSTs signed the consent form, we included their data in our analyses.

Second, PSTs completed a pretest. On this pretest, PSTs provided information about how often they wrote in their mathematics classes as a K-12 student and provided examples of their MW. We also asked PSTs to reflect upon the benefits of MW, whether they anticipated asking their future students to engage in MW, and what kinds of things they might ask students to write about in mathematics. PSTs solved a Grade 4 high-stakes test item focused on geometry in which they had to use MW to explain why two students' work examples were correct. Then, PSTs worked on a Grade 3 MW prompt in which they had to write about a pseudo student's mistakes and how they would solve the problem correctly (see Figure 2). We included these two activities so the PSTs could understand the MW activities that students are asked to do in schools. Finally, we provided two real student answers to the Grade 3 MW prompt and asked how the PSTs would assess the student's MW and things they would do in the classroom to respond to the student's MW. Finally, PSTs filled out survey questions about their mathematics, writing, and MW self-efficacy.

Third, PSTs read two articles. The first, titled "Why Should Students Write in Math Class" (Casa et al., 2017), reviewed the four types of MW (i.e., exploratory, informative/explanatory, argumentative, mathematically creative). The second, "Integrating Writing and Mathematics" (Wilcox & Monroe, 2011) provided suggestions for MW which included learning logs, think-write-share, note-making, shared writing, class books, and alphabet books. After reading both articles, PSTs responded to three open-ended questions: What is mathematical writing? Why is mathematical writing important for learning mathematics? What is one strategy you may use in your future classroom to foster students' mathematical writing? These prompts were intended to encourage PSTs to engage with the content.

Next, PSTs completed steps four and five according to their randomly assigned order (a) step 4 then step 5 (i.e., treatment) *or* (b) step 5 then step 4 (i.e., control). In step four, PSTs were introduced to the idea that MW can be used as a form of assessment. To expand upon this idea, PSTs watched a video that provided the following information. In the first few minutes of the video, two hosts (co-authors of this chapter) described the interest in MW and explained that about half of the states in the United States require students to write about mathematics on high-stakes mathematics tests. Then, the hosts shared the benefits of MW. Benefits for teachers included allowing teachers to understand student-level communication, providing teachers with a method for multi-disciplinary learning, helping teachers understand student thinking, and allowing teachers to use problems with multiple entry points and multiple ways to solve a problem. Benefits for students included allowing students to communicate about mathematics, engaging students in writing to learn, allowing students to generate and edit ideas, and connecting mathematics to real life. Then, the video hosts proposed the idea that MW involves several skills, including writing skills and mathematics knowledge, with these relying on foundational skills such as oral language, reading, cognition, and self-regulation.

After a brief review of the four types of MW, the video hosts shared different ways to score MW, such as using a holistic rubric, an analytical rubric, or elements scoring. With a holistic rubric, student MW is scored on a scale (e.g., 1 [no explanation] to 5 [excellent explanation]) to describe the quality of the student's MW sample. With an analytical rubric, student MW is scored on scales (e.g., 1 to 5 points) across several categories (e.g., writing, mathematics understanding, mathematics vocabulary). Analytical scoring is likely more helpful for making instructional decisions than a holistic rubric. With elements scoring, student MW is scored as 1 (present) or 0 (not present) on elements of a MW sample. For example, "identified the pseudo student used the wrong operation" or "identifies the math is correct." Elements scoring may be just as helpful as analytical scoring but may take longer to score. In the module, we wanted to expose PSTs to different methods for the scoring of MW but then we provided an in-depth explanation of scoring MW using an analytical rubric. We focused on this type of MW scoring based on recommendations from Namkung et al. (2020).

To help PSTs understand analytical rubric MW scoring, the hosts introduced the analytical scoring rubric seen in Figure 1. With this rubric, the hosts described how to assess students' MW according to five categories important for mathematics and writing: mathematics content, mathematics vocabulary, mathematics-writing organization, mathematics-writing grammar, and clarity and precision. Following the introduction of a MW rubric, the hosts then provided examples for scoring two MW samples for the rest of the video.

Fifth, PSTs scored two MW samples according to the MW rubric introduced in the video (and seen in Figure 1). On these samples, PSTs saw two Grade 4 student responses to a MW prompt, The prompt featured the fraction representations for four pseudo students. The prompt asked the writer to write about the mistakes of the pseudo students and how they would help the pseudo students solve the problem correctly. Approximately half of participants were randomly selected to score the two student samples

Figure 1. MW scoring rubric from the MW module

	5	4	3	2	1
Math content	Complete understanding of math concepts and procedures	Adequate understanding of math concepts and procedures	Partial understanding of math concepts and procedures	Limited understanding of math concepts and procedures	No understanding of math concepts and procedures
Math vocabulary	Accurate use of formal math vocabulary	Adequate use of formal math vocabulary	Use of formal and informal math vocabulary	Use of informal math vocabulary only	No use of (or incorrect use of) math vocabulary
Writing organization	Clear progression of ideas, effective transitions, and fluent sentences	Adequate progression of ideas, adequate transitions, and fluent sentences	Weak progression of ideas, ineffective transitions and connectors	Limited progress of ideas, limited use of transitions and connectors	No progression of ideas, limited use of transitions and connectors
Writing grammar	Free of grammatical errors	1 or 2 grammatical errors	Several errors, but they do not interfere with comprehension	Many errors interfere with comprehension	Grammatical errors make comprehension impossible
Clarity and precision	Conveys math concepts and procedures clearly and precisely	Conveys math concepts and procedures clearly but no quite precisely	Only some concepts and procedures are conveyed clearly and precisely	Most concepts and procedures conveyed are unclear and imprecise	Completely lacks clarity and precision

prior to watching the video (i.e., control), but were given access to the video after completing the task, while the remaining PSTs watched the video prior to scoring the student samples (i.e., treatment).

Sixth, PSTs answered questions about the module. These questions included a self-reflection piece on scoring the two MW samples and then general feedback for the module. Seventh, PSTs completed a posttest containing the self-efficacy questions as well as required them to generate their own MW response to an elementary grade MW prompt.

Overview of Findings

To understand the impact of participation in the MW module for PSTs, we administered survey questions and activities at pre- and posttest. On questions about self-efficacy, PSTs' self-efficacy for teaching MW increased from an approximate average of 3 on a five-point Likert scale at pretest to a 4 at posttest. In contrast, PSTs' self-efficacy for teaching writing and mathematics demonstrated an approximate score of 4 on the five-point Likert scale at pre- and posttest. However, self-efficacy was higher for teaching both writing and mathematics than for MW, before and after the module, despite the gains in MW self-efficacy.

We also explored whether the module had an impact on PSTs' anticipated future mathematics instructional practices (Smith et al., 2021; Smith et al., in preparation). Aligned with previous research (Kenney et al., 2014; Kuzle, 2013), we saw that exposing PSTs to MW lead to an increase in the frequency that PSTs would ask their students to write in mathematics. Prior to the module, the median frequency PSTs

Figure 2. Sample argumentative MW prompt

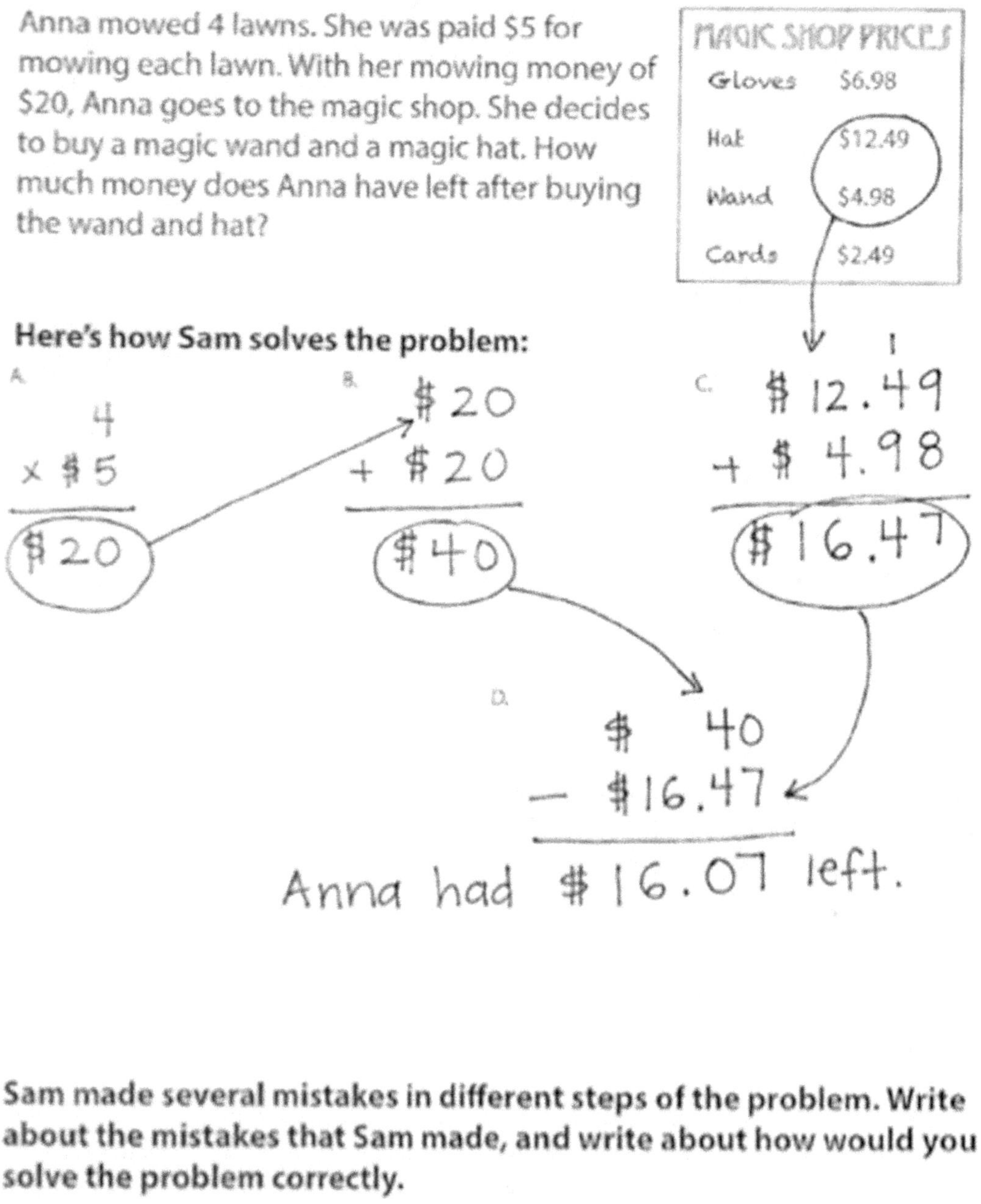

planned to incorporate MW was once a week, but after the module the planned frequency was several times a week. Across the module, explanatory MW was the most common type of writing activity PSTs planned to use with students. We also noted an increase in the frequency that PSTs planned to have students evaluate and critique another person's MW (i.e., argumentative MW) after the module. Critiquing mathematical reasoning in MW was not an explicit focus of the module, but PSTs were asked to respond to an argumentative MW task (see Figure 2) during pre- and posttest, and to evaluate a student's written response to such a task. Thus, the activities we asked the PSTs to engage in influenced their planned future instruction even though there was no explicit explanation of their value.

In the video lesson, PSTs received training on how to evaluate students' MW using a rubric (see Figure 1) adapted from Namkung et al. (2020) and then were asked to apply this rubric to assess two students' MW samples (i.e., Bill and Ted). Preliminary findings indicated that PSTs did not reliably score the two student MW samples regardless of condition (treatment, control), as compared to an expert's scoring. Further, PSTs' mean scores for both MW samples often exceeded researcher scores across the five dimensions included in the rubric (Smith & Smith, 2024). Said another way, PSTs evaluated the two student's MW samples as demonstrating greater mathematical content knowledge, mathematics vocabulary usage, writing organization, and clarity and precision than did the expert scorers.

Finally, PSTs also identified next instructional steps based on the MW samples scored with the rubric. Generally speaking, PSTs' instructional decisions indicated greater attention to writing and language features and less attention to the mathematical content. Moreover, the instructional steps identified were often vague or lacked specificity. For example, when PSTs identified they would focus on mathematics vocabulary, they did not specify which vocabulary they would target or how they would support its use. PSTs' comments included, "I would help Bill using math vocabulary," "Teach him proper formal math vocabulary", and "I will go over mathematical language that they can use for this concept." Overall, access to the video did not appear to significantly impact PSTs' ability to apply the provided rubric when scoring student MW samples or the instructional decisions they communicated in response to the MW samples.

CONCLUSION AND RECOMMENDATIONS FOR MATHEMATICS METHODS INSTRUCTORS

Although there is rather limited research to guide instructors in integrating MW into their mathematics content or methods courses, we pull from the research reported here to provide some general guidance. First, mathematics methods instructors should anticipate that many of their PSTs have limited prior experience with MW and may struggle generating their own MW. As such, instructors should develop PSTs' background knowledge of MW early in the course, the various subgenres and purposes, similar to how the MW module did as described prior. However, requiring PSTs to simply read content describing the subgenres and purposes of MW is not enough to impact PSTs' own MW or influence their future instructional decisions related to MW. Mathematics methods instructors might also consider training PSTs to use a rubric similar to that presented in Figure 1 then utilizing that rubric for peer-review, self-assessment, and assessing sample student work. Strong instruction would also structure opportunities for revision based on peer-review and self-assessment, as well as reflection on the MW process. These practices should not only support PSTs' own MW competencies, but also provide a model for effective MW process instruction they can use in their future classrooms.

Further, mathematics methods instructors should include multiple opportunities for PSTs to compose MW responses across genres. This could provide PSTs with the experience of writing in mathematics as well as support their understanding of the cognitive benefits of writing. Teacher preparation programs should simultaneously support PSTs' mathematics *and* writing competencies alongside each other. For

Figure 3. Sample four-square graphic organizer

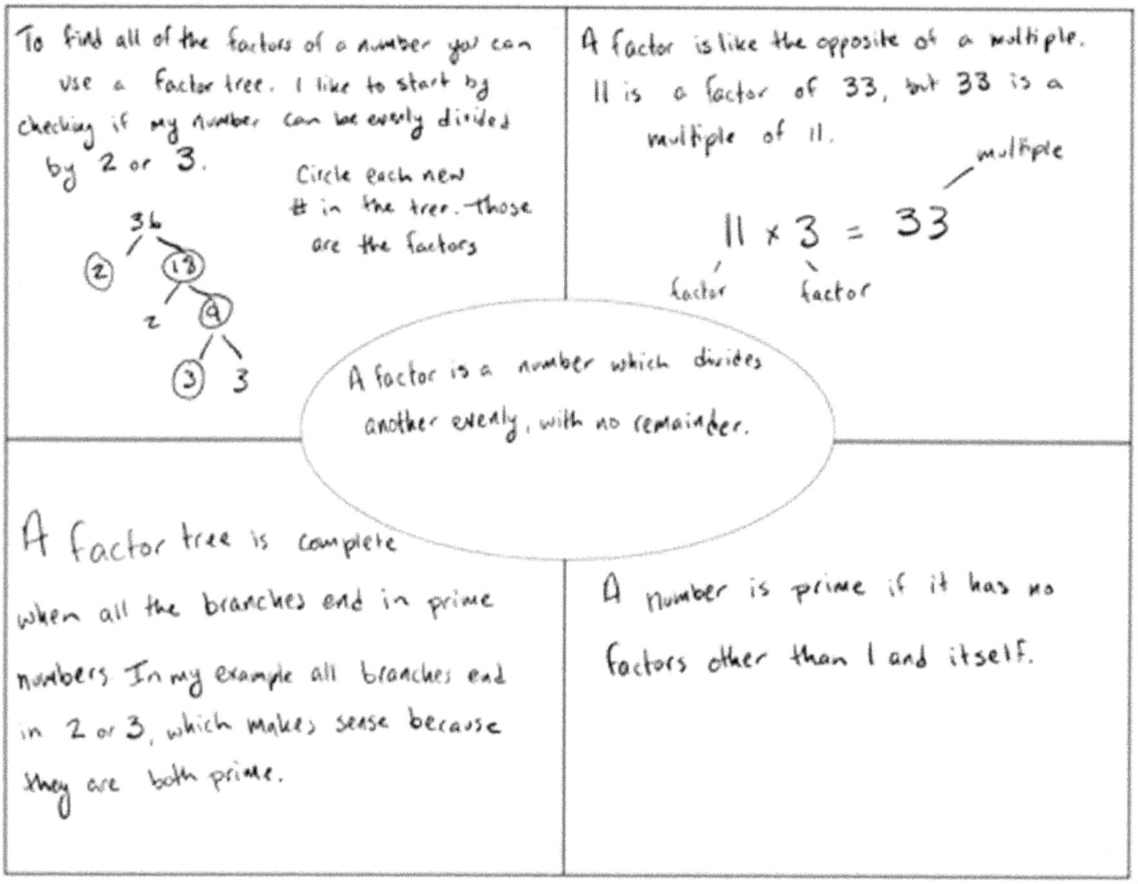

instance, instructors could incorporate mnemonic strategies (e.g., PSOLVE, Hauk & Isom, 2009) or graphic organizers in conjunction with their mathematics or mathematics methods instruction while teaching PSTs to compose their own MW responses. Figure 3 shows a simple four-square graphic organizer being used to craft a definition of a factor. The four-square graphic organizer centers a topic sentence or main idea, surrounded by four areas for writers to add examples, explain, and clarify.

As PSTs develop their own MW skills, including self-assessment, mathematics methods instructors may begin engaging PSTs in exercises wherein they assess sample student MW using the same rubric and generating instructional responses. Ideally, the MW aligns with the mathematics content being addressed (e.g., single step change word problems or multi-step word problems involving multiplication and division). PSTs should be given opportunities to work collaboratively in revising their own MW as well as in developing instructional responses to sample student MW. While instructors may begin with explanatory MW, they should be sure to transition to and provide many opportunities for PSTs to engage in argumentative MW in order to develop PSTs' own mathematical reasoning as well as to promote their thinking as teachers of mathematics. In short, PSTs should engage in both explanatory and argumentative MW across the course with an initial focus upon explicitly teaching PSTs how to compose and self-assess their own MW and later how to teach and assess their future students' MW.

The MW problems presented should include the mathematics content being addressed in the course. In this way, MW is both supporting the PSTs' learning of the content while providing models of how they can support the learning of their future students using MW.

Beyond writing and assessing explanatory and argumentative MW, PSTs should engage in exploratory and creative MW across the course. Exploratory MW could include a regularly updated journal wherein students reflect upon what they have learned, record specific strategies and justifications they have learned, refine definitions of mathematics vocabulary, and specifically describe their own dispositions toward mathematics (as both a learner and teacher of mathematics). For creative MW, PSTs could regularly engage in generating their own MW prompts integrating the content standards being addressed or journaling activities they may ask students to engage in. PSTs with interest could even draft their own mathematical storybooks, connecting content standards to the ways families use math in their lives every day, potentially enabling them to frame instruction through engaging and culturally relevant narratives (An et al., 2019).

The key takeaways from the research are that MW can support meaningful changes in PSTs' own competencies as well as their future instructional practices, but they often enter their college preparation coursework with limited prior experience and competencies in MW. Therefore, mathematics methods instructors will need to strategically build PSTs' own MW competencies by modeling best practices and then transition to guiding PSTs in designing and making their own instructional decisions related to MW. While a one-week lesson/unit or MW module is not enough to accomplish all of these goals, it could be a good way to start a course in order to build background knowledge and draw attention to MW as a tool to support mathematical learning. As PSTs engage in MW practices across the semester on a consistent basis, they will be more likely to translate these practices to their future classrooms.

FUTURE RESEARCH DIRECTIONS

Future research should longitudinally examine what PSTs identify as anticipated MW instruction with what they actually do in practice. This line of research could examine PST practice during student teaching as well as in their future classroom. Such research may also examine facilitators and barriers to implementing PSTs' planned MW instruction (e.g., curriculum, district professional learning). Future research should also work to identify the key instructional practices, beyond building background knowledge via assigned readings, that improve current PSTs' own MW competencies as well as promote high-quality instructional practices once they are in the classroom. There needs to be more research identifying ways to integrate MW within mathematics content and methods courses in ways that support PSTs' own MW competencies and mathematics content knowledge that simultaneously provide models of high-quality MW instruction. Finally, beyond our own study, we were not able to identify any research that examined PSTs' instructional decision-making related to MW and there is limited research on the assessment of MW. It is critical that the field moves beyond acknowledging that MW can support mathematical learning to understanding how teachers can best evaluate MW to inform meaningful instructional decisions, which can then be integrated into PSTs' education.

REFERENCES

Abidin, Z. (2018). Mathematical communication characteristics of pre-service primary school teacher in explaining the area of trapezoid reviewed from school origin. *Journal of Research and Advances in Mathematics Education*, *3*(2), 118–126. https://eric.ed.gov/?id=EJ1283918. doi:10.23917/jramathedu.v3i2.6784

An, S., Tinajero, J., Tillman, D., & Kim, S. J. (2019). Preservice teachers' development of literacy-themed mathematics instruction for early childhood classrooms. *International Journal of Early Childhood*, *51*(1), 41–57. doi:10.100713158-019-00232-9

Arsenault, T. L., Powell, S. R., Hebert, M. A., King, S. G., Lin, X., & Lang, D. (2023). Mathematics-writing profiles for students with mathematics difficulty. *Reading and Writing*, *36*(8), 2025–2052. doi:10.100711145-022-10375-2

Bicer, A., Lee, Y., Perihan, C., Capraro, M. M., & Capraro, R. M. (2020). Considering mathematical creative self-efficacy with problem posing as a measure of mathematical creativity. *Educational Studies in Mathematics*, *105*(3), 457–485. doi:10.100710649-020-09995-8

Boaler, J. (2002). The development of disciplinary relationships: Knowledge, practice and identity in mathematics classrooms. *For the Learning of Mathematics*, *22*(1), 42–47.

Bruun, F., Diaz, J. M., & Dykes, V. J. (2015). The language of mathematics. *Teaching Children Mathematics*, *21*(9), 530–536. doi:10.5951/teacchilmath.21.9.0530

Bundock, K., Hawken, L. S., Kiuhara, S. A., O'Keeffe, B. V., O'Neill, R. E., & Cummings, M. B. (2021). Teaching rate of change and problem solving to high school students with high incidence disabilities at tier 3. *Learning Disability Quarterly*, *44*(1), 35–49. doi:10.1177/0731948719887341

Casa, T. M., Firmender, J. M., Cahill, J., Cardetti, F., Choppin, J. M., Cohen, J., Cole, S., Colonnese, M. W., Copley, J., DiCicco, M., Dieckmann, J., Dorl, J., Gavin, M. K., Hebert, M. A., Karp, K. S., LaBella, E., Moschkovich, J. N., Moylan, K., Olinghouse, N. G., & Zawodniak, R. (2016). *Types of and purposes for elementary mathematical writing: Task force recommendations.* Math Writing. https://mathwriting.education.uconn.edu/wpcontent/uploads/sites/1454 /2016/04/Types_of_and_Purposes_for_Elementary_Mathematical_Writing_for_Web-2.pdf

Casa, T. M., MacSwan, J. R., LaMonica, K. E., Colonnese, M. W., & Firmender, J. M. (2019). An analysis of the amount and characteristics of writing prompts in Grade 3 mathematics student books. *School Science and Mathematics*, *119*(4), 176–189. doi:10.1111sm.12333

Cobb, P., & Hodge, L. L. (2002). A relational perspective on issues of cultural diversity and equity as they play out in the mathematics classroom. *Mathematical Thinking and Learning*, *4*(2-3), 249–284. doi:10.1207/S15327833MTL04023_7

Freeman, B., Higgins, K. N., & Horney, M. (2016). How students communicate mathematical ideas: An examination of multimodal writing using digital technologies. *Contemporary Educational Technology*, *7*(4), 281–313. doi:10.30935/cedtech/6178

Graham, S., Hebert, M., Paige Sandbank, M., & Harris, K. R. (2016). Assessing the writing achievement of young struggling writers: Application of generalizability theory. *Learning Disability Quarterly*, *39*(2), 72–82. doi:10.1177/0731948714555019

Graham, S., Kiuhara, S. A., & MacKay, M. (2020). The effects of writing on learning in science, social studies, and mathematics: A Meta-Analysis. *Review of Educational Research*, *90*(2), 179–226. doi:10.3102/0034654320914744

Hauk, S., & Isom, M. A. (2009). Fostering college students' autonomy in written mathematical justification. *Investigations in Mathematics Learning*, *2*(1), 49–78. doi:10.1080/24727466.2009.11790290

Hebert, M. A., & Powell, S. R. (2016). Examining fourth-grade mathematics writing: Features of organization, mathematics vocabulary, and mathematical representations. *Reading and Writing*, *29*(7), 1511–1537. doi:10.100711145-016-9649-5

Helstad, K., Solbrekke, T. D., & Wittek, A. L. (2017). Exploring teaching academic literacy in mathematics in teacher education. *Education Inquiry*, *8*(4), 318–336. doi:10.1080/20004508.2017.1389225

Hughes, E. M., & Lee, J.-Y. (2019). Effects of a mathematical writing intervention on middle school students' performance. *Reading & Writing Quarterly*, *36*(2), 1–17. doi:10.1080/10573569.2019.1677537

Hughes, E. M., Riccomini, P. J., & Lee, J.-Y. (2020). Investigating written expressions of mathematical reasoning for students with learning disabilities. *The Journal of Mathematical Behavior*, *58*, 100775. doi:10.1016/j.jmathb.2020.100775

Ivanič, R. (1998). *Writing and identity*. John Benjamins. doi:10.1075wll.5

Kenney, R., Shoffner, M., & Norris, D. (2013). Reflecting to learn mathematics: Supporting pre-service teachers' pedagogical content knowledge with reflection on writing prompts in mathematics education. *Reflective Practice*, *14*(6), 787–800. doi:10.1080/14623943.2013.836082

Kenney, R., Shoffner, M., & Norris, D. (2014). Reflecting on the use of writing to promote mathematical learning: An examination of preservice mathematics teachers' perspectives. *Teacher Educator*, *49*(1), 28–43. doi:10.1080/08878730.2013.848002

Kenney, R. H., & Montan, N. (2015). Using writing to encourage PSMTS' reflections on ambiguity in mathematical language. *North American Chapter of the International Group for the Psychology of Mathematics Education.* https://files.eric.ed.gov/fulltext/ED584314.pdf

Kim, H., & Oh, Y. (2014). The effect of essay writing-centered mathematics teaching on problem solving and mathematical disposition. *Communications of Mathematical Education*, *28*(1), 131–154. doi:10.7468/jksmee.2014.28.1.131

Kiuhara, S. A., Rouse, A. G., Dai, T., Witzel, B. S., Morphy, P., & Unker, B. (2019). Constructing written arguments to develop fraction knowledge. *Journal of Educational Psychology*, *112*(3), 584–607. doi:10.1037/edu0000391

Kuzle, A. (2013). Promoting writing in mathematics: Prospective teachers' experiences and perspectives on the process of writing when doing mathematics as problem solving. *CEPS Journal*, *3*(4), 41–59. doi:10.26529/cepsj.222

Magiera, M. T., & Zambak, V. S. (2020). Exploring prospective teachers' ability to generate and analyze evidence-based explanatory arguments. *International Journal of Research in Education and Science, 6*(2), 327–346. https://eric.ed.gov/?id=EJ1250671. doi:10.46328/ijres.v6i2.765

Marks, G., & Mousley, J. (1990). Mathematics education and genre: Dare we make the process writing mistake again? *Language and Education, 4*(2), 117–135. doi:10.1080/09500789009541278

McCarthy, D. S. (2008). Communication in mathematics: Preparing preservice teachers to include writing in mathematics teaching and learning. *School Science and Mathematics, 108*(7), 334–340. doi:10.1111/j.1949-8594.2008.tb17846.x

Murphy, P., & Hall, K. (Eds.). (2008). *Learning and practice: Agency and identities*. Sage.

Namkung, J. M., Hebert, M., Powell, S. R., Hoins, M., Bricko, N., & Torchia, M. (2020). Comparing and validating four methods for scoring mathematics writing. *Reading & Writing Quarterly, 36*(2), 157–175. doi:10.1080/10573569.2019.1700858

National Center for Education Statistics. (2023). Mathematics Performance. *Condition of Education*. U.S. Department of Education, Institute of Education Sciences. https://nces.ed.gov/programs/coe/indicator/cnc

National Council of Teachers of Mathematics. (2000). *Principles and standards for school mathematics* (Vol. 1). NCTM.

National Governors Association Center for Best Practices & Council of Chief State School Officers. (2010). *Common Core State Standards mathematics*. SSO.

Powell, S. R., & Hebert, M. A. (2016). Influence of writing ability and computation skill on mathematics writing. *The Elementary School Journal, 117*(2), 310–335. doi:10.1086/688887

Powell, S. R., & Hebert, M. A. (2022). Thinking beyond symbols: Writing and reading in mathematics. In Z. A. Philippakos & S. Graham (Eds.), *Writing and reading connections: Bridging research and classroom practice*.

Powell, S. R., Hebert, M. A., Cohen, J. A., Casa, T. M., & Firmender, J. M. (2017). A synthesis of mathematics writing: Assessments, interventions, and surveys. *Journal of Writing Research, 8*(3), 493–526. doi:10.17239/jowr-2017.08.03.04

Powell, S. R., Hebert, M. A., & Hughes, E. M. (2021). How educators use mathematics writing in the classroom: A national survey of mathematics educators. *Reading and Writing, 34*(2), 417–447. doi:10.100711145-020-10076-8

Salinas, T. M. (2004). Effects of reflective notebooks on perceptions of learning and mathematics anxiety. *PRIMUS (Terre Haute, Ind.), 14*(4), 315–327. doi:10.1080/10511970408984096

Sezer, R. (2010). Effects of reflective writing in mathematics methods courses on pre-service teachers' perceptions. *Eurasian Journal of Educational Research, 40*(1), 188–206.

Smith, E., & Smith, R. A. (2024). *Elementary preservice teachers' assessment of and instructional decisions for mathematical writing* [Report presentation]. *Annual Conference of the Association of Mathematics Teacher Educators*, Orlando, FL, United States.

Smith, E., Smith, R. A., Powell, S., & Hebert, M. (2021). Examining the value, importance, self-efficacy and prior experiences of mathematical writing for preservice teachers. In D. Olanoff, K. Johnson, & S. M. Spitzer (Eds.) P*roceedings of the forty-third annual meeting of the North American Chapter of the International Group for the Psychology of Mathematics*. PMENA.

Smith, E., Smith, R. A., Price, M., & Powell, S. (2023). Assessing future teachers' mathematical writing: An assessment tool and insights. In C. Cobbs and D. Kombe (Eds.) *Proceedings of the 50th Annual Conference of the Research Council on Mathematics Learning*. https://www.rcml-math.org/assets/Proceedings/RCML%202023_Final_Proceedings.pdf

Smith, E., Smith, R. A., Price, M., Powell, S., & Hebert, M. (2023). *An examination of preservice teachers' experiences with, beliefs about, and instructional practices for mathematical writing*.

Stonewater, J. K. (2002). The mathematics writer's checklist: The development of a preliminary assessment tool for writing in mathematics. *School Science and Mathematics*, *102*(7), 324–334. doi:10.1111/j.1949-8594.2002.tb18216.x

Swanson, H. L., Kong, J. E., Moran, A. S., & Orosco, M. J. (2019). Paraphrasing interventions and problem-solving accuracy: Do generative procedures help English language learners with math difficulties? *Learning Disabilities Research & Practice*, *34*(2), 68–84. doi:10.1111/ldrp.12194

Wall, J., Selmer, S., & Bingham Brown, A. (2016). Assessing elementary prospective teachers' mathematical explanations after engagement in online mentoring modules. *Contemporary Issues in Technology & Teacher Education*, *16*(3), 373–414.

Wilcox, B., & Monroe, E. E. (2011). Integrating writing and mathematics. *The Reading Teacher*, *64*(7), 521–529. doi:10.1598/RT.64.7.6

Zarestky, J., & Bigler, M. (2021). Reflective journaling in mathematics: Insights into the development of future teachers. *New Horizons in Adult Education and Human Resource Development*, *33*(3), 52–65. doi:10.1002/nha3.20334

ADDITIONAL READING

Arsenault, T. L., Powell, S. R., Hebert, M. A., King, S. G., Lin, X., & Lang, D. (2023). Mathematics writing profiles for students with mathematics difficulty. *Reading and Writing*, *36*(8), 2025–2052. doi:10.100711145-022-10375-2

Beaver, C., & Beaver, S. (2011). The effect of peer-assessment on the attitudes of pre-service elementary and middle school teachers about writing and assessing mathematics. *Issues in the Undergraduate Mathematics Preparation of School Teachers, 5*. https://files.eric.ed.gov/fulltext/EJ962630.pdf

Hebert, M. A., & Powell, S. R. (2016). Examining fourth-grade mathematics writing: Features of organization, mathematics vocabulary, and mathematical representations. *Reading and Writing*, *29*(7), 1511–1537. doi:10.100711145-016-9649-5

Hebert, M. A., Powell, S. R., Bohaty, J. J., & Roehling, J. (2019). Piloting a mathematics writing intervention with late elementary students at-risk for learning difficulties. *Learning Disabilities Research & Practice*, *34*(3), 144–157. doi:10.1111/ldrp.12202

Hughes, E. M., & Lee, J.-Y. (2019). Effects of a mathematical writing intervention on middle school students' performance. *Reading & Writing Quarterly, 36*(2), 1–17. doi:10.1080/10573569.2019.1677537

Hughes, E. M., Riccomini, P. J., & Lee, J.-Y. (2020). Investigating written expressions of mathematical reasoning for students with learning disabilities. *The Journal of Mathematical Behavior, 58*, 100775. doi:10.1016/j.jmathb.2020.100775

Namkung, J. M., Hebert, M. A., Powell, S. R., Hoins, M., Bricko, N., & Torchia, M. (2020). Comparing and validating four methods for scoring mathematics writing. *Reading & Writing Quarterly, 36*(2), 157–175. doi:10.1080/10573569.2019.1700858

Powell, S. R., & Hebert, M. A. (2016). Influence of writing ability and computation skill on mathematics writing. *The Elementary School Journal, 117*(2), 310–335. doi:10.1086/688887

Powell, S. R., Hebert, M. A., Cohen, J. A., Casa, T. M., & Firmender, J. M. (2017). A synthesis of mathematics writing: Assessments, interventions, and surveys. *Journal of Writing Research, 8*(3), 493–526. doi:10.17239/jowr-2017.08.03.04

Powell, S. R., Hebert, M. A., & Hughes, E. M. (2021). How educators use mathematics writing in the classroom: A national survey of mathematics educators. *Reading and Writing, 34*(2), 417–447. doi:10.100711145-020-10076-8

Price, M. D., Smith, E., & Smith, R. A. (2023). Explain it to me: Supporting student voice in mathematical writing. In S. Finley, P. Correll, C. Pearman, & S. Huffman (Eds.), *Empowering and engaging students through academic discourse* (pp. 260–280). IGI Global. doi:10.4018/979-8-3693-0543-0.ch014

KEY TERMS AND DEFINITIONS

Argumentative MW: MW which presents an argumentative line of reasoning, typically in response to another's faulty argument. Questions which ask students to critique others' reasoning or process prompt argumentative MW.

Creative MW: Any writing within or about mathematics that seeks to express understanding in creative or novel ways.

Explanatory MW: MW intended to inform; can focus on explaining a math concept in isolation or the concepts used in the process of solving a problem. Common in curriculum and assessment.

Exploratory MW: MW which positions the author as the audience; generally used for metacognitive purposes such as reflection and planning.

Math Teacher Educator: Defined broadly, any instructor or instructional coach who supports the learning and professional development of pre-service of practicing math teachers.

Math Writing: A large category that encompasses many genres of writing in or about mathematics, including both the communication of mathematical ideas and the use of writing to support learning mathematics

Pre-Service Teacher: Someone pursuing K-12 teacher's licensure either through undergraduate or graduate coursework.

Chapter 3
Explanatory Writing: A Window Into K-8 Prospective Teachers' Reasoning About Contextual Problems

Marta T. Magiera
https://orcid.org/0000-0003-2944-9787
Marquette University, USA

Mohammad Al-younes
Marquette University, USA

ABSTRACT

This chapter contributes to broader research on mathematical discourse and its multimodal forms that reveal how students interpret, make meanings in mathematics, and communicate their understandings. Central to this chapter is explanatory writing in mathematics, which targets a specified audience to explain and transmit mathematical ideas. In the context of a mathematics problem-solving course and focusing on contextual problem-solving situations, this chapter illustrates how explanatory writing in mathematics offers a window into prospective elementary school teachers' (PTs') reasoning and communication skills. The chapter highlights PTs' strategic reasoning and sense-making and describes how they communicate their problem solutions. It provides mathematics teacher educators with insights into how PTs think about contextual problem situations, offering directions for planning how to help PTs develop new ways of thinking. This chapter also includes suggestions for future research on PTs' learning to construct written mathematical explanations.

INTRODUCTION

This chapter contributes to broader discussions focused on mathematical discourse as a multimodal practice involving a range of modes of communication used to coordinate and complement the meaning-making processes in mathematics. While *mathematical discourse* encompasses both written and spoken mathematical communication, in this chapter, we limit our discussion to only written communication. We situate this chapter in the context of elementary teacher education and focus on prospective elementary

DOI: 10.4018/978-1-6684-6538-7.ch003

teachers' (PTs') explanatory writing in contextual problem-solving situations, that is, writing through which PTs explain and justify their problem solutions. This work contributes to the knowledge about the significance of engaging PTs in mathematical writing and illustrates implementing and assessing mathematical writing in a mathematics course for elementary school teachers.

The focus on explanatory writing that engages PTs in describing and explaining their reasoning in problem-solving contexts is warranted for several reasons. First, focusing on mathematical explanations supports PTs' mathematical development as problem solvers. Formulating problem explanations helps "building and connecting knowledge in a systematic, structured way by linking an explanandum (i.e., the issue that needs to be explained) to an explanans (i.e., [means] by which the issue is explained)" (Prediger & Erath, 2014, p. 317). Prior researchers also identified that producing (and evaluating) mathematical explanations supports the development of ways of mathematical thinking that promote reasoning and proof (e.g., Bicknell, 1999; Hodds et al., 2014). Written explanations have particular advantages because the act of writing is reflective; thus, it can further enhance understanding (Steele, 2007). Writing increases awareness about the way one conceptualizes mathematics and solves problems (Berkenkotter, 1982; Pugalee, 2001).

Second, the mathematics education community affirms the importance of integrating writing into learning mathematics at all levels (Casa et al., 2016; National Council of Teachers of Mathematics [NCTM], 2000; Santos & Semana, 2015; Taylor & McDonald, 2007). At the same time, many mathematics teachers still consider writing an add-on rather than an integral aspect of mathematics learning and teaching (Sezer, 2010). Researchers document that practicing mathematics teachers are often reluctant and uncomfortable incorporating writing into their instruction (Fukawa-Connelly & Buck, 2010; Kosko, 2016). Engaging PTs in explanatory writing in the problem-solving course can provide them with first-hand experiences with writing in mathematics *as students*. PTs' prior experiences with writing as mathematics students are typically limited, and thus, they might not consider writing in mathematics as a high-leverage pedagogical practice (Kenny et al., 2014; McCarthy, 2010). PTs need to experience writing as *learners* and *doers* of mathematics in a consistent way to help them develop views of mathematical writing as an integral aspect of problem-solving and, thus, the development of problem-solving skills. When PTs experience writing in mathematics as students, they will likely be better positioned to think about writing in mathematics as a pedagogical practice and integrate writing into their future practice as mathematics teachers.

Researchers largely agree that by engaging students in mathematical writing, teachers have the opportunity to understand the depth of their students' mathematical knowledge (Steele, 2007; Teledahl, 2017). Writing can serve as a window into students' knowledge and ideas and, thus, could be used to assess students' mathematical knowledge and understanding. At the same time, research on writing in mathematics primarily addresses writing in K-12 mathematics classrooms. Studies with PTs are scarce overall and predominantly focus on PTs' perceptions of using writing-to-learn activities in mathematics (Craig, 2016; Kenney et al., 2014; Sezer, 2010). Research that examines how PTs communicate their mathematical reasoning in writing is even more scarce (Colonnese & Castro, 2023; Magiera & Zambak, 2020).

This chapter adds to research on PTs' mathematical writing by focusing on how PTs communicate their mathematical reasoning in problem-solving contexts. This chapter also adds to the growing body of research on mathematical writing by providing important pedagogical and theoretical implications. Pedagogically, this chapter illustrates PTs' strategic reasoning and sense-making and describes how they communicate their reasoning in solutions to contextual problems by providing examples of their explanatory writings. The explicit focus on strategic reasoning provides mathematics teacher educators with insights and directions for planning how to help PTs develop new ways of thinking. Theoretically, this chapter contributes to a broader body of research on mathematical discourse and its multimodal forms that reveal how students interpret, make meanings in mathematics, and communicate their understandings in writing.

BACKGROUND

Writing to Communicate Problem-Solving

An essential aspect of problem-solving is constructing and communicating mathematical meaning by sharing and explaining mathematical ideas orally and in writing (Clarke et al., 1993; NCTM, 2000). Both forms of communication support and provide evidence of mathematics learning. Mathematical communication includes providing interpretations and meanings, using representations, and providing justifications (Santos & Semana, 2015).

Communicating problem-solving in writing can take different forms. For example, problem-solvers might write polished explanations to convey arguments for their problem solutions (Craig, 2016; Magiera & Zambak, 2020; Shield & Galbraith, 1998). They might write less structured journals to document their thinking and experiences with problem-solving activities (McCarthy & Faulkner, 2019; Pugalee et al., 1999; Waywood, 1992). They might also write exploratory notes that record how they make sense of problems and mathematics for themselves (Colonnese, 2022).

Proponents of writing to learn mathematics argue that writing fosters the development of reasoning and problem-solving skills because by engaging in mathematical writing, students gain better awareness of their thinking (Connolly, 1989; Pugalee, 2001; Swafford & Bryan, 2000). Connolly (1989) described that, by engaging in mathematical writing, students can develop thought processes that support *doing* mathematics. Other researchers shared that writing in mathematics contributes to the development of conceptual and procedural knowledge (Porter & Masingila, 2000) and supports students' metacognitive abilities (Pugalee, 2001).

While having students communicate mathematics in writing might contribute to students' mathematical development, it also gives instructors insights into student learning (Pugalee, 2001; Seto & Meel, 2006). Because students' written responses can provide insights into the quality of student reasoning, Kuzle (2013) advocated that communicating mathematics through writing must be integral to mathematical problem-solving. Written problem solutions can reveal how students understand problem context, what information they identify as essential for solution development, what assumptions they make, what plan(s) they develop, or what conceptual and procedural knowledge they draw upon as they reason about a given problem.

Why Explanatory Writing?

When asked to write problem solutions, most students limit their work to records of their execution of mathematical algorithms and procedures. Usually, students' written responses provide minimal information about their reasoning that supports developing problem understanding and solution strategy. Brozo and Crain (2018) argued that problem-solving should begin with literacy. They described that students must read and explain problems, describe and provide rationales for forming solution steps, and defend their answers and solution approaches. Explanatory writing is suited to document this process. Explanatory writing focuses on explaining and transmitting mathematical ideas (Craig, 2016; Santos & Semana, 2015). In explanatory writing, problem solvers document complete mathematical ideas and present these ideas with a specific audience in mind (e.g., a teacher, problem client, or fellow student). In problem-solving contexts, explanatory writing necessitates that writers elucidate *how* to solve a problem, perform a mathematical procedure, and explain *why* their outcomes are valid (Santos & Semana, 2015; Shield & Galbraith, 1998). Explanatory writing prompts writers to consider what mathematical terminology to use as they communicate their problem understanding and problem solutions. It fosters thinking about what representations might be helpful to present solutions to problems and what models might provide additional insights into solutions and solution development. Mathematical writing for a specified audience (i.e., explanatory writing) prompts writers to consider how someone else can make sense of their ideas. Communicating problem solutions in writing prompts writers to be explicit about aspects of their thinking that they might take for granted otherwise. These aspects add insights into the understanding of a problem solution. Writing problem explanations facilitates clarifying aspects of mathematical thinking that might not be readily apparent to others (Santos & Semana, 2015).

Assessment of Students' Mathematical Writing

Researchers investigating students writing in mathematics have proposed and adapted various approaches and theoretical assessment frameworks. Some assessment methods utilize qualitative research paradigms and delineate students' written responses into conceptual categories that characterize student mathematical writing. Other methods utilize quantitative approaches that reveal how well students communicate mathematics (Clarke et al., 1993; Santos & Semana, 2015; Scheld & Galbraith, 1998). For example, with a focus on mathematical journals, Clarke et al. (1993) developed a scheme for analyzing students' written responses as *recount*, *summary*, and *dialogue*. They defined instances of recounts when students reported on nonmathematical aspects of their learning (e.g., they reported what happened in the classroom without much focus on mathematics). They defined a summary as an instance of writing that captured mathematical description (e.g., how to solve an equation or what a given mathematical term means). Finally, Clarke et al. categorized student journal writing as a dialogue when students interpreted mathematics, constructed meaning for themselves, and made mathematical connections.

Other researchers emphasized the type of information embedded in students' written responses. For example, Schield and Galbraith (1998) categorized students' explanatory writing through the lens of purpose. Their framework for analyzing students' explanations included four aspects of mathematical expositions: *kernel* (presenting main ideas), *goal statements* (presenting concepts or procedures), *justification* (providing justifications), and *links to prior knowledge* (making mathematical connections). Santos and Semana (2015) proposed a framework for analyzing students' reasoning evident in students' explanatory writing. They focused on *task interpretation* (how well students articulate problem goals), *justifications* (whether students' justifications are complete and correct), and *representations* (types of representations students use, representational precision, and completeness). Colonnese and Casto (2023) examined PTs' written responses with a focus on (1) PTs' reasoning and (2) elements of mathematical writing. With attention to reasoning, they analyzed connections PTs make between mathematics and number sentences when asked to explain computational strategies in writing (e.g., what relevant properties on which a given computational strategy hinges PTs identify). They also evaluated PTs' facility with mathematical language (i.e., the overall logic of PTs' written explanations, precision, use of correct mathematical language, and notation).

MAKING REASONING VISIBLE VIA EXPLANATORY WRITING

This chapter reports on a study conducted in the context of a larger research and development project that explored how PTs learn about science, technology, engineering, and mathematics (STEM) integration in a mathematics problem-solving course for elementary education majors. In this project, STEM integration was interpreted as solving real-world problems bound by STEM practices that support solving non-routine and authentic problems. The instructional focus was on guiding students to think out of the box, foster their ability to identify and analyze problems, develop new strategies for problem solutions, and communicate their thinking. The instructional goal was twofold: supporting PTs' development of habits of mind that facilitate problem-solving across STEM disciplines and helping PTs develop as problem-solvers, discoverers, and users of mathematics. With this second goal in mind, we examined *what PTs' explanatory mathematical writing reveals about their reasoning and communication skills in solving contextual problems.*

We discuss data from a cohort of 18 PTs enrolled in the Problem-Solving and Reasoning mathematics course for elementary education majors to document and discuss how explanatory writing can provide insights into PTs' reasoning and development as problem-solvers. The inquiry-oriented course that served as the context for this study was designed to engage PTs in the processes of mathematical discoveries and problem-solving, capitalizing on PTs' authentic experiences with problem-solving. At the beginning of the semester, PTs were introduced to Pólya's (1945) problem-solving heuristics (i.e., problem understanding, designing a plan, implementing the plan, checking and interpreting solutions) to provide a framework for their learning about habits of mind and practices that support STEM-based problem-solving. To develop language for thinking about problem-solving practices in which they engaged, PTs also discussed habits of mind that support STEM-based problem-solving activities. For example, the habit of perseverance, being open to diverse ideas, the habit of thinking systematically, analyzing and revising problem solutions, and communicating, to name some. They also explicitly discussed several general problem-solving heuristics (e.g., strategies for organizing problem information such as drawing diagrams, analyzing patterns, making systematic lists, identifying sub-problems, or solving simpler-related problems to gain entry points into more complex problems).

PTs solved a wide range of problems by working in small groups or individually. Class discussions and problem presentations followed individual and group problem-solving sessions. Through a collective examination of the merits of the presented problem explanations, the whole class, as a group, played an important role in deciding whether the proposed solutions were valid. The goal of in-class presentations was two-fold. First, helping PTs understand different solution approaches and strategies for the problems they solved, providing opportunities for asking questions, building on one another's ideas, and experiencing a wide range of reasoning. Second, fostering PTs' ability to communicate and justify findings and developing their understanding of good explanations for solutions to class problems. During class presentations, the instructor focused PTs' attention on features of explanations that make them relevant or effective, such as completeness of reasoning and clarity of evidence provided to support given assertions, level of evidence credibility, and the correctness of mathematical language and notation.

All problems presented in class were non-routine and open-ended contextual problems. All problems allowed for multiple entry points and solution strategies. While all problems were contextual, they ranged from more structured to ill-defined. All problems were open to interpretation, had multiple entry points, and were accessible to PTs. None of the problems required any algebraic manipulations, and all promoted mathematical reasoning and sense-making rather than the application of algebraic manipulations and previously known procedures.

Writing Tasks and In-Class Assessments

Writing Tasks

The focus on writing complete problem explanations was consistent throughout the semester. PTs were expected to communicate their complete mathematical solutions in writing in a way that provides someone not involved in solving the problem with a full understanding of the problem situation and the problem solution. They were also consistently engaged in writing personal problem-solving reflections about their thinking in the context of the problems they solved. In their reflections, PTs documented "rough draft" aspects of their solution approaches: Initial ideas, questions they had, mathematical insights they developed in the context of a given problem, connections to other problems they recognized, difficulties they experienced, and ways in which they overcame their difficulties. Reflections accompanying written solutions provided insights into the problem-solving processes in which PTs engaged, their initial ideas, understandings, solution approaches, or any practices in which they engaged.

Explanatory writing about one's solutions to mathematical word problems poses many unique challenges. To explain their solutions to contextual problems, writers need to articulate their understanding of the context of the problem, what information they deem relevant, and what mathematical interpretations of the problem situation they make to develop their solution strategy. They need to explain their decisions as they justify their strategies or results. Moreover, the problem-solving processes in which one engages are often messy, unlike explanatory writing, which produces a more polished story explaining the problem solution. To support PTs in organizing their written explanations, we used Pólya's (1945) problem-solving framework discussed in class as a structure for PTs' explanatory writing. Figure 1 illustrates the relationship between Pólya's and explanatory writing frameworks. With this structure in mind and our focus on communicating mathematics, we provided PTs with the following general directions for writing up solutions to class and homework problems:

Include a concise restatement of the problem in your own words. Provide enough information so that someone reading your work can understand what you were asked to do without referring to the original problem. Present your solution in an organized format. Using complete sentences, explain in detail what you did to solve the problem and why. Use visual representations (such as, pictures, diagrams, or tables, whenever appropriate). Be detailed about your work and answer using complete sentences. If your solution strategies include any equations or expressions, explain how you derived them and why they are valid. Be sure to explain what each variable represents. Your solution should be written so that it is convincing to someone else—even someone who initially disagrees with your answer.

In addition to documenting and explaining problem solutions, we required that PTs reflect on their problem-solving experiences. To do so, we provided PTs with the following general guidelines for writing their reflections:

For each problem, include a brief reflection on your problem-solving experiences. Briefly discuss what worked, what did not work, and what you did when you got stuck. What relationships did you notice between the work you did for this problem and your work for other problems you have solved? What are you still working on to understand better?

By engaging PTs in reflecting on their problem-solving experiences, we wanted them to gain awareness of their thinking and activities in which they engaged in the context of each problem.

Figure 1. Relationship between Pólya's problem-solving framework and the explanatory writing framework

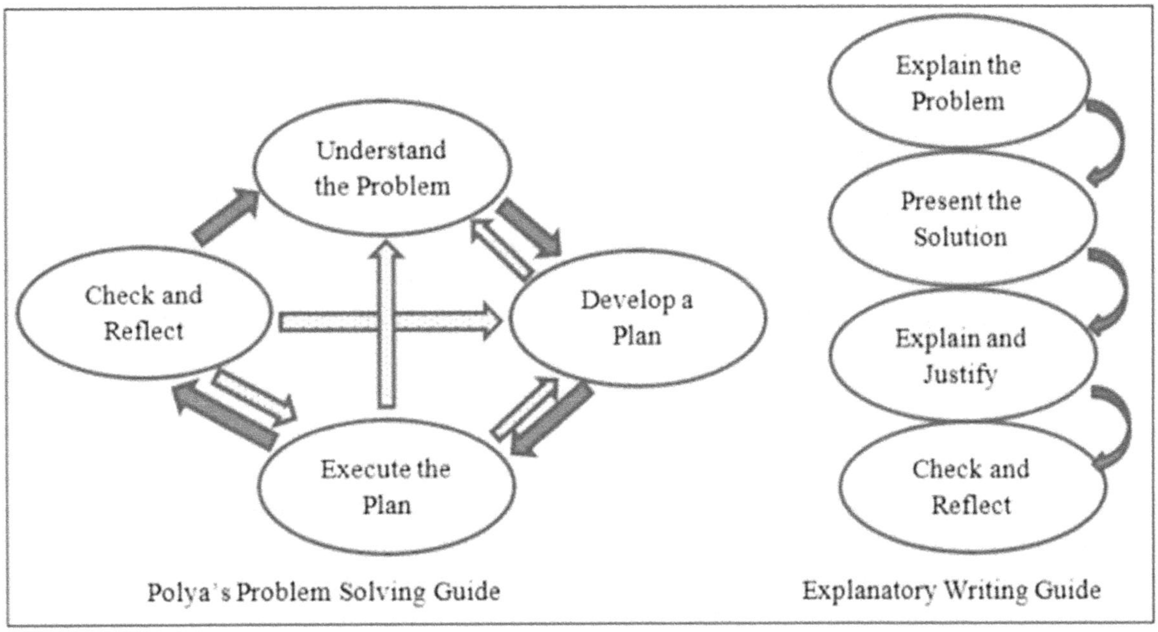

Because in our work with PTs, our primary focus was on their reasoning and problem-solving skills, we did not limit PTs on how much time they could spend on each problem. PTs could take as much time as they needed to solve each problem and write an explanation for their solution. The only requirement was to present a complete written solution that others not directly involved in solving the problem could understand, as explained in the general directions for their mathematical writing. In this chapter, we discuss PTs' individual written responses to two problems (Figure 2), which we present in the order in which PTs solved them.

We selected these two problems because both problems engaged PTs in reasoning about existential statements. Both problems facilitated sense-making and reasoning but differed progressively in complexity (e.g., the complexity of interpretation and logical complexity). Both problems facilitated proof by exhaus-

Figure 2. Problems[1] for which PTs' written responses are discussed in this chapter

Corporation Meeting. Dayna was in charge of a large corporate meeting. She wanted everyone to work in groups, and she really wanted all the groups to be the same size. She started out by organizing everyone into groups of 6, but one person was left out and didn't have a group. So she started over again and organized everyone into groups of 7, but again there was one person left out. So she tried again, this time organizing everyone into groups of 8, and this time there were five people left out. So finally, she organized everyone into groups of 5, and nobody was left out. There were less than 100 people at the meeting. How many people were there?

Stamps: The Tarnovik Republic prints stamps in the denominations 17 cents, 9 cents, and 5 cents. What is the largest amount of postage that cannot be made using these denominations?

tion and had the potential to engage PTs in reasoning focused on refutations. The Corporation Meeting problem fostered thinking about a number, or numbers, that meet a set of given conditions. The problem could be solved by examining and carefully checking a finite number of possibilities. Nevertheless, to develop a resourceful problem solution, PTs needed to reason about divisibility and make connections between multiples and remainders. They needed to recognize that a number with remainder x is x more than a multiple of that number. Moreover, they also had to develop a strategy for solving the problem rather than rely on a known, previously learned strategy. Like most problems that PT discussed in class, this problem could be solved without using symbolic or algebraic manipulations.

The Stamps problem also fostered thinking about numbers that meet a set of given conditions. This time, in addition to reasoning about multiples of given numbers, PTs needed to interpret and systematically reason about linear combinations of a given set of numbers. As with the Corporation Meeting problem, they had to develop a novel strategy rather than rely on previously known strategies.

While both problems engaged PTs in reasoning about existential statements, they had different logical structures. The problem space for the Corporation Meeting problem was finite, while the problem space for the Stamps problem was infinite. In other words, PTs could solve the Corporation Meeting problem and justify their solution by carefully checking which number within a finite set of numbers (whole numbers less than 100) meets all given conditions. However, without an upper limit, the same was not true for the Stamps problem. The difficulty of the Stamps problem was knowing how to think about "the largest impossible case" and where to start.

Class Assessment Strategies

Formative assessment strategies included sharing with PTs supporting documents that encompassed general guidelines for explanatory writing (as illustrated in directions for writing up problem solutions shared with PTs and discussed earlier) and sharing assessment criteria for their written responses, which focused on (1) mathematical content and (2) communication. The assessment of mathematical content addressed how well PTs demonstrated their problem understanding and if their solution was correct and complete. The assessment of communication addressed the clarity with which PTs communicated and justified their reasoning behind the problem solution. The expectations for communicating problem solutions were systematically reinforced in class through instructor modeling and providing oral and written feedback. The instructor's feedback included positive comments on anything that was well and clearly explained and comments on aspects of PTs' explanations that could be improved or extended. For example, "*Why did you decide to start with multiples of 5?*", "*Great strategy. Explain how you decided on numbers in each group.*" "*How can you be certain that every number beyond 21 would work?*" The goal of the feedback was to bring PTs' attention to aspects of their explanatory writing that needed improvement and, at the same time, highlight the positive aspects of their written solution presentations. Similarly, feedback during oral presentations focused on clarifying explanations, including the need to communicate problem understanding and assumptions on which the developed solution strategy rests.

Data Analysis

We recognize the complexity related to the assessment of problem-solving and reasoning in the context of explanatory mathematical writing. Guided by research on the assessment of mathematical writing (Albano et al., 2023; Numkung et al., 2019; Santos & Semana, 2015), we analyzed PTs' written responses for evidence of communicating an understanding of the problem situation and for evidence of strategic reasoning. For each problem, using qualitative content analysis and open coding (Saldaña, 2016), we identified and cataloged how PTs made sense of the problems and what strategies they developed. We also analyzed and tracked how PTs communicated their problem understanding and solution strategies (e.g., representations they used, completeness and precision of their procedures descriptions, and justifications). We systematically compared and contrasted written responses across the entire data corpus until the final list of characteristics of PTs' problem explanations was developed.

In our assessment of *overall* communication of problem solutions, we focused on how well each written solution articulates a cohesive and complete chain of reasoning and documents the flow of ideas that connect to one another in terms of (1) description of procedures and (2) their justifications. With a focus on descriptions of procedures and justifications, we assessed the overall quality of PTs' written explanations using a 3-point scale (*competent*, *emerging*, and *insufficient*).

We categorized written solution reports that were clear, logically organized, and included a comprehensive description of the problem and the solution strategy as *competent* in describing the problem solution. We categorized written reports that included only a partial description of the problem or a partial explanation of the solution steps as *emerging* in communicating information about the problem solution. Finally, we categorized written reports that included mostly mathematical work with very limited or no explanations of the problem and solution steps as *insufficient* in communicating a solution strategy.

Similarly, we assessed each written response, focusing on the quality of justifications. If the PTs included comprehensive and valid justifications for the reasoning behind the procedure used to solve the problem, we categorized the explanation as *competent* in providing justifications. If, in the written solution report, PTs justified only selected steps of the procedure or provided only partial justifications for the solution procedure and results, we categorized the explanation as *emerging* in terms of providing justifications. Finally, we categorized written explanations that included very limited or no justifications as *insufficient* in terms of justifications.

To identify possible patterns across PTs' problem explanations in terms of the quality of their descriptions of procedures and justifications, we tabulated PTs' written responses based on the overall quality of their explanations for *how* to solve the problem (explaining how) and *why* the solution is correct (justifying). In the Results section that follows, we further discuss our assessment and illustrate with data examples.

RESULTS

The analysis revealed a broad range of approaches, reasoning strategies, and ways in which PTs communicated their problem solutions. Below, we discuss aspects of PTs' reasoning identified across the analyzed written responses, focusing on sharing ways of understanding, strategic reasoning, and overall communication.

Transmitting Problem Information

Contextual problem situations often offer an overabundance of information, making it difficult to process everything at once. Problem solvers (writers) can demonstrate their understanding of the problem (or can help someone else to understand the problem) by describing known and unknown elements, quantities that are known, whether there are exceptions, assumptions that need to be made, and describing what is being asked. Our analysis revealed what aspects of contextual problem situations PTs articulated in their written solution reports and how they conveyed the problem information in their explanations. We identified two ways in which PTs communicated problem information in their solution reports (1) *summarizing* or (2) *interpreting*. We also identified different levels of precision and completeness in articulating problem information. For example, while most PTs included complete information regarding conditions described in each problem statement, only a few included information about problem goals. The lack of PTs' attention to articulating problem goals was particularly noticeable when PTs communicated problem information using shortcuts or iconic representations. Table 1 summarizes how PTs communicated problem information in their written solutions. It also shows that in their reports, some PTs included very limited or no information about the problem they solved.

Table 1. Conveying information about the problem

Evidence of Problem Understanding	Description	Problem one	Problem two
		# PTs* (n = 18)	# PTs (n = 18)
Summarizing	Reporting the statement of the problem in a direct way (shortcut or paraphrasing)	10	9
Interpreting	Interpreting the statement of the problem with a focus on embedded mathematical meaning	2	3
No information	The written solution provides very limited or no information about the problem statement	4	5

Note. *Two PTs did not complete the first problem, and one PT did not complete the second problem

Summarizing

One way in which PTs presented problems they solved was by extracting contextual information from the problem statement. As Table 1 shows, most PTs expressed the problem information subject to their writing in a direct form by briefly restating the problem statement. By doing so, they communicated with the reader of their explanation aspects of the problem that were readily apparent in the problem statement. We coded this way of presenting the problem information as *summarizing*. To do so, some PTs used various shortcuts to highlight the essential problem facts, (Figure 3). Some PTs briefly paraphrased the problem statement (Figure 4), or used iconic representations to convey problem information (Figure 6). (All excerpts are verbatim).

Figure 3. Direct shortcut description of the problem statement (PT #4)

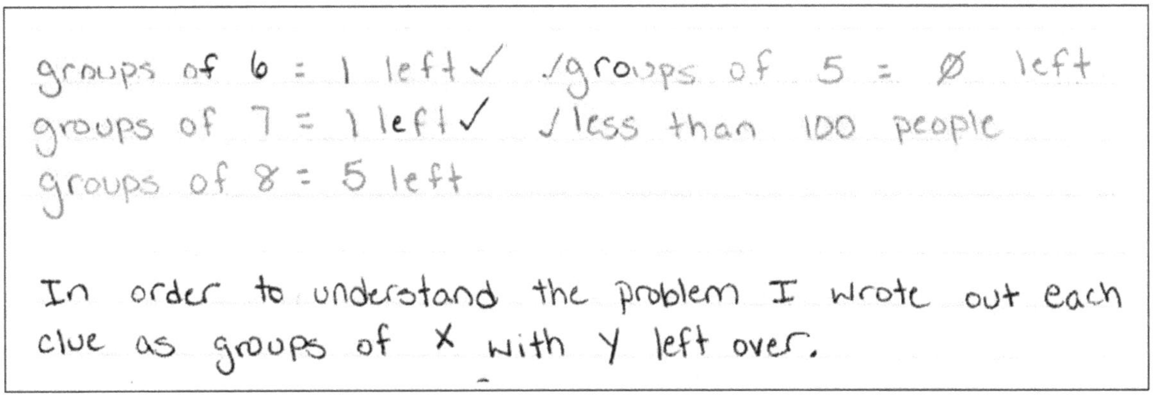

groups of 6 = 1 left ✓ ✓groups of 5 = ∅ left
groups of 7 = 1 left ✓ ✓less than 100 people
groups of 8 = 5 left

In order to understand the problem I wrote out each clue as groups of X with Y left over.

Figure 4. Direct paraphrasing of the problem statement (PT #12)

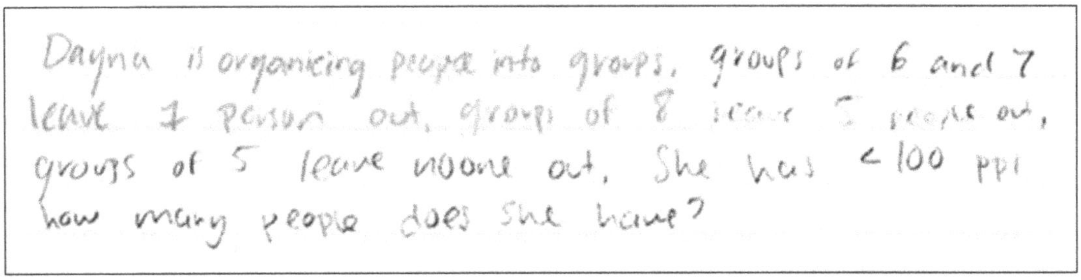

Dayna is organizing people into groups, groups of 6 and 7 leave 1 person out, groups of 8 leave 5 people out, groups of 5 leave noone out. She has < 100 ppl how many people does she have?

Interpreting

Some PTs presented the contextual problem information, clarifying the problem statement to bring its central mathematical meaning. Their written responses communicated aspects of the problem that were not readily apparent in the problem statement. We coded this way of presenting the problem information as *interpreting*. PTs who interpreted the problem communicated problem information beyond the mere summary of the problem statement. They shared mathematical insights that revealed their mathemati-

Figure 5. Example of interpreting problem statement mathematically (PT #1)

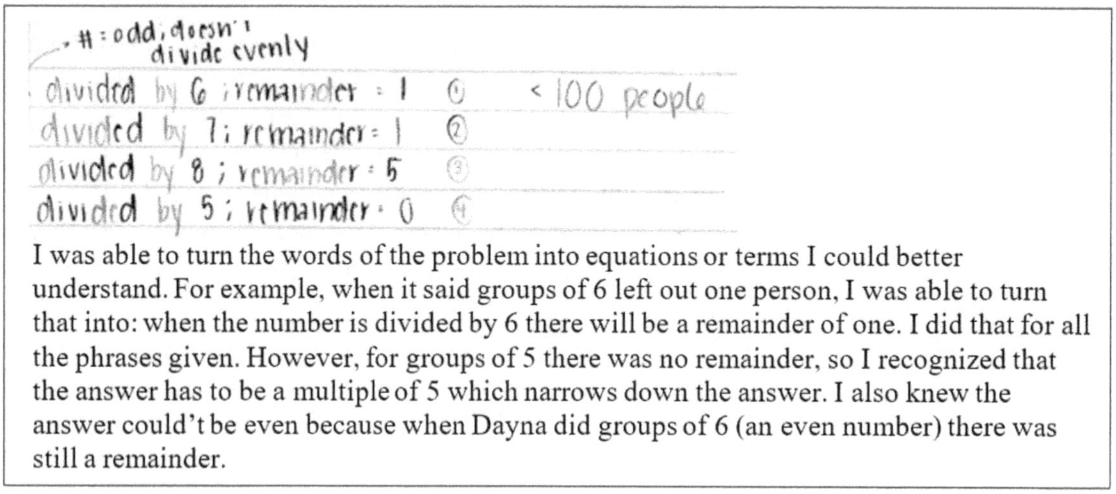
#: odd; doesn't divide evenly
divided by 6; remainder = 1 ①   < 100 people
divided by 7; remainder = 1 ②
divided by 8; remainder = 5 ③
divided by 5; remainder = 0 ④

I was able to turn the words of the problem into equations or terms I could better understand. For example, when it said groups of 6 left out one person, I was able to turn that into: when the number is divided by 6 there will be a remainder of one. I did that for all the phrases given. However, for groups of 5 there was no remainder, so I recognized that the answer has to be a multiple of 5 which narrows down the answer. I also knew the answer could't be even because when Dayna did groups of 6 (an even number) there was still a remainder.

Figure 6. Interpreting problem statement (PT #11)

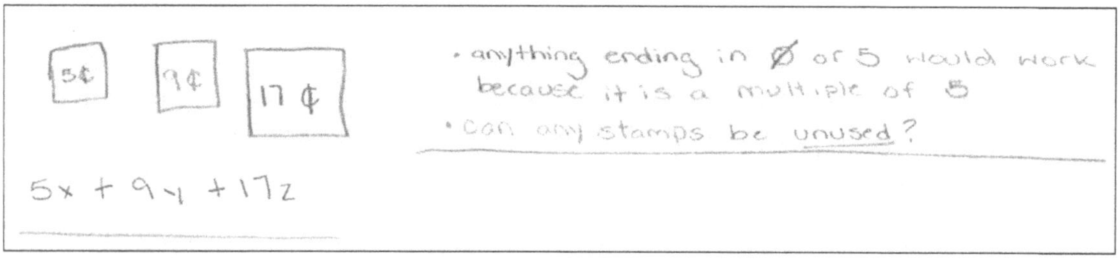

cal understanding of the problem, ideas about a potential solution, or solution strategies. Their written explanations revealed how they reinterpreted and mathematized the contextual situation presented in the problem statement.

Unlike the solution excerpts in Figures 3 and 4 above that conveyed only the direct information about the problem (i.e., contextual information), the explanations included in Figures 5 and 6 reveal that PTs shared their mathematical interpretations of the problem statement.

Figure 6 provides another illustration of *interpreting* by an excerpt from a PT's solution to the Stamps problem. The PT presented information about the problem using iconic representations and articulated their understanding that the problem involves reasoning about a linear combination of the given values. They also documented that they considered the implications of using the 5¢ stamp, revealing their reasoning about multiples of five, and communicated what they still sought to understand.

Strategic Reasoning

We operationalized PTs' strategic reasoning in terms of evidence of successful strategies for solving each problem. In problem-solving, strategic competence refers to the ability to represent the problem mathematically and formulate an effective strategy for solving the problem. Within PTs' written responses

Table 2. Summary of PTs' strategic reasoning (corporation meeting problem)

Strategic reasoning	Strategy description	# of PTs (n = 18)	Solution success
Searching within all multiples of five	Recognizing that the problem solution lies within the list of multiples of five and reducing the list of multiples of five using modified criteria	7	yes (5) no (2)
Searching within odd multiples of five	Recognizing that the problem solution lies within the list of odd multiples of five and searching within this list for numbers that meet the remaining conditions	1	yes
Checking multiple lists	Using the meaning of the remainder and generating lists of numbers that meet each of the specified conditions and seeking commonalities across them	4	yes
Systematic checking	Systematically checking all numbers between one and 100 for the given sets of conditions	2	yes
Least common multiples	Generating and checking the list of least common multiples	1	yes
Unsystematic checking	Unsystematic checking of the problem conditions for a subset of numbers <100	1	no

Note. *Two PTs did not complete this problem

to the first problem (Corporation Meeting), we identified five distinct ways of strategic reasoning (Table 2). We describe and illustrate the three most prevalent: Searching within the multiples of five, reasoning about commonalities across multiple lists, and systematic checking.

Table 2 shows that most PTs successfully solved this problem. Two of the unsuccessful solutions based on reasoning about all multiples of five included careless arithmetic errors or were incomplete (not checking all conditions) rather than a lack of strategic approach and problem understanding. The unsuccessful solution that stemmed from the unsystematic checking lacked a strategic approach and focused on guessing and checking a limited number of possibilities.

Searching Within Multiples of Five

This reasoning strategy was the most prevalent. PTs who used this strategy recognized that the solution could possibly exist within the set of multiples of five and searched for a solution by systematically reducing the list of multiples of five. To do so, they applied *modified* criteria to eliminate unwanted possibilities. The modified criteria stemmed from their reinterpretation of the problem conditions. For example, instead of reasoning about numbers that generate the remainder of one when divided by six (as provided in the problem statement), PTs reasoned about numbers divisible by six and reduced the initial list of possibilities by eliminating all multiples of six (Figure 7). They continued this reasoning in successive steps, considering other conditions included in the problem statement.

PT #12 followed describing their strategy and justification:

I started by writing out a list of all the numbers that were divisible by 5 that were less than 100. Then I eliminated multiples of 6, 7, 8 because I knew that there was a remainder. I knew that a number was 1 more than a multiple of 6 so I eliminated all the numbers that were not that. I did the same for numbers 1 above multiples of 7. That left me with 85, which I confirmed that it was 5 more than a multiple of 8.

Figure 7. Documenting reasoning focused on reducing the list of multiples of five (PT #12)

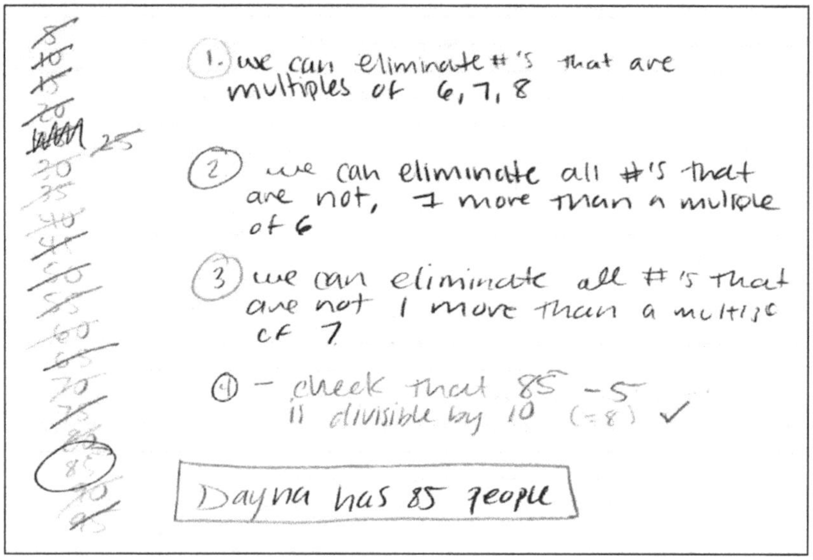

Once I figured out that I needed to eliminate numbers that were not, multiple of 6 and 7, the problem got a lot easier. (PT #12)

Two PTs explicitly recognized and expressed in their written reports that the solution would lay within the *odd* multiples of five, but only one of them used this fact as a starting point for strategy development. The other PT solved the problem using the generating multiple lists strategy described next.

Generating Multiple Lists and Checking for Commonalities.

PTs who developed this strategy reasoned about the meaning of remainders and generated lists of numbers that produced the given remainders. They then examined the generated lists for common elements to identify the problem solution. The excerpt in Figure 9 illustrates this strategy and this line of reasoning.

Figure 8. Documenting reasoning about common elements across multiple lists (PT #3)

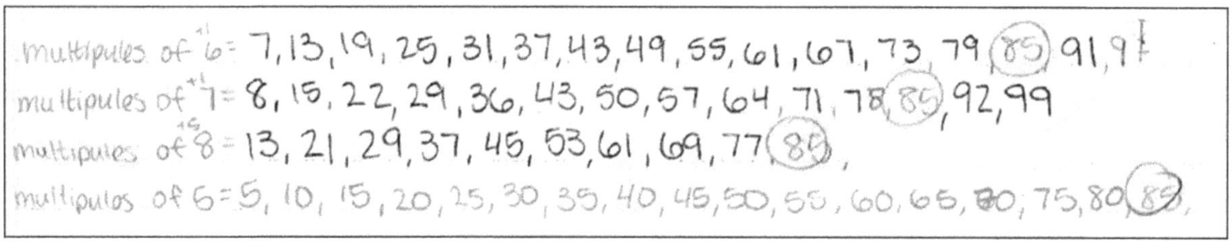

PT #3 further explained:

First, I listed all the multiples of 6 and added 1 to each number. Then I did the same thing with 7 and added 1 to each number as well. Then I did that to 8 and added 5 to each number and then I wrote out all multiples of 5, and I didn't need to add anything to that. Seeing all the numbers was so helpful to me. I looked at every number, and I got rid of the ones that weren't in all 4 [lists]. Finally, I got to the number 85. To double-check, I divided 85 by 6, 7, & 8. For each answer, I added the correct number for how many people there were left out. For each one, they all came out to 85, so there were 85 people in the meeting. (PT #3)

Systematic Checking

PTs who used this strategy recognized that the problem solution would lie within the multiples of five and generated complete lists of multiples of 6, 7, and 8 (accounting for the respective remainders) while simultaneously checking each successive value against the list of multiples of 5. Included in Figure 9 is PT's #4 solution, followed by an accompanying explanation to illustrate this strategy:

Figure 9. Systematic checking of all possible cases (PST #4)

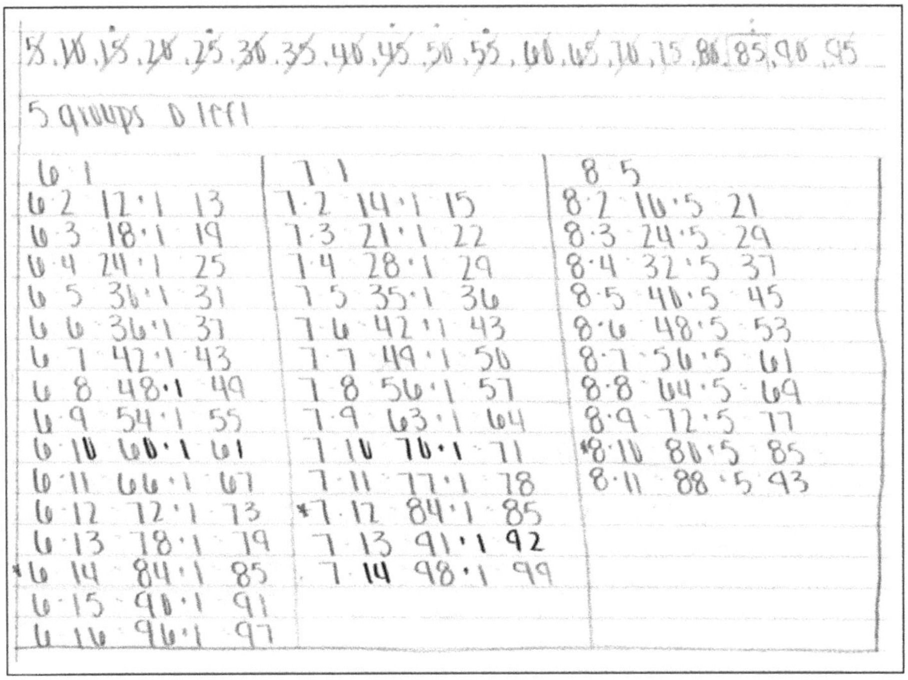

5, 10, 15, 20, 25, 30, 35, 40, 45, 50, 55, 60, 65, 70, 75, 80, 85, 90, 95

5 groups 0 left

6·1	7·1	8·5
6·2 12·1 13	7·2 14·1 15	8·2 16·5 21
6·3 18·1 19	7·3 21·1 22	8·3 24·5 29
6·4 24·1 25	7·4 28·1 29	8·4 32·5 37
6·5 30·1 31	7·5 35·1 36	8·5 40·5 45
6·6 36·1 37	7·6 42·1 43	8·6 48·5 53
6·7 42·1 43	7·7 49·1 50	8·7 56·5 61
6·8 48·1 49	7·8 56·1 57	8·8 64·5 69
6·9 54·1 55	7·9 63·1 64	8·9 72·5 77
6·10 60·1 61	7·10 70·1 71	*8·10 80·5 85
6·11 66·1 67	7·11 77·1 78	8·11 88·5 93
6·12 72·1 73	*7·12 84·1 85	
6·13 78·1 79	7·13 91·1 92	
*6·14 84·1 85	7·14 98·1 99	
6·15 90·1 91		
6·16 96·1 97		

PT #4 followed describing her reasoning:

When I understood what I had to do I started to do calculations. The approach I took was first writing out all multiples of 5. Then I started with the group of 6 and with 1 person left out. I started by multi-

plying 6 by 2 which gave me 12 and I added 1 because 1 person was left. It gave me 13 which wasn't on the numberline above so I moved on to the next number. Then I did 6 times 3 plus 1 to get 19 which also was not on the numberline. Then I did 6 times 4 and add 1 to get 25 and that was on the number line so I put dot above it because it could be a possibility. I continued doing this until I reached under 100. ..I started with 8 times 2 and add 5 to get 21. I did this until I got 93. When I finished doing all the calculations, I found out that 85 was the only one that worked because it was the only one that had 3 dots. So there were 85 people in the meeting.

Table 3 summarizes PTs' reasoning about the second problem (Stamps). Figures 10 and 11 include excerpts from PTs' written responses.

Table 3. PTs' strategic reasoning (stamps problem)

Strategic reasoning	Strategy description	# of PTs* (n = 18)	Solution success
Reasoning about infinite patterns of one's place values	Recognizing that if a number ending in x is possible, any number $10k + x$ (by adding groups of two five's) is also possible. Identifying infinite possible patterns	4	yes (4)
Generating a finite list of possible denominations and checking possibilities	Generating an initial finite list of possible denominations and checking and eliminating possibilities	11	yes (11)
Constructing unsystematic lists or guessing and checking	Guessing and checking in an unsystematic way	2	no (2)

Note. *One PT did not complete this problem

Table 3 shows that all but two PTs were successful in generating the correct numerical answer for this problem. PTs' unsuccessful solutions stemmed from a lack of a systematic strategic approach and a focus on unsystematic guessing and checking.

Reasoning About Infinite Patterns of One's Place Values

This strategy stemmed from reasoning about patterns of linear combinations of 17, nine, and five and recognizing that any combination of given stamp denominations could be indefinitely extended to form a sequence of numbers sharing the same one's place value. The solution excerpt in Figure 10 illustrates reasoning about infinite sequences of possible combinations, each sequence ending in the same digit. Recognizing the infinite nature of these sequences, PT #13 organized information about possible denomination values in a 10-by-10 array (considering possible postage values between zero and 99), reasoned about sequences of possible combinations of the stamps and eliminated those possibilities from her array.

PT #13 further explained: "After many tries, I found that my answer is 21. I slowly removed rows from these patterns like 14, 24, 34, 44…. or 22, 32, 42, 52…and so on. From there, I was left with small numbers that I had to manually check off from my list."

Figure 10. Documenting reasoning about sequences of one's place values (PT #13)

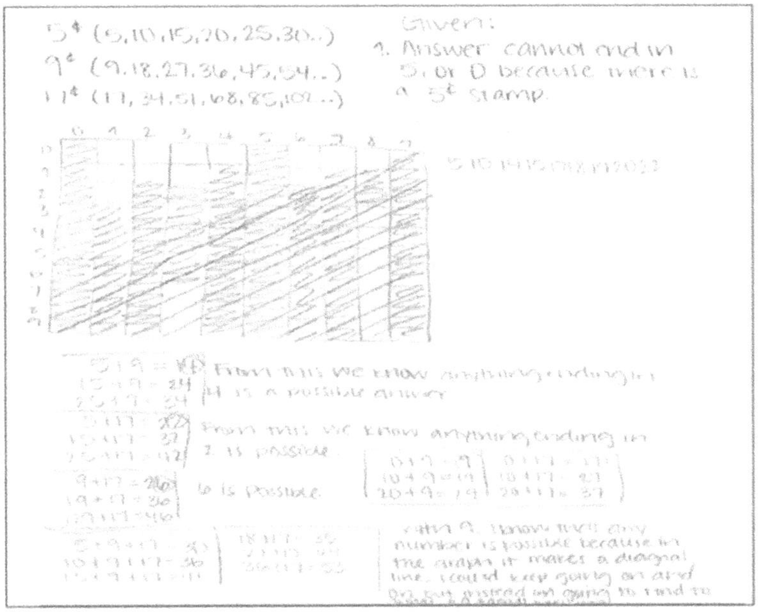

Generating a Finite List of Possible Denominations and Checking Possibilities

This strategy was by far the most frequent. To solve the problem, PTs generated an initial list of possible stamp denominations and continued checking possible combinations of the given stamp values against their initial list. All PTs who used this strategy constructed and checked sufficiently long lists of possible

Figure 11. Reasoning about a finite number of possible denominations (PT #6)

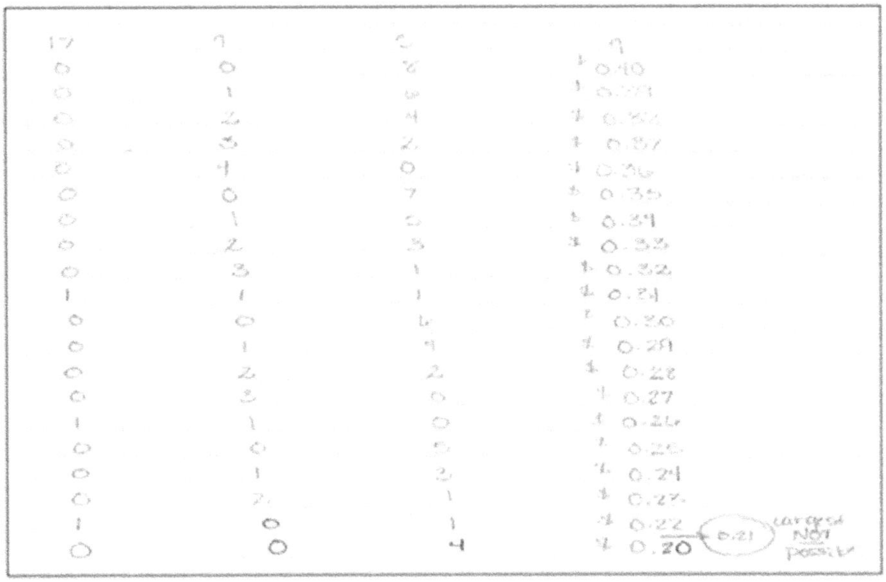

postage values to discover the correct answer of 21. However, none of the PTs who used this strategy justified that *all* values beyond the maximum value in their list were possible. Their justifications were limited to only the results observed within their initial finite lists. Figure 11 and the accompanying written explanation below illustrate this line of reasoning.

PT #6 described their thinking:

I was first confused about how I should begin. I first started by writing down important parts of the problem, like denominations for the stamps as well as the job to find the largest amount of postage that cannot be made. I worked my way up to 0.20 and 0.30 cents, combining the denominations in different ways to reach different answers. However, this left me even more confused so I thought I have to approach the problem in a different way. I decided to approach the problem backwards. I knew that the total amount I get from combining the denominations is infinite so I needed to start smaller. I thought that starting at 0.40 cents was a good jumping-off point. I began combining 0.17, 0.9, and 0.5 cents in various ways to get 0.39, cents, 0.38 cents, and so on. As I did so, I was looking at the combinations and their totals, ensuring that I was getting a number that was chronological to the one above it and worked in a coherent sequence. I saw this to be true until I reached 0.23 cents. I then tried to make 0.22 cents, and this was possible, so I moved to 0.21 cents, but that was impossible. So I tried 0.20 cents, and it was possible. So I knew my answer had to be 0.21 cents as the largest not possible.

Communicating Thinking

As the solution excerpts from both problems illustrate (Figures 7 – 11), PTs used multiple forms of representations to communicate their reasoning, their understanding of the problem, and the solution strategies they devised. In their problem explanations, PTs use different cognitive tools to communicate their reasoning, combining tabular, numerical, verbal, and iconic representations (and, in a few cases, used symbolic notations, as shown in Figure 6). They used written language (to a different extent and with varying levels of precision) to provide meanings for the representations through which they

Table 4. Overall assessment of mathematical communication in PT's written solutions

Features of Mathematical Communication	Description	Problem one	Problem two
		PTs* $n = 18$	PTs $n = 18$
Communicating how	*Competent*: Gives a comprehensive and clear description of strategy (how the problem was solved).	6	5
	Emerging: Gives a partial description of how the problem was solved.	7	11
	Insufficient: Gives a limited explanation of the steps of the solution.	3	1
Providing justifications	*Competent*: Gives comprehensive and valid justification for the reasoning behind the procedure and the result.	8	4
	Emerging: Gives partial justification (justifies some steps of a procedure or partially justifies results).	3	8
	Insufficient: Gives a limited justification or does not provide justification.	5	5

Note. *Two PTs did not complete the first problem, and one PT did not complete the second problem

Figure 12. An example of a competent explanation (PT #6)

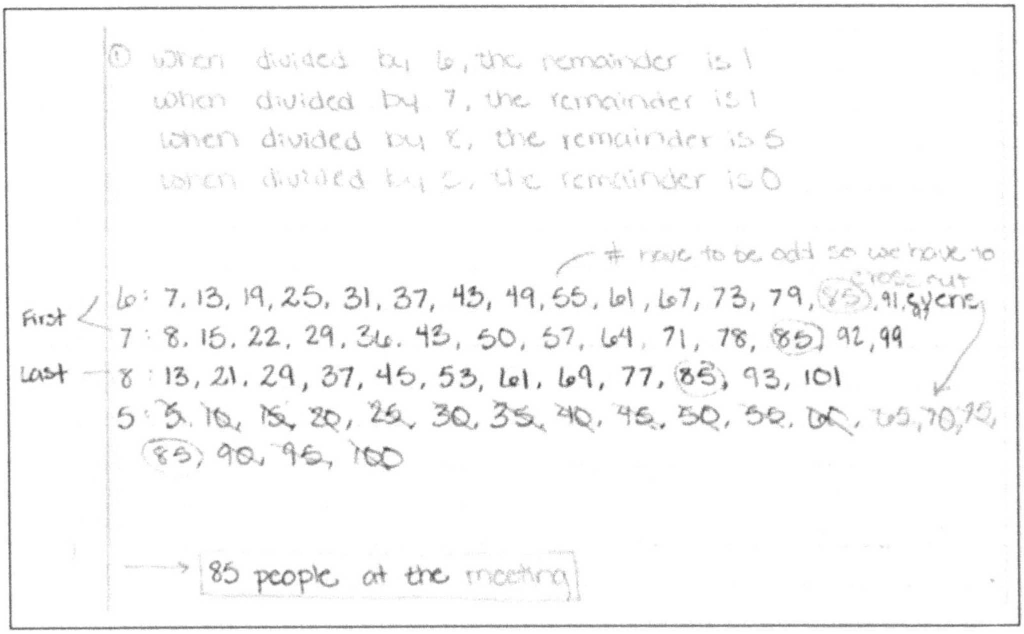

communicated their reasoning. Included in Table 4 is a summary of the overall characteristics of PTs' written responses, aggregated by the level of clarity of their explanations of solution procedures and completeness of justifications.

We illustrate a *competent* problem explanation using PT #6 response (Figure 12) to the Corporation Meeting problem.

PT #6 included description of their strategy:

I first wrote out the size groups and then how many people were left over when these groups were made. I then made lists of the multiples of these numbers but added however many people were left out in correlation with that number to the multiples. For example, if I was doing multiples of 6 and within the problem, it said one person was left out, my list would be 7, 13, 19, 25, and so on. I repeated this process with 7, 8, and 5. I used the multiples of 5 as my point of comparison since this was the only number of people per group that ended up with no leftover people. From there, I knew that the number of people I was looking for had to be odd because the first list I was making was all odd. So, I crossed off all the even numbers. From there, I tried to find the numbers that corresponded within the list of multiples of 5 that were left after crossing off all the even numbers. Here, I would look between all the lists I had and found that the only number that corresponded to all lists was 85. I then checked 85 with all three group sizes to make sure they all corresponded to their leftover amounts and found that 85 was the correct answer.

We assessed this explanation as *competent* in both aspects (i.e., communicating how and providing justifications) because PT articulated information about the problem, elucidated how each list was generated, was transparent in describing each list, describing how she recognized that the problem solution would be odd, and how they searched across the lists to find the answer to the problem. PT #6 also justified their choice of numbers within each list and included justification for their result.

In contrast, the earlier discussed response of PT #4 (Figure 9) provides an example of an *emerging* explanation in communicating how and an *insufficient* explanation in providing justifications. PT#4 described how they generated lists of numbers by showing all calculations, but their explanation of their procedure was unclear. For example, the written solution was not transparent about what the "dot notation" documented and how the PT reasoned to assign "dots" to the entries in generated list of multiples of five. The explanation was limited to describing what calculations were completed to generate the various lists of numbers and was limited in terms of describing any decisions the PT made to identify the problem solution from the generated lists. PT #4's attempt to justify their final result ("because it was the only one that had 3 dots") was also limited.

Table 4 illustrates that explaining and justifying problem solutions might not be easy for PTs, despite the fact that most PTs could generate the correct numerical answers for both problems. Less than half of the PTs competently described their strategies for the first problem, and an even smaller number of PTs provided competent explanations of strategies they developed to solve the second problem. While most of the PTs demonstrated sufficient mathematical knowledge and problem-solving skills (i.e., the ability to interpret the contextual situation mathematically and devise and implement a solution), only a few PTs thoroughly articulated their reasoning and strategies in writing. Similarily, in their written responses most PTs provided only partial justifications, and a few did not attempt to justify and exclusively focused their explanations on articulating what they did to solve the problem.

While we cannot establish any direct relationship between the requirement of writing complete problem explanations and PTs' problem-solving and reasoning skills, many of the PTs indicated that writing about and reflecting on the problems they solved helped them organize what they knew and think about their own understanding. Our data suggest that writing might have forced PTs to confront their understanding and prompted them to work towards achieving deeper levels of understanding needed to develop and adequately describe their solution strategies. We hypothesize that writing might have helped PTs orient themselves to each problem. PTs frequently commented on difficulties they had designing a strategy for solving each problem. At the same time, many PTs acknowledged that writing helped them to organize their thinking and supported their problem understanding. Comments such as the following were frequent across the analyzed responses:

I didn't have a clear idea on how to start this problem. I thought that maybe drawing circles that represent people would be a good start. However, this plan failed quickly. ....I first wrote out the size of the groups and then how many people were left over once these groups were made. ... Reflecting on the process as a whole, I am very pleased with the result. I wasn't entirely sure how to solve this problem at first, but I worked to find a systematic way of solving the problem in its entirety. (PT #6)

The same PT commented on her solution to the second problem:

Much of what I understood from the problem reading did't make sense. However, as I started to write down the important part of the problem, it all started to unfold. I went through a trial and error in my first attempt but I was quick to find a solution that worked. Reflecting on this problem as a whole, I found great difficulty in developing the procedure.

FUTURE RESEARCH DIRECTIONS

This work provides a good starting point for further research to investigate how explanatory writing contributes to PTs' reasoning and problem-solving skills. The data presented here show that writing is a challenging cognitive process in problem-solving contexts that requires careful examination of the thinking one chooses to articulate.

As Table 4 shows, not all of the PTs in this study provided complete descriptions of *how* they solved the problem, and only a few included complete justifications. At the same time, the data show that most PTs successfully reasoned about each problem situation to solve unfamiliar contextual problems. Almost all PTs identified relevant mathematical concepts and developed strategies for solving each problem. This underscores that mathematical reasoning and communication of that mathematical reasoning are symbiotic but different sets of skills. Analyzing students' responses to non-contextual tasks, other researchers reported similar results. For example, Evens and Houssart (2004) observed elementary school students who successfully reasoned to generate problem solutions but had difficulty explaining their reasoning in writing. Colonnese and Castro (2023) made a similar observation while studying the written responses of prospective elementary school teachers and their reasoning about computational tasks. Further research needs to explore effective ways in which mathematics educators can support PTs in developing an understanding and knowledge of how to write effective mathematical explanations for contextual problems.

With a focus on PTs' development as problem solvers and doers of mathematics, we examined what PTs' written explanations reveal about their reasoning rather than what their written explanations lack. Thus, we did not catalog any errors nor focused our analysis on levels of precision in using mathematical language within and across the PTs' responses (e.g., precision in the use of mathematical symbols, terms, notation, and phrases PTs used to connect ideas or their use of grammar). However, like other scholars, we recognize that the effectiveness of written mathematical explanations relates to the choices of presented information *and* the overall quality of writing and precision of the mathematical language used (Powell & Hebert, 2016). Future research that explores these features of PTs' explanatory responses can contribute to building a more robust understanding of how PTs communicate and explain solutions to contextual problems in writing. Understanding and classifying errors within PTs' written mathematical explanations can contribute to building interventions focused on supporting PTs' learning to explain. Kuzle (2013) argued that PTs might need more focused instruction on mathematical writing, including stating the problem information and the problem goal, linking and explaining representations, and integrating mathematical notation with words. Understanding different types of errors in PTs' mathematical writing can provide a step toward designing appropriate interventions to support PTs' mathematical writing to explain and justify their reasoning.

Working with a group of elementary PTs, Magiera and Zambak (2020) identified several criteria that PTs used when evaluating student-generated written explanatory arguments in problem-solving contexts. These included explanation correctness, overall organization, clarity of assumptions-mathematical foundations, presence of justifications, explanation generality, and explanation communicative power, which PTs interpreted through the lens of their ability to understand the student's explanation. Future research also needs to examine how PTs interpret and view features of effective problem explanations in their own work. Examined from both a learner's and a future teacher's perspective, how PTs perceive effective explanations in problem-solving contexts can add to the knowledge of student writing in mathematics.

CONCLUSION

This chapter explored and illustrated how explanatory writing in the context of a problem-solving course for future elementary school teachers provides insights into PTs' reasoning in contextual problem situations. By presenting PTs with contextual problems, which they had to interpret mathematically, and analyzing PTs' written explanations of their problem solutions, this work brings new perspectives to the research on students' writing in mathematics. With a focus on how PTs communicate their understanding of the problems they solve, we discussed PTs' strategic reasoning in contextual problem situations that require developing new approaches and ways of thinking. We also described the overall nature of PTs' written responses with attention to how they articulate their reasoning about solution procedures they devised, how well they justify, and what representational choices they make to communicate their solutions.

The results reported in this chapter add to prior research that describes how PTs communicate their mathematical reasoning in writing (Albano et al., 2023; Colonnese & Casto, 2023; Taylor & McDonald, 2007). First, our analysis revealed that PTs took two approaches to communicate contextual problem information in their explanatory mathematical writing. One approach emphasized providing information about the problem that was readily apparent in the problem statement (*summarizing*). Another approach emphasized sharing insights about the mathematical interpretation of the contextual situation and articulating underlying mathematical concepts not explicit in the problem statement (*interpreting*). PTs who interpreted problems in their written explanations rendered the underlying mathematical meaning of the contextual problem situation and communicated that meaning to the intended readers. We identified that summarizing problem context was far more frequent across the PTs' written responses than providing explicit interpretations of that content with a focus on mathematical meaning. Given the centrality of explanations in teaching and learning (Charalombous et al., 2011), this result brings a question of how to support PTs so they emphasize meaning-making in all aspects of their written explanations. While in this study, we did not provide PTs with opportunities to think about explanatory mathematical writing from the perspective of a reader (e.g., engage PTs in analyzing written responses), we believe that PTs might benefit from analyzing written problem solutions to help them focus on conveying meaning in their own mathematical expositions.

In this chapter we reveal how PTs express their reasoning and share their understanding of contextual problem situations in writing. The results of the analysis of PTs' strategic reasoning in the context of both problems draw attention to and raise questions about PTs' understanding of the logical structure of both problems. We could only hypothesize that lack of such understanding could contribute to difficulties in providing complete and valid justifications we observed across PTs' responses. Problem one (Corporation Meeting) was set within a finite problem space, and thus could possibly be solved by reasoning about all possible cases. We illustrated this reasoning with an excerpt included in Figure 9. However, problem 2 (Stamps) was set within an infinite problem space. Thus, reasoning about a finite number of possible cases in the context of that problem (even if the number of cases examined was sufficient enough to generate the correct numerical result) could not lead to providing valid justification. As the presented data illustrate, more than 50% of PTs reasoned about and analyzed a finite number of cases while responding to the Stamps problem. Drawing PTs' attention to the logical structure of the problems they solve might prove helpful in supporting PTs' strategic knowledge. Even if PTs can solve mathematics problems, mathematics teacher educators need to assess their conceptual knowledge for mathematical reasoning through mathematical writing. Emphasis on effective communication in mathematics and the application of effective mathematical reasoning warrants the focus on developing instructional interventions that address these areas simultaneously.

ACKNOWLEDGEMENT

This research was partially supported by the National Science Foundation, grant NSF IUSE #2142646. Opinions, findings, conclusions, or recommendations are those of the authors and do not necessarily reflect the views of the funding agency.

REFERENCES

Albano, G., Swidan, O., & Pierri, A. (2023). A model for analyzing the explanatory writing of undergraduate students when solving mathematical tasks. *International Journal of Mathematical Education in Science and Technology*, *54*(2), 180–194. doi:10.1080/0020739X.2021.1949757

Berkenkotter, C. (1982). Writing and problem solving. In T. Fulwiler & A. Young (Eds.) *Language connection: Writing and reading across the curriculum.* (pp. 33—44). National Council of Teachers of English. https://wac.colostate.edu/books/language_connections/chapter3.pdf

Bicknell, B. (1999). The writing of explanation and justification in mathematics: Differences and dilemmas. In J. Truran & K. M. Truran (Eds.), *Proceedings of the 22 Annual Conference of the Mathematics Education Research Group of Australasia* (pp. 75—83). Mathematics Education Research Group of Australasia.

Brozo, W. G., & Crain, S. (2018). Writing in math: A disciplinary literacy approach. *The Clearing House: A Journal of Educational Strategies, Issues and Ideas*, *91*(1), 7–13. doi:10.1080/00098655.2017.1342435

Casa, T. M., Firmender, J. M., Cahill, J., Cardetti, F., Choppin, J. M., Cohen, J., & Zawodniak, R. (2016). *Types of and purposes for elementary mathematical writing: Task force recommendations*. Mathwriting. http://mathwriting.education.uconn.edu

Charalombous, Y. C., Hill, H. C., & Ball, D. (2011). Prospective teachers' learning to provide instructional explanations: How does it look and what it might take? *Journal of Mathematics Teacher Education*, *14*(6), 441–463. doi:10.100710857-011-9182-z

Clarke, D. J., Waywood, A., & Stephens, M. (1993). Probing the structure of mathematical writing. *Educational Studies in Mathematics*, *25*(3), 235–250. doi:10.1007/BF01273863

Colonnese, M. W. (2022). Exploratory writing to support mathematical sense making. *Mathematics Teacher PK-12. Learning and Teaching*, *115*(4), 265–271.

Colonnese, M. W., & Casto, A. R. (2023). Analysis of pre-service teachers' written mathematical reasoning. *Mathematics Teacher Education and Development*, *25*(1), 23–42.

Connolly, P. (1989). Writing and the ecology of learning. In P. Connolly & T. Vilardi (Eds.), *Writing to learn mathematics and science* (pp. 1–14). Teachers College Press.

Craig, T. S. (2016). The role of expository writing in mathematical problem solving, *African Journal of Research in Mathematics. Science and Technology Education*, *20*(1), 57–66. doi:10.1080/10288457.2016.1147803

Evens, H., & Houssart, J. (2004). Categorizing pupils' written answers to a mathematics test question: "I know but I can't explain". *Educational Research, 46*(3), 269–282. doi:10.1080/0013188042000277331

Fukawa-Connelly, T., & Buck, S. (2010). Using portfolio assessments to assess students' mathematical thinking. *Mathematics Teacher, 103*(9), 649–654. doi:10.5951/MT.103.9.0649

Hodds, M., Alcock, L., & Inglis, M. (2014). Self-explanation training improves proof comprehension. *Journal for Research in Mathematics Education, 45*(1), 62–101. doi:10.5951/jresematheduc.45.1.0062

Kenny, R., Shoffner, M., & Norris, D. (2014). Reflecting on the use of writing to promote mathematical learning: An examination of preservice teachers' perspectives. *Teacher Educator, 49*(1), 28–43. doi:10.1080/08878730.2013.848002

Kosko, K. (2016). Writing in mathematics. A survey of K-12 teachers' reported frequency in the classroom. *School Science and Mathematics, 116*(5), 276–285. doi:10.1111sm.12177

Kuzle, A. (2013). Promoting writing in mathematics: Prospective teachers' experiences and perspectives of writing when doing mathematics as problem-solving. *Center for Educational Policies Study Journal, 3*(4), 41–59. doi:10.26529/cepsj.222

Magiera, M. T., & Zambak, V. C. (2020). Exploring prospective teachers' ability to generate and analyze evidence-based explanatory arguments. *International Journal of Education and Science, (IJRES), 6*(2), 327—346.

McCarthy, D. (2010). Communicating in mathematics: Preparing prospective teachers to include writing in mathematics teaching and learning. *School Science and Mathematics, 108*(7), 334–340. doi:10.1111/j.1949-8594.2008.tb17846.x

McCarthy, L. A., & Faulkner, M. (2019). Integrating writing and mathematics: Journaling to increase learning and enjoyment while reducing anxiety. *Teaching Mathematics and its Applications: An International. The Journal of IMA / Islamic Medical Association of North America, 39*, 145–159.

National Council of Teachers of Mathematics. (2000). *Principles and standards for school mathematics*. NCTM.

Numkung, J., Herebrt, M., Powell, S., Hoins, M., Bricko, M., & Torchia, M. (2019). Comparing and validating four methods for scoring mathematical writing. *Reading & Writing Quarterly, 36*(2), 157–175. doi:10.1080/10573569.2019.1700858

Pólya, G. (1945). *How to solve it*. Princeton University Press. doi:10.1515/9781400828678

Porter, M. K., & Masingila, J. O. (2000). Examining the effect of writing on conceptual and procedural knowledge in calculus. *Educational Studies in Mathematics, 42*(2), 165–177. doi:10.1023/A:1004166811047

Powell, S. R., & Hebert, M. A. (2016). Influence of writing ability and computation skill on mathematics writing. *The Elementary School Journal, 117*(2), 310–335. doi:10.1086/688887

Prediger, S., & Erath, K. (2014). Content, interaction, or both? Synthesizing two German traditions in a video study on learning to explain the mathematics classroom microcultures. *Eurasia Journal of Mathematics, Science and Technology Education, 10*(4), 313–327. doi:10.12973/eurasia.2014.1085a

Pugalee, D. K. (2001). Writing, mathematics, and metacognition: Looking for connections through students' work in mathematical problem solving. *School Science and Mathematics*, *101*(5), 236–245. doi:10.1111/j.1949-8594.2001.tb18026.x

Pugalee, D. K., DiBiase, W., & Wood, K. (1999). Writing and the development of problem solving in mathematics and science. *Middle School Journal*, *30*(5), 48–52.

Saldaña, J. (2016). *The coding manual for qualitative researchers*. Sage Publications Inc.

Santos, L., & Semana, S. (2015). Developing mathematics written communication through expository writing supported by assessment strategies. *Educational Studies in Mathematics*, *88*(1), 65–87. doi:10.100710649-014-9557-z

Seto, M., & Meel, E. D. (2006). Writing in mathematics: Making it work. *PRIMUS: Problems, Resources, and Issues in Mathematics Undergraduate Studies*, *16*(3), 204–232. doi:10.1080/10511970608984147

Sezer, R. (2010). Effects of reflective writing in mathematics methods courses on pre-service teachers' perceptions. *Eurasian Journal of Educational Research*, *40*, 188–206.

Shield, M., & Galbraith, P. (1998). The analysis of student expository writing in mathematics. *Educational Studies in Mathematics*, *36*(1), 29–52. doi:10.1023/A:1003109819256

Steele, D. (2007). Understanding students' problem-solving knowledge through their writing. *Mathematics Teaching in the Middle School*, *13*(2), 102–109. doi:10.5951/MTMS.13.2.0102

Swafford, J., & Bryan, J. (2000). Instructional strategies for promoting conceptual change: Supporting middle school students. *Reading & Writing Quarterly*, *16*(2), 139–161. doi:10.1080/105735600278006

Taylor, J. A., & McDonald, C. (2007). Writing in groups as a tool for non-routine problem-solving in first year university mathematics. *International Journal of Mathematical Education in Science and Technology*, *38*(5), 639–655. doi:10.1080/00207390701359396

Teledahl, A. (2017). How young students communicate their mathematical understanding. *International Journal of Mathematical Education in Science and Technology*, *48*(4), 555–572. doi:10.1080/0020739X.2016.1256447

Waywood, A. (1992). Journal writing and learning mathematics. *For the Learning of Mathematics*, *12*(2), 34–43.

ADDITIONAL READING

Graham, S., Kiuhara, S. A., & MacKay, M. (2020). The effects of writing on learning in science, social studies, and mathematics: A meta-analysis. *Review of Educational Research*, *90*(2), 179–226. doi:10.3102/0034654320914744

Guce, I. K. (2017). Mathematical writing errors in expository writings of college mathematics students. *International Journal of Evaluation and Research in Education*, *6*(3), 233–242. doi:10.11591/ijere.v6i3.8549

Hallman-Trasher, A., Rhodes, G., & Schultz, K. (2020). Supporting mathematics teacher educators' practices for facilitating prospective teachers' mathematical explanations in content courses. *The Montana Math Enthusiast, 17*(2&3), 883–906. doi:10.54870/1551-3440.1506

Kline, S., & Ishii, D. K. (2008). Procedural explanation in mathematics writing. A framework for understanding college students' effective communication practices. *Written Communication, 25*(4), 441–461. doi:10.1177/0741088308322343

Kulikovich, J. MMason, LBrown, S. (2008). Evaluating fifth- and sixth-grade students' expository writing: Task development, scoring, and psychometric issues. *Reading and Writing, 21*(1-2), 153–175. doi.org/10.1007/s11145-007-9068-8 doi:10.100711145-007-9068-8

Lee, G., Lim, C., & Leong, L. (2020). Use mathematical writing as a practical approach to increase students' problem solving skills: A case study. *The Montana Math Enthusiast, 17*(1), 239–272. doi:10.54870/1551-3440.1486

Morgan, C., Craig, T., Schuette, M., & Wagner, D. (2014). Language and communication in mathematics education: An overview of research in the field. *ZDM Mathematics Education, 46*(6), 843–853. doi:10.100711858-014-0624-9

O'Halloran, K. L. (2008). Mathematical discourse: Language, symbolism and visual images. *Continuum.*

Powell, S. R., Hebert, M. A., Cohen, J. A., Casa, T. M., & Firmender, J. M. (2017). A synthesis of mathematics writing: Assessments, interventions, and surveys. *Journal of Writing Research, 8*(3), 493–530. doi:10.17239/jowr-2017.08.03.04

Teuscher, D., Hodges Kulinna, P., & Crooker, C. (2015). Writing to learn mathematics: An update. *The Mathematics Educator, 24*(2), 56–78.

KEY TERMS AND DEFINITIONS

Contextual Problems: Mathematical problems with experientially real contexts. The problems may provide a basis for connecting informal and formal knowledge of mathematics.

Explanatory Writing in Mathematics: Writing that targets a specified audience (e.g., a teacher, problem client, or fellow student) focused on describing, explaining, and transmitting mathematical ideas.

Mathematical Discourse: Verbal, visual, and written communication that facilitates the exchange of ideas among a mathematics classroom community members.

Problem Explanation: Communicates the meaning of a problem by providing grounds, evidence, and reasons for why a statement being explained (a problem solution) is true (or not).

Strategic Reasoning: The ability to represent the problem mathematically and formulate an effective strategy for solving the problem.

ENDNOTE

1 Problems are selected from curricular materials from: Johnson, K., Herr, T., & Kysh, J. (2018). *Crossing the River with Dogs: Problem Solving for College Students,* (3rd Edition). Hoboken, NJ: John Wiley & Sons.

Chapter 4
Enhancing Preservice Teachers' Mathematical Knowledge Through Writing:
From Peer Review to Inter-Institutional Conversations

Christopher Dennis Nazelli
https://orcid.org/0000-0002-4511-5712
Wayne State University, USA

S. Asli Özgün-Koca
Wayne State University, USA

Deborah Zopf
Henry Ford College, USA

ABSTRACT

In this chapter, the authors present and analyze a mathematical writing activity for preservice elementary school teachers with a novel peer review component. The products of this activity, referred to as "journals", are detailed representation-rich explanations of problems solved in small groups. These journals were exchanged across institutions, and the addition of a computer-mediated communication platform transformed the peer reviews into inter-institutional conversations. An analysis of the journals and the conversations they sparked reveal four opportunities to develop mathematical knowledge for teaching. Evidence, in the form of journal and review excerpts as well as two case studies, suggests that the process enabled preservice teachers to deepen their understanding of foundational concepts from the elementary curriculum, to communicate mathematics more effectively to others, to better make sense of the mathematical thinking and writing of others, and to incorporate their colleagues' suggestions into their future writing. Potential future research paths are discussed.

DOI: 10.4018/978-1-6684-6538-7.ch004

INTRODUCTION

In this chapter we describe a mathematical writing activity designed to enhance preservice elementary school teachers' mathematical knowledge for teaching (MKT) (Ball et al., 2008). The addition of an inter-institutional exchange of this written work enriched the learning opportunities for both parties involved in the peer-review process; and, after shifting to an online platform, the process evolved into a mathematical conversation grounded in their written work. We will identify four key opportunities to learn within this activity and present evidence of growth observed in our students' writing and their ability to offer, accept, and use critique of mathematical explanations and representations.

BACKGROUND

The relationship between oral and written mathematical communication and mathematical understanding has been highlighted by numerous scholars. Hiebert and his colleagues (1997) emphasized reflection and communication as major components of learning mathematics with understanding. Accordingly, the Standards for Mathematical Practice from the Common Core State Standards and the key standards documents of the National Council of Teachers of Mathematics (NCTM) encourage mathematics teachers to support written and oral communication of all learners of mathematics (National Governors Association Center for Best Practices, and Council of Chief State School Officers, 2010; NCTM, 1989, 2000, 2014).

NCTM's (2020) Standards for the Preparation of Middle Level Mathematics Teachers expect that "Candidates [will] organize their mathematical thinking and use the language of mathematics to express ideas precisely, both orally and in writing to multiple audiences" (p. 18). Therefore, it is important that preparation programs for preservice teachers (PSTs) provide rich mathematical writing experiences. Writing in mathematics can help PSTs "consolidate their thinking because it requires them to reflect on their work and clarify their thoughts about the ideas" and communicate them (p. 61). Through communication, ideas become public. We can "challenge each other's ideas and ask for clarification and further explanation" which might, in turn, result in additional reflection (Hiebert et al., 1997, p. 5). Thus, we believe that mathematical writing is an important activity to support PSTs as they transition from learning mathematics for personal use to learning mathematics for others. Expressing their ideas clearly in writing helps PSTs build their mathematical knowledge for teaching, but the act of reading, interpreting, and reflecting on the mathematical ideas of others offers the potential for additional growth as future teachers.

The APEX Cycle and Inter-Institutional Conversations

Content courses are critical sites for the development of mathematical knowledge and practices that teachers will use when teaching children (Ball, 2000). As part of a large Noyce Mathematics Fellows project, we redesigned the Mathematics for Elementary Teachers courses at a large urban public university and a nearby college to improve students' MKT (Ball et al., 2008). This conceptualization of the knowledge needed to teach mathematics encapsulates both content and pedagogical knowledge. The domains of MKT that are relevant to our work are Common Content Knowledge, Specialized Content Knowledge, and Knowledge of Content and Teaching. Common Content Knowledge includes facts, skills, and understandings that are drawn upon by all users of mathematics. Specialized Content Knowledge is the

Figure 1. The APEX cycle

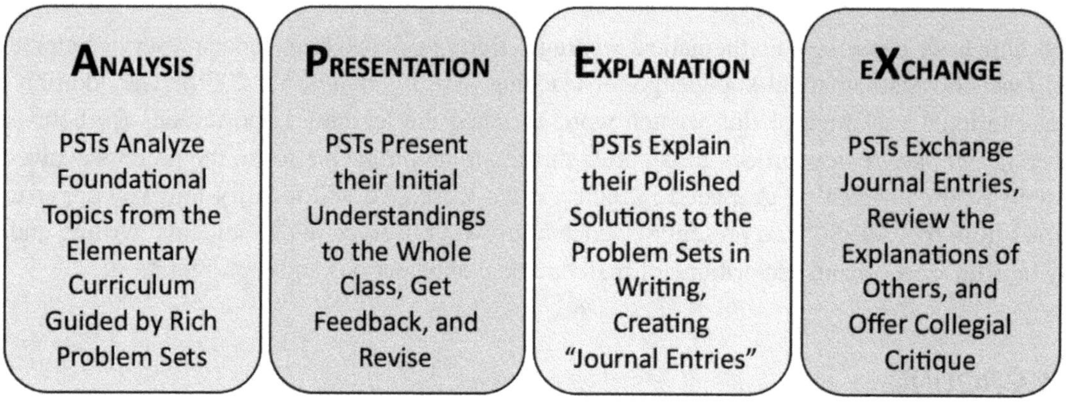

knowledge used by teachers, for example, when responding to student thinking. Knowledge of Content and Teaching represents the knowledge of how to effectively convey mathematics through examples, representations, and discussions (Ball et al., 2008).

Our students engaged with mathematics through a sequence of problem analysis, presentation, explanation, and exchange, that we have come to call the APEX Cycle (Figure 1). We present PSTs with a problem set designed to deepen their understanding of the fundamental concepts that undergird the mathematics they will teach to children. The PSTs analyze the problem set in small groups, first drafting their responses to the problem sets independently, then working with their group to compare their work and share ideas. This phase of the cycle requires the PSTs to "…work backward from mature and compressed understanding of the content to unpack its constituent elements. Knowing for teaching requires a transcendence of the tacit understanding that characterizes and is sufficient for personal knowledge and performance" (Ball, 2000, p. 245). Following this, the PSTs' thinking is elicited via whole-class presentations. The presentation begins with PSTs offering their (often unfinished or unpolished) solutions, and the class offer recommendations for improvement or alternative solution strategies. This phase answers the call of the Conference Board of Mathematical Sciences (2012) to have PSTs experience mathematics as a dynamic field and have opportunities "to construct viable arguments, to listen carefully to other people's reasoning, and to discuss and critique it" (p. 33). After class, the PSTs write their personal solutions to the task in their problem solving "journals". Lastly, the PSTs exchange their journal entries. Each PST reviews their partner's work and acknowledges writing that communicates the mathematics well, recommends ways to improve, or poses probing questions to the author. As can be seen, the APEX Cycle involves multiple forms of mathematical communication; but, in this chapter, we will focus on the opportunities to learn within the explanation (i.e. journal entry writing) and exchange phases[1]

Hiebert and his colleagues (1997) define mathematical understanding as a complex and dynamic phenomenon: "we understand something if we see how it is related or connected to other things we know" (p. 4). With this definition in mind, they highlight "two processes that play an important role in the making of connections: reflection and communication" (p. 5). Reflection happens when one thinks about their own knowledge, experiences, and even one's thinking. Communication includes "talking, listening, writing, demonstrating, watching, and so on" (Hiebert et al, 1997, p. 5). Reflection and communication are interwoven and interact with each other; it is very difficult to pinpoint where one ends and

Figure 2. Two sample journal pages

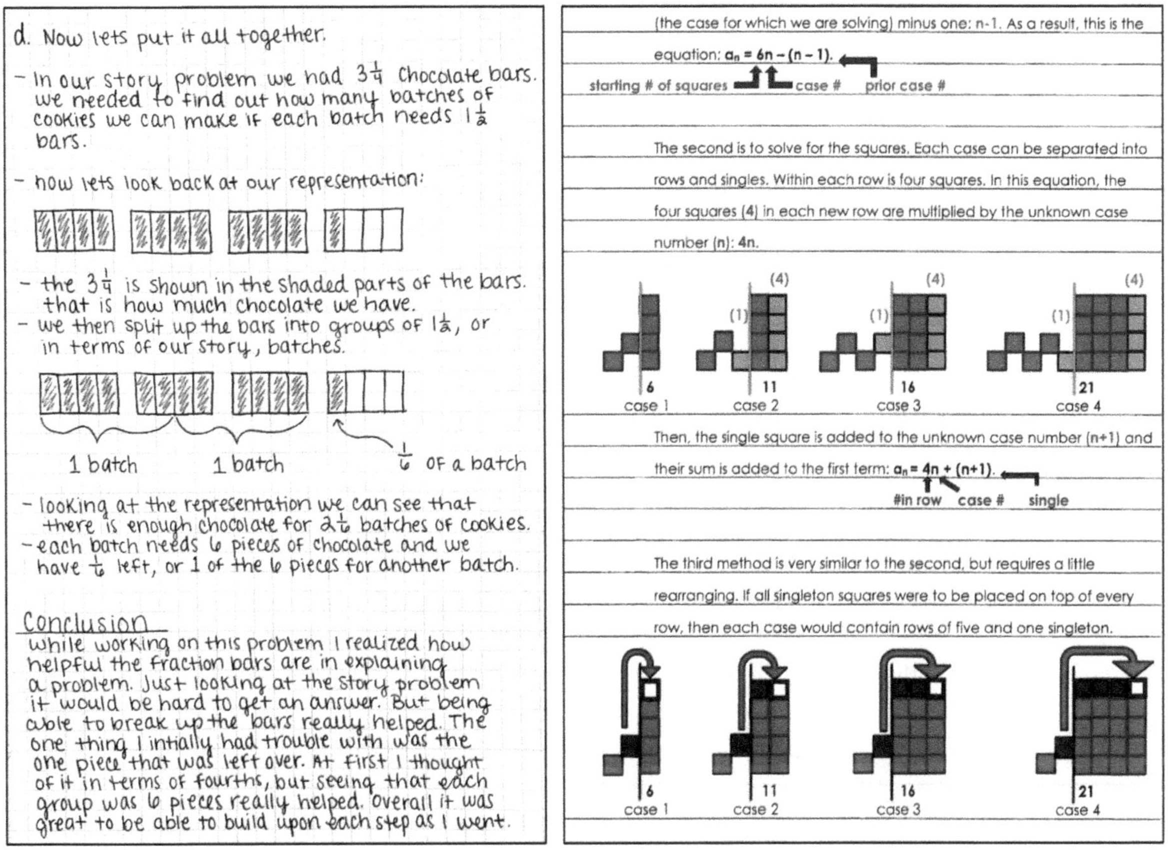

the other starts. One should reflect before communicating with others; and, when communicating orally and in writing, one's thinking becomes visible to oneself. This in turn may trigger further reflection.

We hypothesized that the journal writing and exchange in our courses would contribute to PSTs' learning by activating both reflection and communication in multiple ways. The PSTs are asked to write detailed solutions to the problems that they analyzed (both individually and in small groups) and presented during whole-class discussions. Their written explanations are expected to demonstrate logical thinking, be clear, concise, and completely understandable to anyone with the appropriate mathematics background (e.g. a peer). The PSTs are asked to include explanations and justifications of the steps involved and to highlight the major concepts that they use (Figure 2). As PSTs write their journals, they need to reflect on their understanding and make sure the connections that they have made in their own mind are accessible and visible to others.

The initial purpose for having the PSTs exchange between the campuses was to promote inter-institutional connections, but we also recognized the value in having the PSTs read the mathematical writing of peers who were taught by a different instructor. We theorized that, even though both groups of students were studying the same mathematics using the same journal assignments, the perspectives and explanations could be different and might require a more careful review. The exchanges offered the

PSTs the opportunity to review others' journals, to make sense of and offer critique of the mathematical explanations and representations, and to make suggestions for improvement. These pairings often remained consistent throughout the semester. To support this work, the instructors offered the following prompts to guide the PSTs' review:

- What part(s) of this journal do you feel is(are) done well? Please be as specific as possible.
- What part(s) of this journal do you feel could be improved? Please offer specific suggestions for improvement.
- What are the important mathematical ideas that you see used?
- Are these ideas used correctly? Are there ideas that you feel deserve more explanation? If so, what would you suggest adding?
- If you could ask the author one question about this work, what would it be?
- The space below is offered as a place for you to write a friendly note to the author.

When the PSTs completed their reviews, course instructors copied and returned the reviews to the authors. The logistics of transporting physical copies of the journals and reviews back and forth across campuses limited the timeliness of the experience. More importantly, the paper exchange occluded the opportunity for the PSTs to respond to the comments or questions of the reviewers. Therefore, in the third year of our collaboration, we began to exchange journals digitally via Google Docs. The asynchronous communication capabilities of the Google Doc platform transformed the peer review process.

Peer Feedback via Computer-Mediated Communication

Many teacher education programs use peer feedback to help PSTs become reflective practitioners and develop a critical lens for teaching and learning. However, providing and receiving constructive and formative critical feedback is difficult. Developing these skills requires structure and support (An et al., 2009; Howard et al., 2010; Lim et al, 2021). Computer-mediated communication (CMC), described by Lee & Oh (2017) as "an umbrella term that encompasses various forms of human communication through networked computers, which can be synchronous or asynchronous and involve one-to-one, one-to-many, or many-to-many exchanges of text, audio, and/or video messages" (p. 1), has affordances that can be leveraged to support PSTs engaging in peer feedback. An et al. (2009) shared that "as a great deal of research has indicated, asynchronous CMC can promote knowledge construction, critical thinking, and problem solving, through interaction and communication with fellow students and instructors" (p. 749). Howard et al. (2010) stated that "asynchronous CMC often reduces fear of humiliation, allows for less stressful preparation of comments, and provides equity in participation, allowing students who may otherwise be hesitant to become active" (p. 91). Absent the expectation of an immediate reply, asynchronous online environments offer participants additional time to reflect not only on the feedback that they received but also the response they will provide (Singer & Zeni, 2004; Sutherland et al., 2010).

The asynchronous CMC capabilities of Google Docs transformed the peer review process into mathematical conversations. These conversations took place within the comments posted inside the Google Document version of the PSTs' journal entries (Figure 3). We will explore these entries and the conversations that they spark in subsequent sections.

Figure 3. Screenshot of a journal with Google Doc comments

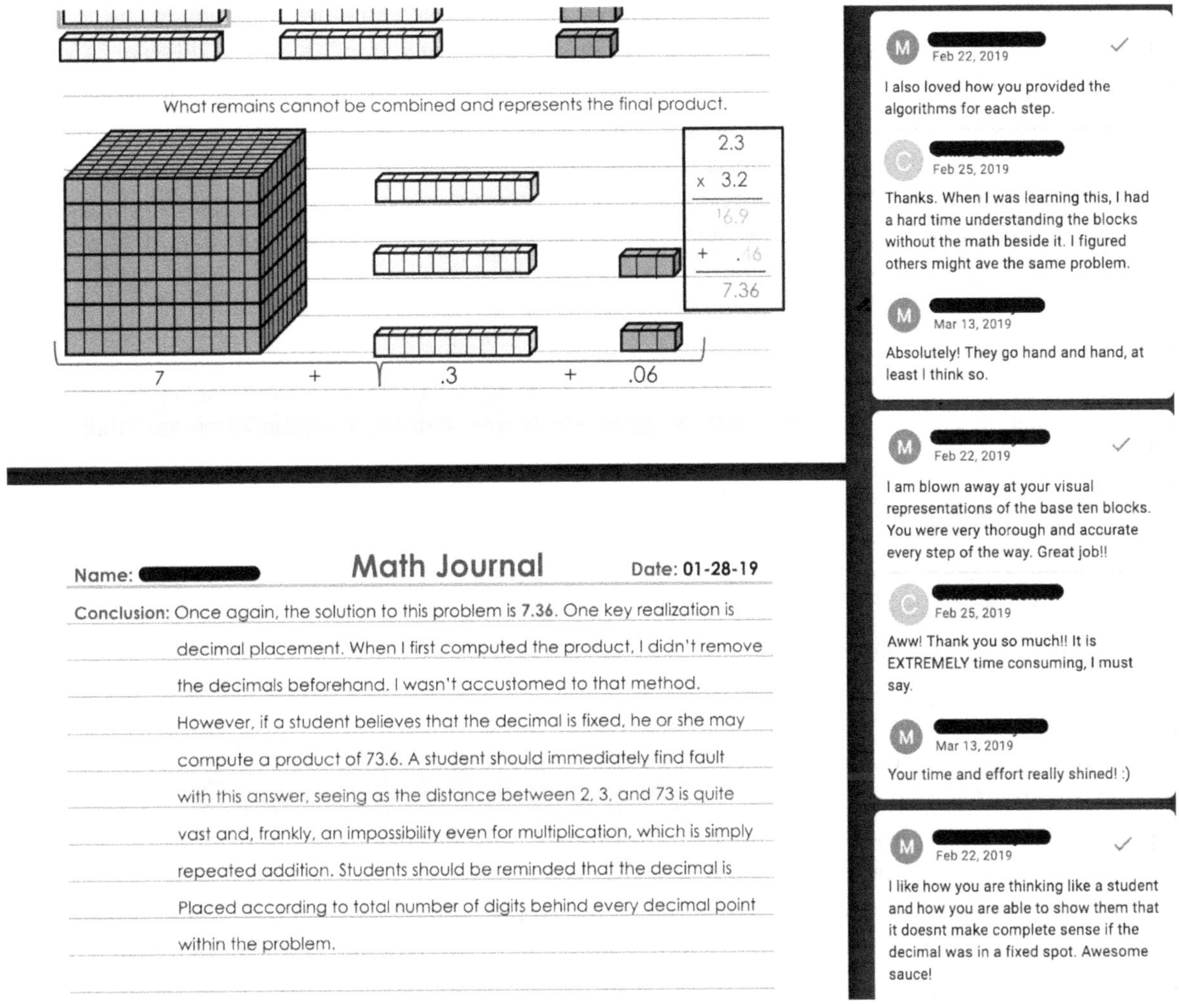

Action Research

As stated above, the support of the Noyce Foundation, initiated in 2016, enabled us to redesign the mathematics content courses at our partner institutions. The original objectives of our collaborative effort were to:

1. Design curricula for two mathematics for elementary teachers courses at a large urban public university and a local two-year college that explore mathematics used for teaching in a deep and meaningful way.
2. Develop teaching strategies that engage pre-service teachers in mathematics used for teaching in a deep and meaningful way.

The APEX cycle was developed as part of this course redesign, and its facilitation was refined over the next seven years. We view this process of refinement as a form of action research. Action research at the higher-education level for teacher education has been described "as a means to achieve a broader understanding of teaching situations and to improve the quality of teaching" as an extension from its general definition "as the systematic reflection of practitioners on action" (Krainer, 2006, p. 213). We were guided by Erbilgin's (2019) notions of intrinsic and extrinsic gains of teacher educators' practitioner research, as we examined our "own teaching practices [and/]or prospective teachers' learning, and viewed such research efforts as a valuable professional learning experience" (p. 29). In this chapter, we focus on the journal writing- and digital-exchange-related components to answer the following research question:

How does the writing and digital exchange of journal entries contribute to the development of PSTs' MKT?

Methods

The Mathematics for Elementary School Teachers courses were taught more than 28 times, collectively, at both institutions over a seven-year period. PSTs typically completed eight journal entries per semester in each course, approximately 224. Journal entries were exchanged across institutions twice (on average) each semester.

Following Price's (2001) definition of action research as "inquiry that is systematic, intentional, collaborative, and democratic in intent and process" (p. 43), we collected data from various sources at both institutions in order to improve different aspects of APEX Cycle. Data included:

- Instructional packages themselves for document analysis
- Course grades such as final exam and/or journal assignments grades
- *PST journal entries*
- *Journal exchange reviews and responses (Google Doc Comments)*
- *Anonymous PST reflections on the APEX Cycle*

To answer the research question, we concentrated on the italicized sources. Using a grounded theory constant comparative method (Strauss & Corbin,1997), we analyzed and coded the 378 PST journal entries and the digital exchange interactions (2500 individual Google Doc comment/response interactions) created during the four-year period from 2018-2022. In addition, we analyzed the anonymous PST reflections that we solicited at various points throughout the semester. Differences in coding were resolved through discussion. Analytic themes were generated, and connections between the themes coalesced into broader categories that we term "opportunities to learn." We paid special attention to negative cases (i.e cases no fitting into established themes) to ensure the trustworthiness of the study).

Results

We begin the presentation of our results with what students recognized about their own learning within the cycle. We will then introduce and explore the four opportunities to learn that our analysis of the PSTs journal entries and inter-institutional conversations brought to light. We conclude with two case studies that reveal the interrelated nature of these opportunities.

Anonymous written self-reflections capture the growth that the PSTs perceived through writing and reviewing mathematical journals. One PST shared that

Writing the journals has given me such a better understanding of each of these concepts. I feel like I've been able to master the art of using both words and pictures/diagrams to explain my own thought process which I know will be helpful to me as a future educator. I've learned a lot about myself and my thought process which is important. The journals were a great help, as I've become more comfortable with my ability to explain mathematical concepts in detail.

The notion of deeper learning through writing and increased comfort in explaining one's own complex mathematical ideas was echoed in other reflections and in the work that the PSTs produced and discussed. These discussions, grounded in the journals exchanged across institutions, were opportunities for additional and different learning. One PST reflected that she

Learned how different two [explanations] of the same thing could be and that there are different ways to get the same answer. There were concepts in math I didn't know, and the journal exchange made me more aware of that and that I should break down the problem more.

Working to understand different ways of thinking about new mathematical ideas is challenging. Noticing how other writers effectively (or ineffectively) present their thinking and getting feedback from their peers enabled the PSTs to make adjustments to their own thinking and writing. Another PST wrote,

Having someone else evaluate my work and provide suggestions for improvement allows me to see someone else's way of thinking, and may be a way to get new ideas I have never thought of before...In the future, going beyond comparing what I wrote to my peers, I should think deeper about the content and maybe look at important concepts that...I missed in my own journal to help my peer and provide a perspective they might not have seen.

These new ideas that PSTs "have never thought of before" can be incorporated into future mathematical writing. In the next section, we present evidence of four opportunities to learn that we detect within the journal entry writing and exchange phases of the APEX Cycle. We include illustrative examples from PST journal entires and reviews that were noted during data analysis.

Opportunities to Learn

Through our analysis of the PSTs' journals, the inter-institutional conversations about the entries, and solicited PST self-reflections, we found that the writing and exchange of the journal entries enabled PSTs to build their MKT in multiple ways. Writing, exchanging, and engaging in conversations about journals proffer four opportunities to learn, or OTLs (Figure 4):

1. Writing mathematics to deepen one's understanding
2. Writing for someone else to communicate one's thinking
3. Reading someone else's journal entry, making sense of others' thinking, and providing feedback
4. Improving one's future mathematical writing

Figure 4. Opportunities to learn

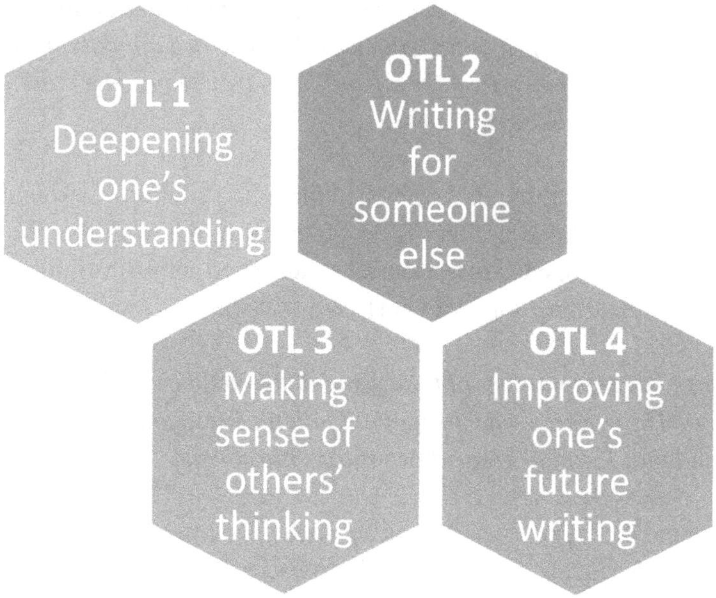

We view OTL 1 and OTL 2 as opportunities to learn along the Common Content Knowledge and Knowledge of Content and Teaching domains of MKT, as the PSTs must understand the mathematics deeply to write clear explanations of the problem sets. As they write their journal entries, the PSTs must consider which representations will best convey their thinking to the reader. As the PSTs read the journal entries of their peers, they must make sense of their peer's thinking expressed in writing. In cases when that thinking might not be clear, the PST reviewer must attempt to fill in missing pieces or interpret imprecise representations. Thus, OTL 3 represents an opportunity to deepen their Specialized Content Knowledge (Ball et al., 2008). Additionally, we ask PSTs to ask clarifying questions of the authors which, again, draws upon this specialized knowledge base. The introduction of Google Docs as a CMC platform opened the doors for peers across institutions to engage in an authentic mathematical dialogue. These conversations offered additional access points to the first three opportunities to learn as well as enriched OTL4, as PSTs frequently mentioned that they planned to incorporate the suggestions and critique offered by their reviewer into their future journal entries. Frequently, we were able to see evidence that the PSTs did just that. In the next sections, we will present representative examples where we see evidence of our students taking advantage of all four of the OTLs.

OTL 1: Deepening One's Understanding

One journal activity, which we will refer to as the Whole Number Multiplication Activity, concentrated the PSTs' attention on the conceptual underpinnings of the standard whole-number multiplication algorithm. Students explored how the computation of 24x58 rests on the distributive property, how it can be represented using an area model, how the multiplication can be done using the expanded algorithm (partial products), and how the standard algorithm compresses the work done in this expanded algorithm. In addition, students were asked to diagnose a common student error and suggest ways to address the

Figure 5. Uma and Shanna's comments

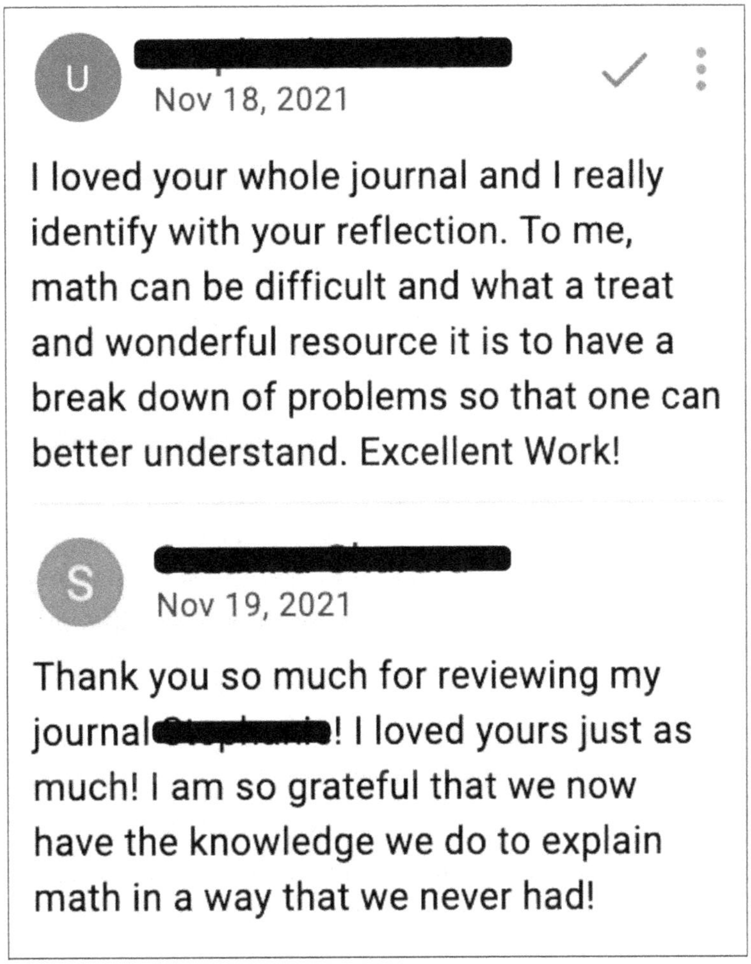

misconception. One college PST, Shanna (Note: all student names are pseudonyms throughout. Within the Google Doc comments, the names have been redacted and pseudonyms are maintained and represented using first initials), wrote in her journal entry's conclusion,

I always enjoy completing journals as they strengthen my knowledge and understanding of each concept. Explaining each problem into (sic) words not only teaches someone else, but breaking it down that finely also helps you when completing the problem. I also enjoyed how this journal showcased how many different ways you can get the same [answer]; I even found myself being able to explain the methods to my younger sister.

Shanna's reflection resonated with her review partner, Uma, a university PST, who left a comment within the Google Doc (Figure 5). Uma describes her partner's mathematical writing as a "wonderful resource…so that one can better understand."

Figure 6. Organizing representations to help own understanding

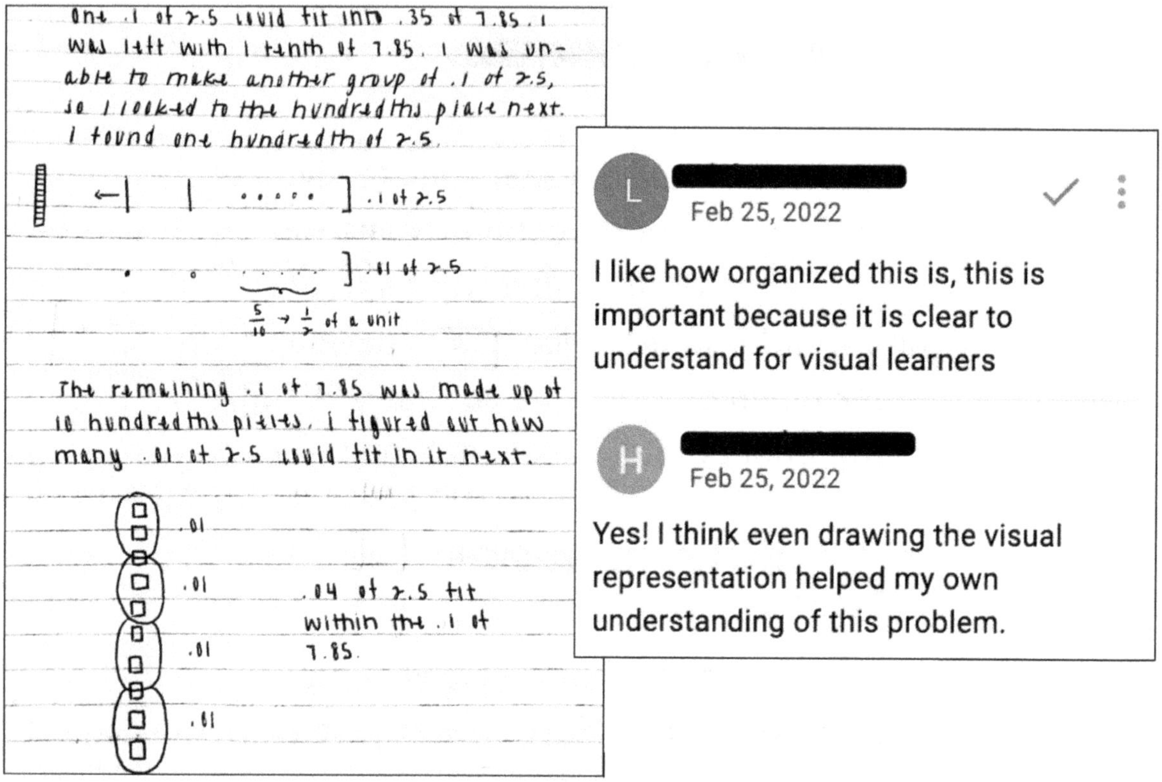

Many PSTs found physical manipulatives to be helpful in solving the mathematical problems in the activities. For one such activity, the Decimal Operations Activity, the PSTs were asked to use base-10 pieces to model the quotient 7.85 divided by 2.5. The journal entries that the PSTs created often included drawings of these pieces to help explain their thinking process. Lisa, in her review of Helena's journal, commented that the organization of Helena's representation-rich entry would be accessible to visual learners (Figure 6). Helena's response to this comment shows that drawing the representation also helped her own understanding of the problem.

Erica, a university PST, and Sandy, a college PST, reviewed each other's journals three times over the two-semester sequence of courses. In her conclusion on the Whole Number Multiplication Activity journal entry, Erica writes, "Not only did I do it algebraically, there was a visual representation which helped me understand it even better. It was nice to be able to relearn a skill that I learned when I was younger in different ways" (Figure 7). In Erica's review of Sandy's journal entry, her final comment shows a recognition of how much she has learned about mathematics through her own writing. Erica shares, "I agree…I learned so much from [the activity], and writing a journal on it also allowed for me to think deeper about the lesson."

As the previous examples show, the OTLs within the explanation phase are often intertwined. Writing can help deepen one's own understanding of a mathematical idea; and writing for someone else often requires detailed explanations that anticipate how the reader may interact with the writing. This leads us to our second OTL.

Figure 7. Deepening the understanding of writing and mathematics

Reflection
So far I believe this is my favorite journal we have done. I liked being able to work independently on it, but also able to discuss and bounce ideas off of my breakgroup classmates. Not only did I do it algebraically there was a visual representation which helped me understand it even better. It was nice to be able to relearn a skill that I learned when I was younger in different ways.

E
Nov 15, 2021

I agree, having the visual helped me grasp the concepts more as well. This activity was for sure my favorite as well. I learned so much from it, and writing a journal on it also allowed for me to think deeper about the lesson. You have done a really good job with thisn journal also!

OTL 2: Writing for Someone Else

As they write for someone else, the PSTs must organize their entries so that their thinking is visible and understandable. Building up arguments and representations in steps and in a specific order, while giving the reader time and space to digest them, are important considerations. These written explanations may need to be organized differently than the way we think of them for ourselves. Returning to the Whole Number Multiplication Activity, and to an earlier journal of Helena's, we see an example of this type of reorganization. In Helena's journal, she models the product of 24x58 using an area model (Figure 8). A different reviewer than in the previous example, Emily, also makes note of Helena's representation (and her helpful color coding) in her comment. Emily writes that Helena's representation "...makes it easier to understand. Especially as a visual learner, this helped me." From Helena's response, we see that she has adapted her original thinking to help her reader better understand. She writes, "When I was solving this problem, I did not organize it in this way, but I found it helpful for my journal explanation." This short interaction over this journal entry, the first exchange for both the author, Helena, and the reviewer, Emily, shows evidence of *both* PSTs learning how to write effectively for others.

Helena found success in this first exchange; and when she reads that her explanation helped Emily understand, Helena happily exclaims, "I was hoping that [my explanation] would have that effect!" Others found writing for others more difficult.

In Erica's journal reflection on the same activity (Figure 9), she stresses that "the most challenging part about this worksheet for me was not actually solving the problems, but being able to fully understand each detail of each problem so I could effectively teach it to a student..." Having her colleague Sandy share that she also struggles with moving beyond *doing* mathematics to *explaining* mathematics goes beyond simple collegial commiseration. Sandy comments, "This is where I sometimes struggle too. I

Figure 8. Organizing writing to help others understand

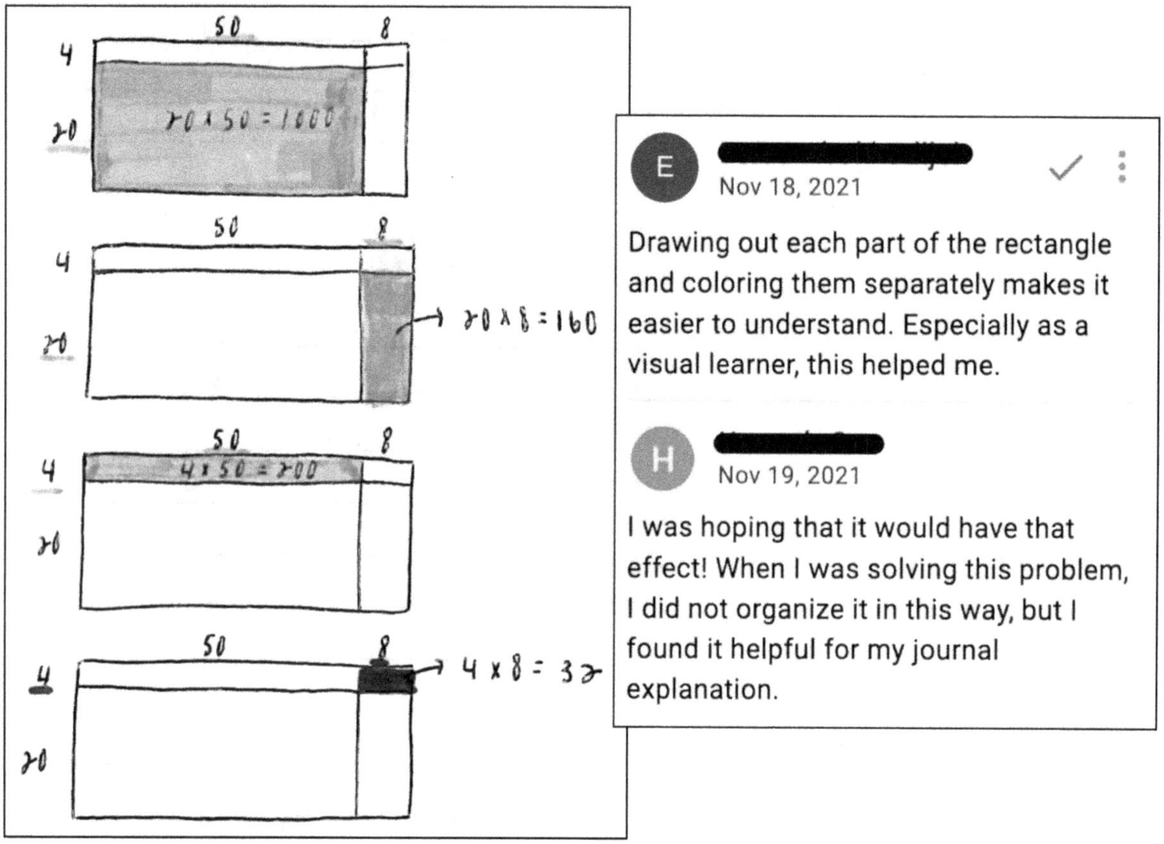

Figure 9. Discussing the challenges and value of writing mathematics

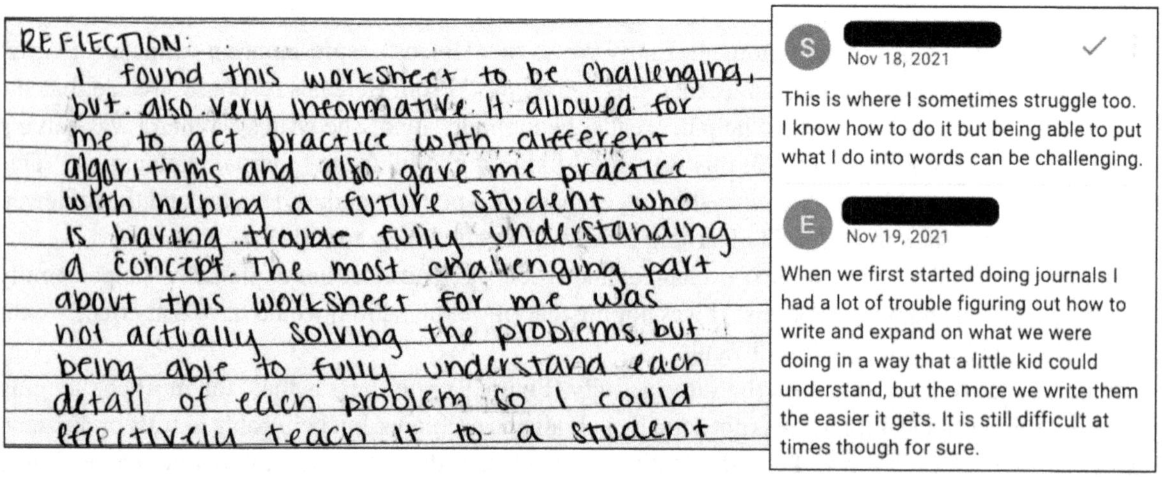

know how to do it, but *being able to put what I do into words can be challenging*" (emphasis added). Such interactions help Erica see that she is not alone, and that the difficulty is inherent in the activity of writing mathematics and not a reflection of her ability. She responds,

When we first started doing journals I had a lot of trouble figuring out how to write and expand on what we were doing in a way that a little kid could understand, but the more we write them the easier it gets. It is still difficult at times though for sure.

OTL 3: Making Sense of Others' Thinking

As mentioned in the previous section, the PSTs found writing for others difficult. It can also be difficult to make sense of the thinking of others as expressed in their writing. This difficulty is compounded by the variety of solution paths a reviewer can encounter. Many PSTs commented on the different ways of approaching the same problem. What is intriguing is how positively the PSTs approached the challenge that this variety in thinking and representing presented. We conjecture that this is based on one of the core values of the redesigned courses at both institutions: an appreciation of different ways of thinking leading to freedom from prescribed mathematical methods (e.g. knowing only the invert and multiply algorithm for division of fractions). This extends to different explanations as well.

Returning to Sandy and Erica's first exchange, we find them discussing the opportunities to learn, both in terms of mathematical content and in writing, involved in the review process of Erica's journal (Figure 10). Sandy voices the value of seeing different approaches to the same problem: explaining why we place a 0 in the ones place when moving on to the tens digit of the multiplier. She writes, "It is interesting because we both explained in different ways but got the same point across." Erica replies that, "I really liked reading through your journal because it allowed for me to reflect and see a different way of explaining." This is a considerable shift from many PSTs' earlier experiences with mathematics, which often involved a procedural understanding of a single prescribed solution path.

In other cases, reviewers had difficulty understanding the authors' explanations. We find that critical feedback by the reviewer sometimes enabled the author to confront incomplete or fragile understanding of the mathematics. In another exchange of the Decimal Operations Activity journals, Martha explained and represented the division of 7.85 by 2.5. Martha's explanation of the quotative model of division and base-10 pieces opened up her thinking to her review partner, Julia. After noting that there were three groups of 2.5 in the dividend 7.85, Martha then considered what one tenth of a group of 2.5 would be (visually, in terms of the base-10 pieces) in order to continue the division computation. Julia, unable to follow Martha's explanation, comments,

This is where I get lost. I see what is left over but am confused about what happened to the 2 longs and 5 little cubes. I see how they match up with the row above it but am unsure about what you did with them

In her reply, Martha laments that, while doing the problem, it made sense to her; but explaining her understanding proved challenging.

This part was a struggle for me. When we did it in class, we had the manipulatives in front of us and it made sense. When it came time for me to draw it, I had a hard time putting it into words. The only way

Figure 10. Discussing different mathematical approaches

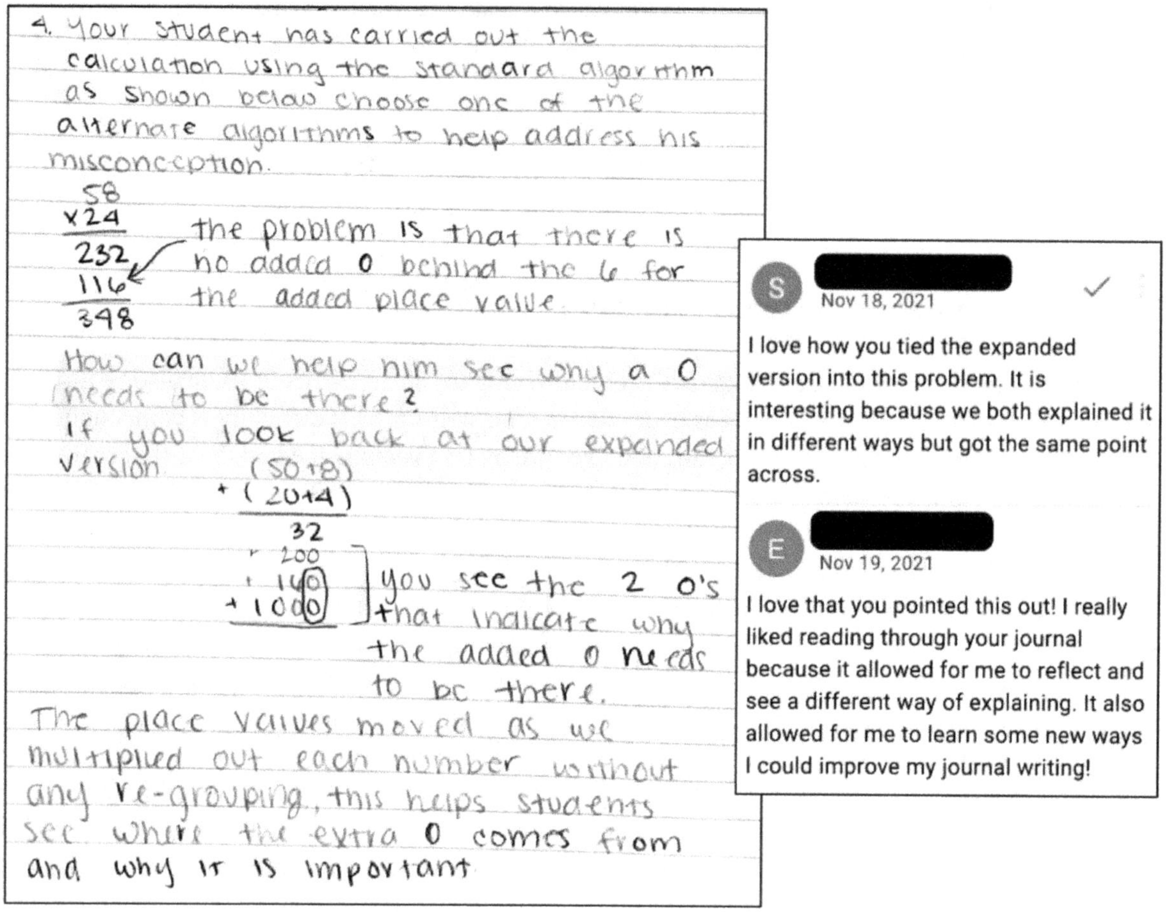

I know how to explain it is to look at the pieces left as a slice out of the 2.5. One long is a tenth of a flat and one unit is a tenth of a long

This short conversation captures the key distinction between knowing for one's self and knowing for others, a component of the specialized content knowledge domain of mathematical knowledge for teaching (Ball et al, 2008). In addition, it allows Martha to confront the depth of understanding that she will need as she moves into her classroom.

Other reviewers highlighted gaps in the written explanations by asking questions of the author. These questions frequently opened up opportunities for the author to reply with clarifying comments, additional explanation, or completely new information. Thus, these dialogues were particularly rich opportunities for learning for both the reviewer and the author.

In another exchange based on the Decimal Operations Activity, Lisa tried to make sense of Melanie's explanation of 2.3 x 3.2. Melanie conceptualized this as 2.3 copies of 3.2, and, after drawing 2 full copies using base-10 pieces, she explained what .1 copy of 3.2 would look like. Like Julia, the reviewer in the previous vignette, this explanation of tenths of a figure was difficult for Lisa to understand and

Figure 11. Asking for clarification and receiving a thoughtful, targeted response

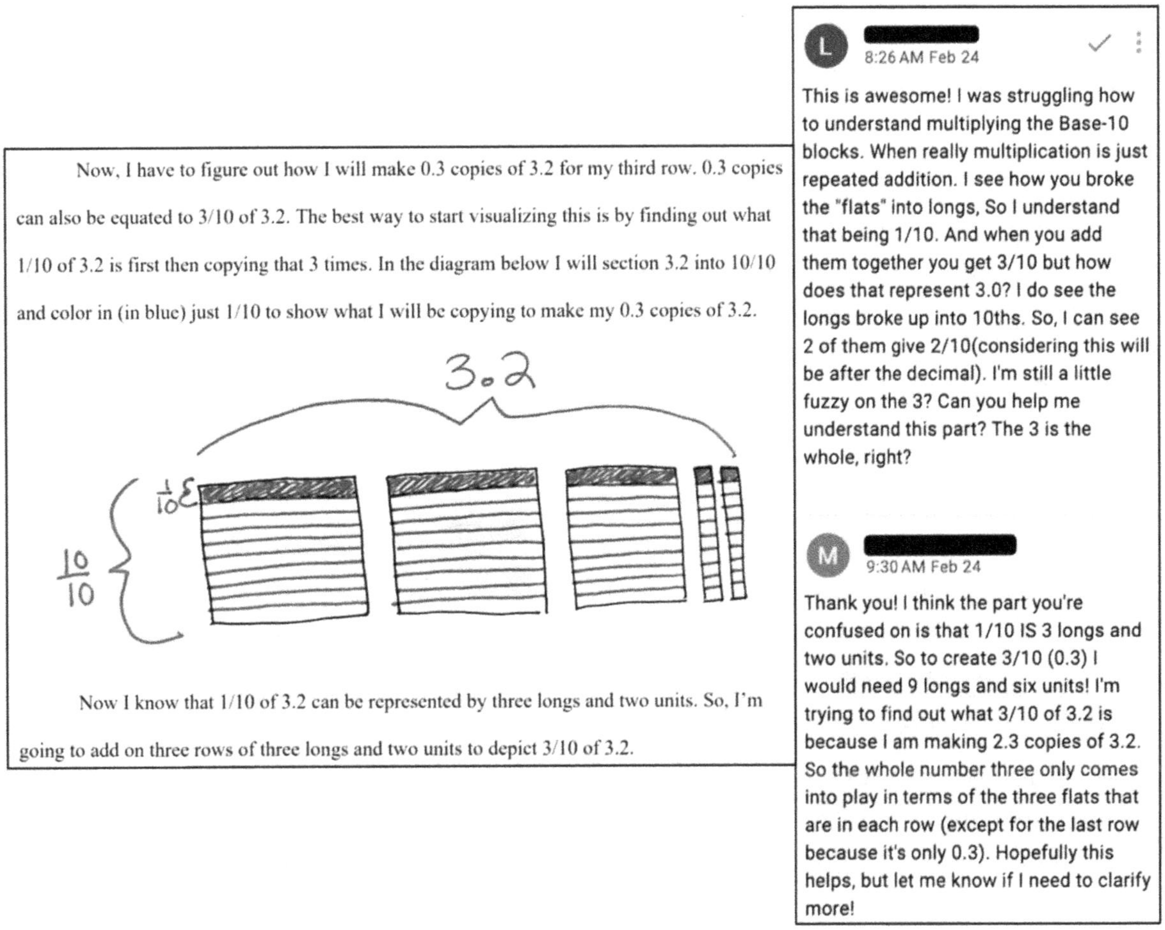

Now, I have to figure out how I will make 0.3 copies of 3.2 for my third row. 0.3 copies can also be equated to 3/10 of 3.2. The best way to start visualizing this is by finding out what 1/10 of 3.2 is first then copying that 3 times. In the diagram below I will section 3.2 into 10/10 and color in (in blue) just 1/10 to show what I will be copying to make my 0.3 copies of 3.2.

Now I know that 1/10 of 3.2 can be represented by three longs and two units. So, I'm going to add on three rows of three longs and two units to depict 3/10 of 3.2.

L 8:26 AM Feb 24

This is awesome! I was struggling how to understand multiplying the Base-10 blocks. When really multiplication is just repeated addition. I see how you broke the "flats" into longs, So I understand that being 1/10. And when you add them together you get 3/10 but how does that represent 3.0? I do see the longs broke up into 10ths. So, I can see 2 of them give 2/10(considering this will be after the decimal). I'm still a little fuzzy on the 3? Can you help me understand this part? The 3 is the whole, right?

M 9:30 AM Feb 24

Thank you! I think the part you're confused on is that 1/10 IS 3 longs and two units. So to create 3/10 (0.3) I would need 9 longs and six units! I'm trying to find out what 3/10 of 3.2 is because I am making 2.3 copies of 3.2. So the whole number three only comes into play in terms of the three flats that are in each row (except for the last row because it's only 0.3). Hopefully this helps, but let me know if I need to clarify more!

she asked for clarification. Unlike what occurred in the previous vignette, Melanie was able to further explain her representation and explanation (Figure 11).

In Melanie's response, we see her process Lisa's misconception and offer an additional explanation to help Lisa better understand the work. Melanie realizes that Lisa is mistaking the three longs (with each long representing 0.1, colored blue in the top row figure) in the blue-colored row as the three tenths in the multiplier 2.3. Melanie replies,

I think the part you're confused on is that 1/10 IS 3 longs and two units. So to create 3/10 (0.3) I would need 9 longs and six units! I'm trying to find out what 3/10 of 3.2 is because I am making 2.3 copies of 3.2

Lisa's efforts to make sense of Melanie's writing presented an additional opportunity for Melanie to practice a key element of teaching: responding to a learner's thinking. This is something that all teachers must do, often on-the-fly; but here, the computer mediated communication may have provided enough additional time for Melanie to organize and produce this thoughtful, targeted response (note that Melanie's response came just over an hour after Lisa's comment). As the PSTs reviewed their peers' journals, they frequently noticed, and commented on, elements of organization and mathematical explanations that helped them understand the author's thinking. Thus, the exchange supported a fourth learning opportunity.

OTL 4: Improving One's Future Mathematical Writing

The exchange phase offers the authors an opportunity to receive critique and use it to improve subsequent written work. In addition, we find evidence that the reviewers also learn a great deal. Many of the conversations sparked by the exchange mention the value of encountering new mathematical terminology, new representations, and new formatting ideas in the author's writing.

In an exchange based on the Whole Number Multiplication Activity, we see both directions of this learning. Erin, the reviewer, offers critique that the author, Monica, graciously acknowledges and accepts. Later in the conversation, Erin comments that she will incorporate "regrouping" (a more conceptual description of what occurs in the multiplication algorithm) in place of "carrying", as well as Monica's thoughtful formatting, into her own work (Figure 12). In this short exchange, we see elements of nascent collegial collaboration that, it is hoped, the PSTs carry into the field.

Figure 12. Author and reviewer learning for improving future journals

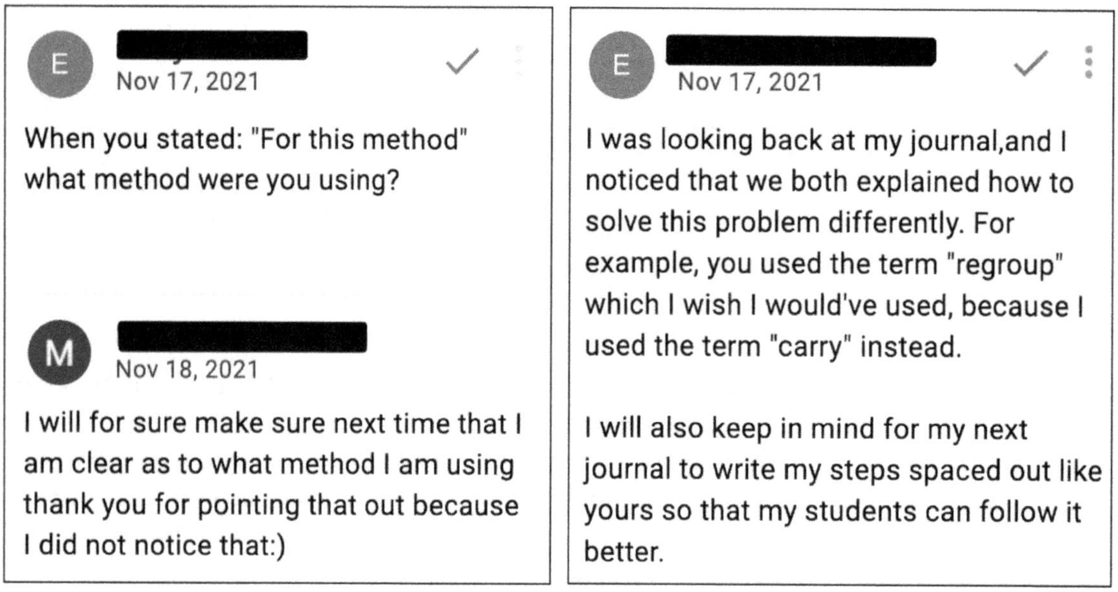

Recall that we saw Julia having some difficulty understanding Martha's explanation of division of decimals. In particular, Julia expressed some confusion over the use of base-10 pieces to represent the division. Despite this, reviewing Martha's work was a rich opportunity to learn for Julia–one that she took advantage of. At the end of Martha's journal, Julia comments,

You did a great job on this journal, Martha, and truly helped me see how you can use base 10 blocks to break down both the multiplication and division problems and I plan on using your methods when

making changes to my journal. It was great working with you again and hope to work with you again in the future!

Julia's instructor, seeing the potential in allowing students to incorporate immediately the learning from the review process, added a revision step to the journal writing process. Julia saw the value of this opportunity to build on her "first draft" and reflected on it in her final version. She shares, "Seeing the way she used base 10 blocks to break the problem down helped me do the same...this is when I learn the most…reviewing another student's work and sharing comments with each other".

This trial implementation of a journal revision by the college instructor, like many phases of the APEX Cycle, was based on reflective iteration and holds great promise in terms of operationalizing the learning exemplified by Julia. It serves as a powerful reminder for us as we consider other tweaks and supports that may allow us to wring even more learning from this activity.

Having examined the four OTLs across multiple exchanges between different pairs of PSTs, we now present how two individual PSTs engaged with these opportunities. We posit that a deeper analysis of their two cases will help the reader understand and appreciate how the multiple layers of mathematical writing can deepen students' MKT.

Emily's Case

Emily was a community college PST studying Pre-Elementary Education and intending to complete her teacher education program at a four-year university. She was in the top 5% of her class. In addition to being mathematically capable, Emily was always eager to participate in class activities, engage with other students, and contribute to the class' understanding of mathematical content. As depicted in Figure 2, there were multiple OTLs for Emily within the mathematical writing and reviewing tasks. Furthermore, in this case, we will demonstrate that these OTLs were not sequential and isolated, but rather interwoven and interdependent.

Even though we see multiple data points indicating that Emily's mathematical understanding deepened as she was writing her journals (OTL 1); her understanding was also advanced when she was reading and responding to her reviewers' comments. Emily was part of three journal exchanges (one in the Fall and two in the Winter Semester) and, in her reflection on the first exchange, she wrote, "After completing problems in this journal, I was able to better understand how I can solve and explain multiplication problems in four different ways." In the same journal exchange Emily and her reviewers, Helena ("H") and Rasheeda ("R"), also acknowledged that they learned from reviewing each other's journals (Figure 13). In these exchanges, we saw multiple OTLs being realized. Emily took great care to make her mathematical thinking visible to others (OTL 2), such as using and describing the correct mathematical terminology, and explaining her mathematical work clearly in order to help her reviewers learn (OTL 3). We observed that Emily's careful and detailed explanation of her mathematical thinking helped her reviewers see different methods (parts a and d of the figure) and different ways of explaining (parts b, c, and e of the figure).

Similarly, as Emily was reviewing others' journals, she learned from them (OTL 3) (Figure 8). Here, Emily shared that the author (Helena) created a representation (an area model) that helped her understand the multiplication concept better. We find multiple comments showing that she valued her fellow PSTs' ways of thinking and explanations as they contributed to her learning: "I did not learn it this way, this

Figure 13. Journal exchange dialogues

helped me understand it better!" Moreover, some conversations were purely mathematical as the PSTs were trying to make sense of each other's mathematical thinking. In one instance, Emily, as the reviewer, wrestled with her partner Lupe's explanation of the division algorithm for numbers involving decimals. Emily puts forth the notion of multiplying both the divisor and dividend by 10:

Emily: Yes, multiplying it by 10/10 is like multiplying it with 1 whole. You can also add on to this by saying how multiplying each by 10 is the same as moving the decimal point one place to the right in both the divisor and dividend, this helps explain why the standard algorithm works.

Lupe: That is true, this part I was having a little more trouble with because I wasn't so sure if i was explaining it right but this helps. Thank you.

In her second semester, Emily pushed her fellow PSTs' mathematical thinking by asking multiple why and how questions in her requests for more explanation. These online dialogues were examples of mathematical communication. Some of the questions that Emily posed were:

- Why did you choose this way to solve the problem?
- How did you come up with this formula?
- Can you elaborate more, do you mean the right angles?
- How did this help you understand it better?
- Can you explain why you chose to make three triangles?

In the first exchange of the first semester, one of her reviewers suggested that Emily use color coding to show the connection between the visual and numerical representations. Emily acknowledged this suggestion and stated that color coding would help readers to understand her work. We treat the color coding used to connect the area model and the numerical representation in her first journal exchange (Figure 14 part a) of the following semester as evidence of Emily taking advantage of OTL 4. For this work, her reviewer commented that "This is so interesting. I am assuming you multiplied then drew the boxes. This is such a cool way to do this. I have never seen it done this way. This is great work, and way less confusing than when I first learned multiplying the base 10 blocks. really great work!" As Emily made her mathematical work more accessible (OTL 2), it created more opportunities for her reviewers to learn from it (OTL 3). In another interaction in the first journal exchange, Emily was encouraged to provide more explanations in an organized way: "If you were provided the opportunity to revise this journal, you could incorporate more of the step-by-step structure from above...These steps are important. However, I feel it is almost more important to explain what prompted your process here or why you solved the problem in this way. I think this part could benefit from a discussion of place value." Again, Emily accepted this suggestion as evidenced in her subsequent journal where she provided steps with more explanations (Figure 14 part b).

Figure 14. Attending to comments from previous exchanges

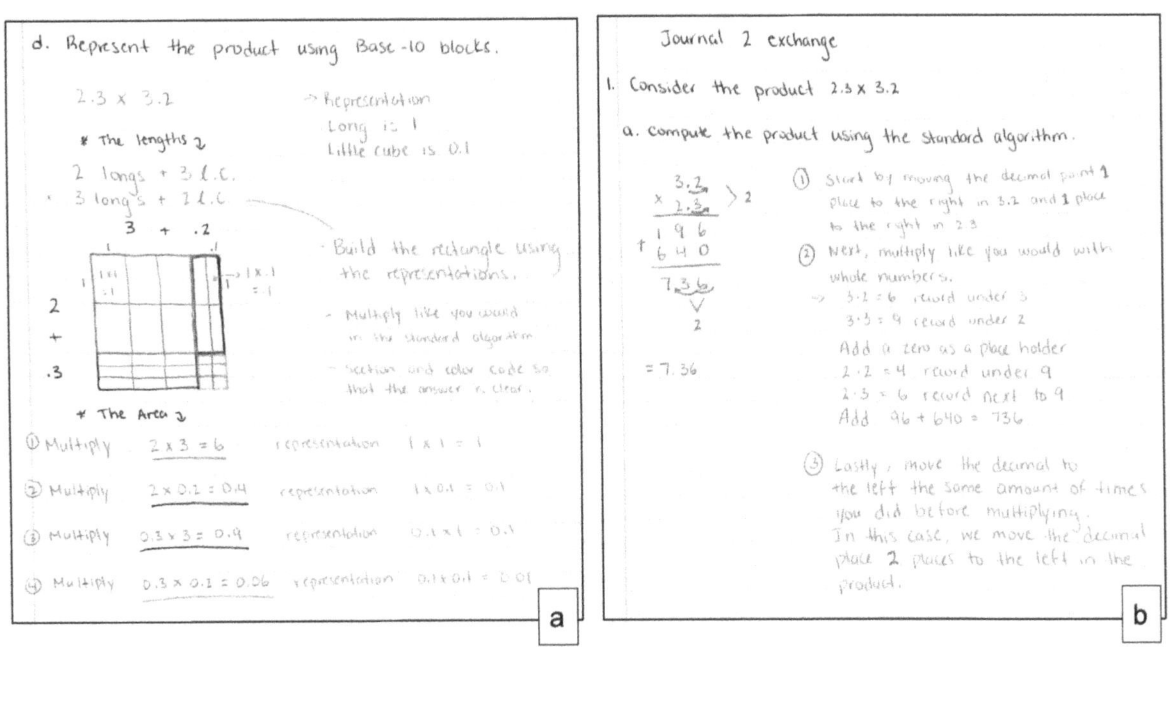

As mentioned above, Emily was confident mathematically and comfortable enough socially to share her ideas in the classroom setting. But seeing Emily improve and deepen her mathematical knowledge demonstrated that writing is a tool that can support all levels of understanding. In the following case, we follow a student whose mathematical knowledge had more room to improve; and mathematical writing supported her in doing so.

Mari's Case

Mari was a community college PST who shared that she struggled with math. During the early weeks of semester, she often worked quietly in her group and had difficulty completing her journals. The challenges she faced extended beyond journal writing to the course content in general. For the initial journal exchange, Mari and her classmates worked on a problem set, shown in Figure 15, that involved pattern recognition and the connection between geometric and algebraic representations (Boaler, 2016).

Figure 15. The initial journal exchange activity

Worksheet 2

1. Consider the following sequence of shapes:

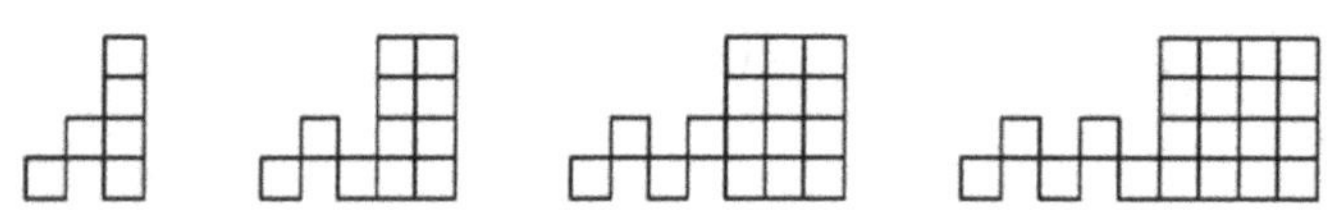

   a. How do you see the shapes growing?
   b. Can you predict what the 100th case would be?
   c. What about the nth case? Connect your formula to the figures.

2. Consider the following sequence of shapes:

   a. How do you see the shapes growing?
   b. Can you predict what the 100th case would be?
   c. What about the nth case? Connect your formula to the figures.

3. If you painted a large cube made up of $n \times n \times n$ tiny cubes, how many of the tiny cubes will have paint on them? Connect your formula to the figures.

   The case where $n = 5$ is shown here:

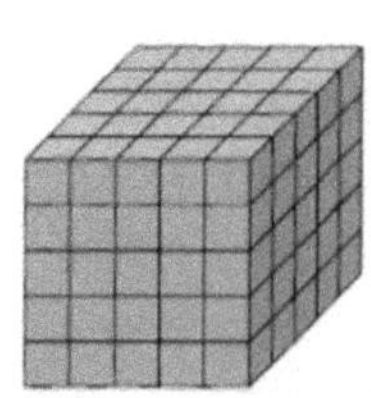

In her journal, Mari completed the problems with almost no explanation (Figure 16). There was no connection between the geometric figures and the numbers recorded in the table, and there was no explanation of how she arrived at the formulas representing the general terms. Shanell begins her review with a positive comment noting that Mari's tables "help visualize how each new figure is changing." She follows with a recommendation for the third problem suggesting that Mari "(t)ry adding a visual of what the cube actually looks like and break it down so that you can show how you got your numbers." Shanell presses further and recommends that Mari add a formula and "connect it back to the figure." Shanell's review guides Mari to include more visual support to help a reader make sense of her mathematical reasoning as she solves the problem. Over the course of the semester, Mari exchanged journals with another university PST, Amanda, in addition to Shanell. We track the learning of all three PSTs through these subsequent journals.

In the next exchange, based on the Whole Number Multiplication Activity, Mari includes a detailed picture (Figure 17) to support her explanation. Amanda joins Shanell in reviewing Mari's journal. Amanda recognizes that Mari's drawing includes lines to show longs and cubes, the base-10 pieces that have been frequently mentioned in the previous vignettes. Mari notes that "the smaller lines help show more of how I broke it down and they also help label the place values … the longs being 10's, the cubes being the ones and so on." Mari appears to address the importance of breaking down the explanation and addressing the needs of a learner, key facets of writing for someone else (OTL2). We view Mari's visual, although still mathematically weak (note the transposition of the 50 and 20 in her figure), as progress in addressing Shanell's earlier recommendation.

Both Shanell and Amanda comment on the detail of the explanation demonstrating Mari's improved attention to detail and effort to explain mathematics to her reader (Figure 18). Shanell highlights Mari's organization and clear steps. She also commends Mari for the "(v)ast improvement from the last journal."

Figure 16. Mari's first journal entry with reviewer comments

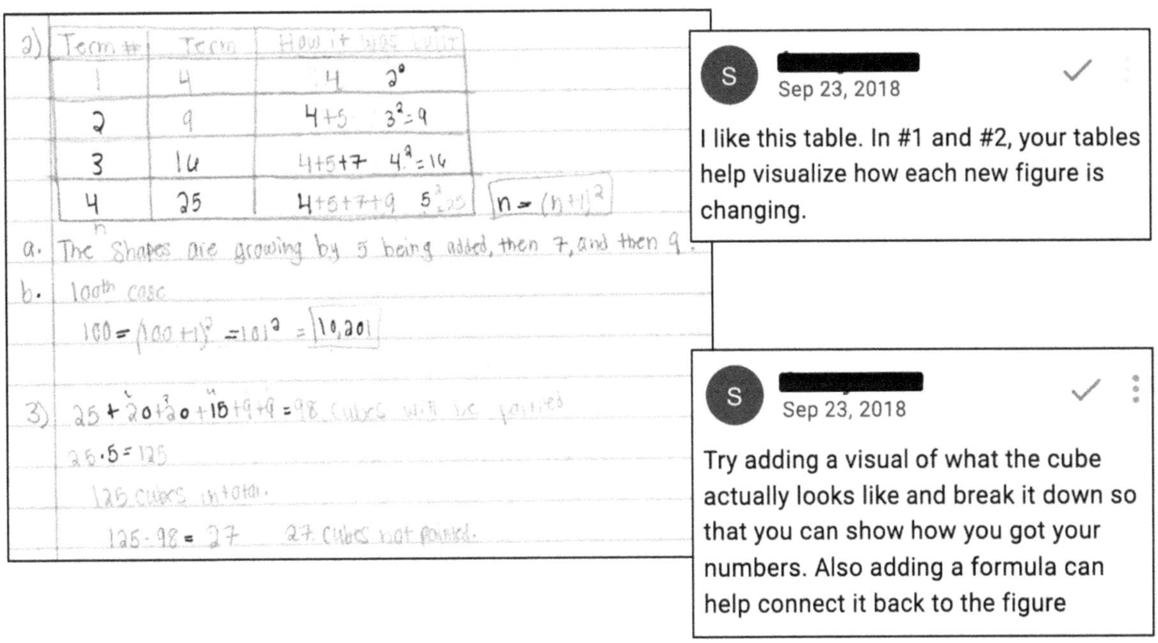

Figure 17. Mari's representation

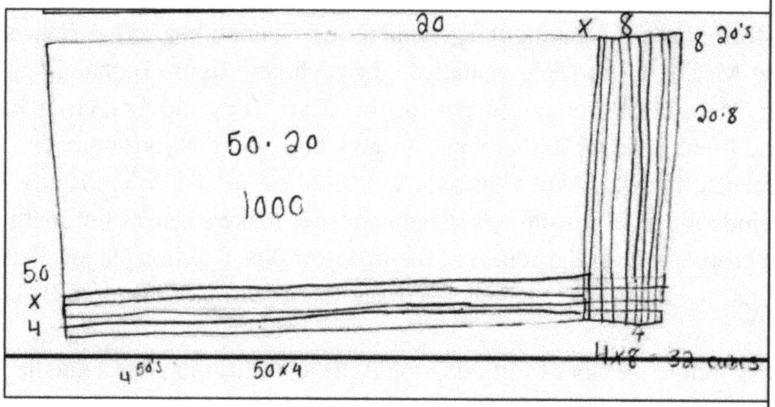

A
Nov 11, 2018

I like how you used smaller lines in this! It shows more in depth of how your broke it down.

M
Nov 13, 2018

The smaller lines do help show more of how i broke it down and they also help label the place values as well such as longs being 10's , the cubes being the ones and so on.

Mari notes that listing steps helps her write an explanation *and* helps with better understanding. She notes "the explanations do help in making it easier to go step by step and understand everything. I also find that when I list the steps it makes it easier to put it into an explanation and it also helps with better understanding how to solve the problem." That is, Mari's writing has deepened her own understanding, and she has improved her ability to write for others (OTL 1 and 2).

Figure 18. Improvement and understanding

A
Nov 11, 2018

I really like this explanation. It goes into a lot of detail and it makes it easy to understand.

S
Nov 12, 2018

I love how you explained the problems in 1 and 2 and how they are similar, but it is more broken down into detail. I love how organized you are. Great job listing the steps on how to solve the equation. Making a list of steps, especially in notes is a great idea to do for your students to thoroughly understand how to solve the product as well in #3. Vast improvement from the last Journal I reviewed!

M
Nov 13, 2018

the explanations do help in making it easier to go step by step and understand everything. I also find that when i list the steps it also makes it easier to put it into an explanation and it also helps with better understanding how to solve the problem.

Amanda continues to address the expanded algorithm explanation, noting its affordances. Mari responds that this "method does help in showing exactly what's happening in each step and how everything is being distributed…" and that it helps students in identifying place values. Additionally, Amanda notes that the expanded form "shows the student step by step how to multiply each number while focusing on the place values." Mari agrees, and notes that "it helps students in identifying place values." In these two short exchanges, Mari shows a deepening awareness of how to write mathematics for someone else (OTL 2).

Shanell's and Amanda's reviews of Mari's journal demonstrate thoughtful analysis of the mathematics and the communication of that mathematics. It appears to have presented an opportunity to make sense of Mari's thinking (OTL 3), one on which both reviewers capitalized. In addition, we posit that their positivity and encouragement of her peers helped Mari feel successful as she worked on her mathematical writing and incorporated their helpful suggestions.

Equally important to the development of Mari's MKT is her review of the university PSTs' work. During these reviews, we see examples of all four opportunities to learn. Examining Amanda's journal from the same activity, Mari notes, "The method that you used to check [the multiplication in the problem] is a really good way to check your answer and it's really organized. That is something I never thought of." Amanda emphasizes that "I really like organizing in math because it helps me follow what I am doing better! It is very helpful." We view these comments as evidence of OTL 1. Later in this journal, Amanda uses color to make visible the individual steps in the expanded form of the multiplication. When Mari notes the affordances of this color coding, Amanda responds that she "likes adding color because it is easier to follow. It also helps a reader understand what I am doing." Mari incorporates this strategy into her later work. Thus, we have both PSTs taking advantage of OTLs 2 and 4.

Mari also reviewed Shanell's journal in which she provided a detailed analysis and repair of a student error. Mari notes that "I like how you showed exactly what mistakes were done and how they were fixed because it helps show the correct process… and what could be done differently." In this comment, Mari also refers to a later portion of Shanell's journal in which she explains, in detail, the multiplication

Figure 19. Mari's review of Shanell's analysis of a student error

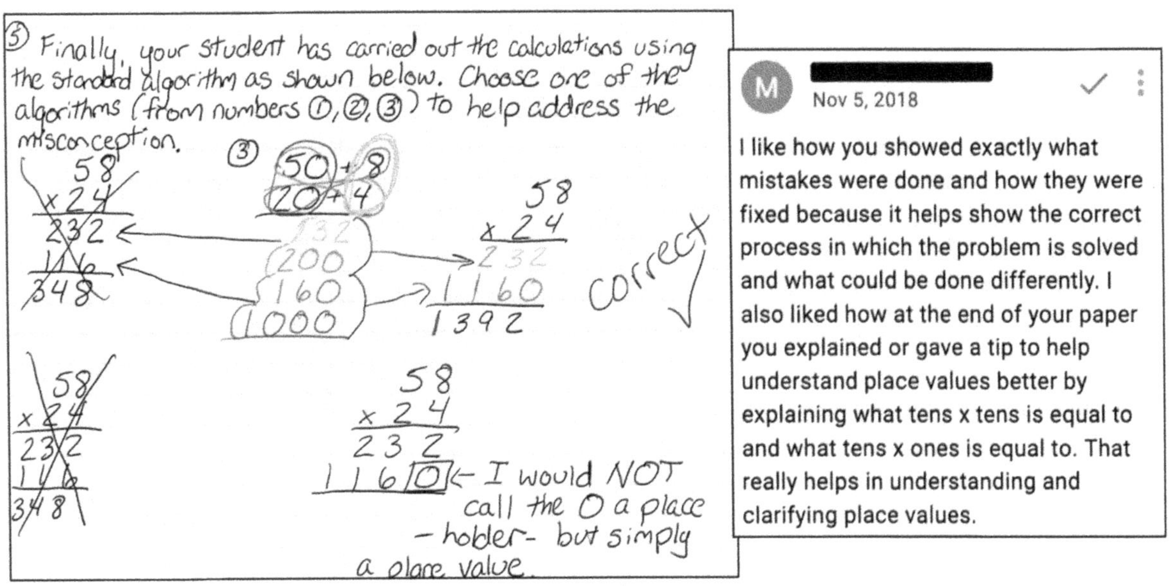

Figure 20. Mari's final representation

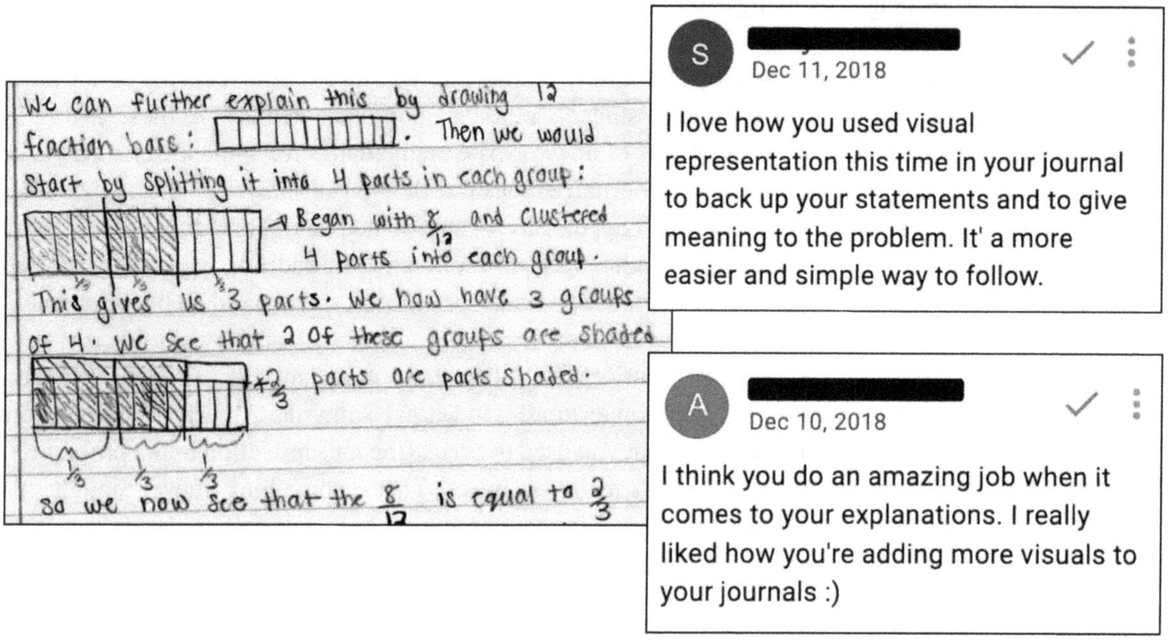

process. Shanell works carefully, place value by place value, using color to make connections to the numerical representation. Mari notes, "That really helps in understanding and clarifying place values." That is, Mari has been successful at understanding Shanell's thinking (OTL 3) (Figure 19).

For the final exchange of the semester, based on an activity dealing with rational numbers, Mari needed to justify that 8/12 and 2/3 are equivalent fractions. Interestingly, Mari chose to use a representation (fraction bars) to support her more thorough explanation. This time, the representation is clear, built up in several steps, and annotated extensively (Figure 20). Shanell comments that the representation gives more meaning to the problem, and Amanda highlights Mari's growth by remarking, "I really liked how you're adding more visuals to your journals." Mari's developing MKT is documented as she moved from a very terse explanation of her problem solving in the first exchange to a more detailed representation-rich explanation of the mathematics in her third journal.

Evidence from the three journal exchanges (and, although not shown here, two additional exchanges in the second course) indicate that Mari developed her MKT through writing and reviewing journals, and she became more confident in her ability to write about mathematics for herself and others. Mari's mathematical development extended beyond journal writing as her work on assignments, quizzes and tests improved throughout the courses. Taken together, these two case studies show that the OTLs were accessible to students with very different mathematical backgrounds and confidence levels. Mari's case raises an important consideration: can this mathematical writing activity be particularly helpful and productive for PSTs who need the most support? We discuss this question, and others, in the next section.

DISCUSSION

Mathematical Knowledge For Teaching

As a response to our research question, we find that journal writing and exchange presented multiple opportunities to learn along key domains of MKT. The writing of the journal itself challenged and supported PSTs to reflect on and communicate their own understanding. According to their reflections, mathematics has been a subject that many of our PSTs have merely endured (and, in some cases, dreaded) during their studies and one that they worry about teaching effectively. They have often experienced mathematics as a set of rules and algorithms to be followed without understanding. In this context, one might think that writing about mathematics would represent an unwelcome extra burden; yet we found the opposite. The APEX Cycle supported this writing process for the PSTs by first allowing them to collaborate in solving rich tasks in their groups during the Analysis Phase, and then to observe their peers' thinking during class presentations. Following the Analysis and Presentation phases, the PSTs were better equipped to reflect on their thinking and express it through writing. This made the process less focused on mathematical performance and allowed them to focus on their writing skills. Making sense of someone else's mathematical thinking and providing constructive suggestions for improvement are not straightforward tasks. To do so, PSTs draw on their Specialized Content Knowledge (Ball et al., 2008). As the various vignettes show, the PSTs' faced challenges while doing so, but they responded skillfully. Mari's case confirms that these learning opportunities are available and accessible to students at all levels; and they may be even more helpful to those students who have not had previous success in mathematics. With its emphasis on understanding and communication rather than rule-following and performance, the mathematical writing and inter-institutional conversations were inviting to all students and provide the time and space to gradually improve their understanding of key mathematics with the help of their peers.

Collegial Collaboration Through Writing

Moreover, we believe that writing for someone other than the course instructor and reviewing the journals of their peers made the learning discussed above a truly collaborative process. Even though contributing to someone else's mathematical knowledge for teaching was not an explicit goal when writing their journals, PSTs appreciated receiving positive comments about how their explanations contributed to their peer's mathematical understanding. The journal exchange gave these PSTs a chance to act as knowledgeable others, perhaps for the first time in mathematics, and experience the authentic demands and practices of teaching with the luxury of time to reflect and plan their responses. These conversations, as we saw, were sites of important layers of learning. We find students able to pose focused mathematical questions to each other, encourage each other as they struggle together to master the mathematics and the art of explaining through writing, and validate each other's successes. The two case studies show how productive these partnerships can be, as PSTs formed collegial relationships grounded in their mathematical writing. Maintaining the review groupings for multiple exchanges deepened these relationships, and often created an opening for more candid critique and sustained encouragement (the cases involving Emily, Mari, and their review partners, are excellent examples of this). By normalizing the cooperative practice-based learning early in the PSTs' preparation, we see this activity as a way to counterbalance the strong individualism and isolationism that pervades teaching in the United States.

We believe that gradual, collaborative, career-long improvement is important for all teachers; and this writing and activity represents an early experience with exactly this type of learning.

Implications for Using Writing as a Teaching and Learning Tool

Even though our study was situated at the higher-education level, we believe that the following implications (lessons learned) are valuable for educators at all levels. Deliberate support might be needed for the learners of mathematics as they

- decide what constitutes a thorough mathematical explanation,
- critique someone else's mathematical writing, and
- value mathematical writing.

As mentioned above, the careful design of the APEX Cycle supported learners of mathematics by balancing collaborative and individual work opportunities. Allowing them to work on the task in a group, they were able to hear the initial, unpolished thoughts of others. Moreover, they listened actively to the whole-class presentations (curated carefully by the instructor) and saw the use of different representations. They had individual time to rethink, reorganize, and write their explanations in their journals. Finally, reading the journals of others and communicating with them in a one-to-one asynchronous format allowed them to see more examples of mathematical explanations. As educators, we found that we needed to be patient for learners of mathematics to see the value of mathematical writing as it enhances their understanding. For some, it took an entire semester (or two), while others shared that they recognized and appreciated it immediately. We see the grouping assignments as a lever with potential to enhance this learning. In Mari's review group, having students with similar mathematical understanding levels, as well as the social skills to offer critique with tact and praise with authenticity, helped Mari extensively. We are aware of the variety of strategies for grouping students to support this type of work; and we see our set of scaffolding prompts for the review process working in concert with those assignments to create a productive setting.

An additional element that we have piloted recently within the APEX Cycle is a set of intramural (typically within a class, as our institutions typically offer only one section of the course semester) journal exchanges. This increased the number of journal exchanges in which students could participate and gave them a chance to critique their classmates' journals before critiquing a journal of someone that they do not know. This raises questions about the exchanges. What are the affordances of exchanging across institutions rather than simply among one's own classmates? Does this lack of social contact add to or detract from the quality of the conversations? This is one phase of the APEX Cycle that we will explore more deeply with future iterations.

Future Research

There are multiple facets of the APEX Cycle that deserve further study. For example, the role of the mathematical tasks themselves, and what features make them most productive for this type of writing, can be better explicated. The alignment of the values and goals of the course (e.g. discussing multiple solution paths and building connections between different representations) and the tasks seems to be an important consideration. In addition, an exploration of the advantages and disadvantages of having the PSTs at partner institutions work on *different* problems involving the same concepts may be worth undertaking. Making sense of both an unfamiliar problem and a peer's explanation of it is certainly more challenging than the activities discussed in this chapter, where the PSTs worked on the same problems at both institutions. One exchange with this modification was piloted, and the small sample of PST feedback suggested that it was difficult to juggle both the new mathematical ideas and someone's explanation of it simultaneously. It would be helpful to know if working on different tasks at the partner institutions hinders the mathematical conversations too much, or if this struggle can be productive in helping the PSTs build even deeper mathematical knowledge for teaching.

The writing activities described in this chapter took place within mathematics content courses for preservice elementary school teachers. Most students take these courses early in their preservice preparation programs, before methods courses, field experiences, and, of course, their initial teaching placements. Therefore, exploring ways to reinforce and extend the learning that we've observed here, in the classes and experiences that follow, seems warranted. Specifically, identifying connections between the learning that takes place within the APEX Cycle and the specific instructional moves that teachers must make in the classroom may allow teacher educators to better help PSTs operationalize this knowledge. We also wonder how mathematical-pedagogical writing and its exchange in methods courses (or student teaching) might further grow the PTS's MKT. Also, if the exchange of mathematical writing involved PSTs and *elementary students*, how might that affect the opportunities to learn? Having to consider student background knowledge when writing, and to make sense of *children's* thinking pose new challenges and opportunities to learn. Anticipating student responses to the writing draws on an important domain of the mathematical knowledge of teaching as conceptualized by Ball and her colleagues (2008). This adjustment would situate OTLs 2 and 3 even closer to the classroom. Tracking this transition from the content courses to the classroom might give us a better understanding of how to better bridge preservice and inservice mathematical communication through writing.

CONCLUSION

As mathematics educators, we need to reflect on our practice and continually improve the teaching and learning environments that we create. When we began our curriculum redesign work, our intention was to create mathematics courses for preservice elementary teachers that engaged PSTs in activities that would deepen their mathematical knowledge for teaching. As our work evolved, three components contributed to the PSTs learning of mathematics: the use of writing as a learning tool, inter-institutional collaboration, and the introduction of CMC. As noted, the PSTs' journal writing engaged them deeply with mathematics as they worked to communicate the content to other learners. Inter-institutional collaboration benefited our work in two ways. First, it allowed us to draw on our varied experiences, to learn from each other, and to build on each other's work. Through this, the APEX Cycle, with its multiple opportunities to learn, was conceived, enacted, and improved over time. One of the key improvements to the cycle was the implementation of the CMC system in our third year that created a platform for the PSTs to engage in inter-institutional conversations. As opposed to earlier one-time, one-way peer reviews of each other's written mathematical work, with no opportunity to respond, the PSTs were now able to engage in true mathematical dialogues grounded in their written work. This leads to the second benefit.

We see the conversations across the institutions as communication across two mathematical communities, sometimes with different language, representations, and points of emphasis. This often required

PSTs to offer a more careful explanation of mathematical terms and solution paths to their peers at the partner institution, as they could not rely on a single, shared classroom experience. This added another layer of learning to the cycle by challenging PSTs to rethink and rewrite their mathematical explanations. It also fostered relationships among the PSTs at the different institutions that, in turn, allowed them to engage in deeper collegial critique of their mathematical writing.

Engaging in this action research provided an important opportunity for us to improve the ways in which we utilize writing as an effective tool to teach mathematics. As previously mentioned, many aspects of this process could be further studied in order to modify and improve it for different learners and environments. We look forward to continuing and expanding this work.

DEDICATION

We, Christopher Nazelli and S. Asli Özgün-Koca, dedicate this chapter to our colleague, coauthor, and friend, Deborah Zopf, in honor of her retirement. Deb's dedication to her students and the field of teacher education have touched countless educators, ourselves included, and she will continue to serve as an inspirational model even as she takes a well-deserved break to spend more time with her beloved family (especially her grandchildren) and to tend to her wonderful garden.

I, Deborah Zopf, dedicate this chapter to my colleagues, coauthors and friends, Christopher Nazelli and S. Asli Özgün-Koca, in gratitude for their ongoing collaboration as we designed and delivered mathematics courses for preservice elementary teachers. Their amazing abilities to study our work have made this chapter possible. I will forever be indebted to their willingness to work with me and enrich my life—both professionally and personally.

ACKNOWLEDGMENT

This work has been supported by two National Science Foundation grants: Division of Undergraduate Education, Student Success Through Evidence-Based Pedagogies (SSTEPs) (DUE-1524878) and Robert Noyce Teacher Scholarship Program, Noyce Mathematics Fellows, TeachDETROIT (DUE-1540819).

REFERENCES

An, H., Shin, S., & Lim, K. (2009). The effects of different instructor facilitation approaches on students' interactions during asynchronous online discussions. *Computers & Education*, *53*(3), 749–760. doi:10.1016/j.compedu.2009.04.015

Ball, D. L. (2000). Bridging practices: Intertwining content and pedagogy in teaching and learning to teach. *Journal of Teacher Education*, *51*(3), 241–247. doi:10.1177/0022487100051003013

Ball, D. L., Thames, M. H., & Phelps, G. (2008). Content knowledge for teaching: What makes it special? *Journal of Teacher Education*, *59*(5), 389–407. doi:10.1177/0022487108324554

Boaler, J. (2016). *Mathematical mindsets*. Wiley.

Conference Board of the Mathematical Sciences (2012). *The mathematical education of teachers II.* American Mathematical Society.

Erbilgin, E. (2019). Two mathematics teacher educators' efforts to improve teaching and learning processes: An action research study. *Teaching and Teacher Education*, *78*, 28–38. doi:10.1016/j.tate.2018.11.005

Hiebert, J. (1997). *Making sense: Teaching and learning mathematics with understanding*. Heinemann.

Howard, C. D., Barrett, A. F., & Frick, T. W. (2010). Anonymity to promote peer feedback: Pre-service teachers' comments in asynchronous computer-mediated communication. *Journal of Educational Computing Research*, *43*(1), 89–112. doi:10.2190/EC.43.1.f

Krainer, K. (2006). Action research and mathematics teacher education. *Journal of Mathematics Teacher Education*, *9*, 213–219. doi:10.100710857-006-9009-5

Lee, E., & Oh, S. Y. (2017). *Computer-mediated communication.* Oxford Bibleographies. https://www.oxfordbibliographies.com/display/document/obo-9780199756841/obo-9780199756841-0160.xml

Lim, W., Son, J. W., & Kang, S. H. (2021). How reducing discomfort impacts peer assessments of preservice teachers. *Sustainability (Basel)*, *13*(11), 6435. doi:10.3390u13116435

National Council of Teachers of Mathematics. (1989). *Curriculum and evaluation standards.* NCTM.

National Council of Teachers of Mathematics. (2000). *Principles and standards for school mathematics.* NCTM.

National Council of Teachers of Mathematics. (2014). *Principles to actions: Ensuring mathematical success for all.* NCTM.

National Council of Teachers of Mathematics. (2020). *Standards for the preparation of middle level mathematics teachers.* NCTM. https://www.nctm.org/uploadedFiles/Standards_and_Positions/NCTM_Middle_School_2020_Final.pdf

National Governors Association Center for Best Practices, & the Council of Chief State School Officers. (2010). *Common Core State Standards for mathematics.* NCTM. http://www.corestandards.org/assets/CCSSI_Math%20Standards.pdf

Price, J. N. (2001). Action research, pedagogy and change: The transformative potential of action research in pre-service teacher education. *Journal of Curriculum Studies*, *33*(1), 43–74. doi:10.1080/00220270118039

Singer, N. R., & Zeni, J. (2004). Building bridges: Creating an online conversation community for preservice teachers. *English Education*, *37*(1), 30–49.

Strauss, A., & Corbin, J. (1997). *Grounded theory in practice*. Sage Publications.

Sutherland, L., Howard, S., & Markauskaite, L. (2010). Professional identity creation: Examining the development of beginning preservice teachers' understanding of their work as teachers. *Teaching and Teacher Education*, *26*(3), 455–465. doi:10.1016/j.tate.2009.06.006

ENDNOTE

1 For more information on the other phases of the APEX Cycle or details about the content courses, please contact the first author.

Chapter 5
Dialogically Focused Secondary Mathematics Teachers' Orientations Toward Mathematical Writing:
A Multiple-Case Study

Ethan P. Smith
Washington State University Tri-Cities, USA

ABSTRACT

Because mathematics classrooms entail expectations for students to communicate understanding through both writing and spoken discourse, it is important to consider the place of writing in the context of such discourse. This chapter details a case study centered around three dialogically focused high school mathematics teachers and their orientations toward mathematical writing. Teachers in this study were first asked to describe their orientations toward mathematical writing. Then, they were asked to reflect on samples of student writing and recordings from their classrooms to further understand how they orient toward students' mathematical writing. Findings suggested that these teachers' orientations toward mathematical writing are embedded in broader views that they hold toward discourse, feedback, and student and teacher self-efficacy around writing. Findings also indicated the value of using the construct of teacher noticing to help investigate and strengthen teachers' orientations toward mathematical writing.

INTRODUCTION

Fostering students' effective communication of mathematical ideas is a critical goal of mathematics education. For instance, The National Council of Teachers of Mathematics (2000) describes the importance of challenging students "to think and reason about mathematics and to communicate the results of their thinking to others orally or in writing" (p. 60), while the Common Core State Standards Initiative (2010) includes an explicit call for mathematically proficient students to be able to "Construct viable arguments

DOI: 10.4018/978-1-6684-6538-7.ch005

and critique the reasoning of others" (p. 6). Ultimately, students must grapple with multiple language demands in the math classroom that extend and connect across different modes of language, including reading, writing, speaking, and listening (Aguirre & Bunch, 2012). This suggests the importance of mathematics teachers attending to both written and spoken language in the classroom.

Although math students do grapple with both written and spoken language, attention toward spoken discourse in the classroom has been more prevalent in mathematics education research (Morgan, 1998; Pugalee, 2004). Additionally, many mathematics teachers are unfamiliar with even the idea of writing-to-learn mathematics and associated writing activities (Teuscher et al., 2015). Indeed, one of the most prominent instructional models in mathematics education has been referred to as *dialogic instruction*, which emphasizes the importance of student exploration with and discourse around novel mathematics tasks (Munter et al., 2015). However, even mathematics teachers whose instruction is dialogically focused often center their classroom discourse around the perceived strength of students' written work (Smith, 2021), so it is not as if students' writing and speaking in school mathematics are diametrically opposed. To this end, understanding the orientations of dialogically focused mathematics teachers could better situate the role of mathematical writing in such classrooms.

This chapter focuses on how a group of three secondary mathematics teachers with dialogically focused instruction oriented themselves toward mathematical writing. Using a multiple-case study approach (Yin, 2018), I drew on interview data and recordings of classroom instruction to better understand how these teachers described their orientation toward mathematical writing, and how such descriptions related to their broader views on teaching and learning mathematics. By using video clips from these teachers' classrooms, I also address how these teachers interpreted samples of students' written work in order to shape (van Es & Sherin, 2021) their noticing of student thinking. I employed these approaches in order to answer the following two questions: (1) How does a group of secondary mathematics teachers with dialogically focused instruction describe their orientation toward mathematical writing? and (2) What are some of the ways that these teachers use students' mathematical writing to help shape their understanding of student thinking?

BACKGROUND

Dialogic instruction (Munter et al., 2015) is a prevalent approach toward mathematics instruction that emphasizes the importance of discourse and mathematical talk in supporting students' learning of mathematics. Because classroom discourse can support students' engagement with each other's ideas and learning in mathematics (Webb et al., 2014), dialogic instruction offers promise in addressing the language demands of the classroom and supporting broader educational policy goals around fostering communication in mathematics. Such instruction entails the use of high cognitive demand tasks, opportunities for students to discuss different strategies or representations, and student discourse (Henningsen & Stein, 1997; Munter et al., 2015), but the role of mathematical writing in the context of such classrooms is less understood. Indeed, classroom spoken discourse has long received more attention compared to students' written work in mathematics education research (Morgan, 1998; Pugalee, 2004).

This lack of clarity in the literature on the role of writing within the context of dialogic instruction gives rise to concerns about how mathematical writing is understood and addressed by educators in the classroom. Mathematical writing can consist of not only the written word (e.g., the construction of words, phrases, and sentences) but also mathematical symbolism and visual imagery (Casa et al.,

2016; O'Halloran, 2008). However, traditional characterizations of writing in mathematics messaged to educators have often emphasized unisemiotic expressions of natural language rather than multisemiotic expressions of writing across different sign systems (Siebert & Draper, 2008). The potential discordance in how mathematics teachers may or may not conceive of different types and purposes of mathematical writing thus warrants careful consideration.

Additionally, teachers' understanding and use of mathematical writing in the classroom may depend on the content and grade level being taught. Indeed, Powell and colleagues (2021) found that secondary mathematics teachers reported focusing less on explanatory writing and more on argumentative writing compared to elementary mathematics teachers. Given that Casa et al. (2016) focus their recommendations specifically on types of mathematical writing in the elementary grades, and given the increasing importance and prevalence of disciplinary literacy (e.g., literacy skills specialized within mathematics) as students advance into the secondary grades (Shanahan & Shanahan, 2008), writing in secondary mathematics is an especially compelling context for further investigation.

As noted by Remillard and Bryans (2004), orientations arise at the intersection of beliefs, perceptions, and practices. Therefore, while interviews and surveys of teachers might help shed light on how teachers think about and report using mathematical writing (e.g., Powell et al., 2021), direct evidence from the classroom – and teachers' interpretations of that evidence – may help provide a more thorough picture of those teachers' orientations toward mathematical writing. To this end, the construct of teacher noticing could be a valuable lens through which to investigate such orientations.

van Es and Sherin (2021) describe how teacher noticing addresses "the invisible, moment by moment attention and sensemaking teachers engage in as they develop classrooms organized around student thinking" (p. 17). Therefore, the conditions and ways in which teachers notice students' mathematical writing in their classrooms could also indicate how they understand the purposes of such writing. Of particular interest is the dimension of noticing that van Es and Sherin (2021) describe as *shaping*, which "involves teachers constructing interactions, in the midst of noticing, to gain access to additional information that further supports their noticing" (p. 23). Because such shaping can entail spoken discourse around a students' written work, the sort of mathematical talk that drives dialogic instruction may offer further evidence of teachers' noticing around students' mathematical writing.

Such in-the-moment noticing could be indicated by teachers' use of *judicious telling*, where they "mediate between the accepted knowledge and methods of mathematics and the particular intellectual communities of their classrooms [in order to] both value their students' constructive activity and introduce elements of accepted mathematics when appropriate" (Smith, 1996, p. 397). Because there is an integrated relationship between what teachers notice and how they decide to respond to students (Jacobs et al., 2010; van Es & Sherin, 2021), instances where teachers build off of students' written work to engage in judicious telling could hint at what aspects of perceived student understandings and perceived mathematical norms those teachers had noticed and chosen to respond to.

Additionally, teacher noticing could be a helpful construct for considering not only the attention of teachers toward students' writing in-the-moment, but also how teachers notice aspects of students' mathematical writing while viewing recordings of classroom instruction. Indeed, previous research on teacher noticing has focused on how teachers can improve aspects of their noticing through professional learning that incorporates video viewings of classroom instruction (Jacobs et al., 2010; Sherin & van Es, 2009; van Es & Sherin, 2002). As such, what teachers notice about students' mathematical writing could help indicate their orientations toward mathematical writing and suggest future avenues for strengthening similar teachers' focus on and use of mathematical writing.

METHODS

I employed a multiple-case study approach (Yin, 2018) to answer the research questions pertaining to this chapter, with each of the three teachers who were interviewed and observed for this study representing a case. These three participants were recruited because their instruction had previously been found to be dialogically focused (Smith, 2021). This entailed instruction that had been found to include the use of (1) high-level, open-ended tasks, (2) opportunities for sharing multiple representations or strategies (e.g., graphs, tables, etc.), and (3) student discourse (Henningsen & Stein, 1997; Munter et al., 2015).

A previous project – the Secondary Mathematics in-the-moment Longitudinal Engagement Study (SMiLES) – included these three aspects of dialogic instruction as part of rubrics that the researchers (including myself) had used to code recordings of classroom instruction. Using data from SMiLES, I previously identified a subset of teachers from that study whose observed instruction strongly demonstrated these aspects of dialogic instruction (Smith, 2021). I had also found that these same teachers, although coded as dialogically focused, often focused their feedback to students on the students' written work, rather than the students' spoken discourse around that work. For these reasons, I felt that such teachers would be strong candidates for the present study.

I ultimately recruited three participants – pseudonyms Mrs. Taylor, Mrs. Hudson, and Mrs. Barnett. Mrs. Taylor, whose Geometry classroom I observed, identified as white and reported having 30 years of teaching experience. Mrs. Hudson, whose Algebra II classroom I observed, identified as Black and reported having 16 years of teaching experience. Mrs. Taylor, whose Honors Pre-Calculus classroom I observed, identified as mixed race (Asian and Caucasian), and reported having nine years of teaching experience. All three participants also identified as female, and all taught at the same secondary level school in a mid-Atlantic U.S. state. I was keen to observe different secondary mathematics courses in order to observe a range of mathematics topics and how writing arose across a variety of contexts. However, these different course choices also arose out of consultation with the participants, with each of them indicating that these particular courses would provide an accurate representation of the ways in which they typically "engage students with their written work."

Each teacher was interviewed in the Spring of 2022 to gain insight into their described orientations toward mathematical writing. The questions in this interview were adapted from Powell et al. (2021) and Cantrell et al. (2008). Questions from the former source (which entailed a survey of mathematics teachers) were adapted to be more open-ended for the interview format, such as asking participants "How frequently do your lessons require students to write? What do those lessons typically look like? Why is that?" Because Cantrell et al. (2008) had investigated teachers' overall content-specific literacy practices, items from this source were adapted to focus specifically on writing. For instance, "How do you help your students become more literate in (state relevant subject area)?" (Cantrell et al., 2008, p. 94) was adapted to "Specifically, how do you help your students become better writers in your classroom?"

Similar to Kauffman et al. (2002), I analyzed these interviews through a multistage coding scheme. This began by creating a narrative profile of each participants' responses by "summarizing prominent topics, uncovering emergent themes, noting memorable responses, and describing overall tone" (Kauffman et al., 2002, p. 277). In particular, these profiles addressed distinctions in how the participants described mathematical writing as it related to both the teaching and learning of mathematics. Again following Kauffman et al. (2002), I then coded subthemes of the participants responses based on the literature – in particular, the recommendations about mathematical writing from Casa et al. (2016) – specifically accounting for the ways in which the teachers referenced how mathematical writing supported the goals of reasoning and communicating.

In a subsequent interview, participants were given an opportunity to member-check (Creswell & Clark, 2017) their profiles, where they were asked about any aspect of these profiles that they would amend, remove, or add in order to more accurately capture their orientations toward writing in mathematics. As noted by Saldaña (2013), such a process can serve as a validation tool during the coding of interview data. The member-checked profiles of these teachers' orientations toward mathematical writing informed my findings around the first research question.

In these later interviews, I also asked teachers to interpret instructional artifacts from their classroom collected from lessons that I had observed. These artifacts included both samples of students' written work and associated video clips of the teacher interacting with students' writing. Similar to the approach of Sherin et al. (2009), the interview questions that followed the showing of the video clip began with a more generalized prompt, "What stands out most to you from this interaction?" After this, the teachers were asked about their recollected intentions with this sort of interaction, and the extent to which this sort of affordance was typical in their instruction. As such, this interview allowed for the teacher's stimulated recall (Lyle, 2003) of the event, serving as "an indirect method of obtaining evidence of cognitive activity" (p. 872) – i.e., an imperfect but insightful look into the teacher's recollection of their thinking during the interaction. Selected teacher responses to some of these interactions inform my findings around the second research question.

RESULTS

Below I describe the results from the study, starting with the teachers' described orientations toward mathematical writing. These results indicate that the teachers recognized mathematical writing as supporting students' mathematical reasoning and communication in line with the recommendations from Casa et al. (2016) regarding elementary mathematical writing. However, the findings also suggest that these teachers' perceptions about mathematical writing – and instructional opportunities and challenges associated with such writing – are embedded in broader beliefs that they held toward teaching and learning mathematics. I then focus on the results from the post-observation interviews conducted with each teacher. These results show how the teachers appeared to hold a stance of inquiry (van Es & Sherin, 2021) toward students' mathematical writing. The findings also suggest differences in how these teachers interacted with students' writing in the classroom and suggest opportunities for how similar teachers might better engage students in discourse around their written work.

Described Orientations Toward Mathematical Writing

Casa et al. (2016) centered their recommendations on elementary mathematical writing around the goals of supporting both communication and reasoning in mathematics. Additionally, Powell et al. (2017) described how communicating through writing can not only serve to benefit discourse and argumentation in mathematics, but also can serve as a tool for assessing students' mathematical knowledge and competence. All three of the teachers who participated in this study alluded to both goals of communication and reasoning when describing writing in mathematics. However, there were differences in how the

Table 1. Teachers' views on writing as a means of communicating mathematical ideas

Teacher	Described Views and Aligned Quotes
Mrs. Taylor	Writing can be used for explaining and justifying to others, which means that it is important to be comprehensible to readers. *"...if somebody else came and looked at this [written justification], would they understand what it was that you meant?"* If students are able to share their writing with others, it allows them to better understand the correctness and comprehensibility of what they wrote. *"I think one of the other things that has to happen is that the kids have to share what they wrote. If they just write it down, who's gonna see it? They don't know whether they did it right, whether they did it wrong, they don't know whether what they wrote makes sense, because some kids are just writing stuff to get it down on paper so Mrs. Taylor leaves them alone."*
Mrs. Hudson	Students can sometimes feel self-conscious or uncomfortable with sharing their writing. However, by encouraging writing opportunities early and frequently, students become more comfortable communicating their thinking. *"I would also say encourage it often, as far as writing or expressing themselves in some way mathematically...In the beginning, I anonymize their names, so the fear goes away of maybe knowing if 'I'm good,' if 'I'm correct or not,' or, if 'this doesn't sound good, but at least I can display it to the entire class at one time.'... just being comfortable with it in the beginning, they were like, 'I don't really like this,' but then at the end now they're like, 'okay, I'm gonna see what I have to say.' And they don't feel limited, you know, to what they want to say there too."*
Mrs. Barnett	Mathematics and writing are both forms of communication. We write in mathematics as a way to communicate mathematical ideas with one another. *"I think again, writing, it's a way of communicating and, I mean, very literally, like equations can be written. Graphs, you know, we can write on a graph. So, I believe that all of those forms of communication fall under that category of writing, especially again, in this context of, in math. Because I wanna communicate something to you. I could draw a picture, or I could, again, write out an equation, things like that. So, I think that they are very much all kind of intertwined."* The precision of written language is not as important as its comprehensibility to whomever is reading what was written. *"...so I'm thinking, the terminology does not have to be perfect. It doesn't have to be, like, fancy. In fact, sometimes when there is fancy jargon, it gets lost in translation. It's the ability to communicate an idea to someone else."*

teachers understood these goals and how they related to their approaches to supporting student writing in their own classrooms.

A summary of the participants' views on writing as a means of communicating mathematical ideas is shown in Table 1. While each teacher described writing as broadly serving a role in communicating, their descriptions suggested different purposes for writing to communicate with others in the context of learning and teaching mathematics. For instance, Mrs. Taylor, who spoke repeatedly about writing in the context of explaining and justifying, described how students' writing needed to be comprehensible to its audience, and that having students' share such writing with their peers could help them to get feedback about the correctness of what they wrote.

Mrs. Hudson also put forward a different perspective about the challenges and affordances of asking students to communicate through writing. She described how student comfort in sharing their ideas with one another through writing takes time to develop, and that one of the fears that students have is about whether or not they are "correct or not." Therefore, while she did describe writing activities as a way to strengthen students' precision of language (in particular, their use of academic vocabulary) over time, she spoke about how this was accomplished by pushing students to want to communicate different ideas even if they are not sure about the correctness of such ideas (i.e., students feel comfortable to "throw different things out" through "rough draft" writing activities). Thus, the goal of such writing in the classroom would presumably not always center on the correctness of what was written, as Mrs. Hudson hoped to move students beyond such a focus on their writing.

Table 2. Teachers' views on writing as a means of reasoning with mathematical ideas

Teacher	Described Views and Aligned Quotes
Mrs. Taylor	Writing is a way to explain and justify mathematical reasoning and helps students to process such thinking. *"...where I'm trying to push them [my students] now is, as they're getting ready now to go into Algebra II, getting them, pushing them into doing that, thinking themselves, because I think that writing part helps them to process the thinking."* Writing helps students to develop generalizations of mathematical ideas beyond being able to produce solutions to specific problems with procedures and algorithms. *"The algorithms I think sometimes gets in their [students'] way. And I even with solving equations and things like that, I try to get them out of that algorithm...so that as they're going along, if they're processing that in their head now, then when they have to write out a generalization later, they have a little bit more idea of what's going on, hopefully.*
Mrs. Hudson	Writing, with a focus on using academic vocabulary, helps develop students' conceptual understanding of mathematical ideas. *"I remember when I first started teaching, I thought vocabulary wasn't important...So I would just kind of dummy down the vocab sometimes and [I eventually realized] that it makes more sense to include it, because when it's seeing the connections [that] are being made across the board, [that is] when that vocab comes in."*
Mrs. Barnett	Writing is a way for students to tangibly reflect on and revise their reasoning with mathematical ideas over time. *"...if they [students] have those kind of not quite put together ideas...we could kind of just revise or restructure the way that you were initially saying that. So again, I want them to write down what they're thinking and then we can refine it as we go."*

Mrs. Barnett, on the other hand, spoke about the importance of comprehension with students' writing similar to Mrs. Taylor. However, Mrs. Barnett also described how "fancy jargon" can actually get "lost in translation." In this way, Mrs. Barnett focused more on communication as a means to facilitate the sharing of ideas, while Mrs. Taylor emphasized an additional role of assessment for such communication.

The teacher participants also had different stated views on how writing was a means of reasoning with mathematical ideas, as shown in Table 2. Mrs. Taylor spoke about how writing, in the form of written explanations or justifications, helped students to process their thinking and could help them develop conceptual understanding of mathematics topics better than if they were simply working through procedures or algorithms to come up with solutions. Mrs. Hudson also saw writing as a way to support students' conceptual understanding of mathematics topics, but specifically because it helped students draw connections by considering the vocabulary that they used in their writing. Therefore, while Mrs. Taylor spoke broadly about how writing helped students to "process the thinking" of the mathematics – in essence, writing as a metacognitive activity – Mrs. Hudson specifically saw word parts and etymology as a way to draw connections for students to other sign systems.

Compared to the other two participants, Mrs. Barnett spoke sparingly about the use of writing as a tool of reasoning – her responses largely centered on the role of writing as a tool to support student communication of mathematical ideas. However, she did discuss how writing can be helpful as a way for students to have a tangible artifact of their thinking. In this way, as students "revise or restructure" their original, "not quite put together ideas," written work could help indicate how students' thinking has changed over time. Mrs. Barnett's responses indicated an idea of writing as a form of rough draft thinking (Jansen, 2020) to be refined over time, rather than as a summative artifact of student learning.

Teachers were then pressed specifically about what writing skills they felt students needed to develop to be successful in mathematics (see Table 3). Perhaps unsurprisingly given the dialogic focus of these participants, they each spoke about the value in developing skills that would serve future discourse or outside interpretation of the writing. Mrs. Taylor spoke of this skill in argumentative terms, describing the importance of writing well-defined justifications for "somebody who's looking at this later." Mrs.

Table 3. Teachers' descriptions of writing skills required for students to succeed in mathematics

Teacher	Described Views and Aligned Quotes
Mrs. Taylor	Students should be able to explain and justify via writing in ways that are comprehensible to a reader. *"I think the biggest obstacle for the writing is the kids knowing what they're thinking...you need to explain, you need to justify, you need to tell me, in writing, [for] somebody who's looking at this later, what it is that you know. And if you can explain it to this person, you can write it down and that's where you need to start. And then we can go from there and make it better and look at it later.*
Mrs. Hudson	Students should be able to use academic language precisely in their writing in order to be both accurate and comprehensible. *"Well, some general things like knowing certain terms, like basic things, like the difference between the word 'and' and 'or.' We were talking about that in Statistics, just what are the differences and what are the similarities actually breaking down English words and translating and see[ing] how they affect the math world. So, when I think about writing, I think about just the common or basic things of knowledge that we can incorporate into math and turn it into some kind of math talk, or vocab."*
Mrs. Barnett	Students should be able to broadly communicate ideas via writing in ways that are comprehensible to a reader. *"I need to be able to transport what I'm thinking in my brain to you in some way, either through words or a picture or an equation. So as far as writing skills, people need to be able to put down their thinking in order for someone else to understand it."*

Hudson did not speak specifically about argumentation, but instead emphasized the importance of using precise academic language as a way to prepare for future discourse or "math talk." Mrs. Barnett also did not reference potential argumentation, but rather focused on clarity – the writer is ultimately trying to "transport what I'm thinking in my brain to you in some way," so students needed to be able to "put down their thinking in order for someone else to understand it." While these responses were in line with the teachers' broader views of writing in math, they also suggested an ultimate aim of such writing as communicating ideas to others, rather than simply reasoning for oneself.

Although the participants all expressed general support of mathematical writing as a valuable instructional tool, they also each identified challenges that they associated with implementing mathematical writing in their own classrooms. As shown in Table 4, these challenges included the time-intensive nature of providing feedback to students' writing (Mrs. Taylor), the need to build students' self-efficacy with mathematical writing over time (Mrs. Hudson), and the teachers' own low self-efficacy around supporting students' mathematical writing (Mrs. Barnett). These challenges were both in line with the teachers' descriptions of their perceived goals with mathematical writing and indicated how their orientation toward mathematical writing was rooted in broader views that they held toward learning and teaching mathematics.

Mrs. Taylor discussed concerns about the amount of time required to provide students with feedback on their writing. Given her aforementioned emphasis on an evaluative purpose of writing, it is understandable that she would see feedback as an important component of any sort of writing-based activity. She described how writing is "the piece that gets pushed to the side when I'm running out of time" in part because of the amount of instructional time that it takes to provide students with feedback on their writing, and that she perceived this as being a reason overall why "teachers can pull away from writing" in math.

While Mrs. Hudson did not explicitly name any challenges that she had with using writing in her mathematics classes, her responses again referenced the importance of building student comfort with writing over time. She stated that writing "should happen all the time" once students "are getting more comfortable," indicating some degree of scaffolding required in successfully implementing writing-based

Table 4. Teachers' perceived challenges of using mathematical writing in their instruction

Teacher	Described Views and Aligned Quotes
Mrs. Taylor	Although writing can be used for learning, it is difficult to make time for it. Because it is important for students to get feedback on their writing to know if it "makes sense" or is correct, it is difficult to use writing in the classroom as it is time intensive. *"So, I've been more worried about getting through the curriculum than spending more time writing, because that does take time and that's where, unfortunately, I think teachers can pull away from writing more because it does take time. And how do I get to see 30 kids' writing? You know? I can look over their shoulders. I can ask what they're saying and things like that. But, you know, I can really only, unless I'm collecting everything and reading everything, I can't – it's almost impossible to do that. So that's one of the hard parts about having the kids write in mathematics is that I'm not taking papers home and looking at them like an English teacher is, I'm 'are you getting it? Are you not?' I don't know."*
Mrs. Hudson	Students need to feel comfortable about their writing in order to engage with activities that include writing. Because of this, building from "rough draft" writing activities to more frequent, "formal" writing activities can help attend to students' growing comfort with writing in math. *"I think they [writing activities] should go throughout [the learning sequence]. I'm a day to day, always sharing in any kind of way, whether it's, like I said, from rough draft to more formal. I think once you see a building process and they [students] are getting more comfortable, I think it [writing] should happen all the time, just so that the comfort is there and that they know the expectation is to go from one type of talk from math to another."*
Mrs. Barnett	Writing was something that Mrs. Barnett did not feel that she does frequently, and when she would engage students in writing this was usually only done at the end of a lesson for students to summarize what they learned. *"So, honestly, I have to be completely honest with you. I haven't been doing [writing] as much and that's why I was feeling a little insecure about being a part of this study, but I'm hoping that it'll encourage me to have this more as a structure in my teaching. A lot of times I would have students write like a summary of a lesson or even just like a, 'this is what I took away from the lesson at the end.' I mean, trying to put the understanding of the lesson into a sentence or into words, that [is something that] I want to be able to do more. I feel like mostly it would be in like summarizing though as a part of the lesson."*

activities. Her responses spoke to the importance of building classroom expectations and norms about how writing fits in with other sorts of activities, such as mathematical talk.

Mrs. Barnett was candid in describing her hesitation to participate in this study because of how infrequently she felt that she used writing in her classroom. However, she did then describe how she tried to build in summative activities where students write "a summary of the lessons" or a response about what students "took away from the lesson at the end." Because she had also discussed different types of and places for writing in the learning sequence earlier in her interview, these responses may have indicated that her perceived difficulties with using writing in the classroom were particularly in relation to this more summative type of writing activity.

In sum, these interview responses illustrated how, although the participants held broad similarities in their described goals and forms of mathematical writing, they also held meaningful differences in how they oriented themselves toward writing as teachers of mathematics. While each teacher mentioned how writing can serve as a means of communicating mathematical ideas, Mrs. Taylor focused on the element of feedback and assessment of student writing, Mrs. Hudson focused on using writing as a tool to build student comfort with communicating in math class, and Mrs. Barnett focused on how different sign systems can allow for more flexibility in how students communicate mathematical ideas. The teachers all described how writing can support student reasoning with mathematics, but Mrs. Taylor focused on writing as a way to support students' explanations and argumentation, Mrs. Hudson focused on writing as a way to improve students' understanding of academic vocabulary, and Mrs. Barnett focused on writing as a tool for revision.

The teachers also each described different perceived challenges to using writing as a tool for learning in mathematics. This included the time intensive nature of providing feedback to students' writing (Mrs. Taylor), the time it takes to build students' comfort with writing in math (Mrs. Hudson), and the difficulty of integrating writing throughout a lesson (Mrs. Barnett). Such responses showed the complexity of how writing relates to the broader ideas of teaching and learning mathematics, and how lines at times were blurred between how these teachers considered student writing versus spoken discourse in their own instruction.

Mathematical Writing and Teachers' Noticing of Student Thinking

In recognizing that teacher orientations fall at the intersection of beliefs, perceptions, and practices (Remillard & Bryans, 2004), I also sought other ways for these teacher participants to express their orientations toward mathematical writing. To this end, I brought instructional artifacts – samples of students' written work and video clips of teachers interacting with students' writing – to a subsequent interview with each teacher. I asked the participants to interpret and recall their thinking regarding these artifacts, with two aims in mind. First, I wanted to understand how these teachers' responses regarding these artifacts were in line with their described orientations toward writing from previous interviews. Second, I wanted to better understand how these teachers engaged in a process of noticing (Jacobs et al., 2010; Sherin & van Es, 2009; van Es & Sherin, 2002, 2021) with these different instructional artifacts.

Below I describe how each teacher participant engaged in a video viewing activity that involved samples of students' writing and a video clip of the teacher's interaction with students' writing. Each participant watched a video clip while also having available to them pictures of students' writing captured during or soon after the clip was recorded. The teachers' responses indicated some distinctions in what (and how) they noticed regarding students' writing, and the extent to which the participants appeared to be shaping (van Es & Sherin, 2021) their understanding of students' thinking amidst their noticing during and after watching the video clip.

Mrs. Taylor observed a video clip from one of her lessons where (according to a pre-observation survey she completed) she had tasked students with creating "a poster where they must use coordinate proofs to determine the type of quadrilateral they were given." Students (in groups of four to five) were given four ordered pairs and asked to "determine the type of quadrilateral *ABCD*" from the given coordinates. Students were also asked to explain on their poster "How can you create parallel lines," "How can you create sides that are perpendicular," and "How can you create sides that are the same length?"

The video that Mrs. Taylor viewed involved her talking with a student (pseudonym Erika) in a group that had been assigned the coordinates of a parallelogram. The poster that the students were creating, as captured shortly after this interaction occurred, is shown in Figure 1. The transcript of the exchange that occurred during the video clip is provided below.

Mrs. Taylor: Um, so, you...just to clarify, right? So, you have the slopes and then (pointing at the poster paper) you said AB is parallel to DC and AD is parallel to BC. How do you know they're parallel?

Erika: Oh, because they're -

Mrs. Taylor: (interrupting Erika) Ah, right? And then, just another little bit to say, how do you know that AD is congruent to DC? Those should have line segments [a vinculum] over the top of them, also.

Figure 1. The parallelogram "proof" poster created by Erika's group

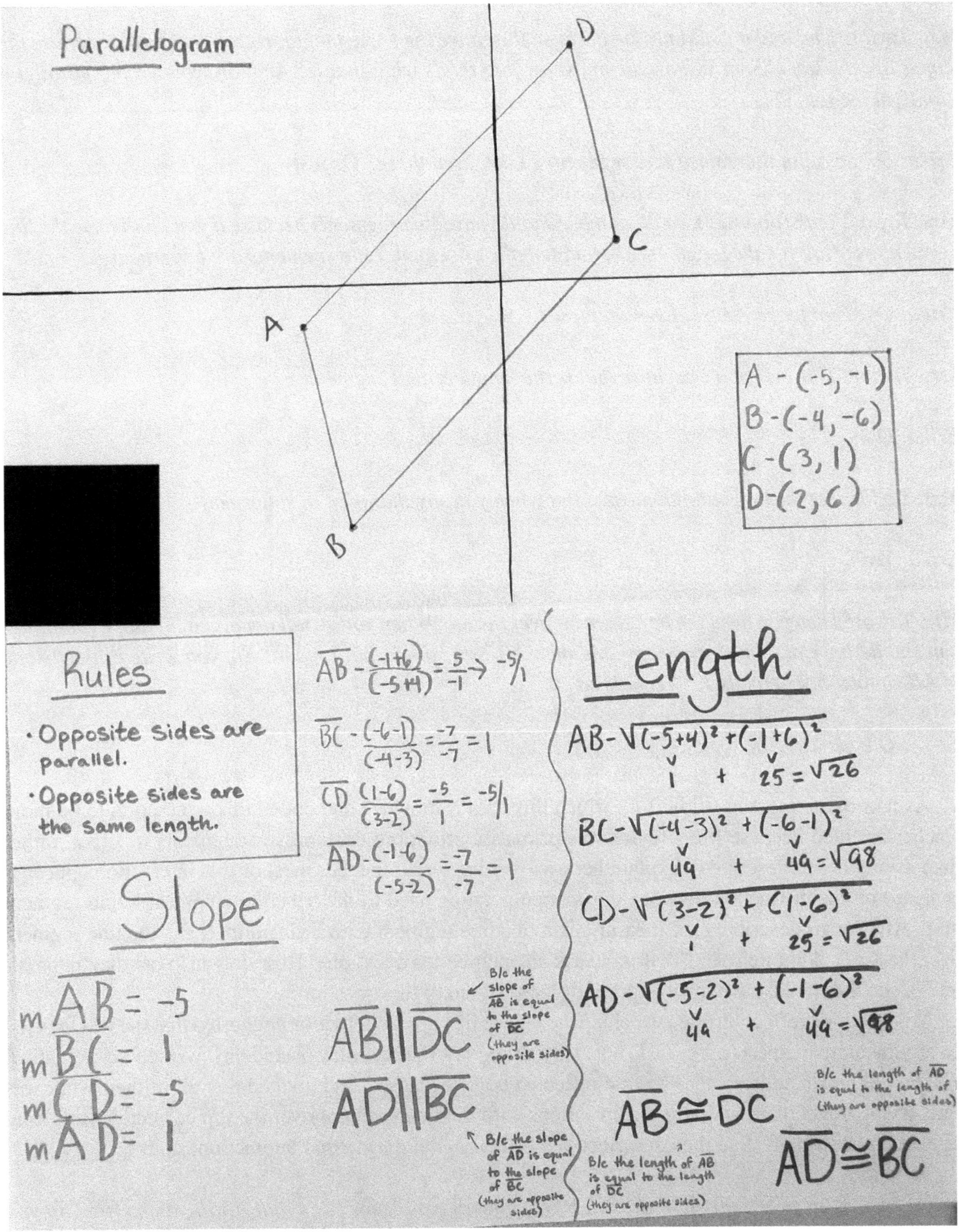

Erika: Ok.

Mrs. Taylor: Um, and it could just be because they have the same length, right? And that's because the slopes are the same. So just, how do you know that that's what that is? And, oh, you wrote "parallelogram" at the top. Ok.

Erika: So, do all of these need line segments? Like, even these? Or just -

Mrs. Taylor: Yeah (pointing), so all of those should have line segments because if you don't have the line segment over top it's the length of AB, and lengths are equal. Line segments are congruent.

Erika: Ok. That makes sense. Even on these?

Mrs. Taylor: That you don't because that is the length of that segment.

Erika: Ok.

Mrs. Taylor: So, that's what that means. But when you say that they're congruent -

Erika: Yeah.

Mrs. Taylor: Then you need the line segment over the top. When you're talking about the slope (pointing) you should have m – lower case m – and then AB, you know what I mean? So, you write it like "slope of AB equals -5," "equals 1," "equals so on."

Erika: Oh, ok.

As part of an ongoing study from which this data is drawn, I had coded this interaction to be more teacher-oriented because Mrs. Taylor was primarily providing declarative statements to Erika, rather than asking her to describe or explain her own written work. Indeed, most of this interaction appeared centered on the distinction between the written notation used to describe the length of a line segment (e.g., AB) versus describing the measurement of a line segment with a vinculum (i.e., "the line segment over the top"). Although Mrs. Taylor did ask an open-ended question ("How do you know they're parallel?"), she did not give Erika an opportunity to respond to this question.

Mrs. Taylor's reflection upon watching the video clip of this interaction suggested that she was focused on argumentative aspects of the task, but also that she saw the accuracy of students' written mathematical notation as worth addressing. She first reflected how "because I had noticed that all of their work was fine, trying to get them to do their conventions with the line segment over the top correctly…was kind of a secondary thing." She then described her primary goal during this interaction, stating:

I say, "you need to tell me, how do you know, how do you know? It's that justification. How do you know? You're telling me that this is true, how do you know that it's true?" And so that's what all of that was, and that's what I was trying to get to, I think, when I went over to their group."

In this way, Mrs. Taylor hoped to push her students from a more descriptive "proof" about their quadrilateral to instead actually justifying their conclusions in writing.

With regard to Mrs. Taylor's noticing around students' written justifications, there appeared to be limited shaping around this aspect of the writing. van Es and Sherin (2021) describe how, with shaping, "part of a teacher's noticing may involve asking a student a question or looking more closely at what a student has written to access further information for the teacher to notice" (p. 23). Because Mrs. Taylor did not give Erika an opportunity to actually respond to the question about how she knew that the opposite sides of her quadrilateral were parallel, this interaction may not have allowed for Mrs. Taylor to gather more information about Erika's thinking.

However, her focus on the group's written notation could suggest that – although she described this aspect as "a secondary thing" – Mrs. Taylor was shaping her understanding of the students' thinking by more closely investigating the symbolic conventions of the students' written work. This could also be considered an instance of judicious telling (Smith, 1996, p. 397), which would be in line with Mrs. Taylor's reflection that, during this interaction, she was "going through [in her] head – 'How is it that I can tell them what they need to do without telling them what they need to do?'" I have previously described how similar examples of judicious telling can be a helpful tool for strengthening students' literacy skills in mathematics (Smith et al., 2023). Therefore, although this interaction did not entail the use of questioning to gather further information about a student's thinking, Mrs. Taylor's descriptive feedback as she observed Erika's poster indicated the focus of her noticing on the perceived accuracy of students' writing conventions.

Similar to Mrs. Taylor, I recorded instruction from Mrs. Hudson's classroom and interviewed her about her interactions with students' writing. Mrs. Hudson's responses to one of these interactions illustrated how she used short written responses from students in a whole class setting to check students' understanding and to build student comfort with sharing their own writing in a math classroom setting. While her interactions with students' writing were mostly evaluative or descriptive (similar to Mrs. Taylor), she also engaged in some brief questioning with a student (pseudonym Emory) that appeared intended to check a potential error with students' explanations.

Prior to this interaction, Mrs. Hudson had asked students to solve $6^m \blacksquare 6^m \blacksquare 6^m = 6^{21}$ for m and had told students to "explain your reasoning" for their solution. Students then input brief explanations using an application on their individual laptop computers that allowed Mrs. Hudson to share student responses on her projector. Mrs. Hudson referred to this as a "Shout it Out" activity, and enacted similar activities in every one of her lessons that I observed. In her interview, she had stated that this activity was the primary way that she asked students to engage in writing in her classroom, and that she saw this activity as an opportunity to strengthen students' academic vocabulary and build their comfort with engaging in writing in mathematics. Several of the responses that Mrs. Hudson pulled up whole group and commented on are shown in Figure 2, and the transcript of her interactions with these students' writing is provided below.

Mrs. Hudson: So, let's look at number 5, this one's a little different. You're multiplying 3 numbers. The good thing is that the bases were all the same, but let's see what we got for m. "I got m = 7 because I divided 21 by..." – I think they meant to put 3 – "...and got 7 so m is equal to 7." "Once again adding, 7 + 7 + 7 is 21." "Divide the exponent of 21 by 3 and to get 7." Good. Oh, let me go back. "When integers are multiplied the exponents are added, so 21 divided by 3 is 7." Question – why in this problem

Figure 2. Several student responses shared by Mrs. Hudson during the "Shout it Out" activity

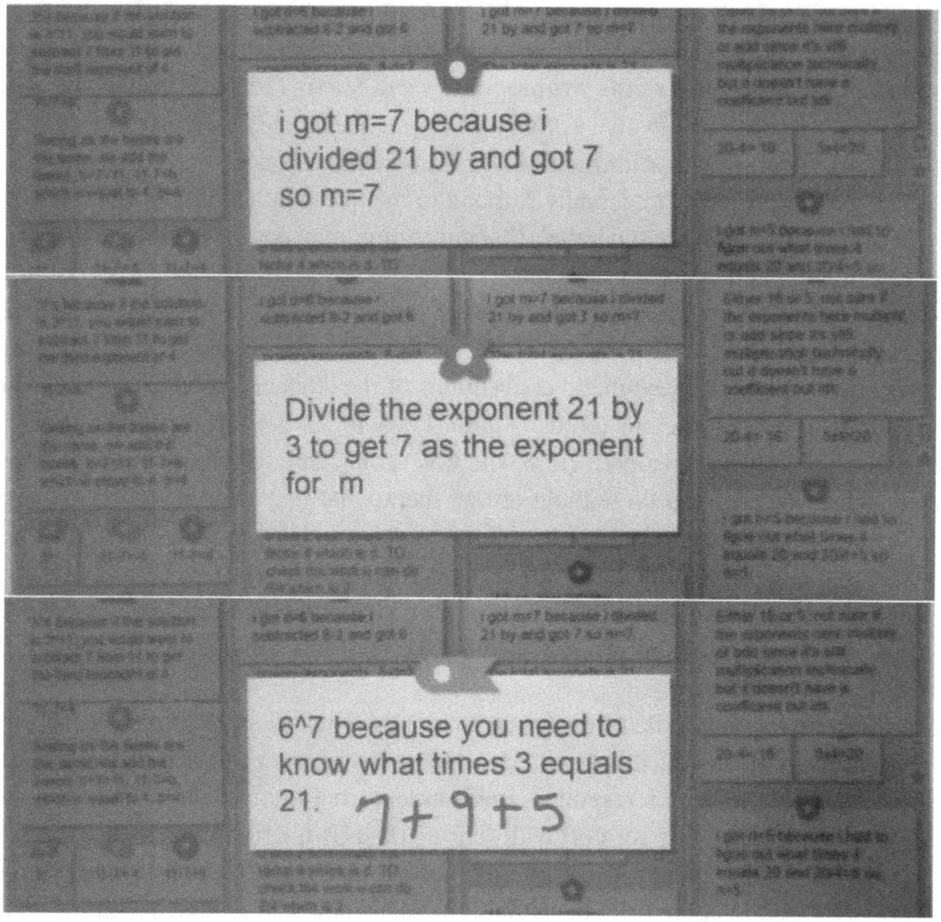

couldn't I put, like…7, let's say, I don't know, 9, and…5. [Writes "7 + 9 + 5" on the Board] Why can't I do 7 and 9 and 5? They equal 21.

Emory: Because they're all the same [unknown variable] so they should be one [value].

Mrs. Hudson: Right. That's the big piece I wanted you all to point out there – if they were different variables, you could have any combination that would create 21, but because they're the same they were all 7s. Alright, so just be careful when they're different variables.

Unlike Mrs. Taylor, Mrs. Hudson here did not appear focused on the precision of the explanations for most of these selected responses. Although most of the students only described the procedures that they performed to find that $m = 7$ (rather than explaining their reasoning), Mrs. Hudson chose to simply restate the students' written responses or to provide brief affirmation (e.g., "Good"). When asked about her interpretation of this interaction, Mrs. Hudson stated the following:

...most of them [students] are used to shouting out or saying their thoughts in this platform just because they know it's going to be shared or they can check in, it's a good way to check in to see if they're on the right page. They don't have to feel like they'll be criticized by responding, so I don't think this is any different than a typical day to be able to share out their ideas and thoughts and some misconceptions that come out in this platform.

This response indicated that Mrs. Hudson's focus in this activity was not only on checking for student misconceptions, but also on building students' comfort with having their writing being shared publicly. This is in line with her described orientation, where she repeatedly emphasized the importance of students' comfort with writing as a factor that impacted how she approached writing in her classroom.

However, Mrs. Hudson did also use this interaction to shape her understanding of students' thinking as well. In her interview, she reflected on how she "thought that there would be students at this question that were going to give me different combinations of [numbers with a sum of] 21." By asking students "Why can't I do 7 and 9 and 5?" Mrs. Hudson appeared to be seeking some student recognition that, because the missing exponents are all equal to m, that they should all be equal to the same value. The whole group context of this question also may have also meant that it served as a sort of judicious telling (Smith, 1996) for the rest of the students listening to their peer's response and Mrs. Hudson's rephrasing of that response.

Finally, I also interviewed Mrs. Barnett about her interactions with students' writing. One of Mrs. Barnett's interactions involved whole class feedback that she provided about students' responses to a writing prompt. Because, like Mrs. Hudson, this was a whole class interaction, I asked Mrs. Barnett to reflect on the recording of this moment in part to contrast her responses with those of Mrs. Hudson.

This interaction occurred near the beginning of a lesson that focused on the inverse relationship between exponential and logarithmic functions. Students were shown a graph of a logarithmic function, an exponential function, and the line $y = x$, and asked to write about "What do you notice about these functions?" This graph, and some of the students' responses to this exploratory writing prompt, are shown below in Figure 3. After students had written down their responses, Mrs. Barnett displayed the students' responses and summarized some of what she was noticing about their explanations.

Mrs. Barnett: So, friends, looking at this first graph, some of us say "whoa, they're reflections of each other across a line in the middle of the two functions." Ooh, I wonder if a lot of people said that too. "Mirrored over the line of $y = x$." Ooh, someone even said that they're inverse functions. A lot of you are talking about the behavior of these curves, like "one of the functions is increasing while the other is decreasing, however they never intersect one another." How interesting. A lot of you just said that they are inverses or they're reflections, there is a line of symmetry...But that's very good – these are key ideas to notice, friends, that these two functions, they are symmetrical to one another about this line. This line is $y = x$, and it's true about any function and its inverse that the graph is going to be symmetrical over the line $y = x$."

Similar to the interaction from Mrs. Taylor's classroom, Mrs. Barnett did not engage students in questioning around their writing. Rather, she appeared to be restating selected students' responses and using these to synthesize the important understandings that she had intended for students to consider during this activity. Similar to Mrs. Hudson, Mrs. Barnett also affirmed several students' responses, even though there appeared to be a range of specificity in these students' written descriptions about the graph.

Figure 3. An exploratory writing task with students' responses from Mrs. Barnett's classroom

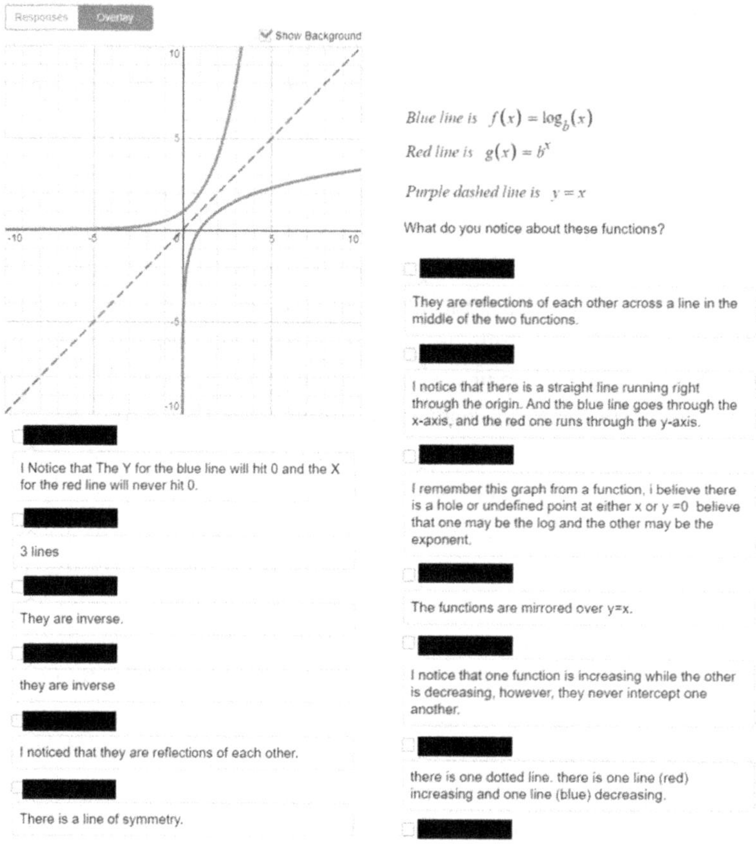

By seemingly synthesizing her thoughts about the students' responses verbally, it could have been that Mrs. Barnett was engaged in shaping amidst her noticing. However, Mrs. Barnett also reflected in her interview that "I wish I had asked the students to make sense of each other's responses instead of reading through them myself" so that it would have seemed "less like I'm just looking through and recounting what they said." This could have suggested that Mrs. Barnett held a stance of inquiry (van Es & Sherin, 2021) toward students' thinking about this task, and that she may have preferred more student-oriented discourse around students' writing.

By reflecting on samples of students' written work and recordings of their instruction, these teachers described some of the ways that they used mathematical writing to shape their understanding of students' thinking. Mrs. Taylor reflected on the delicate balance she was hoping to strike between engaging in judicious telling (Smith et al., 2023; Smith, 1996) and simply "telling them [students] what to do" with their writing. Mrs. Hudson indicated that, while she saw writing as a way to build students' academic vocabulary and address misconceptions, her "Shout it Out" writing activities were also intended to help build a classroom community that valued writing and made it safe for students to share their mathematical writing with others. Mrs. Barnett indicated how, although her interactions (similar to the other participants) appeared more teacher-oriented, she idealized greater student engagement with their own and their peers' writing in the classroom. These sorts of responses showed some of the ways that these teachers' described orientations toward mathematical writing translated into classroom practice, and suggested challenges and opportunities that these teachers faced in shaping their understanding of students' thinking based on students' written work.

DISCUSSION

These results illustrated different types of mathematical writing that these dialogically focused teachers reported promoting through their tasks and their intended rationales for students' writing. The distinctness of each teacher's described orientation toward mathematical writing demonstrated how, although writing in math may broadly be understood to support student reasoning and communication (Casa et al., 2016), there are different ways to approach such goals in the classroom. Mrs. Taylor focused primarily on writing as a means to support student reasoning and communication through the process of students explaining and justifying their mathematical thinking. Mrs. Hudson, on the other hand, saw writing as a tool for strengthening the precision of students' communication of mathematical ideas with academic vocabulary, as well as strengthening their reasoning by connecting that vocabulary to other representations of such ideas. Mrs. Barnett, who described mathematics as ultimately a form of communication, saw both writing and engaging in mathematical thinking and learning as being inherently "intertwined." Although these different descriptions ultimately aligned with broader goals of mathematical writing to support reasoning and communication, the distinctness of each teachers' responses suggested substantial differences in how these teachers considered and attended to writing in their classrooms.

The findings also suggested that these dialogically focused teachers approached writing as an embedded practice within their instruction rather than as a standalone activity. Writing was not only described as a means for students to communicate and reason with mathematical ideas, but also as a means to support students' self-efficacy as mathematical thinkers and doers and to strengthen classroom discourse. These responses extended beyond the scope of teacher surveys about mathematics writing such as that used by Powell et al. (2021), where teachers were surveyed on explicit Likert-scale agreement questions such as "Students understand math better when they write math explanations" (p. 428). The interview responses from the teachers in the present study indicated that teachers' orientations toward writing in mathematics should also be considered in the context of their broader beliefs about instruction and student learning, as the teachers reported their writing practices and beliefs to be strongly influenced by such broader views.

This study also indicated some of the perceived challenges that these teachers felt regarding mathematical writing, particularly in regard to supporting student learning through writing. Mrs. Taylor, for instance, described how writing was often too time intensive for her to use frequently. This was in part because of the value that she placed on students receiving feedback on their writing: If she did not have time to provide individual feedback to each student's written work (which she indicated that she did not), then she was not inclined to use writing in her classroom. This was in line with the findings from Teuscher et al. (2015), who also found that teachers were less inclined to use writing-to-learn strategies in mathematics because of how much class time it was perceived to require. The present study added some potential context to these feelings of time constraints. It was not the enactment of actual writing activities that Mrs. Taylor perceived as being too time intensive, but rather the feedback with which she felt compelled to provide students in response to their writing.

Mrs. Barnett, however, described a different challenge in that she did not feel efficacious with regards to using writing in her teaching. This appeared more in line with the findings from Powell et al. (2021), who found that teachers with higher self-efficacy toward using mathematics writing were more likely to report using such writing activities in their classrooms. However, given Mrs. Barnett's broader described orientation of both writing and mathematics being "intertwined" forms of communication that transpire across multiple semiotic systems, it seemed that Mrs. Barnett was perceiving her low self-efficacy with writing in part because she was considering writing as solely focused on the written word. This is worth considering in the context of studies like that of Powell et al. (2021), especially given that they more narrowly define mathematics writing as "a writing activity in which students write about mathematics concepts or procedures" (p. 418). It stands to reason that, if teachers such as Mrs. Barnett came to associate writing in mathematics as attending to connections across multiple semiotic systems (i.e., a multisemiotic approach; O'Halloran, 2008), she may have reported more self-efficacy in attending to writing in her teaching.

Mrs. Hudson, on the other hand, did appear confident in her attention to student writing in her classroom, especially through her references to her "Shout it Out" activities. However, she did note an overarching challenge of needing to foster students' comfort with sharing their thinking through writing over time. Her approach to building that student comfort – by making frequent use of "Shout it Out" activities that involve the public, whole-class sharing of and teacher feedback on students' written solutions and explanations to instructional tasks – appeared generally in line with the recommendation from Powell et al. (2021) that teachers make regular use of writing across a variety of writing activities (e.g., exploratory or explanatory writing tasks). However, while Powell et al. focused more on the connections between the teacher's self-efficacy toward mathematics writing and the frequency of their use of writing activities, Mrs. Hudson's responses ask us to consider the importance of attending to *students'* self-efficacy with writing in mathematics. As it was with Mrs. Hudson, students' comfort (or perceived comfort from the view of the teacher) with writing may play an important role in determining when and under what conditions mathematics teachers like Mrs. Hudson are likely to integrate writing activities into their teaching.

Finally, the recordings of these teachers' interactions with students' writing in the classroom, and the teachers' reflections on those recordings, illustrated some of the ways that these teachers engaged in noticing around students' writing – and how such noticing aligned with the participants' described orientations toward mathematical writing. The teachers' responses suggested that their distinct views on the purposes of mathematical writing impacted what aspects of students' writing they attended to and interpreted, but also that there were limitations in how they shaped (van Es & Sherin, 2021) their understanding of student thinking amidst their noticing in-the-moment. Participants' reflections during their video viewing activities also indicated a recognition and desire among the participants for better incorporating students' thinking in their interactions with students' writing. These findings are informative in how they suggested opportunities to further relate teachers' described orientations toward mathematical writing to their observed instruction, especially in the context of teacher noticing.

The different vignettes of students' written work, transcripts of the teachers' interactions with that written work, and the teachers' responses to those artifacts showed different approaches to responding to students' mathematical writing. Mrs. Taylor used judicious telling (Smith et al., 2023; Smith, 1996) to provide a student with concrete, corrective feedback about their writing conventions regarding mathematical symbolism, and also pushed the student to strengthen their written justifications. Mrs. Hudson and Mrs. Barnett, on the other hand, publicly displayed and commented on a range of student written responses whole group, potentially as a means to affirm students' effort with their writing regardless of the accuracy or precision of those responses. Mrs. Hudson also engaged in questioning with students around the meaning of their written work, which indicated her use of shaping (van Es & Sherin, 2021) amidst her noticing. Mrs. Barnett similarly commented on how she would have liked to have engaged students in questioning about their peers' responses.

Given the importance of mathematical talk and collaboration in dialogic instruction (Munter et al., 2015), it is perhaps unsurprising that these dialogically focused teachers were grappling with when and how to engage students as co-participants in their discourse around the students' writing. The selected examples in this chapter show a range of opportunities and challenges that teachers face in strengthening students' mathematical writing while also fostering student ownership over their learning. These findings also suggest that a framework of teacher noticing (van Es & Sherin, 2002, 2021) could be helpful in better understanding how teachers attend to students' mathematical writing, and in supporting teachers to strengthen their noticing practices and thus better achieve their stated goals regarding students' mathematical writing.

FUTURE RESEARCH DIRECTIONS

Although the nature of this case study limits the generalizations of any findings, these results do offer implications for future research. In particular, the described orientations of these teachers toward mathematical writing suggests that broader teacher beliefs toward teaching and learning mathematics warrant closer examination in relation to teachers' beliefs about – and instruction involving – mathematical writing. Additionally, the teachers' reflections on their interactions with students' writing provide insight into how investigations of teachers' use of mathematical writing could benefit from employing a noticing framework (van Es & Sherin, 2002, 2021) to better understand how teachers attend to, interpret, and respond to such writing in the classroom.

In line with Casa et al. (2016), the teachers in this study broadly cited mathematical writing as a means to support goals around mathematical reasoning and communicating. However, these teachers' specific instructional approaches also drew on other beliefs that they referenced regarding feedback, student self-efficacy with writing, and the teachers' own self-efficacy with supporting mathematical writing. Although past studies have focused on teacher beliefs specifically about mathematics writing (Powell et al., 2021) and writing-to-learn mathematics activities (Teuscher et al., 2015), future research should consider how broader beliefs about teaching and learning mathematics impact the ways in which teachers approach mathematics writing and the frequency with which they engage in writing activities. Similarly, because the teachers in this study taught at the secondary level and had been found to be dialogically focused (Munter et al., 2015), future research might consider the extent to which different kinds of instructional orientations among teachers – and different grade levels being taught – might lead to differences in how teachers orient toward and enact different writing activities in the classroom.

The observed interactions with students' writing and the participants' reflections on instructional artifacts from their classrooms also indicate future directions of research around mathematics writing. An initial analysis of selected interactions from these teachers' classrooms suggested different ways that teachers notice and respond to students' writing in-the-moment. However, the teachers' reflections on these artifacts in subsequent interviews also helped to connect the teachers' observed actions in the classrooms to the teachers' instructional goals and recollected intentions with their instruction. Therefore, future research concerning teachers' enactment of mathematical writing activities – or teachers' interpretations of those activities – could benefit from drawing on a noticing framework (e.g., van Es & Sherin, 2002, 2021). Similar to Jacobs et al. (2010), such future research could better understand differences in the strength of teachers' noticing and the potential of using noticing activities as a form of professional learning for teachers. In the context of mathematical writing, this direction of research could examine the ways that teachers do or do not shape (van Es & Sherin, 2021) their noticing through interactions with students or the extent to which such teachers' noticing attends to different purposes of mathematical writing, such as those laid out by Casa et al. (2016).

CONCLUSION

This chapter examined how three dialogically focused secondary mathematics teachers described their orientation toward mathematical writing, and how they described using writing to shape their understanding of students' thinking. I found that the teachers' described orientations toward mathematical writing drew on other goals that these teachers shared regarding teaching and learning mathematics. Namely, the teachers emphasized goals around classroom discourse as well as student (and teacher) self-efficacy as important drivers in determining the form and frequency through which they used writing in their own classrooms. I also illustrated a variety of ways in which teachers appeared to be shaping their understanding of students' thinking – and focusing their responses to students – during writing activities, both in the classroom and during subsequent interviews.

This investigation suggests that teachers' orientations toward mathematical writing are interwoven with other beliefs about teaching and learning. Therefore, future research could benefit from further considering the ways that mathematical writing does (or has the opportunity to) align with different approaches toward teaching mathematics, or how writing might be situated as a tool for addressing broader challenges facing mathematics education. Such an approach could both validate teachers in their currently held orientations toward mathematical writing while also pushing them to refine and strengthen their understanding and use of mathematical writing in the classroom.

REFERENCES

Aguirre, J. M., & Bunch, G. C. (2012). What's language got to do with it?: Identifying language demands in mathematics instruction for English language learners. In S. Celedón-Pattichis & N. Ramirez (Eds.), *Beyond good teaching: Advancing mathematics education for ELLs* (pp. 183–194). National Council of Teachers of Mathematics.

Cantrell, S. C., Burns, L. D., & Callaway, P. (2008). Middle-and high-school content area teachers' perceptions about literacy teaching and learning. *Literacy Research and Instruction, 48*(1), 76–94. doi:10.1080/19388070802434899

Casa, T. M., Firmender, J. M., Cahill, J., Cardetti, F., Choppin, J. M., Cohen, J., Cole, S., Colonnese, M. W., Copley, J., DiCicco, M., Dieckmann, J., Dori, J., Gavin, M. K., Hebert, M. A., Karp, K. S., LaBella, E., Moschkovich, J. N., Moylan, K., Olinghouse, N. G., & Zawodniak, R. (2016). *Types of and purposes for elementary mathematical writing: Task force recommendations.* https://mathwriting.education.uconn.edu.

Common Core State Standards Initiative. (2010). *Common Core State Standards for Mathematics*. Core Standards Initiative. http://www.corestandards.org/assets/CCSSI_Math%20Standards.pdf

Creswell, J. W., & Clark, V. L. P. (2017). *Designing and conducting mixed methods research*. Sage publications.

Henningsen, M., & Stein, M. K. (1997). Mathematical tasks and student cognition: Classroom-based factors that support and inhibit high-level mathematical thinking and reasoning. *Journal for Research in Mathematics Education*, *28*(5), 524–549. doi:10.2307/749690

Jacobs, V. R., Lamb, L. L., & Philipp, R. A. (2010). Professional noticing of children's mathematical thinking. *Journal for Research in Mathematics Education*, *41*(2), 169–202. doi:10.5951/jresematheduc.41.2.0169

Jansen, A. (2020). *Rough draft math: Revising to learn*. Stenhouse Publishers.

Kauffman, D., Johnson, S. M., Kardos, S. M., Liu, E., & Peske, H. G. (2002). "Lost at sea": New teachers' experiences with curriculum and assessment. *Teachers College Record*, *104*(2), 273–300. doi:10.1111/1467-9620.00163

Lyle, J. (2003). Stimulated recall: A report on its use in naturalistic research. *British Educational Research Journal*, *29*(6), 861–878. doi:10.1080/0141192032000137349

Morgan, C. (1998). *Writing mathematically: The discourse of investigation*. Routledge.

Munter, C., Stein, M., & Smith, M. (2015). Dialogic and direct instruction: Two distinct models of mathematics instruction and the debate(s) surrounding them. *Teachers College Record*, *117*(11), 1–32. doi:10.1177/016146811511701102

National Council of Teachers of Mathematics. (2000). *Principles and standards for school mathematics*. NCTM.

O'Halloran, K. L. (2008). *Mathematical Discourse: Language, Symbolism and Visual Images*. A&C Black.

Powell, S. R., Hebert, M. A., Cohen, J. A., Casa, T. M., & Firmender, J. M. (2017). A synthesis of mathematics writing: Assessments, interventions, and surveys. *Journal of Writing Research*, *8*(3), 493–530. doi:10.17239/jowr-2017.08.03.04

Powell, S. R., Hebert, M. A., & Hughes, E. M. (2021). How educators use mathematics writing in the classroom: A national survey of mathematics educators. *Reading and Writing*, *34*(2), 417–447. doi:10.100711145-020-10076-8

Pugalee, D. K. (2004). A comparison of verbal and written descriptions of students' problem solving processes. *Educational Studies in Mathematics*, *55*(1), 27–47. doi:10.1023/B:EDUC.0000017666.11367.c7

Remillard, J. T., & Bryans, M. B. (2004). Teachers' orientations toward mathematics curriculum materials: Implications for teacher learning. *Journal for Research in Mathematics Education*, *35*(5), 352–388. doi:10.2307/30034820

Saldaña, J. (2013). The coding manual for qualitative researchers. *Sage (Atlanta, Ga.)*.

Shanahan, T., & Shanahan, C. (2008). Teaching disciplinary literacy to adolescents: Rethinking content-area literacy. *Harvard Educational Review, 78*(1), 40–59. doi:10.17763/haer.78.1.v62444321p602101

Sherin, M. G., Linsenmeier, K. A., & van Es, E. A. (2009). Selecting video clips to promote mathematics teachers' discussion of student thinking. *Journal of Teacher Education, 60*(3), 213–230. doi:10.1177/0022487109336967

Sherin, M. G., & van Es, E. A. (2009). Effects of video club participation on teachers' professional vision. *Journal of Teacher Education, 60*(1), 20–37. doi:10.1177/0022487108328155

Siebert, D., & Draper, R. J. (2008). Why content-area literacy messages do not speak to mathematics teachers: A critical content analysis. *Literacy Research and Instruction, 47*(4), 229–245. doi:10.1080/19388070802300314

Smith, E. P. (2021). Semantically linked syntactic literacy affordances in secondary mathematics. In D. Olanoff, K. Johnson, & S. Spitzer (Eds.), *Proceedings of the forty-third annual meeting of the North American Chapter of the International Group for the Psychology of Mathematics Education* (pp. 1469-1478). Philadelphia, PA.

Smith, E. P., Kelly, J., Sappington, S., Warren, K., & Jansen, A. (2023). Using judicious telling to strengthen literacy in mathematics. *Mathematics Teacher: Learning and Teaching Pre-K–12, 116*(2), 115-123.

Smith, J. P. (1996). Efficacy and teaching mathematics by telling: A challenge for reform. *Journal for Research in Mathematics Education, 27*(4), 387–402. doi:10.5951/jresematheduc.27.4.0387

Teuscher, D., Kulinna, P. H., & Crooker, C. (2015). Writing to learn mathematics: An update. *The Mathematics Educator, 24*(2).

van Es, E. A., & Sherin, M. G. (2002). Learning to notice: Scaffolding new teachers' interpretations of classroom interactions. *Journal of Technology and Teacher Education, 10*(4), 571–596.

van Es, E. A., & Sherin, M. G. (2021). Expanding on prior conceptualizations of teacher noticing. *ZDM Mathematics Education, 53*(1), 17–27. doi:10.100711858-020-01211-4

Webb, N. M., Franke, M. L., Ing, M., Wong, J., Fernandez, C. H., Shin, N., & Turrou, A. C. (2014). Engaging with others' mathematical ideas: Interrelationships among student participation, teachers' instructional practices, and learning. *International Journal of Educational Research, 63*, 79–93. doi:10.1016/j.ijer.2013.02.001

Yin, R. K. (2018). *Case study research and applications: Design and methods*. SAGE Books.

Chapter 6
Mathematical Writing, Teacher Noticing, and Text Clubs

James P. Bywater
https://orcid.org/0000-0001-6053-6382
James Madison University, USA

Sarah Lilly
University of Virginia, USA

Jennifer L. Chiu
University of Virginia, USA

ABSTRACT

This chapter explores how using explanatory mathematical student writing artifacts can be used to position teachers to attend to, interpret, and respond to student mathematical thinking from an iterative and engaged stance. This chapter argues that as conceptualizations of teacher noticing shift from a view that is linear and detached to a view that is iterative and engaged, a corresponding shift should occur in the learning to notice activities that pre-service teachers engage in. The chapter proposes the idea of text clubs, and explores, with two illustrative examples, how pre-service teacher discussions that center explanatory mathematical student writing artifacts offer a potential alternative to the predominant role of video clubs that center teaching video artifacts, and how this approach may support the advancement of new paths for mathematical writing research.

INTRODUCTION

For more than a decade, many mathematics teacher preparation programs have focused on developing pre-service teachers' noticing practice. Teacher noticing is the ability of teachers to see and make sense of events within their classroom (Erickson et al., 1986; Mason, 2002), or more specifically, the ability to attend to, interpret, and decide how to respond to such events (Jacobs et al., 2010). Studies have shown that teacher noticing is an essential practice of teaching (Jacobs & Spangler, 2017), that expert teachers

DOI: 10.4018/978-1-6684-6538-7.ch006

are more skilled at noticing than novice teachers (Mason, 2002), and that a teacher's ability to notice can be developed (van Es & Sherin, 2002). As such, teacher noticing is now a major trend within mathematics education research (Yig, 2022). However, despite the rich body of teacher noticing research, surprisingly little of this literature has involved mathematical writing (some exceptions to this are described in Bywater et al., 2019; Magiera & Zambak, 2021; Monson et al., 2020) despite the fruitful insights into student thinking that mathematical writing - especially explanatory mathematical writing - often afford teachers (e.g., Bicer et al., 2013; Dacey et al., 2018). The predominant artifact used for teacher noticing has been video recordings of classroom teaching, and teachers typically participate in "video clubs" in which they watch video recordings of themselves or others teaching, while reflecting on or discussing what they notice in the video (e.g., van Es & Sherin, 2002). In addition to video, or instead of, "written work" artifacts have been used as the focus of the noticing discussions (e.g., Fernández et al., 2012; Goldsmith, & Seago, 2011; Selmer et al., 2022; Simpson & Haltiwanger, 2017). However, in most studies, mathematical written work artifacts consist of paper on which students write numbers, symbols, and diagrams, but rarely explanatory text. The absence of mathematical explanatory writing within the teacher noticing literature might not be surprising given that explanatory writing still occurs infrequently in mathematics classrooms (Kosko, 2016) despite its well documented benefits (Bicer et al., 2018). Recognizing the importance and absence of explanatory mathematical writing, the objective of this chapter is to explore:

1. How the inclusion of mathematical explanatory writing within teacher noticing practices can help enrich the teacher noticing literature and offer potential solutions to the challenges emerging within teacher noticing research, and
2. How the techniques used to promote teacher noticing can also be used to support the implementation of mathematical writing in classrooms.

BACKGROUND

Explanatory Mathematical Writing

This chapter focuses on *explanatory mathematical writing*. Different types of mathematical writing are identified based on the purpose of the writing activity (Casa et al., 2016). For explanatory mathematical writing, the purpose is to articulate ideas for someone else, typically a teacher or peers. This stands in contrast to other types of writing such as exploratory writing which gives students opportunities to write to themselves to make sense of a problem or their own ideas, argumentative writing which intends to support an understanding of the reasoning and evidence for mathematical concepts, and mathematically creative writing which aims to extend or apply student thinking in new ways. These different types of writing play different roles during instruction. For example, exploratory writing may be more appropriate at the beginning of a problem-solving cycle and explanatory writing near the end. In this chapter, we focus on the student explanatory mathematical writing that occurs after students complete a task or a lesson that provides opportunities for them to explore a mathematical idea (Martin, 2015; McCormick, 2010).

While empirical research about non-mathematical writing is extensive (Juzwik et al., 2006), studies about mathematical writing of any type are limited (Powell et al., 2017). Where they exist, they typically examine the use of journal writing or learning logs at the end of class periods and over multiple weeks, and most have observed that there is a positive impact of writing on students' mathematical conceptual understanding (e.g., Jurdak & Abu Zein, 1998; Kostos & Shin, 2010; Cross, 2009; Atasoy & Adnan, 2020; Tan & Garces-Bascal, 2013; Meel, 1999) but not all (Porter & Massingila, 2000). A meta-analysis of the impact of mathematical writing on standardized test scores, reported an effect size of 0.42 (Bicer et al., 2018), and benefits have also been described in numerous practitioner-based journal articles (e.g., Sanders, 2009; McIntosh & Draper, 2001; Bixby, 2018, Baxter et al., 2002). These positive results have been attributed to the cognitive and metacognitive processes involved in writing. For example, some have argued that the process of writing mathematical explanations encourages students to reflect on what they have done or observed during a task and to consider why processes work or patterns exist (Kenney, 2005). Others argue that explanatory mathematical writing involves processes of self-explanation and helps students connect new and prior knowledge (Chi et al., 1994; Totten, 2005), while others highlight that when students are asked to explain their ideas in writing, particularly when there are a variety of approaches that can be used to solve a problem, they have opportunities to develop their reasoning and the precision with which they use mathematical terminology (McIntosh & Draper, 2001).

While explanatory mathematical writing can support student conceptual understanding, it is particularly valuable for making that understanding visible to others. By reading the explanations written by their students, teachers can get insight into the students' mathematical thinking (Bicer et al., 2013), interpret the reasoning behind every student's work, notice a wide range or complementary set of students' mathematical ideas, and make informed instructional decisions that respond to the needs of their students (Dyer & Sherin, 2016). However, this potential is rarely utilized (Kosko, 2016). Further, skills that are required to attend to, interpret, and decide how to respond to student writing can be developed through structured practice (van Es & Sherin, 2002); however, teachers report insufficient training, time, and confidence for effectively implementing mathematical writing in their classrooms (Powell et al., 2021; Teuscher et al., 2015).

Teacher Noticing

Teacher noticing is commonly described as an activity where teachers identify important events during classroom instruction and then make sense of those events by using contextual information and making connections to broader pedagogical ideas (van Es & Sherin, 2002). Drawing upon Erickson and colleagues' (1986) early work about what teachers pay attention to while they teach, Goodwin's (1994) descriptions of professional vision as the ability to see and make sense of events, and Mason's (2002) view that noticing practice involves being intentionally attentive, early teacher noticing research focused on understanding what teachers attended to, and how they reasoned about those events with the goal of developing the ability to become more attentive to what was most important (Sherin, 2007).

Subsequently, conceptions of teacher noticing became more action-oriented and were expanded to include teacher decision making. For example, Jacobs and colleagues (2010) described teacher noticing as having three components: attending to, interpreting, and deciding how to respond. The first two components, attending to and interpreting, drew upon the established view of teacher noticing, and the third, deciding how to respond, directed the interpretation towards action. Deciding how to respond typically involves deciding how to give feedback to students about their mathematical thinking, and in such cases has much in common with the effective formative feedback strategies such as addressing the features of a student response that relate to the lesson task, including guidance that directs students towards the desired learning goals, and avoiding too much feedback at one time (Bransford et al., 2000;

Hattie & Timperley, 2007; Narciss & Huth, 2004; Shute, 2008). Deciding how to respond may also involve drawing on questioning strategies or talk-moves that help center and explore student ideas (e.g., Chapin et al., 2009; Herbel-Eisenmann et al., 2013), or involve making decisions about future classroom instruction (Kaiser et al., 2015).

Despite the variations in how teacher noticing has been conceptualized and theorized (for a detailed discussion, see König et al., 2022), there has been a consistent emphasis on the centrality of student ideas and the need to consider the "disciplinary substance" (Coffey et al., 2011, p. 1109) of student mathematical thinking. That is, teacher noticing of student mathematical ideas has involved teachers identifying the important mathematical details of a student's strategy or explanation, attending to the subtlety and diversity of the ideas, making mathematical connections between them, interpreting them in the context of the intended mathematics learning goals, and then making decisions about how to respond (Jacobs et al., 2010; Robertson et al., 2016). Research demonstrates that noticing is difficult, and many teachers struggle to notice the diversity and nuance of student ideas in a classroom (e.g., Sherin, Jacobs, & Philipp, 2011). Noticing student mathematical thinking supports, and is supported by, high levels of Mathematical Knowledge for Teaching (MKT; Ball et al., 2005) and detailed understandings of the conceptual development of student mathematical thinking (Carpenter et al., 2003).

Video technologies have been used to support teacher noticing because they allow prior recordings of teaching and learning to be slowed down, paused, or rewatched so that the nuances in what teachers and students do and say during their interaction can be made visible to viewers in ways that are difficult to do during live observations (Santagata et al., 2021). Video recordings can be discussed synchronously by groups of teachers, but video technologies also support asynchronous discussions by allowing moments during the video recordings to be tagged or commented on (Evi-Colombo et al., 2020). Within teacher noticing studies reported in the literature, video clips are more commonly used than recordings of whole lessons, they are typically accompanied with guidance about what pre-service teachers should look for while watching, and pre-service teachers are generally asked to respond to prompts individually (Santagata et al., 2021). However, several studies report groups of teachers participating in open discussions about what they notice while watching teaching videos and these occasions are typically referred to as "video clubs" (e.g., van Es & Sherin, 2002). In general, these studies report that using video in these ways is effective at supporting teachers to attend to and interpret events within the videos (Santagata et al., 2021; Gaudin & Chaliès, 2015). However, given the smaller number of studies that have focused on supporting teachers to decide how to respond to students or to actually respond to students, it is still uncertain whether the skills that teachers develop by watching video recordings transfer to classroom teaching (Santagata et al., 2021).

EXPLANATORY MATHEMATICAL WRITING AND ENGAGED TEACHER NOTICING

Engaged Teacher Noticing

To address these challenges, a number of scholars have explored how teacher noticing might become more engaged with students and their thinking. For example, Scheiner (2021) critiques the linear attend-interpret-respond view of teacher noticing by recognizing that what teachers attend to is often influenced by how they interpret the world, their beliefs about students and learning, and their mathematical understanding. Further, Scheiner (2021) highlights how the linear view of teacher noticing positions teachers as observers, detached from students and the learning environment, and proposes a more broadly situated view of teacher noticing that involves "recurrent interactions" between teachers and students (Scheiner, 2016, p. 237). This view echoes earlier calls for an "enactivist" approach to teaching rather than the more evaluative or interpretive approaches (Davis, 1997; Crespo, 2000), and aligns with the ideas of responsive teaching that highlight the need for an on-going, dynamic process of interpreting student thinking and then actively responding to them (Hammer, 2012; Dyer & Sherin, 2016; Robertson et al., 2016). From these perspectives, teacher noticing is cyclical rather than linear and engaged rather than detached, and teachers are able to take an inquiry stance towards noticing rather than an observational stance (van Es & Sherin, 2021).

At the same time, most teacher noticing research continues to use video artifacts of teaching for developing teacher noticing practice (Weyers et al., 2023). The affordances of video such as the low barriers to recording, viewing, and pausing examples of classroom teaching make it a valuable technology for developing noticing practice. But given the prevailing view that noticing should involve teachers taking an inquiry stance, the constraints of video technologies limits how effective they can be. With recorded video, events can only be observed and not interacted with, and what happens after the events cannot be changed. With video, it is possible to observe consequences of teacher responses, but not change those responses and then observe different consequences. As such, video artifacts align well with linear views of teacher noticing that consist of attending to, interpreting, and deciding how to respond, but less well with the perspective that teacher noticing should be engaged and iterative. We therefore propose that the use of video to support teacher noticing should be supplemented with alternative artifacts that have affordances that support engaged noticing. One alternative approach has been to use interactive teaching simulation artifacts where subsequent events are dependent upon teacher responding decisions such as LessonSketch (Herbst et al., 2013). As a further alternative, we propose the use of another authentic classroom artifact that, like video, is easily recorded and viewed, and is paused for teachers to interpret. However, it is expected that this artifact will be responded to, that students will react to the response, and that the artifact will support a cycle of engaged iterative noticing. The artifact we propose is student explanatory mathematical writing.

Explanatory Mathematical Writing to Support Engaged Teacher Noticing

For students who are developing their mathematical understanding, written explanations can be a more accessible way to share what and how they currently understand an idea (Stylianou & Silver, 2004; Nathan & Koedinger, 2000). Student writing can give teachers important clues about partial understandings that students may have (Dacey et al., 2018); for example, the use of specific vocabulary can help teachers see the precision with which a student understands an idea (Barwell, 2005). Written explanations also enable teachers to see the reasoning of every student, not just the few students who provide verbal explanations of their thinking during classroom discussion; thus, there is the potential for a wider range of students' mathematical ideas to be noticed (Benjamin, 2013). It is not surprising, therefore, that pre-service teachers show greater attention to student ideas when shown artifacts of student written explanation than when shown video recordings (Magiera & Zambak, 2021), and that explanatory mathematical writing artifacts are a rich site for teacher noticing.

Further, given that asking students to write explanations of their mathematical understanding is an activity that can be easily transferred into authentic classroom teaching, for example, during an inquiry activity or as an end-of-task exit ticket, student explanatory mathematical writing artifacts are well aligned with engaged and iterative conceptions of teacher noticing. The teacher noticing skills that pre-service teachers might develop by participating in structured noticing discussions will have been developed in contexts that are very similar to in-service teaching contexts. And the expected outcomes of these discussions will not just be that teachers attend to and interpret the written artifacts, but that they also respond to each student knowing that students will receive those responses.

The techniques used by mathematics methods instructors to develop pre-service teacher noticing can also be used to support the infusion of mathematical writing into classrooms. By shifting the artifacts used to prepare teachers to notice from video artifacts to mathematical writing artifacts, teachers will enter classrooms with greater familiarity with and confidence using writing in their classrooms, and with lower barriers to implementing mathematical writing (Powell et al., 2021). Rather than teachers viewing writing as an extra activity that they have insufficient time for (Powell et al., 2021), the use of written artifacts for noticing has the potential to demonstrate in a new way how writing should be integral and how insights into student thinking may be missed without it.

Together with the widespread use of text-based tools that help teachers to collect student writing, collaboratively discuss the explanations, and respond to students, and with the recent advances in natural language processing technologies, there are multiple innovative ways that tools can be used to support students to write mathematical explanations and for teachers to engage in cycles of noticing and responding.

SOLUTIONS AND RECOMMENDATIONS

Text Clubs

Drawing upon the established practice of using video clubs for supporting and studying teacher noticing, this section explores the potential affordances and limitations of analogous text clubs. In video clubs, groups of teachers watch classroom teaching videos, discuss their interpretations of specific events, and consider how they might respond if they were the teacher (van Es & Sherin, 2010). Similarly, the text clubs that we envisage involve groups of teachers reading student explanatory writing, discussing different interpretations of how students understand the mathematical ideas, and considering how to respond. During this discussion, multiple student explanations might be considered and the diverse and nuanced ways that students describe their thinking within the artifacts can be considered (Magiera & Zambak, 2021; Stockero et al., 2017). However, in contrast to video clubs, text clubs can also be action oriented. The discussions and interpretations that occur during text clubs have the goal of supporting teachers to then write responses to each student that will be subsequently shared with students, initiating new opportunities for noticing.

To illustrate our view of text clubs and how we envision teacher responses being used to engage students in ways that support teachers to take an inquiry stance towards future noticing, we share the following two examples from our prior work. These examples describe professional development and classroom teaching activities that use explanatory writing to iteratively support teachers to notice student thinking. When first published, these examples were not described as text clubs and not framed using the perspective of engaged teacher noticing that we detail above. However, these two examples are shared from this new perspective to articulate our thinking about the intersection of teacher noticing and explanatory mathematical writing in different areas of our work.

In the first example, we share a text club where teachers participate in interpretive discussions of student explanatory writing in order to develop example student responses that they later use to respond to students using an online tool designed to support teacher responding (Bywater et al., 2019; Bywater et al., 2022). In this example, we share how teachers wrote back to students and how these written responses initiated in-class discussions with students about their mathematical thinking that subsequently led to students rewriting their explanations. In the second example, we share how teachers supported students to engage in mathematical writing through a science, technology, engineering, mathematics, and computer science (STEM+CS) project within an elementary science classroom (e.g., Lilly et al., 2023). In this example, teachers and students discussed elements of their mathematical writing within a whole-class discussion and then students utilized their new understandings to revise mathematical and computational models.

Example One: Engaged Mathematics Teacher Noticing Using Technology

Instructional Context

Our first example centers on a group of four geometry teachers who teach at a diverse public high school (ages 14-18 years old) in a small, mid-Atlantic city in the United States. In their classes they implemented a mathematical modeling project that was designed to help students to understand how scale factor changes the dimensions, surface area, and volume of a rectangular prism in the context of candy packaging design. The project began with tasks that involved hands-on and interactive online activities to build understandings of the surface area and volume of rectangular prisms. For example, students were provided with a packet of 12 small rectangular prism shaped candies and asked to find an arrangement that would have the smallest surface area. Afterwards, the project included tasks about scale factor that asked students to identify the relationship between scale factor and a rectangular prism's dimensions, surface area, and volume. These tasks included interactive GeoGebra visualizations and data tables for students to record observations. As shown in Figure 1, students were prompted to write explanations of the mathematical relationships between scale factor and the dimensions, surface area, and volume of a rectangular prism (for details see Bywater et al., 2019).

Teacher Participation in a Text Club Meeting

After the first implementation of the project, the teachers met to discuss the different explanations that students had written about how scale factor impacts dimensions, surface area, and volume. The teachers shared explanations that students had written in each of their classes and the teachers discussed what they noticed and how they might respond to each student's explanation (Bywater et al., 2019). The discussion was open but framed with the question "how would you respond to this student?" Teachers were asked to justify why they intended to respond in a particular way based on their interpretations of the student explanation. These discussions highlighted different interpretations of student thinking, often shifting from interpretations about what teachers saw as missing or incorrect to what assumptions students were making or what partial understandings they held that would lead to the explanation given. Sometimes these conversations were inconclusive, and teachers found it difficult to decide on an interpretation without further information. Having the goal of writing draft responses to students helped focus these discussions, especially when teachers were not sure how to interpret a student response and knew that

Figure 1. Example of a mathematical writing prompt from the scale factor project

Below you can **change the size** of the 5x4x3 shape by changing the **scale factor**.

Because **the shape doesn't change**, the bigger and smaller objects are **similar figures.**

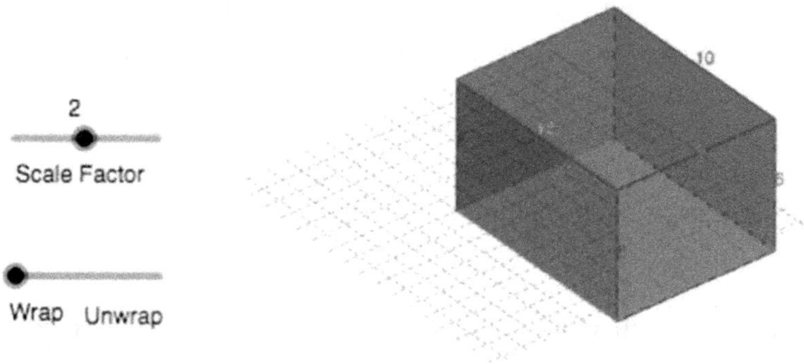

For the shape shown above, **change the scale factor** and write the new sizes for length, width and height in the table.

	Scale Factor = 0.5	Scale Factor = 1	Scale Factor = 2	Scale Factor = 3
Length	2.5	5	10	15
Width	2	4	8	12
Height	1.5	3	6	9

How does the scale factor change the length, width and height of the object?

they could respond by asking clarifying questions. The teachers often found themselves wanting to ask lots of questions, or provide lots of feedback, but we asked teachers to prioritize their responses to include at most two or three follow-up questions. This constraint meant that the discussions also focused on the relative importance of different responses and teachers tended to prioritize questions about key mathematical ideas. In addition, because the teachers were asked to draft their responses in writing, and because they had read and discussed student writing, nuances in vocabulary and mathematical thinking were often discussed together and led to teachers thinking critically about how they phrased their own responses.

Technology-Supported Responding

The responses that were written as part of this text club meeting were not immediately provided back to the students because the text club occurred after the project had ended. However, when the project was implemented the following year with another group of students, the responses that the teachers had drafted were used to create example response recommendations that the teachers could use to assist them when responding to each of the new students' explanations. During the second-year implementation, the teachers used the online Teacher Responding Tool (for details see Bywater et al., 2019) which asks teachers to review each student's written explanation and write responses back to students. For each student explanation, the tool automatically selects the three most relevant example recommendations from those that were written the prior year and shows them to the teacher as they consider how to respond. The tool does not automatically respond to the student for the teachers, but by providing mathematically relevant examples that are based on the text club discussion, the tool aims to remind the teachers of some of the nuanced and prioritized ways of interpreting and responding to students. The tool is intended to be used by teachers between classes so that the written explanations that students write during one class can each be responded to by the start of the next. In this way, the noticing and responding that teachers did while using the tool was engaged with the mathematical thinking of the students in their classroom.

Engaging Students With Teacher Responses

When students read their written responses from their teachers, we observed that they reacted in different ways. All students were asked to read the responses and revise their explanations, and so some students immediately began to revise their explanations based on the questions they were asked. Others returned to the activity from the previous day to explore the scale factor relationship further. For many students, the teacher's response initiated a verbal discussion, either with other students or with the teacher, about the mathematical ideas that the teacher had responded to. For the teachers, this gave them further opportunities to take an engaged inquiry stance towards noticing and responding to student ideas. In the post-study interviews, teachers commented that using the tool to respond to students helped them feel much better prepared for the upcoming class because they were more aware of the different ways that students understood the scale factor relationships (Bywater et al., 2019). In general, the students' revisions showed improvements in their articulation of the scale factor relationships that they were asked to explain (for details see Bywater et al., 2022). These improvements cannot be assigned solely to the teacher responses because the responses initiated further classroom investigations and discussions. However, from our observations, the responses initiated these further investigations and conversations in a way that centered each student's mathematical understanding.

Example Two: Engaged Mathematics Teacher Noticing in an Interdisciplinary Project

Instructional Context

In our second example we consider teacher and student engagement with explanatory mathematical writing in an interdisciplinary project with 5th grade students (ages 10-11 years old) at a school in a mid-size, mid-Atlantic city in the United States. Considering mathematical writing in interdisciplinary learning contexts builds upon research in science where teachers use students' written explanations to support teacher noticing of and responding to students' scientific ideas (e.g., Rodriguez & Zozakiewicz, 2010). For example, Gerard et al. (2010) demonstrated that helping science teachers examine students' written explanations of plate tectonics resulted in teachers' changes to instruction that increased student learning of targeted science concepts. Helping teachers examine students' written explanations and revisions can also result in customizations of curricular units that increase students' self-directed learning (Gerard et al., 2022). Thus, there is opportunity to utilize students' mathematical writing to support teacher noticing within interdisciplinary contexts.

The overarching goal for the project was to support students to redesign their school grounds to reduce water runoff through science, engineering design, mathematics, and computational thinking activities (SPICE; for details see Chiu et al., 2019). The project was aligned to the Next Generation Science Standards (NGSS), that specifically call for teachers to use mathematical thinking within the science and engineering practices (SEPs) to engage students in STEM+CS disciplines. Specifically, NGSS integrates mathematical and computational thinking within the SEPs as well as mathematical concepts within the cross-cutting concepts (CCCs) of scale, proportion, and quantity and patterns. Some of these SEPs, such as developing mathematical models, analyzing and interpreting data, and using mathematics and computational thinking, may seem more obviously dependent on the integration of mathematics. We propose that teachers can also support students' mathematical writing through SEPs such as constructing explanations, engaging in argument from evidence, and obtaining, evaluating, and communicating information. Particularly in NGSS-aligned projects, teachers can support their students to engage in mathematical writing as well as to revise their written work as they develop new understandings based on evidence (Davis et al., 2019; Odden & Russ, 2019). Utilizing mathematical writing through these other STEM+CS disciplines is important as it offers students opportunities to authentically engage in articulating their mathematical understanding through writing as scientists, engineers, and computer scientists do in practice.

Through this project, the teachers supported students to develop mathematical rules from hands-on investigations, use their experiences to write scientific causal or mechanistic explanations about processes, create mathematical models to develop a computational model, and then use the computational model to test different engineering designs and develop arguments for why a specific design best met the project constraints that were supported by evidence and reasoning. The science, engineering, and computational modeling activities, specifically, were designed to support students to develop and explain models, including mathematical models (Lilly et al., 2020). For example, during the project, teachers supported students to conduct hands-on investigations of water run-off in order to construct scientific explanations. These hands-on investigations included students collecting data, making observations, and gathering evidence about water absorption and runoff over time. Specifically, students would simulate a "light" rainfall by pouring a specific amount of water over each surface and collect the amount of water that ran off after specific time intervals to make observations about how different surface conditions (e.g., grass and concrete) affected water runoff. After collecting the data within small groups, and comparing this group data to aggregate classroom data, students wrote explanations of the mathematical rules about the total amount of rainfall, the total amount of water absorbed, and runoff. For example, Figure 2 shows an activity in the Student Guide where there is a table for students to fill in and notice patterns about total rainfall, absorption ratio, and total absorption. Students are prompted to fill in the value of water absorbed after a specific elapsed time with a specific hourly rainfall. For example, after 1 hour of 0.3 inches of rainfall on a surface with an absorption ratio of 0.7, there will be 0.21 inches of rain ab-

Figure 2. Example of a mathematical writing prompt from the SPICE Student Guide

Activity 9.1: Calculating Absorption of Grass

To help you program your model, let's calculate the amount of absorption by grass by hand.

Starting with an hourly rainfall of 0.3 inches per hour for 4 hours, how would you calculate the amount of water absorbed by grass for each hour? Remember, the absorption ratio for grass is 0.70. Fill in the empty boxes in the table below.

Duration (hours)	Elapsed time (hours)	Hourly rainfall (inches)	Total rainfall (inches)	Absorption ratio	Total Absorption (inches)
4 hours	0 hours (rain starts)	0.3 inches	0	0.7	0
4 hours	1 hour	0.3 inches	0.3 inches	0.7	
4 hours	2 hours	0.3 inches	0.6 inches	0.7	
4 hours	3 hours	0.3 inches	0.9 inches	0.7	
4 hours	4 hours	0.3 inches	1.2 inches	0.7	

What pattern or rule do you notice in your table?

____

____

sorbed. After two hours and 0.6 inches of total rainfall, there will be 0.42 inches of rain absorbed. After filling out the table, students are prompted to write about what pattern or rule they notice in the table. Students can answer the additive pattern of 0.3 inches adding to the total amount of rainfall for every hour. Students can also notice the pattern that total water absorbed = total rainfall x absorption ratio.

Having students write about their understanding of the scientific phenomena using mathematical evidence gives insight to the teacher as to what students may or may not understand. If students write about the additive relationship, then teachers can prompt the student to notice or explain the multiplicative relationship with water absorbed. Having the explanation alongside their values can also help teachers diagnose when students might be having a procedural error (e.g., a mistake with multiplying decimals) as opposed to a conceptual error (e.g., understanding that the values need to be multiplied).

Students then abstracted their mathematical rules as a base to program their computational models. For example, students could use a block-based programming environment to translate the rule of "total rainfall = rain duration (hours) x hourly rainfall (inches/hour)" into their computational model (Zhang et al., 2020). The importance of being able to translate their mathematical rule into a computational model is that students can then use the computational model to test each of their different engineering designs for different configurations of surface materials on the school grounds and compare the designs within the project constraints. Using these tests, students then developed written arguments with claims and evidence from their mathematical and computational models for why one of their designs was the best choice based on the project criteria. Finally, the students defended their solutions in a final presentation to their teacher, classmates, and principal.

Teacher Participation in Informal Text Club Interactions

During this project the teachers reviewed a range of student artifacts, including written mathematical explanations. The explanatory writing artifacts were especially useful for helping teachers notice student mathematical ideas and how those ideas impacted their computational or scientific understandings. For example, teachers noticed that some students could explain the mathematical multiplicative relationship clearly and could transition this understanding to a generalizable computational model with variables that represented the scientific concepts. Similarly, student explanatory mathematical writing helped teachers uncover when students may not have understood the scientific concept but did understand the mathematical ideas or computational model. Teachers were able to notice where students were able to calculate values and articulate mathematical rules, so that they were well positioned to respond to students about the application to the scientific concepts and provide more nuanced and tailored feedback to students.

The text club in this example was much more informal than that described in Example 1. The discussions between teachers typically consisted of pairs of teachers sharing what they noticed with each other between classes or at the end of the school day, and the teachers attended to other artifacts in addition to student mathematical writing when noticing. However, the key elements of a text club, as we envisage them, still occurred. That is, groups of teachers read student explanatory writing, discussed different interpretations of how students understand the mathematical ideas, considered how to respond with the goal of responding to students, and responded to students in ways that initiated new opportunities for noticing.

Engaged Teacher Responding

In this project, teachers typically responded to student writing verbally during class rather than writing responses as in Example 1. For example, when a teacher noticed that a student understood the pattern as repeated addition, they were able to respond in ways that allowed the students to build upon this understanding within the computation model that they were developing. Similarly, if students understood the pattern as multiplication, the teachers were able to respond in ways that built on this mathematical understanding, and the alternative ways of understanding pattern and creating models could be discussed. This then helped teachers respond in ways that supported students to make interdisciplinary connections. For example, one teacher said, "So this is just kind of one operation instead of multiple additive operations. So you could have done it either way, which is a really important point when you're coding. There's not always one solution. There's often lots of different solutions. So if you've got here a different way, that's wonderful." This teacher went on to support students to make epistemic connections between being able to write different mathematical models for the same scientific concept and being able to write different code or have multiple solutions (instead of a single correct answer) in computer science (Lilly et al., 2023).

In addition, in this project teachers sometimes responded by adjusting the planned activities. For example, on one occasion the teachers chose to add an activity for one section of students (in which students went outside to take pictures of their water runoff problem) after realizing that students needed to better understand the connection of the project to their own context. The teachers were also involved as co-designers of the curricular unit, and as such, were able to reflect on student work from past implementations during professional learning sessions to make revisions to the unit from year to year.

FUTURE RESEARCH DIRECTIONS

Together these two examples highlight potential avenues for future research and productive approaches to supporting both the implementation of mathematical writing and the adoption of a more engaged conceptualization of teacher noticing.

In the future we envision teacher preparation programs creating text clubs for pre-service teachers who are participating in school placements (e.g., practicums or practice teaching). During text club meetings pre-service teachers would share anonymized examples of student explanatory writing from their school placement and the group would discuss their interpretations of each explanation and draft potential responses. But rather than this being an isolated activity, pre-service teachers would be expected to use the draft responses to respond to each student explanation, to use their responses as new opportunities to further elicit explanations from students at their placement, to notice new mathematical ideas, and continue to respond. We anticipate that these iterative cycles would help promote an engaged view of noticing and at the same time promote the implementation of mathematical writing activities, but further research is needed to illuminate this approach.

In addition, given the concerns about the transfer of teacher noticing skills to authentic classroom contexts, the adoption of student explanatory writing and text clubs have the potential to bridge this gap. As researchers consider ways to measure and support the transfer of teacher noticing skills, text clubs present an alternative model for transferring these skills that might be studied because they support iterative engagement with student thinking in ways that video does not. At the same time, given that teacher noticing is theoretically evolving towards a more iterative, engaged conceptualization, artifacts that have affordances that better align with this new perspective are important for researchers to consider when developing future studies.

Many teachers indicate that they receive insufficient preparation for implementing and assessing mathematical writing within their instruction (Powell et al., 2021), and so the use of mathematical writing artifacts within text clubs can help address this need. As such, we see the potential for productive collaborations between teacher educators who are interested in promoting teacher noticing and teacher educators who are interested in promoting mathematical writing. By engaging with student written explanations and iteratively noticing and responding to student mathematical ideas, pre-service teachers are likely to be better prepared to use mathematical writing during their instruction.

Mathematical writing can also support learning within and across different disciplines. Our examples demonstrate how supporting students to engage in mathematical writing provided the glue among science, engineering, and computational contexts. Such experiences are important to enable all students the opportunity to authentically engage, and be supported in, interdisciplinary work. Our work showed that there were differences in teachers' verbal support in inclusive class settings as teachers worked to meet their expectations of diverse students' needs (Lilly et al., 2023). We propose here that, by introducing additional opportunities for teachers to support students in mathematical writing, and by helping teachers to develop their skills in noticing and giving feedback, the teachers in such projects may be able to support students more equitably as well as build authentic connections between STEM+CS disciplines.

CONCLUSION

This chapter explored how the inclusion of explanatory mathematical writing can support teacher noticing practices in mathematics classrooms. The chapter puts forth that students' mathematical explanations can serve as fruitful resources and provide insight into students' mathematical thinking. Prior research has mainly focused on video clubs to support teacher noticing and responding. This chapter presented two examples of how teachers used mathematical writing in their classrooms and how potential "text clubs" can help teachers respond to students' mathematical ideas in mathematics and interdisciplinary settings.

ACKNOWLEDGMENT

This research was supported by the National Science Foundation [grant numbers DRL- 1253523, DRL-1503170, and DRL-1742195].

REFERENCES

Atasoy, E., & Adnan, B. A. K. İ. (2020). Investigation of students' cognitive learning in mathematics lessons supported with writing activities. *Turkish Journal of Computer and Mathematics Education, 11*(2), 528–583. doi:10.16949/turkbilmat.703648

Ball, D. L., Hill, H. C., & Bass, H. (2005). Knowing mathematics for teaching: Who knows mathematics well enough to teach third grade, and how can we decide? *American Educator*, (Fall), 14–46. http://hdl.handle.net/2027.42/65072

Barwell, R. (2005). Ambiguity in the mathematics classroom. *Language and Education, 19*(2), 117–125. doi:10.1080/09500780508668667

Baxter, J. A., Woodward, J., Olson, D., & Robyns, J. (2002). Blueprint for writing middle school mathematics. *Mathematics Teaching in the Middle School, 8*(1), 52–56. doi:10.5951/MTMS.8.1.0052

Benjamin, A. (2013). *Math in plain English: Literacy strategies for the mathematics classroom*. Routledge. doi:10.4324/9781315855127

Bicer, A., Capraro, R. M., & Capraro, M. M. (2013). Integrating writing into mathematics classroom to increase students' problem solving skills. *International Online Journal of Educational Sciences*, *5*(2), 361–369.

Bicer, A., Perihan, C., & Lee, Y. (2018). The impact of writing practices on students' mathematical attainment. *International Electronic Journal of Mathematics Education*, *13*(3), 305–313. doi:10.12973/iejme/3922

Bixby, M. M. (2018). Effective and efficient use of math writing tasks. *Mathematics Teacher*, *112*(2), 143–146. doi:10.5951/mathteacher.112.2.0143

Bransford, J. D., Brown, A. L., & Cocking, R. R. (2000). *How people learn*. National Academy Press.

Bywater, J. P., Chiu, J. L., Hong, J., & Sankaranarayanan, V. (2019). The teacher responding tool: Scaffolding the teacher practice of responding to student ideas in mathematics classrooms. *Computers & Education*, *139*, 16–30. doi:10.1016/j.compedu.2019.05.004

Bywater, J. P., Lilly, S., & Chiu, J. L. (2022). Examining technology-supported teacher responding and students' written mathematical explanations. *Journal of Mathematics Teacher Education*, 1–23.

Carpenter, T. P., Franke, M. L., & Levi, L. (2003). *Thinking mathematically: Integrating arithmetic and algebra in elementary school*. Heinemann.

Casa, T. M., Firmender, J. M., Cahill, J., Cardetti, F., Choppin, J. M., Cohen, J., & Zawodniak, R. (2016). *Types of and purposes for elementary mathematical writing: Task force recommendations*. Math Writing. https://mathwriting.education.uconn.edu/

Chapin, S. H., O'Connor, C., & Anderson, N. C. (2009). *Classroom discussions: Using math talk to help students learn*. Math Solutions.

Chi, M. T. H., de Leeuw, N., Chiu, M., & LaVancher, C. (1994). Eliciting self-explanations improves understanding. *Cognitive Science*, *18*(3), 439–477. doi:10.1016/0364-0213(94)90016-7

Chiu, J. L., McElhaney, K., Zhang, N., Biswas, G., Fried, R., Basu, S., & Alozie, N. (2019). *A principled approach to NGSS-aligned curriculum development: A pilot study* [Paper presentation]. *NARST Annual International Conference*, Baltimore, MD.

Coffey, J. E., Hammer, D., Levin, D. M., & Grant, T. (2011). The missing disciplinary substance of formative assessment. *Journal of Research in Science Teaching*, *48*(10), 1109–1136. doi:10.1002/tea.20440

Crespo, S. (2000). Seeing more than right and wrong answers: Prospective teachers' interpretations of students' mathematical work. *Journal of Mathematics Teacher Education*, *3*(2), 155–181. doi:10.1023/A:1009999016764

Cross, D. I. (2009). Creating optimal mathematics learning environments: Combining argumentation and writing to enhance achievement. *International Journal of Science and Mathematics Education*, *7*(5), 905–930. doi:10.100710763-008-9144-9

Dacey, L. S., Salemi, R. E., & O'Connell, K. H. (2018). *Why write in math class? K-5*. Stenhouse Publishers.

Davis, E. A., Zembal-Saul, C., & Kademian, S. M. (2019). Working towards a vision of sensemaking in elementary science. In E. A. Davis, C. Zembal-Saul, & S. M. Kademian (Eds.), Sensemaking in elementary science: Supporting teacher learning (pp. 1–11). Routledge.

Davis, B. (1997). Listening for differences: An evolving conception of mathematics teaching. *Journal for Research in Mathematics Education*, *28*(3), 355–376. doi:10.2307/749785

Dyer, E. B., & Sherin, M. G. (2016). Instructional reasoning about interpretations of student thinking that supports responsive teaching in secondary mathematics. *ZDM Mathematics Education*, *48*(1-2), 69–82. doi:10.100711858-015-0740-1

Erickson, F. (1986). *Teachers' practical ways of seeing and making sense: A final report*. Office of Educational Research and Improvement.

Evi-Colombo, A., Cattaneo, A., & Bétrancourt, M. (2020). Technical and pedagogical affordances of video annotation: A literature review. *Journal of Educational Multimedia and Hypermedia*, *29*(3), 193–226.

Fernández, C., Llinares, S., & Valls, J. (2012). Learning to notice students' mathematical thinking through on-line discussions. *ZDM Mathematics Education*, *44*(6), 747–759. doi:10.100711858-012-0425-y

Gaudin, C., & Chaliès, S. (2015). Video viewing in teacher education and professional development: A literature review. *Educational Research Review*, *16*, 41–67. doi:10.1016/j.edurev.2015.06.001

Gerard, L., Bradford, A., & Linn, M. C. (2022). Supporting teachers to customize curriculum for self-directed learning. *Journal of Science Education and Technology*, *31*(5), 660–679. doi:10.100710956-022-09985-w

Gerard, L. F., Spitulnik, M., & Linn, M. C. (2010). Teacher use of evidence to customize inquiry science instruction. *Journal of Research in Science Teaching*, *47*(9), 1037–1063. doi:10.1002/tea.20367

Goldsmith, L. T., & Seago, N. (2011). Using classroom artifacts to focus teachers' noticing. In M. G. Sherin, V. R. Jacobs, & R. A. Philipp (Eds.), *Mathematics teacher noticing: Seeing through teachers' eyes* (pp. 169–187). Routledge.

Goodwin, C. (1994). Professional vision. *American Anthropologist*, *96*(3), 606–633. doi:10.1525/aa.1994.96.3.02a00100

Hammer, D., Goldberg, F., & Fargason, S. (2012). Responsive teaching and the beginnings of energy in a third grade classroom. *Review of Science. Mathematics and ICT Education*, *6*(1), 51–72.

Hattie, J., & Timperley, H. (2007). The power of feedback. *Review of Educational Research*, *77*(1), 81–112. doi:10.3102/003465430298487

Herbel-Eisenmann, B. A., Steele, M. D., & Cirillo, M. (2013). (Developing) teacher discourse moves: A framework for professional development. *Mathematics Teacher Educator*, *1*(2), 181–196. doi:10.5951/mathteaceduc.1.2.0181

Herbst, P., Aaron, W., & Chieu, V. M. (2013). LessonSketch: An environment for teachers to examine mathematical practice and learn about its standards. In D. Polly (Ed.), *Common core mathematics standards and implementing digital technologies* (pp. 281–294). IGI Global. doi:10.4018/978-1-4666-4086-3.ch019

Jacobs, V. R., Lamb, L. L. C., & Philipp, R. A. (2010). Professional noticing of children's mathematical thinking. *Journal for Research in Mathematics Education, 41*(2), 169–202. doi:10.5951/jresematheduc.41.2.0169

Jacobs, V. R., & Spangler, D. A. (2017). Research on core practices in K-12 mathematics teaching. In J. Cai (Ed.), *Compendium for research in mathematics education* (pp. 766–792). National Council of Teachers of Mathematics.

Jurdak, M., & Abu Zein, R. (1998). The effect of journal writing on achievement in and attitudes toward mathematics. *School Science and Mathematics, 98*(8), 412–419. doi:10.1111/j.1949-8594.1998.tb17433.x

Juzwik, M. M., Curcic, S., Wolbers, K., Moxley, K. D., Dimling, L. M., & Shankland, R. K. (2006). Writing into the 21st century: An overview of research on writing, 1999 to 2004. *Written Communication, 23*(4), 451–476. doi:10.1177/0741088306291619

Kaiser, G., Busse, A., Hoth, J., König, J., & Blömeke, S. (2015). About the complexities of video-based assessments: Theoretical and methodological approaches to overcoming shortcomings of research on teachers' competence. *International Journal of Science and Mathematics Education, 13*(2), 369–387. doi:10.100710763-015-9616-7

Kenney, J. M. (2005). *Literacy strategies for improving mathematics instruction.* Association for Supervision and Curriculum Development.

König, J., Santagata, R., Scheiner, T., Adleff, A. K., Yang, X., & Kaiser, G. (2022). Teacher noticing: A systematic literature review of conceptualizations, research designs, and findings on learning to notice. *Educational Research Review, 36*, 100453. doi:10.1016/j.edurev.2022.100453

Kosko, K. W. (2016). Writing in mathematics: A survey of K-12 teachers' reported frequency in the classroom. *School Science and Mathematics, 116*(5), 276–285. doi:10.1111sm.12177

Kostos, K., & Shin, E. K. (2010). Using math journals to enhance second graders' communication of mathematical thinking. *Early Childhood Education Journal, 38*(3), 223–231. doi:10.100710643-010-0390-4

Lilly, S., Fick, S. J., Chiu, J. L., & McElhaney, K. W. (2020). Supporting elementary students to develop mathematical models within design-based integrated science and mathematics projects. In *Proceedings of the 14th international conference of the learning sciences*. International Society of the Learning Sciences.

Lilly, S., McAlister, A. M., Fick, S. J., & Chiu, J. L. (2023). A comparison of elementary teachers' verbal supports for students in inclusive and general classroom contexts during an NGSS-aligned science, engineering, and computer science unit. *Science Education, 107*(4), 853–883. doi:10.1002ce.21788

Magiera, M. T., & Zambak, V. S. (2021). Prospective K-8 teachers' noticing of student justifications and generalizations in the context of analyzing written artifacts and video-records. *International Journal of STEM Education, 8*(1), 1–21. doi:10.118640594-020-00263-y

Martin, C. L. (2015). Writing as a tool to demonstrate mathematical understanding. *School Science and Mathematics, 115*(6), 302–313. doi:10.1111sm.12131

Mason, J. (2002). *Researching your own practice: The discipline of noticing*. Routledge. doi:10.4324/9780203471876

McCormick, K. (2010). Experiencing the power of learning mathematics through writing. *Issues in the Undergraduate Mathematics Preparation of School Teachers: The Journal, 4*, 1–8.

McIntosh, M. E., & Draper, R. J. (2001). Using learning logs in mathematics: Writing to learn. *Mathematics Teacher, 94*(7), 554–557. doi:10.5951/MT.94.7.0554

Meel, D. E. (1999). Email dialogue journals in a college calculus classroom: A look at the implementation and benefits. *Journal of Computers in Mathematics and Science Teaching, 18*(4), 387–413.

Monson, D., Krupa, E., Lesseig, K., & Casey, S. (2020). Developing secondary prospective teachers' ability to respond to student work. *Journal of Mathematics Teacher Education, 23*(2), 209–232. doi:10.100710857-018-9420-8

Narciss, S., & Huth, K. (2004). How to design informative tutoring feedback for multi-media learning. In H. M. Niegemann, D. Leutner, & R. Brunken (Eds.), *Instructional design for multimedia learning* (pp. 181–195). Waxmann.

Nathan, M. J., & Koedinger, K. R. (2000). An investigation of teachers' beliefs of students' algebra development. *Cognition and Instruction, 18*(2), 209–237. doi:10.1207/S1532690XCI1802_03

Odden, T. O. B., & Russ, R. S. (2019). Defining sensemaking: Bringing clarity to a fragmented theoretical construct. *Science Education, 103*(1), 187–205. doi:10.1002ce.21452

Porter, M. K., & Masingila, J. O. (2000). Examining the effects of writing on conceptual and procedural knowledge in calculus. *Educational Studies in Mathematics, 42*(2), 165–177. doi:10.1023/A:1004166811047

Powell, S. R., Hebert, M. A., Cohen, J. A., Casa, T. M., & Firmender, J. M. (2017). A synthesis of mathematics writing: Assessments, interventions, and surveys. *Journal of Writing Research, 8*(3), 493–530. doi:10.17239/jowr-2017.08.03.04

Powell, S. R., Hebert, M. A., & Hughes, E. M. (2021). How educators use mathematics writing in the classroom: A national survey of mathematics educators. *Reading and Writing, 34*(2), 417–447. doi:10.100711145-020-10076-8

Robertson, A. D., Atkins, L. J., Levin, D. M., & Richards, J. (2016). What is responsive teaching? In A. D. Robertson, R. E. Scherr, & D. Hammer (Eds.), *Responsive teaching in science and mathematics* (pp. 1–34). Routledge.

Rodriguez, A. J., & Zozakiewicz, C. (2010). Facilitating the integration of multiple literacies through science education and learning technologies. In A. J. Rodriguez (Ed.), *Science education as a pathway to teaching language literacy* (pp. 23–45). Sense Publishers. doi:10.1163/9789460911316_005

Sanders, C. V. (2009). Exploring and writing geometry. *Mathematics Teacher, 102*(6), 432–439. doi:10.5951/MT.102.6.0432

Santagata, R., König, J., Scheiner, T., Nguyen, H., Adleff, A. K., Yang, X., & Kaiser, G. (2021). Mathematics teacher learning to notice: A systematic review of studies of video-based programs. *ZDM Mathematics Education*, *53*(1), 119–134. doi:10.100711858-020-01216-z

Scheiner, T. (2016). Teacher noticing: Enlightening or blinding? *ZDM Mathematics Education*, *48*(1-2), 227–238. doi:10.100711858-016-0771-2

Scheiner, T. (2021). Towards a more comprehensive model of teacher noticing. *ZDM Mathematics Education*, *53*(1), 85–94. doi:10.100711858-020-01202-5

Selmer, S., Lindstrom, D., & Lampen, E. (2022). A Case Study of Prospective Teachers Engaged in Professional Noticing of their Students' Mathematical Thinking. *Education Sciences*, *12*(10), 656. doi:10.3390/educsci12100656

Sherin, M. G. (2007). The development of teachers' professional vision in video clubs. In R. Goldman, R. Pea, B. Barron, & S. Derry (Eds.), *Video research in the learning sciences* (pp. 383–395). Routledge.

Sherin, M. G., Jacobs, V. R., & Philipp, R. A. (2011). Situation awareness in teaching: What educators can learn from video-based research in other fields. In M. G. Sherin, V. R. Jacobs, & R. A. Philipp (Eds.), *Mathematics teacher noticing: Seeing through teachers' eyes* (pp. 81–95). Routledge. doi:10.4324/9780203832714

Shute, V. J. (2008). Focus on formative feedback. *Review of Educational Research*, *78*(1), 153–189. doi:10.3102/0034654307313795

Simpson, A., & Haltiwanger, L. (2017). "This is the first time I've done this": Exploring secondary prospective mathematics teachers' noticing of students' mathematical thinking. *Journal of Mathematics Teacher Education*, *20*(4), 335–355. doi:10.100710857-016-9352-0

Stockero, S. L., Leatham, K. R., Van Zoest, L. R., & Peterson, B. E. (2017). Noticing distinctions among and within instances of student mathematical thinking. In E. O. Schack, M. H. Fisher, & J. A. Wilhelm (Eds.), *Teacher noticing: Bridging and broadening perspectives, contexts, and frameworks* (pp. 467–480). Springer. doi:10.1007/978-3-319-46753-5_27

Stylianou, D. A., & Silver, E. A. (2004). The role of visual representations in advanced mathematical problem solving: An examination of expert-novice similarities and differences. *Mathematical Thinking and Learning*, *6*(4), 353–387. doi:10.120715327833mtl0604_1

Tan, T., & Garces-Bacsal, R. M. (2013). The effect of journal writing on mathematics achievement among high-ability students in Singapore. *Gifted and Talented International*, *28*(1-2), 173–184. doi:10.1080/15332276.2013.11678412

Teuscher, D., Kulinna, P. H., & Crooker, C. (2015). Writing to learn mathematics: An update. *The Mathematics Educator*, *24*(2), 56–78.

Totten, S. (2005). Writing to learn for preservice teachers. *The Quarterly, 27*(2), 17-20, 28.

van Es, E. A., & Sherin, M. G. (2002). Learning to notice: Scaffolding new teachers' interpretations of classroom interactions. *Journal of Technology and Teacher Education*, *10*(4), 571–596.

van Es, E. A., & Sherin, M. G. (2010). The influence of video clubs on teachers' thinking and practice. *Journal of Mathematics Teacher Education*, *13*(2), 155–176. doi:10.100710857-009-9130-3

van Es, E. A., & Sherin, M. G. (2021). Expanding on prior conceptualizations of teacher noticing. *ZDM Mathematics Education*, *53*(1), 17–27. doi:10.100711858-020-01211-4

Weyers, J., König, J., Santagata, R., Scheiner, T., & Kaiser, G. (2023). Measuring teacher noticing: A scoping review of standardized instruments. *Teaching and Teacher Education*, *122*, 103970. doi:10.1016/j.tate.2022.103970

Yig, K. G. (2022). Research trends in mathematics education: A quantitative content analysis of major journals 2017-2021. *Journal of Pedagogical Research*, *6*(3), 137–153.

Zhang, N., Biswas, G., McElhaney, K., Basu, S., McBride, E., & Chiu, J. L. (2020). Studying the interactions between science, engineering, and computational thinking in a learning-by-modeling environment. In I. I. Bittencourt, M. Cukurova, K. Muldner, R. Luckin, & E. Millán (Eds.), *Proceedings of the international conference on artificial intelligence in education* (pp. 598-609). Springer. 10.1007/978-3-030-52237-7_48

Chapter 7
Assessing Mathematical Writing:
Comparative Judgment and Professional Learning

Ian Jones
Loughborough University, UK

Jodie Hunter
Massey University, New Zealand

ABSTRACT

The chapter discusses potential professional learning benefits for educators who engage in assessing students' mathematical writing. It draws on interview data from twelve mathematics educators who were experienced in assessing primary students' written responses to free response prompts covering a range of topics. The first stage of the interviews used a stimulated recall protocol that followed a comparative judgment procedure in which each participant was presented with pairs of students' written responses and asked to decide which was 'better'. The second stage was semi-structured with questions about how participants made their comparative judgment decisions, and whether doing so improved their understanding of students' thinking. The findings are that assessing mathematical writing can provide educators with insights into students' representations, underlying ideas and learning trajectories, and can also provide stimulus for changing classroom practice.

INTRODUCTION

It has been well documented in research literature (e.g., Carpenter et al., 1996; Jacobs et al., 2022; Krebs, 2005; Wilson et al., 2014) that there are a range of positive outcomes when educators learn about students' mathematical thinking. These include shifts in pedagogical approaches and beliefs, and the development of educators' own mathematical understanding. There have been varied professional learning approaches to providing educators with opportunities to both understand and learn from examples of students' mathematical thinking. Examples include watching video excerpts of classroom instruction, conducting task-based or clinical interviews of students, and examining students' written work (Krebs, 2005; Wilson et

DOI: 10.4018/978-1-6684-6538-7.ch007

al., 2014). For example, Krebs (2005) investigated how teachers could analyze mathematical writing as a form of professional learning. Teachers examining responses from students in eighth grade in relation to a task that involved generalizing a quadratic pattern. Key findings indicated that engaging in analysis of mathematical writing offered teachers opportunities to both identify the mathematics in the task and how the student responses built on this mathematics. Further opportunities included teachers being able to 'see' aspects of student thinking and understandings that they had not predicted as well as considering the next steps in learning mathematics for the students. A study by Wilson et al. (2014) involved teachers using a learning trajectory to make sense of students' written work. Findings demonstrated how using a learning trajectory approach supports teachers to examine student work beyond correct or incorrect responses and instead understand conceptual development as a continuum.

Our focus in this chapter is to build on this work by examining the potential professional learning benefits for educators who engage with students' mathematical thinking by assessing their mathematical writing. To achieve this, we describe an alternative approach that we have adopted based on comparative judgment, which we introduce below. We frame the chapter in terms of the literature on peer assessment in mathematics education with a focus on the potential learning benefits for assessors, and extend this theoretical framing so that educators rather than peers are the assessors. We argue that this approach has potential both in relation to considering how to assess mathematical writing and through the opportunities for teacher learning and professional development afforded by the approach.

PEER ASSESSMENT

Peer assessment involves students making judgments about the quality of their peers' work (Topping, 2009). Learning benefits have been cited both for the activities of judging quality, and of receiving feedback from peers (Falchikov & Goldfinch, 2000) and it is the first of these that we are interested in here. The learning benefits of samples of others' work for the case of mathematics have been reported across topics and age ranges, from school students learning algebra (Rittle-Johnson & Star, 2007), fractions (Jones & Wheadon, 2015) and mathematical problem solving (Evans & Swan, 2014), to undergraduates learning calculus (I. Jones & Alcock, 2014) and proof (Davies et al., 2020).

Reviews of the evidence of the learning benefits peer assessment highlight the importance of using detailed assessment criteria (e.g. Orsmond et al., 1996; van den Berg et al. 2006). This might involve provided criteria that the students are expected to internalize, or it might be student-generated criteria that are then used for making judgments of peers' work. Our approach challenges this assumption that detailed assessment criteria are indispensable, and argues there are peer assessment contexts in which avoiding detailed criteria benefits learning.

COMPARATIVE JUDGMENT

The comparative judgment approach we present here is simple. It involves presenting two pieces of written work to an assessor (examples are given in Figure 1), and asking them to make a holistic decision about which piece demonstrates the 'better' understanding. This process is usually conducted with several assessors making holistic decisions about many presented pairings until every piece of work has been judged several times by different assessors. Once the judging is complete, a mathematical model is fitted to the holistic decision data in order to produce a unique score for every piece of writing (Pollitt, 2012).

Outside of peer assessment and professional learning contexts, comparative judgment approaches to assessing student writing have been shown to offer benefits compared to traditional criteria-based scoring methods. These benefits mostly pertain to higher consistency of assessors, greater reliability of the resultant scores, and increased validity of assessment outcomes. These benefits are not of direct relevance to the current discussion and we do not raise them again, but interested readers can consult summaries of the evidence for these benefits in Bartholomew and M. Jones (2022), I. Jones and Davies (2023), and Verhavert et al. (2019).

A key benefit, and our focus here, is that comparative judgment lends itself well to peer assessment activities. The application of comparative judgment to peer assessment has been studied for different mathematical topics and age ranges. I. Jones and Wheadon (2015) investigated the outcomes of middle school students judging peers' mathematical writing about placing fractions in value order. They used an experimental design in which one group of students made their assessments using a comparative judgment approach, and the other using an 'absolute' judgment approach. They found that only students in the comparative judgment group produced assessment outcomes that were coherent. Other work has conceptually replicated I. Jones and Wheadon's findings at the university level for the case of undergraduate writing about multivariate calculus (I. Jones & Alcock, 2014) and proof (Davies et al., 2020). Surveys and interviews completed by undergraduates after they had judged peers' mathematical writing suggested that many found peer assessment based on comparative judgment engaging and beneficial for learning (e.g. I. Jones & Sirl, 2017). At the time of writing we are undertaking a further study to investigate the potential benefits on mathematical learning in the context of undergraduate differential calculus (I. Jones, 2021).

PROFESSIONAL LEARNING

Building on these findings, we conjecture that comparative judgment can offer engaging professional development and learning opportunities for educators. To the best of our knowledge there is a paucity of studies that focus on opportunities for teacher learning about student mathematical thinking through the comparison of mathematical writing. However, the small amount of evidence that is available looks promising. I. Jones and Inglis (2015) surveyed educators after they had comparatively judged a mathematics problem-solving exam paper that had been completed by high school students. The educators reported favoring responses that demonstrated knowledge of formal mathematical facts, notation and vocabulary, and responses that were presented with elements of creativity and flair, as well as those responses that had addressed the presented task in full.

Similarly, Hunter and I. Jones (2018) interviewed educators after they had comparatively judged elementary students' mathematical writing based on free response prompts similar to those shown in Table 1. They found that educators favored responses that were mathematically sound, meaningful in terms of the prompt, and that included examples, illustrations, or explanations. The educators also reported that they felt that, compared to traditional tasks comprising closed or short questions, the use of free-response tasks allowed students to better express their thinking. Here we extend this work and present a study to investigate our conjecture in the context of elementary students' mathematical writing.

STUDY CONTEXT

We begin by considering the nature of assessment methods that are typically used in New Zealand elementary school settings for mathematics, and the corresponding opportunities to learn about student thinking. Common assessment tools include Progressive Assessment Tests (PATs) which are standardized multiple-choice item tests. However, a common criticism of such tests is the predominance of short, closed items that emphasize the recall and application of isolated facts and procedures (Berube, 2004). Alternatively, many schools use a form of a diagnostic interview focused on number knowledge and strategies called the Global Strategy Stage (GloSS) assessment. Students are asked to solve numerical tasks while thinking aloud and teachers use a detailed rubric to determine the 'strategy stage' for each student. A limitation with this form of assessment has been a structure which largely assesses students' aptitudes to remember and correctly apply taught strategies to solve problems (Fitzgerald et al., 2021). In addition, this type of approach to categorizing the stage of individual students is time intensive and requires significant periods of teaching time to be used for assessment. Additionally, the rubric used for this only focuses on student use of numerical strategies and offers limited opportunities for teacher learning beyond the focus on number.

Mathematical writing has not typically been used as a form of assessment in elementary schooling in New Zealand. To the best of our knowledge this is the case for many countries around the globe. Such absence of writing as a source of assessment evidence in mathematics classrooms is perhaps linked to concerns in relation to how mathematical writing can be assessed in a fair and reliable way, as discussed earlier and also highlighted by Powell et al. (2017). To address this, recently we have been pioneering the assessment of mathematical writing using comparative judgment with a focus on professional learning (Hunter & I. Jones, 2018). This has included working with teachers and mathematics educators to trial the use of mathematical writing and comparative judgment as a means for assessment, and also in terms of growth of mathematical knowledge for teaching and pedagogical content knowledge as described by Ball at al. (2008).

THE STUDY

We conducted a study using an interview-based structure with twelve experienced mathematics educators as participants. Specifically, we used a two-stage interview process to examine mathematics educators' use of comparative judgment to assess students' mathematical writing. This included a focus on perceptions of the enabling factors as well as the barriers to using this form of assessment. Crucially, we also examined the opportunities for professional development and learning that are supported by assessing mathematical writing via comparative judgment. Full ethical approval for the study was provided by the ethics committee at the second author's institution.

Research Questions

1. What do educators need to know and learn when using mathematical writing for assessment?
2. What opportunities does comparative judgment provide to educators in relation to learning to use writing as mathematical assessment?

Method

The participants were twelve mathematics educators from New Zealand with experience in elementary, middle, or high school teaching. All participants were already familiar with the use of mathematical writing for assessment and with the NoMoreMarking comparative judgment platform. Prior to participating in the interview study, the participants had engaged in a series of workshops to learn about assessing mathematical writing from students and using comparative judgment. The first workshop included working collaboratively in small groups to read students' mathematical writing and then engaging in the pairwise decision-making required for comparative judgment while discussing and justifying which piece of writing showed higher conceptual mathematical understanding. This included assessing students' written mathematical responses to free response prompts across a range of mathematical topics. All participants were asked to reflect at the end of the workshop on the strengths in mathematical thinking that they had noted and the areas which indicated the need for further teaching of specific concepts. The second workshop involved examining students' written mathematical responses and levelling these according to the New Zealand curriculum which uses banded achievement levels representing two years of progress (Ministry of Education, 2007). For example, students in Year One and Two are expected to be working at Level One in the curriculum while students in Year Five and Six are expected to be working at Level Three. In this workshop, the participants engaged in collective discussion of specific student responses to develop a common understanding of what types of responses indicated achievement at specific curriculum levels.

The study involved a three-stage interview. In the first stage, participants used comparative judgment to assess three student responses to five mathematical writing tasks focused on algebra, measurement, number operations, geometry, and statistics as shown in see Table 1. These mathematical writing tasks took the form of free response prompts and were designed for students from Year Three to Year Eight (aged six years to 13 years). Example student responses to the 'Number Operations 2' task in Table 1 can be seen in Figure 1, and to the 'Algebra' task in Figure 2.

The second stage of the interview was semi-structured with twelve questions. The third and final stage of the interview involved a think-aloud review of the pairings and decisions from the first stage of the interview. Here, we focus on responses to the second, semi-structured stage of the interview. Specifically, we draw on responses to five of the twelve questions, as follows:

1. How did you make your decisions?
2. How does comparative judgment compare to forms of assessment that you have used in the past?
3. What benefits do you see from comparative judgment and the moderation process that you have engaged in?
4. Are there any challenges which you see from comparative judgment?
5. After completing the comparative judgments, do you feel that you have a better understanding of the students' mathematical thinking? Why or why not?

Table 1. The six mathematical writing task examples

Algebra: Year Three – Four (aged six through to nine)

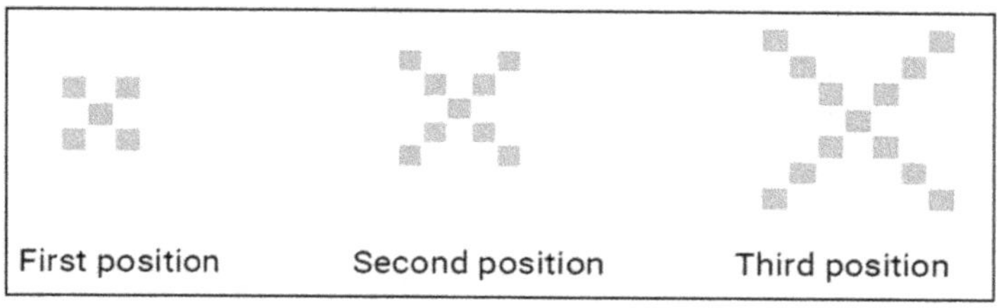

Look at this growing pattern. Can you:

- Show how the number of squares grows using a picture, a table, ordered pairs, and/or a graph.
- Describe the pattern in words and/or numbers.
- Show how you work out the number of squares for the following pattern numbers: Pattern 5; Pattern 10; Pattern 50.
- Write the rule for the number of squares in words or symbols.

Measurement: Year Five – Six (aged eight through to eleven)

Draw different shapes that have a perimeter of 24cm. Which shape would have the largest area? Show how you worked out the area.

Number Operations 1: Year Five – Six (aged eight through to eleven)

The school is going on a trip and has ordered 19 buses. Each bus can take 26 children. How many children can go on the trip?

The library is moving. They have 484 books and can fit 22 books into each box. How many boxes will be needed for the books?

Write your own multiplication or division problems. Show how you would solve them.

Geometry: Year Five – Six (aged eight through to eleven)

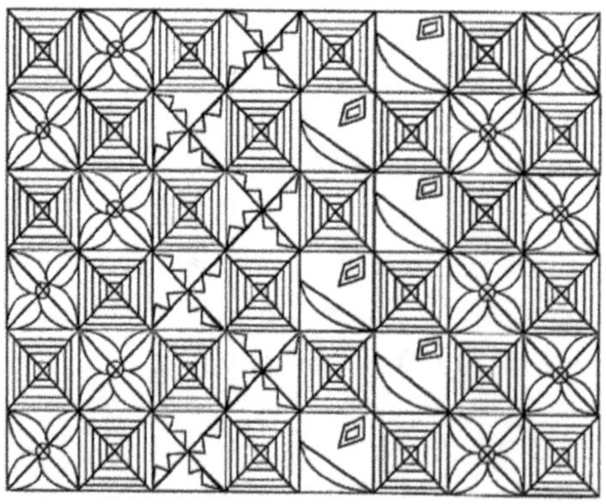

continued on following page

Table 1. Continued

Write a geometrical description of this picture. Use words like reflection, rotation, translation, and symmetry. You may draw or label the picture to highlight parts of your description.

Number Operations 2: Year Seven – Eight (aged ten through to 13)

76 x 15	37 + 43 + 40 + 36	99 ÷ 3 ÷ 3
7 x 86	12 x 22	
6^3	(70 x 5) + (70 x 10) + (6 x 10) + (6 x 5)	
37 + 40 + 36 + 43	99 ÷ 9	6 x 6 x 6
(7 x 90) - (7 x 4)	4 x 66	

Look at the number sentences above:

- Describe what patterns you can find.
- Why do your patterns work?
- Do they work with other numbers? Will they always work? Explain and justify your thinking.

Statistics: Year Seven – Eight (aged ten through to 13)

A company was testing rubber bands to see which were stronger. They recorded in centimetres how far the rubber band stretched before it broke. Here are the results for the tests.

	Length of stretch at breaking point (centimetres)
Blah Bands	50 57 45 55 47 53 54 49 50 46 52 50 52 48 53
Boing Bands	61 36 38 40 61 61 43 38 60 60 42 50 37 39 41

How can you organise and compare this data? Think about things like the mean, mode, median and range.

What statements can you make about the different types of rubber bands?

Which rubber bands do you think the company should sell? Why?

Analysis

All interviews were audio-recorded and transcribed holistically. Analysis included an inductive approach to identify key themes. Multiple levels of coding were used with two over-arching questions used to guide the analysis. These focused on (i) educator knowledge required to accurately assess mathematical understanding represented in mathematical writing, and (ii) professional learning and development opportunities provided by assessing mathematical writing through comparative judgment for educators.

Table 2. Codes generated from inductive analysis of the interviews with descriptions and example excerpts from the transcripts

Code	Description	Example transcript excerpt
Perceptions	Educators' perceptions of comparative judgment in contrast to other assessments	"different than what we do now in assessment … we actually find what the students are thinking"
Student representations	Opportunities to view how students use mathematical representations	"how they've represented, their ability to explain and justify what they've done is really important"
Student ideas	Inferring students' mathematical ideas	"we need to look a little bit more than the surface features. It's not about the answer. We're trying to figure out those thought processes"
Student learning trajectories	Awareness of trajectories for student learning	"it's not just about looking for the answer that you're looking for with the different levels of [mathematics] are"
Changes to classroom practice	Reflecting on changing classroom practice in response to comparative judgment	"I'll be pushing teachers and classrooms to make students … put more justification"

Initial inductive coding focused on analysis of the transcript of the semi-structured interview from each mathematics educator. Table 2 below shows the codes that were developed and examples of coding.

The codes were used to examine, cluster, and integrate emerging themes and to develop the findings presented below. To ensure reliability, both researchers coded the interviews and then discussed any discrepancies in the coding until agreement was reached.

FINDINGS AND DISCUSSION

In this section we present the themes evident from analysis of the teacher interview data and discuss our findings. We begin with a contrast of the typical assessment methods used in New Zealand elementary mathematics classrooms with the opportunities and requirements of using mathematical writing and comparative judgment. We follow this by presenting examples of student mathematical writing to highlight the types of mathematical thinking and ways of working that are evident in the students' responses to the free response prompts shown in Table 1. The following sections then focus on both the requirements on teachers' professional knowledge when using mathematical writing and comparative judgment for assessment, as well as the opportunities for professional development and learning that they reported arising while making comparative judgments.

Perceptions

Participants in this study referred to both PAT and GloSS assessments as frequently being used in their practice to assess mathematical learning. Interestingly, they noted concerns in relation to the limitations that they perceived about these assessment methods. A consistent response from the participants was that both PATs and GloSS are largely oriented to an assessment of the answer provided by the student. Indicative of the common themes from the interview data, one participant stated:

"those were very narrow summative assessments. I had no insight into the child's thought processes at all. On GloSS, I obviously had insight, but I was very much constrained by what the guidelines said was

acceptable. Whilst I could see some students who had really good strategizing, they made a computational error, they got it wrong, and I could not give it to them."

These reflections indicate the difficulties for educators when assessment tools have reduced opportunities for students to demonstrate their thinking and for their teachers to professionally learn from making sense of student reasoning. In contrast, and similar to the teachers in Hunter and I. Jones (2018), these mathematics educators described the use of mathematical writing and comparative judgment as offering opportunities both for students to demonstrate a wider range of mathematical thinking and for teachers to access a broader view of mathematical learning. As one participant said, "we actually find what the students are thinking and the creativity and allowing them to show us everything they do". However, while both mathematical writing and comparative judgment provide opportunities, they also entail specific requirements for teachers as shown in the next section which provides examples of students' mathematical writing.

Student Test Responses

Typical assessment activities focus on correct answers along with rubrics to make sense of student thinking and reasoning. In contrast, our comparative judgment approach uses free response prompts such as those shown in Table 1. This provides students with opportunities to respond with writing complemented with representations in order to present explanations and show creativity. Two example student responses are shown in Figure 1 and 2.

As highlighted in the examples of student responses to the prompt, we can see the multiple layers of student reasoning which are evident. This includes written explanations of mathematical properties using informal mathematical language, for example, describing the use of the distributive property as a "broken down version" (Figure 2). Similarly, evident is justification through the use of examples and informal generalization, for example, for the associative property, "the order doesn't change to answer" (Figure 1). Our comparative judgment approach can stimulate further insights. For example, response (b) in Figure 2 is more comprehensive than response (a) because the student has commented on all the presented number sentences. However, response (a) evidences conceptual understanding by stating, for example, that "having brackets doesn't change the answer" in comparison to response (b) which suggests a procedural view of the same pair of number sentences by saying that "both equal 602".

In contrast to assessments that focus on correct answers or teacher guidance in the form of rubrics, in order to adequately assess these pieces of mathematical writing, teachers both must be prepared and also willing to engage in professional learning. For example, we might hope that educators base their decision about whether (a) or (b) in Figure 1and 2 is 'better' not on presentational style (the use of arrows and connecting lines in (a) to indicate which number sentences are being discussed, versus the use of a key in response (b)) but instead on features pertinent to mathematics, mathematical thinking and practices.

Teacher Preparedness

A theme identified in the analysis of the mathematics educator responses was that the use of mathematics writing and comparative judgment for assessment requires specific preparation. This included two related key teacher actions encompassing anticipation and reference to curriculum documents. Firstly, the mathematics educators engaged in this study highlighted the benefits of anticipating the student

Figure 1. Two example responses to the equations task shown in Table 1. In Stage one of the interview participants were asked to decide which of these two responses is 'better.'

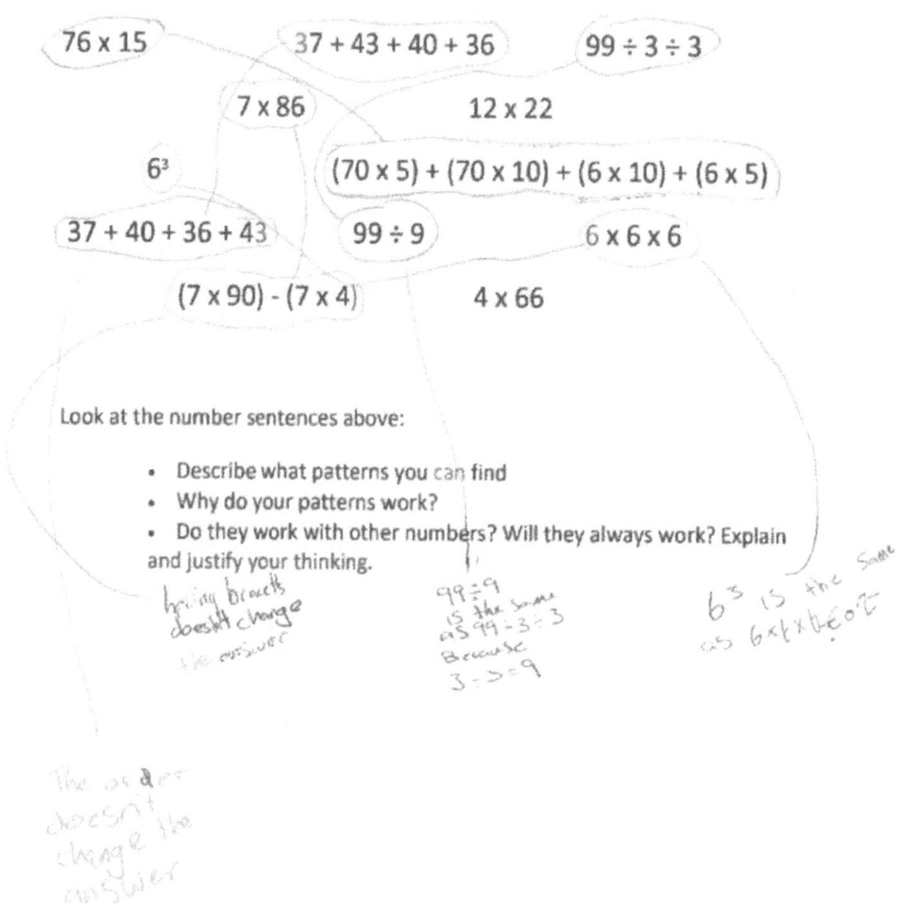

responses to the task prompt prior to undertaking assessment through comparative judgment: "I have to solve the problem myself in order to really understand what I should be looking for when I am doing comparative judgment". Another participant extended this further, referring to the process that she used of anticipating student responses while also referring to the New Zealand curriculum expectations: "You solve it yourself before you go to the elaborations [in New Zealand 'the elaborations' means additional documentation that elaborates on the content of the curriculum] and see okay this is what they're looking for at this level and then try and solve it differently". Previous research studies (e.g., Schoenfeld, 1998; Stein et al., 2008) have emphasized the importance of teachers solving tasks to ensure that they are prepared to recognize, connect and build on student solution strategies as well as addressing student misconceptions. These are also key means for teachers to increase their knowledge of students' mathematical thinking.

Figure 2. Two example responses to the equations task shown in Table 1. In Stage one of the interview participants were asked to decide which of these two responses is 'better.'

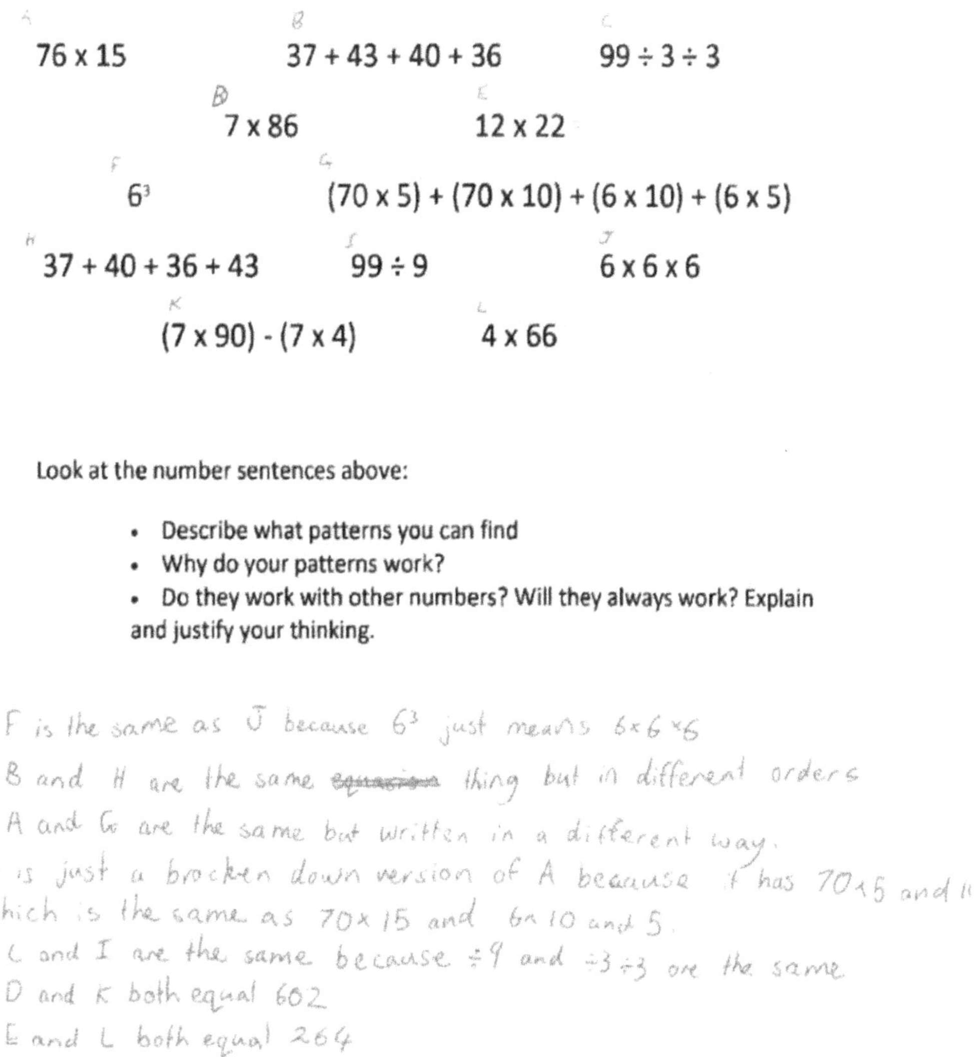

A 76 x 15   B 37 + 43 + 40 + 36   C 99 ÷ 3 ÷ 3

D 7 x 86   E 12 x 22

F 6^3 G (70 x 5) + (70 x 10) + (6 x 10) + (6 x 5)

H 37 + 40 + 36 + 43   I 99 ÷ 9   J 6 x 6 x 6

K (7 x 90) - (7 x 4)   L 4 x 66

Look at the number sentences above:

- Describe what patterns you can find
- Why do your patterns work?
- Do they work with other numbers? Will they always work? Explain and justify your thinking.

- F is the same as J because 6^3 just means 6x6x6
- B and H are the same ~~equation~~ thing but in different orders
- A and G are the same but written in a different way. G is just a brocken down version of A because it has 70x5 and 10 which is the same as 70x15 and 6x10 and 5.
- C and I are the same because ÷9 and ÷3÷3 are the same.
- D and K both equal 602
- E and L both equal 264

Secondly, it is important when using comparative judgment to assess mathematical writing that the assessors are first familiar with the level of content (I. Jones & Inglis, 2015). Similarly, Widjaja et al. (2017) documented teacher professional learning through anticipating student solution strategies during lesson study preparation. We argue that anticipating student responses to assessment tasks also provides opportunities for professional learning in relation to both mathematical content and pedagogical content knowledge (Ball et al., 2008).

Figure 3. Example responses to the algebra task showing student use of mathematical writing and representations

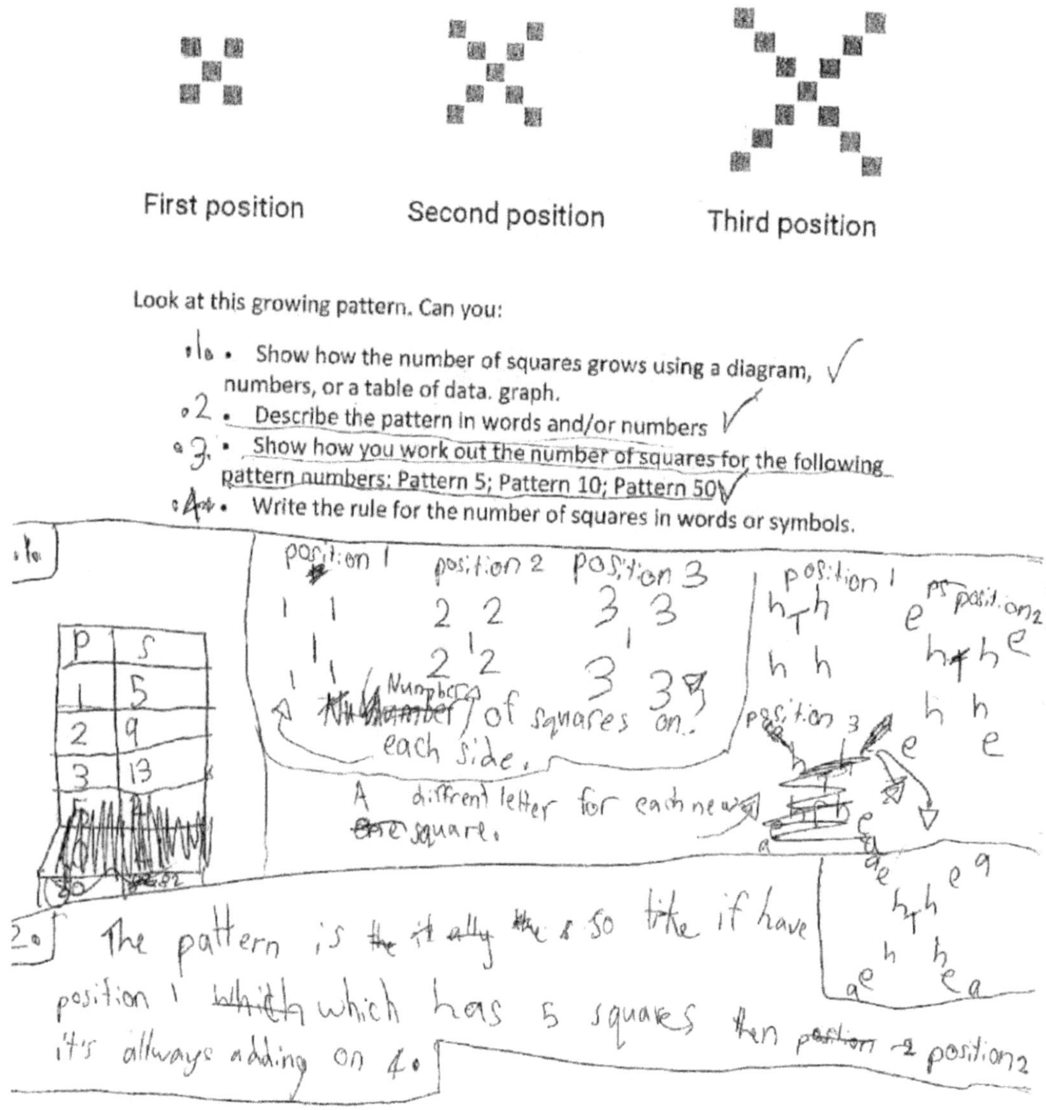

Student Ideas

As shown in Figures 1 and 2, the free response nature of the prompts resulted in students having opportunities to produce creative mathematical writing that included explanation and elaboration of their mathematical ideas. Contrasting this with typical mathematics assessment items, participants identified the need to assess student responses beyond the duality of correct responses or incorrect responses: "I look beyond the right answers into how students explain and justify". The mathematics educators connected both explanations and justification to the need to assess both for procedural and conceptual understanding in mathematics with one noting: "as soon as I see a procedure, I'm thinking, do they actually understand what they're doing here?". Research studies and policy documents (e.g., NCTM, 2014; Rittle-Johnson et al., 2015) have both identified procedural fluency and conceptual understanding as important aspects of developing mathematical proficiency when learning mathematics. A key critique of many assessment formats is that they focus heavily on procedural fluency given that, as Rittle-Johnson and colleagues describe, assessing procedural knowledge involves asking learners to solve a familiar task and examine the accuracy of the answer given or procedure applied. In this study, it appears that the mathematics educators viewed the mathematical writing tasks as an opportunity to examine the development of both conceptual understanding and procedural fluency. However, this also requires teachers to make sense of student explanations and take the time to look through those explanations as windows onto students' understanding. This was reported in the interviews by a participant who stated: "you're actually looking at the thought processes". In similar responses that extended this notion of looking beyond explanations further, a participant described how the process involved "really trying to understand where the kids are coming from", and another participant said "you have to read the students' work in [terms of] how to interpret the answers ... when they give you the solutions". These examples highlight the effort and knowledge base of the teachers required to make sense of the students' mathematical writing. However, also evident is the opportunity for teachers to learn more about student explanations and conceptual understanding, as well as to develop understanding of how conceptual understanding may be represented across different topics within mathematics.

Student Representations

Both the structure of free response prompts which encompass mathematical writing and the use of comparative judgment provide teachers with a window into student thinking. A key affordance for the participants was the opportunity to view and examine students' use of multiple and complementary representations as shown in Figure 3.

The representations that the students developed were used by the teachers to gain insight into their understanding of mathematical concepts. For example, one participant described the multiplicity of representations that could be seen in student responses: "it could be drawing, could be a table … a diagram". Coupled with mathematical writing, these provide rich insights into student thinking and learning.

In some cases, the participants used the representations to cross check student understanding with the answer to the task that they had provided: "representations can actually tell a different story to what a potential answer may be". The theme of representations and writing as an opportunity to clarify student understanding was evident across the participants' responses. In describing the use of comparative judgment to assess student responses to the algebra task, one participant discussed his pairwise decision stating:

"firstly, he [referring to student response although gender was not indicated] represented his work on three different representations. That shows the students' understanding of different representations. The second one [indicating other student response], you've got the rule, and also the table, and the diagram and no, or not much explanation. I chose the one that goes from diagrams to tables and also word form".

These mathematics educators viewed the opportunity to see mathematical representations alongside the writing as an opportunity for their learning: "I've learned about all the different representations that students can use". Again, it was evident that the educators were both assessing students' mathematical learning while also engaging their pedagogical content knowledge, in this case in relation to the use of mathematical representation along with mathematical writing.

Student Learning Pathways

Establishing pathways and trajectories of how students learn in mathematics has gained considerable attention in research studies in recent years (Clements & Sarama, 2004; Siemon et al., 2017; Weber & Lockwood, 2014; Wilson et al., 2014). Curricular documentation can be used as one form of learning about expected progression and development of mathematical ideas. In the New Zealand setting, the curriculum expectations span over a two-year period and are presented in a less detailed form than in many jurisdictions. In recent years, there has been development of curriculum support material including elaborations of the curriculum and learning progression frameworks (LPFs) to provide further detail given awareness that there are differing interpretations of curriculum expectations both between teachers and across schools. This can lead to variability in teaching and opportunities for mathematical learning which is reflected in New Zealand's results in international testing such as PISA and TIMMS which show a large spread of results for students working at the same year level (Caygill & Zhao, 2021). Consequently, an important area of professional development for teachers is developing understanding of the curriculum.

In this study, it was evident that the participants connected their interpretation of students' mathematical writing to the curriculum and related documents, and this supported them both to understand responses and to make sense of the curriculum: "we're having robust discussions about our interpretation of what it says in the curriculum". In a broader sense, the participants acknowledged that using mathematical writing, comparative judgment, and collaborative discussion could be a fruitful form of development: "it helps other teachers … they get to know what the different levels are, particularly if they're moderating and if they're talking together, you know, across levels and stuff like that". In this instance, the participant highlighted opportunities for teachers to learn about different curriculum levels beyond only focusing on the year group that they taught. Similarly, another participant specifically highlighted her own growing knowledge of the curriculum: "it's good for your own content knowledge of the curriculum because it makes you think, what the different levels are within a strand of math. What is level three? What is level four? Then that helps you decipher the different in the trajectory". We argue that assessing mathematical writing using comparative judgment can position teachers to explore their understanding of the curriculum and connect to student responses to further understand the expectations and illustrations of learning.

A key argument in research studies has been that learning pathways and trajectories can be both used to view progress and to guide mathematical teaching and learning (Confrey et al., 2017; Siemon et al., 2017; Wilson et al., 2014). Beyond the development of further understanding of the curriculum, assessing mathematical writing through comparative judgment led to these mathematics educators undertaking a process of developing their own progressions and pathways of learning in relation to the sophistication of student responses. This was described by one participant in general terms as developing a: "trajectory of learning, and think where we are as in, where are these kids on their trajectory? So it's like a sloping number-line and you can see where something is on that number-line". It was evident that the participants were considering mathematical learning as a nuanced pathway that they needed to understand.

Interestingly, many of the mathematics educators specifically explained the progressions that they had developed to guide them in assessing mathematical writing focused on different areas of mathematics. We have used quotes of some of these to illustrate the richness in teacher learning about pathways. For example, in relation to number and operations, one participant discussed her focus on whether students used repeated addition or multiplication which she viewed as more sophisticated. Another participant in considering area measurement described a progression as: "they're [students] using different shapes to try and apply the common thought processes that they have learned... you can see them pushing beyond the regular four-sided shapes. They're trying to apply an understanding of area within a triangle, within a pentagon or hexagon". Finally, in a detailed explanation which connected the earlier examples of using representations and mathematical writing, a participant explained the progression that they had developed for examining algebraic growing patterns: "I'm looking to see if the kids have used a graph or a diagram... I consider a table of data or a graph to be more sophisticated than a diagram so in an order of sophistication, I would look for a drawing, then some numbers in a table of data, and then a graph". In these examples, we see how the mathematics educators engaged in developing their ideas about learner pathways or progressions, and in doing so, also furthered their own professional learning.

Changes to Classroom Practice

Throughout the findings and discussion, we have illustrated both the demands on teachers when using mathematical writing and comparative judgment for assessment and also the opportunities for learning. In this final section, we examine the changes to classroom practice which the mathematics educators described in relation to what they learned through this assessment process. In the first instance, the educators described noticing opportunities to further extend students' mathematical learning. This included identifying missing elements in students' mathematical writing related to their understanding and considering what teaching practice would support further development. For example, in relation to the statistics task, one participant highlighted the demands of making statements about the types of rubber bands and the learning experiences that students needed to have to achieve this:

"Have more of a literacy talk with which ones do you think they should sell?...I've not seen a lot of statements as to why...what can we do to relate that, what they've just presented in the graph to this... push them to say 'why are they the best?'"

Similarly, another participant in referring to the same assessment task, explained how they reflected on the need for students to have adequate time to both learn the concepts but also the need to introduce representation (graphs and co-ordinates) as deliberate acts of teaching. These findings parallel those of Krebs (2005) in showing how examining mathematical writing can also support teachers to consider teaching actions.

Also, evident in the interview data from the educators were the opportunities that assessing mathematical writing using comparative judgment provided to examine the use of explanations and justification as mathematical practices. Again, this was positioned as an area for further development in the classroom. For example, one participant described her realization of the need to balance a focus on students developing a rule for a linear growing pattern with the need for understanding and justification:

"they've got a plus four but then they can't tell you why it is actually multiplied by four... I'll be pushing teachers and classrooms to take students to justify, put more justification and then into representations".

Clearly, the participants engaged in professional learning in relation to students' thinking and use of mathematical practices through the act of reading and assessing samples of writing. Interestingly, we note that for this group of mathematics educators, the requirement for strong understanding of multi-faceted areas of mathematics to undertake comparative judgment appeared to prompt them to engage in further professional development which also was taken into classroom practice.

FUTURE RESEARCH DIRECTIONS

Future research is needed to investigate how the professional learning model presented here might be applied across different mathematical topics and age ranges. A particular interest here is the relative benefits for teacher development at high school rather than elementary school level. The mathematics knowledge and confidence of elementary teachers is known to vary widely (Hill, 2010), especially for those who have specialized in arts and humanities subjects in their own higher education. Conversely, high school mathematics teachers tend to be subject specialists who would consider themselves confident with the curricular topics they teach. To what extent might the nature and depth of the insights reported by elementary teachers also be reported by middle or high school mathematics teachers? Tangential evidence can be found in previous studies (I. Jones et al., 2014; I. Jones & Inglis, 2015), but that work focused exclusively on how professional educators made their comparative judgment decisions rather than what they learned from making decisions. Further theoretical support is provided from peer assessment studies that make use of comparative judgment: given results have been conceptually replicated for the case of school children (I. Jones & Wheadon, 2015) and undergraduates (Davies et al., 2020; I. Jones & Alcock, 2014; I. Jones & Sirl, 2017), we conjecture that the findings reported here in the context of elementary mathematics might translate to middle and high school mathematics.

The data here are grounded in self-reports from stimulated-recall, semi-structured interviews. However, self-report is not a reliable measure of learning (Alcock et al., 2013) and if we wish to know whether comparatively judging students' written mathematical work *causes* professional learning then experimental methods are required. Of particular interest would be to define professional learning in terms of measurable impacts on classroom practice over time for educators who engage regularly with assessing mathematical writing. Such measurement is likely to involve classroom observations, where the focus might be on how teachers respond to students' explanations and representations, as well as measures of students' mathematical achievement with a focus on conceptual understanding.

Should comparative judgment methods as described here prove attractive to teachers, school leaders and others then consideration would need to be given to how they can be practically incorporated within professional development activities. For example, in our experience key to maximizing the types of insights our participants reported in this study is encouraging discussion amongst educators while making comparative judgment decisions. Therefore while comparative judging can be conducted asynchronously and remotely it might be preferable for it take place synchronously and in-person when educator learning is a key goal. Professional development based on comparative judgment might also be enriched with related activities such as educators designing their own free response prompts so that students' responses are better attuned to catching informative student responses in each teachers' particular classroom context.

CONCLUSION

To conclude, we return and address the research questions that motivated this study.

RQ1. What do educators need to know and learn when using mathematical writing for assessment?

Educators need to be familiar with the mathematical task used and should complete it themselves before undertaking assessment activities. It is also beneficial for educators to know the curriculum and understand the expected levels of performance for different year groups. This is unsurprising and consistent with our previous work (Hunter & Jones, 2018; Marshall et al., 2020). Moreover, some participants reported that engaging in the assessment activity helped them strengthen their knowledge and to gain fresh insights as to what it means for students to be at certain levels of performance. Some participants reported further they felt they had a better oversight of students' learning trajectories.

The participants reported other types of learning from undertaking the assessment, which we coded as student representations and student ideas. Some reported learning about the range of representations that students make use of when responded to free response prompts. Crucially, many also talked about seeing through those representations to make inferences about students' underlying mathematical ideas and thinking. Some of the participants commented on how these insights would impact on future classroom practice.

RQ2: What opportunities does comparative judgment provide to educators in relation to learning to use writing as mathematical assessment?

Undertaking comparative judgments of students' writing engaged the participants and helped them to perceive limitations with existing assessment methods. In particular, some commented on how even the relatively rich GLoSS tasks, which involve one-to-one oral assessments with students, could be constrained and frustrated by the associated rubrics. The comparative judgment approach freed the participants from rubrics, allowing them to make holistic assessment decisions based on their own interpretations of students' responses.

We believe the findings reported here are not limited to the particular setting of the study, nor the particular topics and free response tasks used. As summarized above, a corpus comparative judgment studies in the context of mathematical writing have demonstrated the usefulness of the method across a varied contexts, age ranges and mathematical topics and tasks. There is no reason to believe that the potential for professional development and learning we have observed in our data would not translate to other areas of mathematics education practice.

ACKNOWLEDGMENT

This work was supported by the Royal Society of New Zealand Rutherford Discovery fellowship under grant RDF-MAU1903.

REFERENCES

Alcock, L., Gilmore, C., & Inglis, M. (2013). Experimental methods in mathematics education research. *Research in Mathematics Education*, *15*(2), 97–99. doi:10.1080/14794802.2013.797731

Ball, D. L., Thames, M. H., & Phelps, G. (2008). Content knowledge for teaching: What makes it special? *Journal of Teacher Education, 59*(5), 389–407. doi:10.1177/0022487108324554

Bartholomew, S. R., & Jones, M. D. (2022). A systematized review of research with Adaptive Comparative Judgment (ACJ) in higher education. *International Journal of Technology and Design Education, 32*(2), 1159–1190. doi:10.100710798-020-09642-6

Berube, C. T. (2004). Are standards preventing good teaching? *The Clearing House: A Journal of Educational Strategies, Issues and Ideas, 77*(6), 264–267. doi:10.3200/TCHS.77.6.264-268

Carpenter, T. P., Fennema, E., & Franke, M. L. (1996). Cognitively guided instruction: A knowledge base for reform in primary mathematics instruction. *The Elementary School Journal, 97*(1), 3–20. doi:10.1086/461846

Caygill, R., & Zhao, B. (2021). *Progress and achievement and the context of mathematics and statistics learning in New Zealand (English-medium education).* Ministry of Education. https://assets.education.govt.nz/public/Literacy-and-Maths-strategy-development-in-2021/Progress-and-achievement-and-the-context-of-mathematics-and-statistics-learning-in-New-Zealand.pdf

Clements, D. H., & Sarama, J. (2004). Learning trajectories in mathematics education. *Mathematical Thinking and Learning, 6*(2), 81–89. doi:10.120715327833mtl0602_1

Confrey, J., Gianopulos, G., McGowan, W., Shah, M., & Belcher, M. (2017). Scaffolding learner-centered curricular coherence using learning maps and diagnostic assessments designed around mathematics learning trajectories. *ZDM Mathematics Education, 49*(5), 717–734. doi:10.100711858-017-0869-1

Davies, B., Alcock, L., & Jones, I. (2020). Comparative judgement, proof summaries and proof comprehension. *Educational Studies in Mathematics, 105*(2), 181–197. doi:10.100710649-020-09984-x

Evans, S., & Swan, M. (2014). Developing students' strategies for problem solving in mathematics: The role of pre-designed "Sample Student Work". *Educational Designer, 2*(7), 1–31. https://www.educationaldesigner.org/ed/volume2/issue7/article25/

Falchikov, N., & Goldfinch, J. (2000). Student peer assessment in higher education: A meta-analysis comparing peer and teacher marks. *Review of Educational Research, 70*(3), 287–322. doi:10.3102/00346543070003287

Fitzgerald, L., Hunter, J., & Hunter, R. (2021). Shifting teacher practices in relation to grouping: Gap gazing or strengths focused approaches. *Mathematics Teacher Education and Development, 23*(3), 97-110. https://eric.ed.gov/?id=EJ1321057

Hill, H. C. (2010). The nature and predictors of elementary teachers' mathematical knowledge for teaching. *Journal for Research in Mathematics Education, 41*(5), 513–545. doi:10.5951/jresematheduc.41.5.0513

Hunter, J., & Jones, I. (2018). Free-response tasks in primary mathematics: A window on students' thinking. In J. Hunter, P. Perger, L. Darragh (Eds.), *Proceedings of the 41st Annual Conference of the Mathematics Education Research Group of Australasia, 41*, 400–407. Auckland, New Zealand.

Jacobs, V. R., Empson, S. B., Jessup, N. A., Dunning, A., Pynes, D. A., Krause, G., & Franke, T. M. (2022). Profiles of teachers' expertise in professional noticing of children's mathematical thinking. *Journal of Mathematics Teacher Education*, 1–30. doi:10.100710857-022-09558-z

Jones, I. (2021). Improving conceptual understanding through comparing and contrasting students' mathematical writing. *5th Northeastern Research in Undergraduate Mathematics Education Conference*, Rutgers, New Jersey (online event). https://sites.google.com/view/pcrg/events/nerume/5th-ne-rume-conference?authuser=0

Jones, I., & Alcock, L. (2014). Peer assessment without assessment criteria. *Studies in Higher Education, 39*(10), 1774–1787. doi:10.1080/03075079.2013.821974

Jones, I., & Davies, B. (2023). Comparative judgement in education research. *International Journal of Research & Method in Education*, 1–12. doi:10.1080/1743727X.2023.2242273

Jones, I., & Inglis, M. (2015). The problem of assessing problem solving: Can comparative judgement help? *Educational Studies in Mathematics, 89*(3), 337–355. doi:10.100710649-015-9607-1

Jones, I., & Sirl, D. (2017). Peer assessment of mathematical understanding using comparative judgement. *Nordic Studies in Mathematics Education, 22*(4), 147–164. https://hdl.handle.net/2134/28439

Jones, I., & Wheadon, C. (2015). Peer assessment using comparative and absolute judgement. *Studies in Educational Evaluation, 47*, 93–101. doi:10.1016/j.stueduc.2015.09.004

Krebs, A. S. (2005). Analyzing student work as a professional development activity. *School Science and Mathematics, 105*(8), 402–411. doi:10.1111/j.1949-8594.2005.tb18060.x

Marshall, N., Shaw, K., Hunter, J., & Jones, I. (2020). Assessment by comparative judgement: An application to secondary statistics and English in New Zealand. *New Zealand Journal of Educational Studies, 55*(1), 49–71. doi:10.100740841-020-00163-3

Ministry of Education. (2007). *The New Zealand curriculum*. Learning Media.

NCTM. (2014). *Ensuring Mathematical Success for All*. National Council of Teachers of Mathematics.

Orsmond, P., Merry, S., & Reiling, K. (1996). The importance of marking criteria in the use of peer assessment. *Assessment & Evaluation in Higher Education, 21*(3), 239–250. doi:10.1080/0260293960210304

Pollitt, A. (2012). The method of Adaptive Comparative Judgement. *Assessment in Education: Principles, Policy & Practice, 19*(3), 281–300. doi:10.1080/0969594X.2012.665354

Powell, S. R., Hebert, M. A., Cohen, J. A., Casa, T. M., & Firmender, J. M. (2017). A synthesis of mathematics writing: Assessments, interventions, and surveys. *Journal of Writing Research, 8*(3), 493–530. doi:10.17239/jowr-2017.08.03.04

Rittle-Johnson, B., & Star, J. R. (2007). Does comparing solution methods facilitate conceptual and procedural knowledge? An experimental study on learning to solve equations. *Journal of Educational Psychology*, *99*(3), 561–574. doi:10.1037/0022-0663.99.3.561

Rittle-Johnson, B., Schneider, M., & Star, J. R. (2015). Not a one-way street: Bidirectional relations between procedural and conceptual knowledge of mathematics. *Educational Psychology Review*, *27*(4), 587–597. doi:10.100710648-015-9302-x

Siemon, D., Horne, M., Clements, D., Confrey, J., Maloney, A., Samara, J., & Watson, A. (2017). Researching and using learning progressions (trajectories) in mathematics education. In *Proceedings of the 41st Conference of the International Group for the Psychology of Mathematics Education* (Vol. 1, pp. 109–136).

Stein, M. K., Engle, R. A., Smith, M. S., & Hughes, E. K. (2008). Orchestrating productive mathematical discussions: Five practices for helping teachers move beyond show and tell. *Mathematical Thinking and Learning*, *10*(4), 313–340. doi:10.1080/10986060802229675

Topping, K. J. (2009). Peer Assessment. *Theory into Practice*, *48*(1), 20–27. doi:10.1080/00405840802577569

van den Berg, I., Admiraal, W., & Pilot, A. (2006). Design principles and outcomes of peer assessment in higher education. *Studies in Higher Education*, *31*(3), 341–356. doi:10.1080/03075070600680836

Verhavert, S., Bouwer, R., Donche, V., & Maeyer, S. D. (2019). A meta-analysis on the reliability of comparative judgement. *Assessment in Education: Principles, Policy & Practice*, *26*(5), 541–562. doi: 10.1080/0969594X.2019.1602027

Weber, E., & Lockwood, E. (2014). The duality between ways of thinking and ways of understanding: Implications for learning trajectories in mathematics education. *The Journal of Mathematical Behavior*, *35*, 44–57. doi:10.1016/j.jmathb.2014.05.002

Widjaja, W., Vale, C., Groves, S., & Doig, B. (2017). Teachers' professional growth through engagement with lesson study. *Journal of Mathematics Teacher Education*, *20*(4), 357–383. doi:10.100710857-015-9341-8

Wilson, P. H., Sztajn, P., Edgington, C., & Confrey, J. (2014). Teachers' use of their mathematical knowledge for teaching in learning a mathematics learning trajectory. *Journal of Mathematics Teacher Education*, *17*(1), 149–175. doi:10.100710857-013-9256-1

Chapter 8
A Genre-Based Approach to Improving Students' Mathematical Writing: Practical Support for Educators

Lauren Haggis
South Australia Department for Education, Australia

Tutita M. Casa
Department of Curriculum and Instruction, Neag School of Education, University of Connecticut, USA

ABSTRACT

This chapter introduces the resources and practical guides that the South Australian Department for Education has developed to support teachers to teach writing in mathematics classrooms and develop their understanding of the types of writing mathematics students should be proficient in. Developing this understanding has the potential to improve classroom practice and positively impact learning outcomes for students in both writing and mathematics. This chapter may offer insights for other jurisdictions seeking to support the teaching of mathematical writing.

INTRODUCTION

The notion that "writing is another important component of the discourse" (National Council of Teachers of Mathematics [NCTM], 1991, p. 34) is one that continues to be emphasized in mathematics education policies across various nations. The *Australian Curriculum*, for example, calls for learning programs that provide opportunities for students to "communicate their thinking, reasoning and solutions to problems using appropriate mathematical language, notation and symbology" (Australian Curriculum, Assessment, and Reporting Authority [ACARA], 2022b, p. 12). Similarly, in the United States, the *Common Core State Standards for Mathematics* (National Governors Association Center for Best Practices and the

DOI: 10.4018/978-1-6684-6538-7.ch008

Council of Chief State School Officers [NGA Center and CCSSO], 2010b) emphasizes how "students at all grades can listen or read the arguments of others" (p. 7), with the English language arts companion (2010a) promoting writing arguments and informative pieces across subject areas and grades. Similarly, the Canadian province of Ontario notes how "as students engage in reasoning, teachers further encourage them to pose questions, make conjectures, and justify solutions, orally and in writing" (Ministry of Education, 2020, p. 74) to attend to the mathematical processes emphasized in their current curriculum guide.

Despite these policies calling for students to write mathematically, there continues to be a lack of guidance for teachers about how to support students to engage in this practice (Powell et al., 2017). Therefore, Morgan's (1998) longstanding question about whether the writing students complete in mathematics represents their understanding remains. If teachers are unsure about how to introduce students to mathematical writing and help them understand the characteristics of high-quality texts, practitioners and researchers will be limited in relying on students' writing to advance student learning of content and the valued ways of communicating in the mathematics discipline.

One way to support the implementation of mathematical writing is from a policy perspective considering all grade levels. The reach of such policies depends on local governance structures. For example, American states have the constitutional authority to oversee their education systems, and each state dictates the extent to which school districts employ their own policies. Hawaii includes only one school district (Our Statewide School District, n.d.) whereas Connecticut includes multiple districts (ConneCT Kids, n.d.). The sizes of the districts vary, with some only having one elementary and one middle school, and high schoolers going to a regional school alongside students from other districts. In South Australia, the Department for Education is responsible for teaching and learning policy in the approximately 500 public schools in the state. Regardless of the governance structures, educators seeking to develop policy need to consider the implications of writing mathematically within their context to account for the ways in which teachers interpret and implement such guidance. Minimally, policies should account for the age of the students being taught and the teachers' responsibilities for teaching subjects in addition to mathematics where disciplinary-specific writing is also required.

This chapter represents longstanding efforts, beginning in 2014, by the South Australian Department for Education to support the teaching of writing in mathematics classrooms across Years 1 to 10, which are students ranging approximately in ages 5 to 15. To date, four sets of maps have been produced that detail the genres students are most frequently asked to produce. In this sense, genre refers to the goal-oriented social practices that have been developed to help achieve certain writing purposes (Martin & Rose, 2008) rather than a category of literature (e.g., romance, mystery). The mathematics genre maps have been developed to identify writing characteristics critical to the discipline of mathematics and how we envisioned students' writing developing in complexity over the years of schooling. When these efforts began, there was limited literature defining what mathematical writing should entail and a lack of research on how students' writing develops in complexity across grade levels. The department's initial efforts only covered procedural writing and did not expand on the purpose of mathematical writing nor did it surface the link between oral language, writing, and mathematical thinking.

The latest version of the genre map, presented later in this chapter, represents a collaboration between Lauren, who oversees literacy policy for secondary schools in South Australia, and Tutita, a faculty member at the University of Connecticut, who was invited to present at the South Australian Department for Education's online Literacy Summit in 2022. Tutita's presentation emphasized the ways mathematical writing supports the development of students' mathematical reasoning. Lauren and her colleagues began exploring which genres would enable students to engage in and demonstrate their reasoning and continued collaborating with Tutita.

We next share the background of this work, highlight key features of the current genre map, and describe the professional learning provided to teachers as they infuse writing into their mathematics instruction. We hope that the lessons learned by the South Australian Department for Education's ongoing efforts benefit other education authorities and researchers grappling with ways to have their students write mathematically. As education systems worldwide seek effective strategies to enhance teaching and learning outcomes, we hope to contribute to the global exchange of effective practices in mathematical writing.

BACKGROUND

The impetus for the South Australian Department for Education to address writing in mathematics resulted from National Assessment Program – Literacy and Numeracy (NAPLAN) test results indicating that student achievement in writing had steadily declined in South Australia and nationally. Research suggests that this trend has continued given that the average Australian student in 2018 performed nearly 1.5 years behind the average student in 2011 (Thomas, 2019). We surmised that part of the reason for this decline was because students spent little time writing and teachers did not emphasize writing across subjects. Although research has shown that limited time teaching and engaging in writing is a worldwide issue (Graham, 2019), we recognized that having students and teachers spend more time in writing would not be enough. Students need strong writing instruction, aligned with goals for instruction, curriculum content, and the application of effective instructional practices (Bransford et al., 2005).

Decisions About What to Emphasize in Mathematical Writing

Because evidence suggests that writing in mathematics is not only an effective means of communicating understanding but also a tool for enhancing students' reasoning and deepening their comprehension of concepts (Barwell, 2018; Pugalee, 2005), the decision to emphasize this practice across grades became obvious. The department noted the variety of purposes students across all grade levels may be asked to write, including on national assessments that ask them to explain their thinking and justify the strategies they use to solve problems as well as assessments through the senior years of schooling (ACARA, 2022b). Consequently, we adopted a critical perspective to ensure that the type of writing students engage in would develop not only their literacy skills but also their mathematical thinking.

When further considering our approach, we recognized that students would need to understand that when writing mathematically, mathematical symbolic systems, visual images and artifacts other than natural language (Moschkovich, 2007, Barwell, 2020, O'Halloran, 2005) and general writing skills are integrated (Namkung et al., 2020). Our aim was to help teachers see that the genres that are valued in the discourse of mathematics differ from those favored in other learning areas of the curriculum and are unique in relation to language, syntax, and conventions (T. Shanahan & C. Shanahan, 2008, 2012; Snow, 1987). These ways of communicating on paper are reflected across various genres of writing mathematically that go beyond recording computational steps. This includes students analyzing and evaluating potential solutions, constructing and evaluating arguments, using evidence and logic to draw conclusions and solve problems, and identifying alternative explanations and possibilities expected of them on national assessments (ACARA, 2022a). The South Australian Department for Education considered the need for students to record more sophisticated mathematical symbols and visual representations over time.

Decisions About Why to Develop a Genre Map

In South Australia, it has long been recognized that teaching students about genre has several benefits that help deepen their disciplinary understanding; it fosters reflection, facilitates the logical organization of ideas, and enhances students' persuasive abilities (Humphrey & Macnaught, 2016). By understanding the structures and language features of the genres in each learning area, we have observed students gaining confidence in conveying their understanding. Consequently, the department has placed a strong emphasis on enhancing teachers' knowledge and understanding of a genre-based approach to teaching mathematical writing.

In addition to developing mathematical thinking, the South Australian Department for Education took into account that a genre-based approach to mathematical writing has the potential to lead to a more equitable mathematics classroom (Casa et al., 2020). Students come from diverse cultural, linguistic, and socio-economic backgrounds and may not have had exposure to the key genres in mathematics. We recognized that making the literacy demands of mathematics visible to students and teaching genre with an equity lens provides support for students who may not have the semiotic capital, which Duke (2000) described as "knowledge of semiotic systems, including systems of written language" (p. 447), to align their current understandings with the knowledge that schools privilege and helped ensure that all students saw themselves as capable mathematics learners. This goal is also a central commitment to equitable mathematics education in Connecticut (Association of Mathematics Teacher Educators in Connecticut, Associated Teachers of Mathematics in Connecticut, & Connecticut Council of Leaders of Mathematics, 2022). In addition, we anticipated it would contribute to a more inclusive curriculum to help students from various backgrounds see the relevance of mathematical writing to their own lives and empower them to make discerning choices in their independent academic writing. Furthermore, some students are reluctant to engage in classroom discussion due to anxiety or a fear of being wrong in front of their peers and may have a stronger "voice" when asked to write (Casa, Choppin & Moschkovich, 2020). In response, a policy was developed that translated the existing scholarship and research on effective teaching of writing into a range of resources that are accessible to South Australian school leaders and teachers. This included a set of advice papers on writing, an online Literacy Summit with the theme *Learning to Write, Writing to Learn*, and an updated set of disciplinary genre maps.

These resources aim to support teachers of mathematics to attend to writing in their planning and teaching and help them to consider not only students' understanding of mathematical content but also the development of their mathematical practices. ACARA (2022b) expects students to embody inquiry skills, such as developing questions to broaden their thinking and evaluating information and solutions; creating possibilities and considering and evaluating alternatives; analyzing problems, drawing conclusions, and providing reasons based on mathematical evidence; and reflecting on their thinking and evaluating their approaches to mathematical problems. In short, mathematics teachers should teach students "how to read, write and critically discuss the disciplinary genres they are routinely required to read and write in school as a way of developing students' content knowledge and disciplinary literacies" (Gebhard, 2019, p. 205).

THE CURRENT MATHEMATICS GENRE MAP

The following sections outline the key features of the South Australian Department for Education's (2023) current version of the mathematical writing genre map. Throughout, impactful changes from previous versions which would allow others seeking to do similar work in their contexts are shared. Important aspects of the features are highlighted to promote applications elsewhere.

Inclusion of Written Mathematical Texts in Primary School

Previous versions of the genre map were limited to examples of texts students may write in secondary school with a focus on "investigation report." In the current version, sample writing tasks have been included from the early years of schooling to support primary school teachers in South Australia (known as "elementary" teachers in the US and elsewhere) to approach the teaching of writing with a stronger focus on discipline-specific writing practices rather than restricting the teaching of writing to English or literacy lessons. Furthermore, as students can show their mathematical reasoning in a written text through drawings and early mark-making, the genre map supports teachers to provide younger students with opportunities to do so.

Developing control of the discourse of mathematics is a cumulative and recursive process that continues throughout the years of schooling. The development of complexity of the genres from the early years through to senior high school is shown in a summary table, depicted in Figure 1, which helps primary school teachers understand the writing demands of secondary school mathematics and vice versa. The department anticipates that this understanding will help teachers see their students' mathematical success as a combination of both their mathematical understanding and their control of the discourse. Primary school teachers will be better informed to strategically approach preparing their students for secondary school mathematical writing.

There are several anticipated benefits to having primary students engage in mathematical writing. First, it provides teachers with opportunities to gather evidence of student learning in which they demonstrate the key mathematical proficiencies or competencies of understanding, reasoning, and problem-solving. Furthermore, the genre map for primary teachers, seen in Figure 2, intends to help them to design assessment tasks and develop assessment criteria and rubrics relevant to each mathematical writing genre. It helps them to provide specific feedback to students during writing conferences and emphasize the importance of revision and editing in mathematical writing to improve clarity and coherence. Teachers can also use the genre map to design formative assessment tasks that enable teachers to gain insight into their students' mathematical thinking (Casa, et al. 2022) and provide students with practice opportunities that scaffold them toward writing complete texts. Using the genre map also helps schools ensure students are taught the genres critical for high-stakes assessment in the senior years.

Explanation Beyond the Investigation Report Genre

The first genre map that was produced in 2014 identified recounts and evaluations of group tasks, including investigations, as a key genre in mathematics. When the genre map was redeveloped in 2020, the investigation report genre was described as a "macrogenre," which is a more complex text that is made up of multiple genres that are interrelated or embedded (Christie, 2002; Martin, 1995). The embedded genres were identified as procedure, explanation, analysis, and evaluation. Although an investigation

Figure 1. This is the first page of the fourth and most current version of the mathematics genre map showing an overview of example writing tasks in primary/elementary and secondary school (South Australian Department for Education, 2023)

GENRES IN MATHEMATICS

In mathematics, students learn to create texts that explain reasoning and processes, recount problem solving approaches and procedures, and produce and justify solutions. They learn to present and evaluate data and create visual texts such as charts, graphs and pictures.

The following table summarises the task examples in the mathematics genre maps – it is a guide only and the genres are not limited to specific year levels.

	Reception	Year 1	Year 2	Year 3	Year 4	Year 5	Year 6	Year 7	Year 8	Year 9	Year 10
Mathematical explanation	Mathematical explanation	Mathematical explanation	Mathematical explanation	Mathematical explanation	Mathematical explanation	Mathematical explanation	Mathematical explanation				
Mathematical argument (personal)	Argument (personal)	Argument (personal)	Argument (personal)	Argument (personal)	Argument (personal)	Argument (personal)	Argument (personal)	Argument (personal)	Argument (personal)	Argument (personal)	Argument (personal)
Mathematical argument (critique)			Argument (critique)	Argument (critique)	Argument (critique)	Argument (critique)	Argument (critique)	Argument (critique)	Argument (critique)	Argument (critique)	Argument (critique)

LITERACY | GENRES IN MATHEMATICS | https://edi.sa.edu.au/writing   Government of South Australia Department for Education

report is a macrogenre students are likely to encounter at school (Derewianka & Jones, 2016), these texts usually report on an investigation into a topic or issue rather than a mathematical investigation.

Figure 3 shows an example of a prompt used to advance students' writing of investigation reports. Although the aim of this writing type was for students to explain, justify, and evaluate their thinking and not just report on results, the example tasks teachers implemented often did not result in this type of writing from students. What we noticed was that students produced texts in which they wrote procedurally. That is, they recounted the steps taken to reach their solution, and their reasoning was often limited despite using prompts like "investigate."

A further revision of the investigation report tasks resulted in writing prompts that included more detail, such as the one in Figure 4 that represents a revision to the ones depicted in Figure 3. Despite these efforts, students' writing still gave little insight into their mathematical thinking as the prompts did not guide students to evaluate or justify their mathematical thinking. Phrases like "describe the process you might use" did not support students to include their mathematical reasoning. We learned that the wording of the writing prompts is critical for making the expectation for the type of texts students produce clear to teachers.

Figure 2a. The second page of the genre map showing example writing tasks for primary/elementary students
(South Australian Department for Education, 2023)

Mathematics genre map – reception to year 6

LITERACY | DEPARTMENT FOR EDUCATION | 2

Reception	Year 1	Year 2	Year 3	Year 4	Year 5	Year 6
MATHEMATICAL EXPLANATION | Describing and explaining mathematical concepts, processes and strategies						
Mathematical explanation Tell a number story to represent a mathematical relationship, for example '8 kangaroos were drinking at the river and 3 hopped away'. Draw a picture to show what happened. Use numbers to explain your picture.	**Mathematical explanation** Use an everyday object, eg a pencil, to measure and sort a range of objects from shortest to longest. Write some sentences explaining the way you sorted them.	**Mathematical explanation** Sort a collection of 2D shapes based on their features. Write some sentences explaining how you sorted them. Use terms to describe their features, eg opposite, parallel, curved, straight and number of sides.	**Mathematical explanation** Explain different ways a collection of geometric objects, eg cylinders, spheres, prisms and pyramids, can be sorted. Refer to their key features, eg shape, the number of faces, surfaces, edges and/or vertices.	**Mathematical explanation** Explain why it is always true that adding 2 even numbers always equals an even number, adding 2 odd numbers always equals an even number and adding an odd and even number always equals an odd number. Draw and annotate models to support your explanation.	**Mathematical explanation** Explain the connection between multiplication and division as inverse operations. Use examples to support your explanation.	**Mathematical explanation** Using a variety of rectangles, explain the relationship between the side lengths and the area of the rectangles. Determine a formula for the area of a rectangle. Record your data in a table.
MATHEMATICAL ARGUMENT (PERSONAL) | Explaining mathematical reasoning and arguing a mathematical position being taken						
Argument (personal) Make a pattern using a variety of materials. Make the pattern again using different materials. Describe how the patterns are the same. Does your friend think it's the same pattern? Draw pictures and use early forms of writing to convince them that they are.	**Argument (personal)** Give directions for people to go from your classroom to the library. Is your way the same as your classmates' way? Argue that your way is the quickest. Use your directions as evidence.	**Argument (personal)** Is it important to use the same units when comparing the measurements of different things? Use an informal unit, eg shoes, to justify your thinking.	**Argument (personal)** Investigate different ways to estimate the number of buttons in a jar. Recommend the method you think will give the most accurate prediction and provide supporting reasons for your choice. Are any of the methods you thought of unreasonable? Why? Evaluate the accuracy of your method by counting the buttons.	**Argument (personal)** Investigate the different ways a data set can be displayed. Identify the features of each representation. How easily can you compare the data in each representation? Convince me of which representation is most appropriate for the given data set.	**Argument (personal)** Investigate the probability of different outcomes to a game, eg using 2 spinners with equal and unequal coloured regions. Convince me that you can predict the outcomes using your understanding of probability. Play the game multiple times and record the distribution of outcomes. Explain how this relates to equal and unequal outcomes using fractions. Evaluate the outcomes against your calculated predictions.	**Argument (personal)** Which letter would be in the 15th position in the following growth pattern: A B C A B C A B C A B C ...? Identify which letter would be in the 150th position. Where would the 40th C be positioned in the pattern? Determine a rule you could use to work out the position of the Cs and convince me why it works. If the pattern consisted of 50 elements, how many As, Bs and Cs would be used?

GENRES IN MATHEMATICS | RECEPTION to YEAR 10 | https://edi.sa.edu.au/writing

Focus on Explanation Writing

Eventually, Lauren identified explanation as a key genre in mathematics. The verb *explain* appears frequently in the mathematics curriculum documentation of many countries from the early years of schooling right through secondary school. In the United States' English language arts version of the *Common Core State Standards* (NGA Center and CCSSO, 2010a), this type of writing is referred to as *informative/explanatory writing* and requires students to provide information or explanations related to mathematical concepts (Casa et al., 2016; Firmender, Casa & Colonnese, 2017). Research suggests that when students write explanations of their mathematical work and give examples, teachers can gain insight into and can assess students' conceptual understanding and progress (Urquhart, 2009).

Although *explanation* was identified as a key written genre in mathematics, it was important to distinguish it from explanations written in other curriculum areas, such as science and social studies, that teachers and students in South Australia were required to address. These subjects asked students to write a number of different types of explanation: sequential (explaining a phenomenon that involves a linear sequence), cyclical (explaining a cyclical phenomenon), causal (explaining a phenomenon that has an element of causality), system (explaining how a system works), factorial (explaining the factors that lead

Figure 2b. The third page of the genre map showing example writing tasks for primary/elementary students (South Australian Department for Education, 2023)

Mathematics genre map – reception to year 6 (continued)

LITERACY | DEPARTMENT FOR EDUCATION | 3

Reception	Year 1	Year 2	Year 3	Year 4	Year 5	Year 6
MATHEMATICAL ARGUMENT (CRITIQUE) | Critiquing the mathematical reasoning of others and establishing the mathematical validity of arguments						
		Argument (critique) If I begin at the number 13 on a number chart and keep adding 10, the number will always end in a 3. Am I correct? Prove your thinking.	**Argument (critique)** Consider the claim that 15 + 28 = 20 + 23. Do you agree? Why?	**Argument (critique)** Two students ran a 7km race. Student A ran 5/6 of the race and student B ran 9/12 of the race. Each person claims they ran more than the other. Who do you agree with? Why?	**Argument (critique)** A real estate agent is selling a block of land they claim has a perimeter that is numerically equal to its area. Is this possible?	**Argument (critique)** You are playing a card game with a friend. They claim they can beat you by using their knowledge of probability. Is this true or is the card game purely a game of chance? Support your argument with mathematical evidence.

GENRES IN MATHEMATICS | RECEPTION to YEAR 10 | https://edi.sa.edu.au/writing

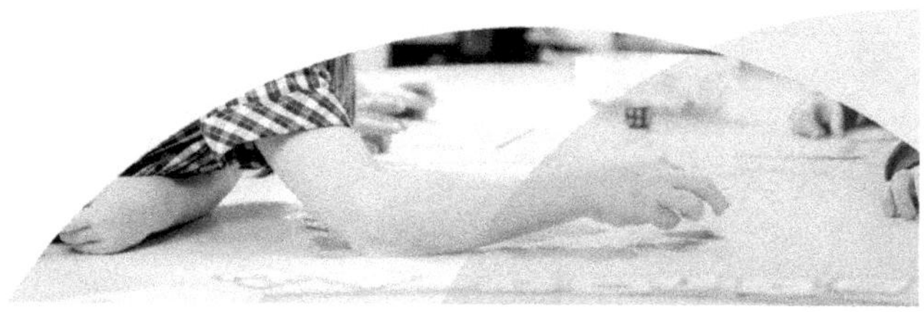

to an outcome), and consequential (to explain the consequences of a particular input (Derewianka & Jones, 2012). Ultimately, we decided to give the genre the name *mathematical explanation* to emphasize

Figure 3. Example investigation report writing prompts in the 2020 version of the mathematics genre map (South Australian Department for Education, 2020)

YEAR 8	YEAR 9	YEAR 10
INVESTIGATION		
Topic: Patterns and algebra, eg Investigate how to find the perimeter of a hexagonal train of any length. 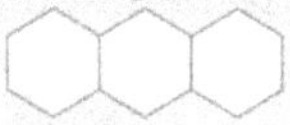	**Topic: Linear and non-linear relations, eg** Investigate the maximum volume of an open box that is made from a piece of cardboard that measures 10 cm by 20 cm.	**Topic: Statistics and probability, eg** Use the points scored by 2 AFL teams in the first 10 rounds to predict the winner when they play each other in round 11. The data can be used in multiple ways to support the prediction.

Australian Curriculum: Literacy in the learning areas: Mathematics

In the Australian Curriculum: Mathematics, students learn the vocabulary associated with number, space, measurement and mathematical concepts and processes. This vocabulary includes synonyms, technical terminology, passive voice and common words with specific meanings in a mathematical context. Students develop the ability to create and interpret a range of texts typical of mathematics ranging from calendars and maps to complex data displays. Students use literacy to understand and interpret word problems and instructions that contain the particular language features of mathematics. They use literacy to pose and answer questions, engage in mathematical problem-solving, and to discuss, produce and explain solutions.

http://tiny.cc/AClitGC

Figure 4. This is an example of an investigation report writing prompt from the third version of the mathematics genre map
(South Australian Department for Education, 2021)

YEAR 7	YEAR 8	YEAR 9	YEAR 10
INVESTIGATION REPORT (This is a macrogenre and may include elements of procedure, explanation, analysis and evaluation)			
Sub-strand: Measurement Design a container to package a 10ml product. Explore using ten 1cm cubes and incorporate the use of isometric dot paper. Describe how you might increase or decrease the prism's surface area and volume and the number of faces, edges and vertices. Include a fully justified recommendation for the best container option. 	**Sub-strand: Patterns and algebra** A path made up of hexagonal pavers requires metal edging around the perimeter. Investigate how the length of metal edging required changes as you add more pavers. How might you predict the perimeter of a hexagon train of 100 hexagons? Describe the process you might use to find the perimeter of a hexagon train of any length. 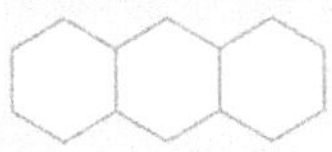	**Sub-strand: Linear and non-linear relations** A packaging company requires a design to maximise the capacity of an open box that will be made from an expensive material. Investigate the different open box options that can be made from a piece of cardboard that measures 10cm by 20cm. Explain the process and justify your recommendation for the open box with the largest volume. 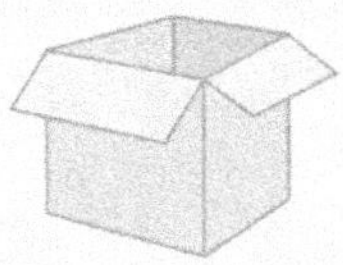	**Sub-strand: Statistics and probability** If the only data you have available is the number of points scored by 2 AFL teams in the first 10 rounds, how might you use this to predict the winner when they play each other in round 11? Discuss how factors such as home/away game, players list, weather, etc might influence your prediction. The data can be used in multiple ways to support the prediction. How confident are you about your prediction?

Notes:

1. **Additional genres:** A broader range of genres than those included in this map may be appropriate for students to use when demonstrating learning in mathematics eg reports of surveys, explanations of processes or evaluation of group teamwork. However, the mathematical investigation is a critical genre for students of mathematics and is the genre most likely to be an extended text. It should be developed over years 7–10, in preparation for the SACE.
2. Genres in mathematics:
   - are informed by social context and therefore will **vary in formality** according to the purpose and context of the task. A formal mathematical investigation report will require technical, abstract and informed language to enable a student to express their thinking at a sophisticated level.
   - may be **macrogenres**: the mathematical investigation is a macrogenre, that is, a combination of genres that may include procedure, explanation, analysis and evaluation, depending on the purpose.
   - may be **multimodal**: genres may be multimodal texts created using various technological platforms. That is, they will integrate a range of spoken, written and visual elements, as appropriate to the purpose and context of the task.

the uniqueness of mathematical writing as students use both written language as well as mathematical symbols and drawings to make their explanations clear to the reader (Casa et al., 2016).

Despite its obvious inclusion in the mathematics curriculum, explanation is not identified as a focus genre in the South Australian Department for Education's mathematics genre map beyond the primary years. This is not to say that students are no longer writing explanations but rather that the genre is more likely to be embedded within the mathematical arguments students write as the emphasis on justification strengthens, as seen in Figure 5 which shows example writing tasks for secondary students.

Focus on Argument Writing

Analysis and evaluation are skills that are identified in a number of national curricula as supporting the development of students' mathematical thinking. The US Standards for Mathematical Practice, for example, recommends that students use writing to "construct viable arguments and critique the reasoning

Figure 5. This is the fourth page of the 2023 genre map showing example writing tasks for secondary students
(South Australian Department for Education, 2023)

Mathematics genre map – years 7 to 10

LITERACY | DEPARTMENT FOR EDUCATION | 4

Year 7	Year 8	Year 9	Year 10
MATHEMATICAL ARGUMENT (PERSONAL) | Explaining mathematical reasoning and arguing a mathematical position being taken			
Argument (personal) Investigate climate or past weather data that interests you from the Bureau of Meteorology website, eg data from 3 different weather stations. Describe and compare trends over time, eg changes in rainfall or temperature. Pose inquiry questions and state a hypothesis giving reasons for your choices. Explain what analysis you will need to do to test your hypothesis. Interpret your statistical results to justify your claim.	**Argument (personal)** You are a designer employed to design packaging for a triangular prism chocolate bar weighing between 200g and 360g using one A4 sheet of cardboard. Write a report for the chocolate manufacturer that includes 2 different designs. Compare the volume of the prisms, the surface area of the required packaging and the perimeter of the nets. Calculate the production cost, including printing. Determine the optimum arrangement for packaging multiple bars for shipping. Calculate the appropriate sale price for the product. Justify all your decisions. Outline any assumptions or limitations to your investigation.	**Argument (personal)** Are all equilateral triangles similar? Explain the theory of similar triangles. Create a data table and use a 3D modelling tool, eg GeoGebra (https://www.geogebra.org), to generate evidence to convince me of your position.	**Argument (personal)** Investigate an aspect of the COVID-19 pandemic. Explore secondary data derived from collating pandemic infection rates (https://ourworldindata.org/covid-cases). Pose an inquiry question and state a hypothesis, giving reasons for your choices. Select and display appropriate data explaining how your chosen displays highlight data trends. Explain the trends in your data set. Interpret the results and justify your claims.
MATHEMATICAL ARGUMENT (CRITIQUE) | Critiquing the mathematical reasoning of others and establishing the mathematical validity of arguments			
Argument (critique) Ethan is playing a game with 2 disks. Each disk is red on one side and green on the other. When Ethan flips the disks, he wins if both disks show the same colour when they land. Ethan thinks he will have more chance of winning if he flips 4 disks and even more if he flips 5 disks. Maria disagrees and says he will have less chance of winning the more disks he flips. Who do you agree with and why? Use evidence to support your argument.	**Argument (critique)** A class of students was seated in a 5 x 5 array. They all wanted to change where they were sitting, but their teacher told them they could only move to a seat in front of, behind or adjacent to the one they were currently in. Is this possible? How do you know? Use evidence to justify your answer.	**Argument (critique)** Eleni says she can create lots of different triangles if she's given the midpoints of the 3 sides. Is she correct? How do you know?	**Argument (critique)** Zoe's ice cream consists of a sphere of vanilla ice cream and a right circular cone that has the same diameter as the sphere. Zoe says that if the ice cream melts, it will exactly fill the cone. Assume that melted ice cream occupies 75% of the volume of frozen ice cream. What is the ratio of the cone's height to its radius? Is Zoe correct in saying that it is possible for a sphere of radius r to sit perfectly inside a cone of radius r? Do the cone and the sphere meet at the equator of the sphere? Clearly document your thinking and justify your results.

This genre map is part of the department's suite of resources designed to support excellence and equity in teaching writing.

Produced by the Department for Education | JUNE 2023 | Images in this resource are copyright to Shutterstock and their submitters and are used under specific license, no third party copying is permitted.

of others" (NGA Center and CCSSO, 2010b, p. 6), while the *Australian Curriculum* encourages students to "be critical thinkers when justifying their choice of computation strategy" (ACARA, 2022b, p. 13).

It was clear that the genre map had to include a focus on the writing of persuasive texts. Importantly, department leaders felt that the arguments students should write in mathematics differ from those in other subject areas. In a mathematical argument, students should state a position and provide argument points just as they would in arguments in other subjects, but they would also include their own mathematical evidence.

As with mathematical explanation, Lauren and her colleagues explored options for the naming of this genre. We considered calling these texts "problem-solution reports" as students were asked to find solutions to mathematical problems, critique and evaluate them, and make a recommendation. This genre, however, is more suited to the exploration of social, economic, and technological problems (Derewianka & Jones, 2012). Furthermore, we wanted to emphasize the element of argument in the macrogenre as it is through developing strong arguments and using mathematical evidence to support their claims, that students consolidate their reasoning (Casa et al., 2016). When producing written arguments, students should include correct mathematical language, logical conclusions, justification of answers and strategies employed, and the use of mathematical evidence to explain their reasoning (Pugalee, 2015).

Therefore, the new genre map specifies two different types of mathematical argument: personal argument and critique. Casa (2022) likens these to *internal* and *external* arguments. The requirement to not only construct their own arguments but also critique the arguments of others is common to a number of mathematics curricula. Mathematical argumentation is a practice standard in the United States (NGA Center and CCSSO, 2010b) and reasoning with an emphasis on justification is a focus of mathematics teaching in the United Kingdom (UK Department for Education, 2014). The Australian Curriculum similarly emphasizes students justifying their mathematical choices (ACARA, 2022b).

Prompt Students to Write Personal Arguments

The current South Australian mathematics genre map positions students to create personal arguments to communicate their mathematical reasoning. These arguments are more than just descriptions of mathematical strategies or recounts of mathematical procedures employed and have a specific structure. Students take a position by making a claim, which is justified through the use of mathematical evidence, and warrants which connect the evidence to the claim. Students write personal arguments using written language and possibly mathematical symbols and drawings (Casa et al., 2016). The wording of the writing prompts, such as the examples presented in Figure 6, guides students to use this structure as they convince their audience using the evidence they uncovered through their mathematical inquiry.

When developing the writing prompts for argument tasks, many of the examples reference "me" as the audience for student writing. This was infused into the prompts with the intention of helping students see the importance of using more academic mathematical language as they are communicating with a mathematically knowledgeable audience.

Prompt Students to Write Critique Arguments

Students are also prompted to write texts that critique the mathematical arguments of others, such as those seen in Figure 7, in the most current iteration of the genre map. They explore claims made about

Figure 6. These are some examples of personal argument writing prompts in the 2023 version of the mathematics genre map
(South Australian Department for Education, 2023)

Argument (personal)
Investigate the different ways a data set can be displayed. Identify the features of each representation. How easily can you compare the data in each representation? Convince me of which representation is most appropriate for the given data set.

Argument (personal)
Are all equilateral triangles similar? Explain the theory of similar triangles. Create a data table and use a 3D modelling tool, eg GeoGebra (https://www.geogebra.org), to generate evidence to convince me of your position.

Figure 7. These are examples of mathematical critique arguments in the 2023 version of the mathematics genre map
(South Australian Department for Education, 2023)

Argument (critique)
Two students ran a 7km race. Student A ran 5/6 of the race and student B ran 9/12 of the race. Each person claims they ran more than the other. Who do you agree with? Why?

Argument (critique)
Ethan is playing a game with 2 disks. Each disk is red on one side and green on the other. When Ethan flips the disks, he wins if both disks show the same colour when they land. Ethan thinks he will have more chance of winning if he flips 4 disks and even more if he flips 5 disks. Maria disagrees and says he will have less chance of winning the more disks he flips. Who do you agree with and why? Use evidence to support your argument.

the results of mathematical inquiries as suggested by the solutions presented to them, including some of which are incorrect. They further analyze and evaluate the strategies employed by using mathematical evidence to justify whether they support or disprove the claims made. The department anticipates that writing arguments of this type will deepen students' understanding of mathematical concepts as they consider ideas other than their own. Again, the wording of the task examples scaffolds the writing of this genre.

TEACHERS' PROFESSIONAL LEARNING OF THE GENRE MAP

The South Australian Department for Education's challenge was to develop resources that clearly focused on the development of mathematical understanding and were meaningful to mathematics teachers while incorporating approaches that we know work when teaching writing. The department recognized the need for teachers to advance their learning of writing across disciplines, including mathematics, to implement the genre maps effectively. We next highlight some key ways that teachers working in South Australia were supported.

Through the online Literacy Summit, teachers were introduced to the three dimensions of writing, including the cognitive, sociocultural, and linguistic perspectives (Myhill, 2022) as shown in Figure 8. Each dimension is an important part of learning to write mathematically. We anticipated that students who experience each dimension will be able to create increasingly complex texts for a range of audiences.

In addition to online professional learning, the department developed a best advice paper that provides fundamental guidance about how to approach the teaching of writing. It is intended to help develop teachers' understanding of how students construct ideas and convey messages in various disciplines. Teachers are supported to plan the teaching of writing using a teaching and learning cycle (Derewianka & Jones, 2012; Rothery 1994), which is a dialogic framework designed to develop students' knowledge of a field as well as scaffold their independent construction of written texts that convey their understanding. It is based on the principle of "guidance through interaction in the context of shared experience" (Rose & Martin, 2012, p. 58) and the gradual release of responsibility (Pearson & Gallagher, 1983). It supports

Figure 8. This diagram shows the cognitive, sociocultural and linguistic dimensions of writing and how they are related
(South Australian Department for Education, 2022)

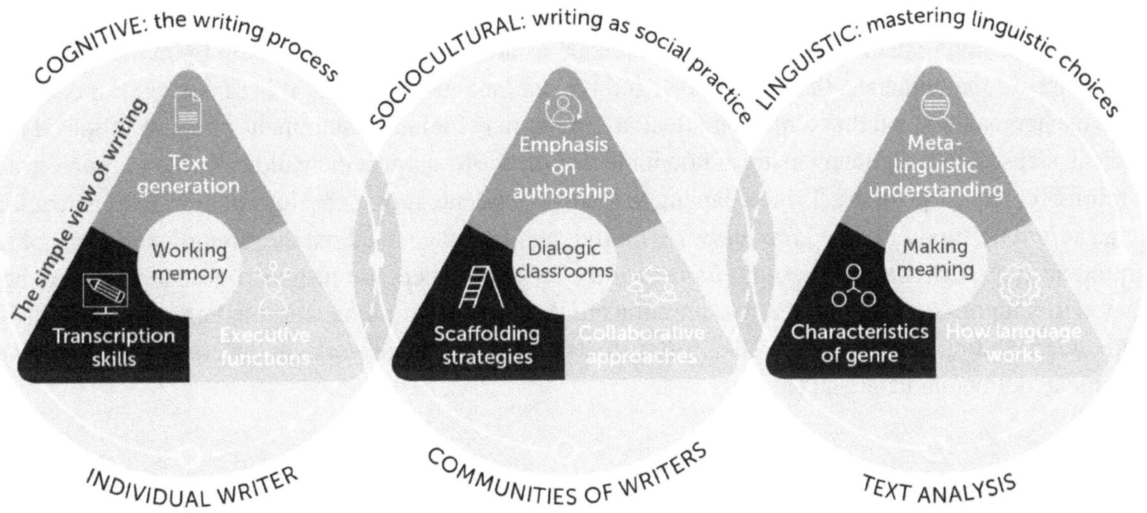

students as they develop an increasing understanding of mathematics concepts and independently write for specific purposes and audiences. The approach is informed by scaffolding and moves students from guided learning to collaborative learning, and then to increasing control of both mathematical content and language in their independent writing.

Additionally, teachers can use the examples in the genre resource paper or they can use the example writing prompts in the genre map to develop model texts that illustrate the key elements of well-written mathematical texts. Deconstructing, analyzing, and discussing these model texts helps students to understand how to structure their own mathematical writing. This understanding helps students to evaluate each other's mathematical writing and provide constructive feedback, as well as reflect on their own writing, identify areas for improvement, and set goals for their mathematical writing.

The teaching and learning cycle provides students with the scaffolding required to produce quality written mathematical explanations and arguments – the genres identified in the genre map as critical for students of mathematics. Styliandes (2019) and Soto-Johnson and Fuller (2012) found that students often create more sophisticated oral arguments than written ones, and the teaching and learning cycle helps students move bridge the gap from oral language to more formal academic written language.

Teachers of mathematics may not have explicitly planned for the teaching of writing in their classes, despite written texts being key forms of assessment in senior high school mathematics subjects. Consequently, the department has developed resources that support mathematics teachers to understand the critical importance of mathematical writing better. By planning the teaching of the key genres in mathematics, students will improve not only their writing but also their understanding of mathematical concepts and mathematical reasoning skills. The resources we developed underwent many iterations as we grappled with the intersection of these two fields.

FUTURE DIRECTIONS

The South Australian Department for Education is currently developing additional resources to help teachers understand that the mathematics genre map and accompanying resources are a guiding framework for teaching mathematical writing. The resources are being designed to help them understand the key genres in mathematics, their structures, and typical language features. It is important, for example, that teachers understand that a mathematical argument may include a statement of claim followed by a logical, step-by-step argument using mathematical evidence to support the validity of their argument and a reinforcement of position. Typical language features students might employ include nominalizations (e.g., *solution*), modality (e.g., *probably, certainly*), and language to adjust the strength and focus of the argument (e.g., *extremely important, increased by 20%*). Teachers are provided guidance to plan how they will support students to clearly communicate their ideas, opinions, and mathematical knowledge and understanding in writing, as well as teach students how language works in different contexts so they can make deliberate language choices.

CONCLUSION

This chapter may stimulate further research and collaboration in the area of integrating writing and mathematics education. Researchers globally may be inspired to investigate and refine the South Australian Department for Education's mathematics genre map and other resources and consider case studies of their application. The resources developed can be adapted and adopted by other education systems promoting international collaboration in the field of mathematical writing. A better understanding of how teachers and students across grades interpret mathematical writing within the context of writing in other disciplines can help advance this work. It is important to consider local contexts, such as if school leaders are working in one school or across them and whether teachers are responsible for teaching writing in other subjects. Importantly, investigations like these will lead to a more comprehensive body of knowledge on the ways teachers can be supported to teach mathematical writing and students can learn this practice.

REFERENCES

Alexander, R. (2008). *Towards dialogic teaching: Rethinking classroom talk* (4th ed.). Dialogos.

Association of Mathematics Teacher Educators in Connecticut. (2022). *Equity in mathematics education: A joint position statement for Connecticut.* Associated Teachers of Mathematics in Connecticut, & Connecticut Council of Leaders of Mathematics. https://portal.ct.gov/-/media/SDE/Board/BoardMaterials120722/Position_Statement_on_Joint_Equity_in_Mathematics_Education.pdf

Australian Curriculum, Assessment and Reporting Authority. (2022a). General Capabilities: Critical and Creative Thinking version 9.0 – About the general capability.

Australian Curriculum, Assessment and Reporting Authority. (2022b). Mathematics F-10 Version 9.0: About the learning area. Retrieved from https://v9.australiancurriculum.edu.au/downloads/learning-areas

Barwell, R. (2018). Writing in mathematics classrooms. In A. L. Bailey, C. A. Maher, & L. C. Wilkinson (Eds.), *Language, literacy and learning in the STEM disciplines: How language counts for English learners* (pp. 101–114). Routledge. doi:10.4324/9781315269610-6

Bransford, J., Darling-Hammond, L., & LePage, P. (2005). Introduction. In L. Darling-Hammond, J. & Bransford (Eds.), Preparing teachers for a changing world: What teachers should learn and be able to do (pp. 35–117). Wiley.

Casa, T., Gilson, C. M., Bruce-Davis, M. N., Gubbins, E. J., Hayden, S. M., & Canavan, E. J. (2022). Maximizing the potential of mathematical writing prompts. *Mathematics Teacher: Learning & Teaching PK-12, 115*(8), 538–550.

Casa, T. M. (2022). *Making things write in maths class.* Presentation to the 2022 South Australian Department for Education's Literacy Summit (online). https://rise.articulate.com/share/L98Q9teAf-10nsBwvtTYurzxKiidxvyfn

Casa, T. M., Choppin, J. M., & Moschkovich, J. N. (2020). *Crafting a research agenda for writing to reason mathematically.* Research Gate. https://www.researchgate.net/publication/343319379_Crafting_a_Research_Agenda_for_Writing_to_Reason_Mathematically

Casa, T. M., Firmender, J. M., Cahill, J., Cardetti, F., Choppin, J. M., Cohen, J., Cole, S., Colonnese, M. W., Copley, J., DiCicco, M., Dieckmann, J., Dorl, J., Gavin, M. K., Hebert, M. A., Karp, K. S., LaBella, E., & Moschkovich, J. N., Moylan, K. Olinghouse, N. G., & Zawodniak, R. (2016). *Types of and purposes for elementary mathematical writing: Task force recommendations.* Elementary Mathematical Writing Task Force. http://mathwriting.education.uconn.edu

Christie, F. (2002). Classroom discourse analysis: A functional perspective. *Continuum.*

ConneC.T. (n.d.). *Connecticut Schools on the Web.* State of Connecticut. https://www.dir.ct.gov/kids/Schools_Libraries/ctschools.asp?ctportalPNavCtr=|27298|#27301

Department for Education South Australia. (2022a). *Episode 4: Learning to write, writing to learn.* Department of Education South Australia. https://www.education.sa.gov.au/schools-and-educators/curriculum-and-teaching/teach-podcast/podcasts/episode-4-learning-to-write-writing-to-learn

Derewianka, B., & Jones, P. (2016). *Teaching language in context* (2nd ed.). Oxford University Press.

Duke, N. K. (2000). For the rich it's richer: Print experiences and environments offered to children in very low- and very high-socioeconomic first-grade classrooms. *American Educational Research Journal, 37*(2), 441–478. doi:10.3102/00028312037002441

Evidence for Learning. (2020). *Improving mathematics in upper primary and lower secondary.* Evidence for Learning.

Firmender, J. M., Casa, T. M., & Colonnese, M. W. (2017). Write on: Reasoning through mathematical writing. *Teaching Children Mathematics, 24*(2), 84–92. doi:10.5951/teacchilmath.24.2.0084

Gebhard, M. (2019). *Teaching and researching ELLs' disciplinary literacies: Systemic functional linguistics in action in the context of U.S. school reform.* Routledge. doi:10.4324/9781315108391

Graham, S. (2019). Changing how writing is taught. *Review of Research in Education, 43*(1), 277–303. doi:10.3102/0091732X18821125

Halliday, M. A. K. (1985). *An introduction to functional grammar.* Hodder Arnold.

Humphrey, S., & Macnaught, L. (2016). Developing teachers' professional knowledge for discipline literacy instruction. In H. de Silva Joyce (Ed.), *Language at work in social contexts: Analysing language use in work, educational, medical and museum contexts* (pp. 68–87). Cambridge Scholars Publishing.

Martin, J. R. (1995). Text and clause: Fractal resonance. *Text & Talk, 15*(1), 5–42.

Martin, J. R., & Rose, D. (2008). *Genre relations: Mapping culture.* Equinox.

Ministry of Education. (2020). *The Ontario curriculum, grades 1–8: Mathematics.* King's Printer for Ontario. https://assets-us-01.kc-usercontent.com/fbd574c4-da36-0066-a0c5-849ffb2de96e/90439c6e-f40c-4b58-840c-557ed88a9345/The%20Ontario%20Curriculum%20Grades%201%E2%80%938%20-%20Mathematics,%202020%20(January%202021).pdf

Moschkovich, J. N. (2007). Examining mathematical discourse practices. *For the Learning of Mathematics, 27*(1), 24–30.

Namkung, J. M., Hebert, M., Powell, S. R., Hoins, M., Bricko, N., & Torchia, M. (2020). Comparing and validating four methods for scoring mathematics writing. *Reading & Writing Quarterly, 36*(2), 157–175. doi:10.1080/10573569.2019.1700858

National Council of Teachers of Mathematics. (1991). *Professional standards for teaching mathematics.* NCTM.

National Council of Teachers of Mathematics. (2000). *Principles and standards for school mathematics.* NCTM.

National Governors Association Center for Best Practices and the Council of Chief State School Officers. (2010a). *Common core state standards for English language arts & literacy in history/social studies, science, and technical subjects.* CCSSO. https://learning.ccsso.org/wp-content/uploads/2022/11/ELA_Standards1.pdf

National Governors Association Center for Best Practices and the Council of Chief State School Officers. (2010b). *Common core state standards for mathematics.* CCSSO. https://learning.ccsso.org/wp-content/uploads/2022/11/ADA-Compliant-Math-Standards.pdf

O'Halloran, K. (2005). Mathematical discourse: Language, symbolism and visual images. *Continuum.*

Our Statewide School District. (n.d.). Hawai'i Department of Education. https://www.teachinhawaii.org/district-profile/#:~:text=On%20the%20island%20of%20O,ll%20find%20Kaua'i%20District

Paxton, M. (2007). Tensions between textbook pedagogy and the literacy practices of the disciplinary community: A study of writing in first year economics. *Journal of English for Academic Purposes, 6*(2), 109–125. doi:10.1016/j.jeap.2007.04.003

Pearson, P., & Gallagher, M. (1983). The instruction of reading comprehension. *Contemporary Educational Psychology, 8*(3), 317–344. doi:10.1016/0361-476X(83)90019-X

Perry, B., & Dockett, S. (2002). Young children's access to powerful mathematical ideas. In L. D. English (Ed.), *Handbook of international research in mathematics education: Directions for the 21st century* (pp. 81–111). Lawrence Erlbaum.

Powell, S. R., Hebert, M. A., Cohen, J. A., Casa, T. M., & Firmender, J. M. (2017). A synthesis of mathematics writing: Assessments, interventions, and surveys. *Journal of Writing Research*, *8*(3), 493–526. doi:10.17239/jowr-2017.08.03.04

Pugalee, D. (2005). *Writing to develop mathematical understanding*. Christopher-Gordon.

Rose, D., & Martin, J. R. (2012). Knowledge about language. In D. Rose & J. R. Martin (Eds.), *Learning to write, reading to learn: Genre, knowledge and pedagogy in the Sydney school* (pp. 206–264). Equinox.

Rossbridge, J., & Rushton, K. (2014). The critical conversation about text: Joint construction. *Primary English Teachers Association Australia Paper 196*. PETAA. https://petaa.edu.au/w/w/Store/Item_Detail.aspx?iProductCode=PP196&Category=PEN

Rothery, J. (1994). *Write it right: Exploring literacy in school English*. Disadvantaged Schools Program, Metropolitan East Region, NSW Department of School Education.

Rothery, J. (1996). Making changes: Developing an educational linguistics. In R. Hasan & G. Williams (Eds.), *Literacy in society*. Addison Wesley Longman.

SACE Board of South Australia. (2015). *Mathematical methods subject outline*. SACE. https://www.sace.sa.edu.au/web/mathematical-methods/stage-2/planning-to-teach/subject-outline

Shanahan, T., & Shanahan, C. (2008). Teaching disciplinary literacy to adolescents: Rethinking content area literacy. *Harvard Educational Review*, *78*(1), 40–59. doi:10.17763/haer.78.1.v62444321p602101

Shanahan, T., & Shanahan, C. (2012). What is disciplinary literacy and why does it matter? *Topics in Language Disorders*, *32*(1), 7–18. doi:10.1097/TLD.0b013e318244557a

Snow, C. E. (1987). The development of definitional skill. *Journal of Child Language*, *17*(3), 697–710. doi:10.1017/S0305000900010953 PMID:2269704

Soto-Johnson, H., & Fuller, E. (2012). Assessing proofs via oral interviews. *Investigations in Mathematics Learning*, *4*(3), 1–14. doi:10.1080/24727466.2012.11790313

Stylianides, A. J. (2019). Secondary students' proof constructions in mathematics: The role of written versus oral mode of argument representation. *Review of Education*, *7*(1), 156–182. doi:10.1002/rev3.3157

Thomas, D. P. (2019). Rapid decline and gender disparities in the NAPLAN writing data. *Australian Educational Researcher*, *47*(5), 777–796. doi:10.100713384-019-00366-8

United Kingdom Department for Education. (2014). *Mathematics programmes of study: Key stages 1 and 2: National curriculum in England*. UKDoE. https://www.gov.uk/government/publications/nationalcurriculum-in-england-mathematics-programmes-of-studyn

ADDITIONAL READING

Campbell, T. G., King, S., & Zelkowski, J. (2021). Comparing middle grade students' oral and written arguments. *Research in Mathematics Education, 23*(1), 21–38. doi:10.1080/14794802.2020.1722960

Humphrey, S., & Macnaught, L. (2016). Developing teachers' professional knowledge for discipline literacy instruction. In H. de Silva Joyce (Ed.), *Language at work in social contexts: Analysing language use in work, educational, medical and museum contexts* (pp. 68–87). Cambridge Scholars Publishing.

Kosko, K. W., & Zimmerman, B. S. (2019). Emergence of argument in children's mathematical writing. *Journal of Early Childhood Literacy, 19*(1), 82–106. doi:10.1177/1468798417712065

Moschkovich, J. (2012). Mathematics, the Common Core, and language: Recommendations for mathematics instruction for ELs aligned with the Common Core. In L. R. Van Zoest, J.-J. Lo, & J. L. Kratky (Eds.). *Proceedings of the 34th annual meeting of the North American Chapter of the International Group for the Psychology of Mathematics Education.* Western Michigan University. https://files.eric.ed.gov/fulltext/ED584919.pdf

Pallanck, J. L., Castro, G. O., Colonnese, M. W. & Casa, T. M. (2020). Improving written mathematical arguments. *Mathematics Teacher: Learning and Teaching PK-12, 113*(11), 910–917.

University of Exeter. (2023). *Writing resources for teachers.* University of Exeter. https://education.exeter.ac.uk/research/centres/languageandliteracy/grammar-teacher-resources/

Urquhart, V. (2009). *Using writing in mathematics to deepen student learning.* McREL International.

KEY TERMS AND DEFINITIONS

Genre: A term used to classify texts with similar purposes, structures, and language features.

Genre map: A graphic representation of genres of writing that evolve across grade levels.

Macrogenre: A text that arises from student inquiry and incorporates more than one genre.

Metacognitive: An awareness and understanding of one's own thinking.

Metalanguage: The vocabulary used to discuss language conventions and use (e.g., grammatical terms such as *sentence, clause, conjunction*).

Scaffolding: Gradually removed support provided to students to develop their learning and independence.

Teaching and Learning Cycle: A framework to support teachers in designing teaching and learning that develops both students' knowledge and their language and literacy skills.

Section 3

Engaging Students in Mathematical Writing

Chapter 9
Supporting Equity Through Mathematical Writing

Christie Sullivan Martin
https://orcid.org/0000-0001-7896-6882
University of South Carolina, USA

Drew Polly
https://orcid.org/0000-0003-2370-4409
University of North Carolina at Charlotte, USA

ABSTRACT

This chapter establishes the relationship between the National Council of Teachers of Mathematics NCTM principles to action, Gutiérrez equity framework, and mathematical writing. The chapter discusses the research that supports mathematical writing as an instructional practice that enhances opportunities for reasoning learning and equity. The chapter provides four cases of activities that embed mathematical writing in the classroom. Each case describes implementation, examples of students' work, discussion of their work, extensions, and outlines the activities connection to the equity framework. The cases are concluded with takeaways that reinforce the purpose of each activity and strategies for implementation. The chapter closes with implications to further support the use of mathematical writing in the classroom and future research direction.

BACKGROUND

The National Council of Teachers of Mathematics' ([NCTM]; 2014) *Principles to Action* shares eight empirically-based mathematics teaching practices that advance the use of effective and equity-based learning opportunities in mathematics. The eight practices highlight effective mathematics instruction that promote reasoning, problem solving, using, and understanding various representations, making connections, engaging in meaningful discourse, and demonstrating evidence of student thinking. In essence, these principles weave together what it means to be mathematically literate. Steen and colleagues (2007) define mathematical literacy as "the capacity to make effective use of mathematical knowledge and un-

DOI: 10.4018/978-1-6684-6538-7.ch009

derstanding in meeting challenges in everyday life" (p. 285). Writing in mathematics has the potential to build understanding (Graham et al., 2020; Pugalee, 2004), increase metacognition (Bangert-Drowns et al., 2004), increase equity (Bieda & Staples, 2020), and support students in becoming mathematically literate. Students' writing in mathematics can serve as a formative assessment tool that allows teachers to be intentional with their instruction.

Furthermore, infusing writing opportunities in the mathematics classroom provides pathways for learners to engage in equity-based mathematics where they have agency and autonomy to select strategies to solve mathematics tasks and then communicate about their reasoning and mathematical ideas (Gutierrez, 2009; 2012; Martin & Polly, 2019). Additionally, allowing students to write about mathematics concepts and their mathematical ideas provides opportunities for students to access and engage with mathematics concepts in multiple ways that may help develop a deeper understanding of concepts and strategies (Hattie et al., 2016). Learners develop a deeper understanding of concepts when they are able to access content in multiple ways (Hattie, 2009; Hattie et al., 2016).

This chapter examines writing in mathematics, using writing as a tool to learn, and writing to increase equity in the classroom. Once these areas are discussed, the authors present examples of how writing can be used in mathematics classrooms to discuss how these examples relate to learning, equity, and formative assessment.

FRAMING EQUITY-BASED MATHEMATICS TEACHING

As we consider how mathematical writing advances both access and equity for students in mathematics, it is beneficial to frame what we mean by access and equity. Gutiérrez (2009; 2012) described four elements of equity-based mathematics instruction (Table 1). The dominant axis consists of access and achievement. It refers to what is needed to be successful in the current environment, Gutiérrez describes the dominant access as "how well students play the game called mathematics." The critical axis includes identity and power, this is where students' resources are highlighted to allow them to "change the game." Table 1 describes these elements in general, it is important to point out that in some of her writing Gutiérrez (2012) has advocated that power includes two aspects student power and agency in the mathematics classroom as well as broader systems-level view of power related to who has power related to educational and societal systems.

Table 1. Framework for equity in mathematics (Adapted from Gutiérrez, 2012)

Dominant Axis	
Access	Students have resources to support mathematics learning, such as high-quality teachers, adequate technology and resources in the classroom, a curriculum aligned to standards, and a classroom environment that invites participation from all learners.
Achievement	Students are active participants in mathematics classes; their test scores and their participation in the math pipeline of mathematics courses provide opportunities to lead to careers in mathematics.
Critical Axis	
Identity	Students view themselves as mathematicians and can see themselves in their curriculum and mathematics that they do. Additionally, the extent to which mathematics provides students with a broader view of the world.
Power	Students have a voice and agency in the classroom.

Since little has been empirically shown about the extent to which writing in mathematics aligns to equity-based mathematics teaching, we aim to use this chapter as a vehicle to show how various writing activities in mathematics classrooms can address the various dimensions of Gutiérrez' (2009; 2012) framework.

SYNTHESIS OF THE LITERATURE

In a large-scale meta-analysis, Graham and colleagues (2020) found that activities that they termed as *writing to learn* led to gains in student achievement. They noted that moderating effects such as writing capabilities, goals, beliefs, knowledge of content, were not able to be examined based on the methods sections of the studies included. They concluded their paper by recommending that writing activities should be used for different purposes and that teachers routinely revisit student work and the success of these activities to ensure that the goals and purposes of writing across content areas are being met. Namkung and colleagues (2020) emphasize that mathematical writing is different from other kinds of writing. Mathematics content knowledge is required in addition to an understanding of how mathematicians convey information.

Writing supports learning and creates an opportunity for students to reflect on their own understanding (Hillocks, 2005). A fundamental part of the writing process is preplanning and monitoring. Brown and Palincsar (1982) suggest that becoming proficient in problem solving requires these skills. Writing is a non-linear process that reinforces revision, rethinking, and moving toward precision in language. Integrating writing in mathematics is an effective tool for developing as a problem solver as it pushes the writer to organize their thoughts and convey them in a coherent manner (Pugalee, 2004). Marilyn Burns states "I can no longer imagine teaching math without making writing an integral aspect of students' learning" (p. 30). There are many ways to infuse writing into the mathematics classroom that address several objectives such as being able to communicate mathematical ideas, develop a deeper understanding of mathematics, and apply problem solving skills.

That meta-analysis supports the ideas that were advanced with the Common Core State Standards for Mathematics (NGA/CCSSO, 2010) that recommended the use of specific Standards for Mathematical Practices (MPs). MP's emphasize behaviors that connect to conceptual understanding (Dacey & Polly, 2012). Writing is a method for learning mathematics and conveys evidence of conceptual learning. NCTM (2000) suggested that teachers create opportunities for students to communicate their mathematical understanding coherently. This practice is reiterated in the third MP that asks students to construct viable arguments.

There are several ways to integrate writing into the mathematics classroom. Colonnese (2020) provided five empirically based instructional guidelines for helping teachers to do so. In this section, we provide a glimpse of research aligned to those guidelines. A mathematics journal is a tool that has been used in the classroom to support writing-to-learn activities. Clarke, Waywood, and Stephens (1993) described students' journal writing in three ways that increase in complexity, beginning with recount, then summary, and finally dialogue. In their experiences with journal writing in mathematics, they had students that would list what occurred in class without connection to content or between content. As students grew in their journal space, they moved to summarizing content. These entries were more developed and relevant to the content. Lastly, students that moved into the phase of the journal that the researchers categorized as dialogue, they would add to the summary by making connections and critically asking questions related to the content.

Kosko and Zimmerman (2019) examined students' writing in kindergarten through third grade. The focus of the study was to explore children's mathematical argumentative writing. Their study also generated a classification scheme to better understand how writing develops in the early grades. Their classification advances in complexity as well, beginning with emergent-mathematical, statement, re-count, procedural, detailing, description, and explanation. The sequence of classifications according to their data shows a correlation with grade level, suggesting that the sequence may be developmental. The researchers conclude that young students can engage in mathematical argumentative writing and having a classification scheme provides insight into their development. Additionally, Goldsby and Cozza (2002) used a journal with eighth-grade students where multiple strategies and explanations were generated and shared with classmates. The students' writing and discussion supported learning.

Writing in mathematics serves several purposes. Students engage in mathematical writing to make sense of content (Ediger, 2006), to develop metacognition (Bangert-Drowns et al., 2004), as a form of assessment (Ntenza, 2006), and to engage in the work of mathematicians (Kline & Ishii, 2008). These experiences are valuable for students to develop as mathematicians and be able to construct viable arguments (Standard for Mathematical Practice 3) while simultaneously attending to precision (Standard for Mathematical Practice 6). There are differences in what it means to use writing as a tool for learning, reflecting on ideas, and increasing understanding versus learning to write well-crafted mathematical text that is aligned with the discipline. There are academic rules that are specific to each discipline and these need to be presented to students (T. Shanahan & C. Shanahan, 2018; T. Shanahan & C. Shanahan, 2008).

Writing as a tool for learning has been part of educational research for decades (Bangert-Drowns et al., 2004; Ediger, 2006;; Liedtke & Sales, 2001; Ntenza, 2006; Pugalee, 2001). Writing is constructed for multiple purposes and serves the writer in different ways. The Elementary Mathematical Writing Task Force (Casa et al., 2016) recommends four types of writing be implemented in the mathematics classroom, including exploratory, informative/explanatory, argumentative, and mathematically creative, which address recommendations put forth by the National Council of Teachers of Mathematics ([NCTM], 2014)the National Research Council (2001), and Common Core State Mathematics Standards (National Governors Association Center for Best Practices & Council of Chief State School Officers, 2010).

EXAMPLES ABOUT HOW WRITING PROMOTES EQUITY-BASED MATHEMATICS TEACHING

The literature discussed in this chapter underscores the research that writing to learn supports student achievement (Graham et.al, 2020), allows students to identify and express their own understanding (Hillocks, 2005), and creating multiple opportunities and purposes for writing increases access to mathematics for students (Colonnese, 2020). The examples presented are directly connected to dimensions of Access, Identity, Achievement, and Power of Gutiérrez' (2009, 2012) framework (see Table 1) that she developed and refined to promote equity-based mathematics teaching. In this section we see examples that include: a) Writers' Workshop in mathematics, b) writing while exploring cognitively-demanding mathematics tasks, and c) brainstorming and graphic organizers.

Writer's Workshop in Mathematics

Description

Martin (2015) adapted the well-known Writer's Workshop (Calkins, 1994) for use in mathematics classrooms. The model has components, such as drafting, revising, and editing that are intended to reflect the writing process, which is not linear and should be adjusted and revisited as needed. While teachers typically implement Writer's Workshop daily, Martin implemented it three times a week. The goal is for learners to have consistent experiences so that the processes become familiar to learners. The Writer's Workshop also includes multiple aspects of Gutiérrez' (2009; 2012) dimensions of equity-based mathematics teaching (see Table 2).

Table 2. Alignment of examples, writer's workshop in mathematics, and Gutiérrez' (2012) equity framework

Dimension	Description of Alignment
Access	Students engaged in various writing activities in a small group setting that included exploring tasks, writing about connections between concepts (e.g., multiplication and divisions), and reflecting on what they learned. Student work was scaffolded by the teacher in small group settings, and by the use of mathematics tools, such as manipulatives, graph paper, etc.
Achievement	The teacher used the Writer's Workshop framework to provide additional learning experiences for foundational mathematics concepts for that grade level.
Identity	Students met in small groups to collaboratively work on various tasks with support. Teachers posed tasks that have a real-life context to more deeply address students' identity and background in the mathematics problems that they are exploring.
Power	Students had power and agency as they solved the multiplication task described above. Students also had power to share what they wanted on the reflection.

Examples of Student Work

The examples provided in Figures 1-3 come from mathematics workshop sessions with third graders focused on multiplication and division. A 10- to 15-minute mini-lesson preceded the students working independently on a task. Each small group had access to their workshop notebook or journal, manipulatives, and whiteboards at their tables. The teacher presented students with writing prompts as they engage in solving mathematical tasks. The students have solved a division problem, "I have 15 pieces of Halloween candy, and I would like to share those pieces equally between myself and two friends" and are asked to write how they thought about solving this problem (informative writing).

Writing to Communicate about Problem Solving Strategies. Figure 1 shows that the student showed a few different computations and summarizes their thinking with a sentence that counting by 5's was going to get me to 15, then counted by three and got 5. The response shows several mathematical connections the student is making that can be built upon as the work of multiplication and division continues. The student writes the division problem $15 \div 3 = 5$ pieces of candy and reports the unit of the quotient. This is important as including the unit is moving toward a level of precision, this will also be important when quotients that include a remainder need to be interpreted. The student includes a list of multiples of five (5, 10, 15). This is interesting since the divisor in the problem written is 3, but the student uses

Figure 1. Writing to explain mathematics concepts

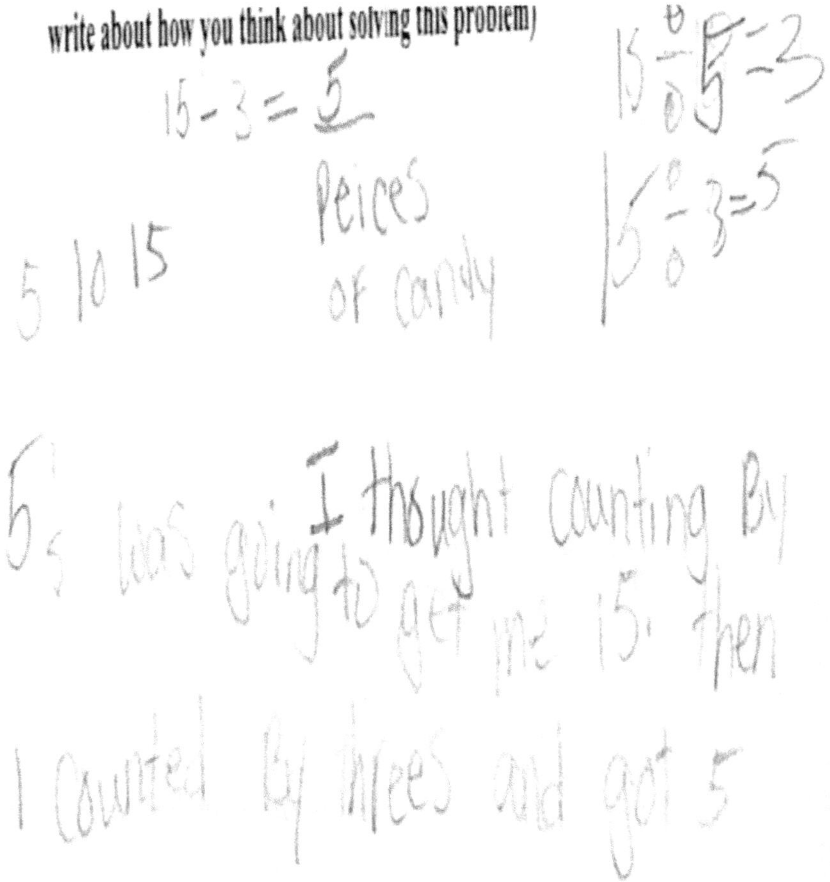

the dividend of 15 and their knowledge of multiples of 5 to solve. The student includes $15 \div 3 = 5$ and $15 \div 5 = 3$ in their work indicating an understanding of fact families. The written expression of thinking and strategy is helpful as we work with students to consider if a strategy will always work, whether there are other ways to solve it, and which method is most efficient.

"I thought counting by 5's was going to get me 15. Then I counted by threes and got 5"

In Figure 2 the third-grade students were prompted to explain the relationship between division and multiplication after completing the task about Halloween candy. The students' work seen in Figure 2, like the previous student whose work is in Figure 1, seems to be using their knowledge of number combinations and fact families to describe the relationship between multiplication and division. The student states that "division begins with the product" followed by the idea that "division ends in small numbers" and "multiplication ends in bigger numbers." The student's statement in this activity highlights what Karp, Bush, and Dougherty (2014) discuss related to "rules" that hold true at a time but soon expire causing confusion. The Achievement dimension of Gutierrez' (2009; 2012) framework advocates that learners

Figure 2. Writing to reflect on their learning

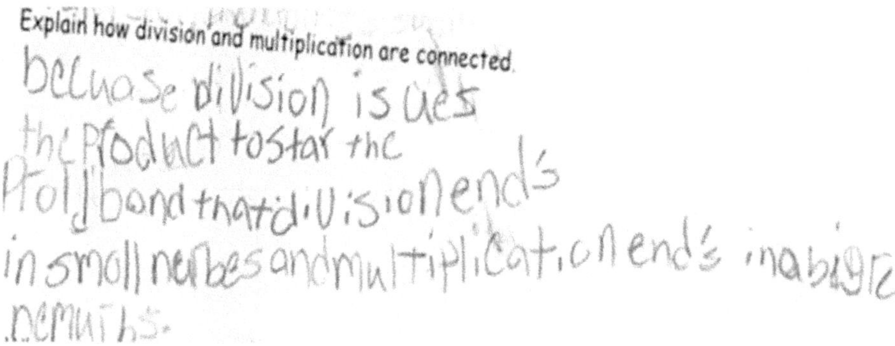

are given access to content that promotes student achievement. Rules that expire like the faulty idea that multiplication makes numbers bigger may help learners in the short-term but will impede achievement and learning with future mathematics concepts. The student's writing in this activity indicates the teacher will need to modify their own discussion of these concepts and attend to precise language. Students' writing here indicates the need for teachers to provide additional support.

"Because division uses the product to start the problem that division ends in small numbers and multiplication ends in bigger numbers."

As part of the Writer's Workshop model, learners reflect on their learning. Figure 3 shows how writing was used to support students' reflection to close the lesson. At the end of the workshop, students reflect on what they learned in the lesson. This offers students a chance to review their work, think through everything that was discussed in the lesson, and clearly articulate something that they learned. This aligns to the Power dimension of Gutierrez' (2009; 2012) model since it provides learners with personal ownership of their own learning.

In Figure 3, the student shows they understand that a number may divide without a remainder or with a remainder. The student takes the additional step of defining what remainder means. The writing is a clear statement of what the student learned. In the mathematics writing workshop model, this reflection time prioritizes students thinking deeply about their work and the embedded mathematics concepts and sharing them with their teacher and classmates (Hattie et al., 2016)

"That some problems might have remainders and some don't and I know what remainder means it means leftover"

Takeaways From Mathematics Writer's Workshop

Infusing aspects of the Mathematics Writer's Workshop in mathematics classrooms provides a structure that many students will be familiar with from their English language arts class, which embeds writing for multiple purposes. Specifically, in the examples discussed above, students wrote during their workshop

Figure 3. Example of a Closing Reflection Prompt

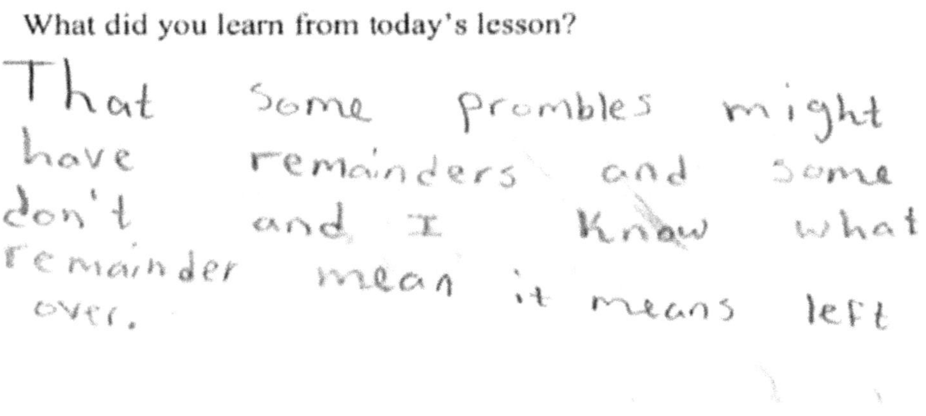
What did you learn from today's lesson?

That some prombles might have remainders and some don't and I know what remainder mean it means left over.

to communicate their problem-solving strategies, to explain mathematics concepts and to reflect on their learning. In the examples above, writing provided equity-based opportunities as seen in Table 2. These writing samples also allowed teachers to get a sense of students' mathematical thinking and provided data for formative assessment.

Writing While Exploring a Cognitively-Demanding Mathematical Task

Description. While exploring cognitively-demanding mathematical tasks teachers can provide opportunities for students to write about their strategies and mathematical ideas. These types of activities align to the Power dimension of Gutierrez' (2009; 2012) framework. While exploring cognitively-demanding tasks students should have autonomy to select their own use of mathematics tools and strategies to help them solve the tasks. In this section, we describe an example from a first-grade classroom. The lesson started when the teacher posed the following task to her first-grade students with an argumentative writing prompt:

Look at the following equations. Determine which is true.

9=14–4 11–5=10–4 15–8=8

Explain how you know the equation is true.

Use pictures, numbers, and/or words to describe your thinking.

Students were given access to unifix cubes, a ten-frame template, paper, and pencil to complete the task. The teacher read the task aloud and then asked two students to revoice what the task was asking them to do. Next, she encouraged them to use their materials to work on their task collaboratively for the next 10 minutes. While students were able to collaborate, each student had to record their own work on their own paper. The student work collected from students indicated that nearly all students explained their thinking in writing as well as with pictures and number sentences. Table 3 provides an alignment of this example to Gutierrez' (2009; 2012) dimensions of equity-based teaching.

Table 3. Alignment of first grade task to Gutiérrez' (2012) equity-based teaching framework

Dimension	Description of Alignment
Access	Students were given a task aligned to grade-level content. Students were provided with access to unifix cubes, a ten frame template, and paper and pencil. Students were asked to revoice the expectations of the task.
Achievement	The task given to students focused on developing fluency, and the meaning of the equal sign. These concepts are critical foundations for future mathematics.
Identity	The task was an open-middle task with one correct answer but multiple entry points or ways to start the task. As a result students could feel successful with at least one of the three tasks. 15-8 = 8 was the first one that most groups explored. There was no real-life context so there was no opportunity for students to see this relate to mathematics in their community or to identify with the context in the task.
Power	Students had power in the classroom as they were able to choose the mathematics tools as well as which of the three parts to explore first.

Figure 4 shows three samples of student work collected as well as what each student said when they were asked to orally explain what they wrote. To represent a range of learners we have provided a range for each level based on the rubric (Toolsforncteachers.com, n.d.). It is worth noting that when this task was posed to students, the teacher encouraged them to use words to explain their answer. This task was posed in January of the school year, and students were working on writing complete sentences in literacy. As evident in all three student work samples, complete thoughts were written down that provided evidence of students' mathematical thinking.

Example Labeled as Developing. The student work labeled as Developing includes an incorrect answer. However, the picture provided shows that the student was able to represent numbers with cubes and thought about crossing out part of the group of 15. The misconception appears to be that the student is not yet able to realize that 15 minus 8 means starting at 15 and removing 8 leaves only 7 remaining. It appears that the list on the right should have started with 9 and since 8 is listed on both the cubes on the left and the list on the right that led to an incorrect answer. The written work with the follow-up oral explanation gives us insight into the students' mathematical thinking. The student was asked in a subsequent activity to solve 15 - 8 = __ using cubes. They were able to determine the correct answer and realized that that equation was incorrect.

Example Labeled as Progressing. The student work and follow-up oral explanation that aligns to the *Progressing* level on the rubric indicates that the student was able to identify the correct equation, but their answer did not reflect much understanding of the equal sign or subtraction. As a result, the teacher posed follow-up tasks that asked the student to identify the correct equation and then explain their actions with unifix cubes. It is worth noting that this student routinely wanted to do math mentally and was reluctant to record their work on paper. While providing students with autonomy aligns to the Power dimension of Gutierrez' (2009; 2012) equity-based framework there sometimes is a need for teachers to be a bit more structured in their request or suggestion for a specific strategy. In this example, the student said that their answer came from their head. The teacher could now respond by asking for them to show their work or by suggesting a specific strategy. If the teacher prompts them to show their work or defend their answer on paper it is a more open question that preserves student autonomy, while their request for a specific strategy, such as a number line or drawings, may inhibit student autonomy.

Example Labeled as Meets Expectations. The student work and oral explanation that aligned to the level of Meets Expectations included the correct answer as well as a written explanation that reflected a clear understanding of what the equal sign means. Additionally, they were able to share their math-

Figure 4. Student work samples from first grade task

Developing:

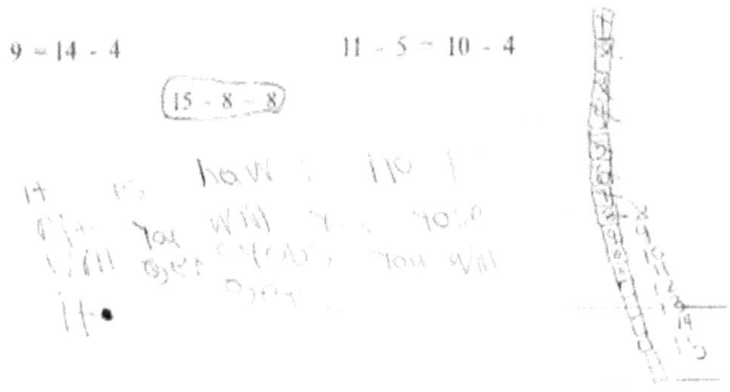

Oral explanation: "It is how I know is right. You will use cubes you will get it."

Progressing:

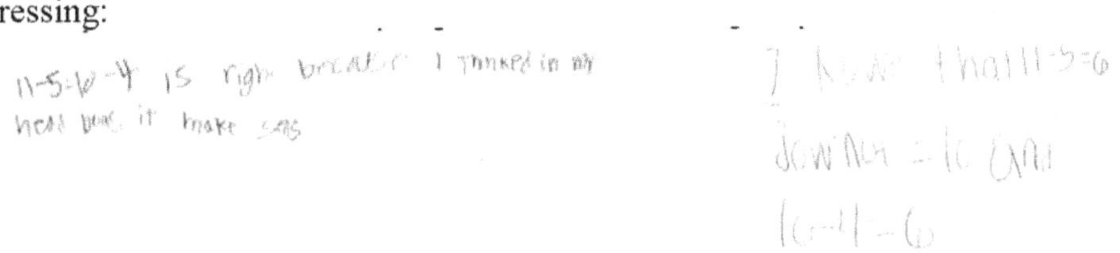

Oral explanation: "11 minus 5 is six minus 4 is right because I think in my head and it makes sense. I know 11 minus 5 is 6 and 10 minus 4 is 6 too."

Meets Expectations:

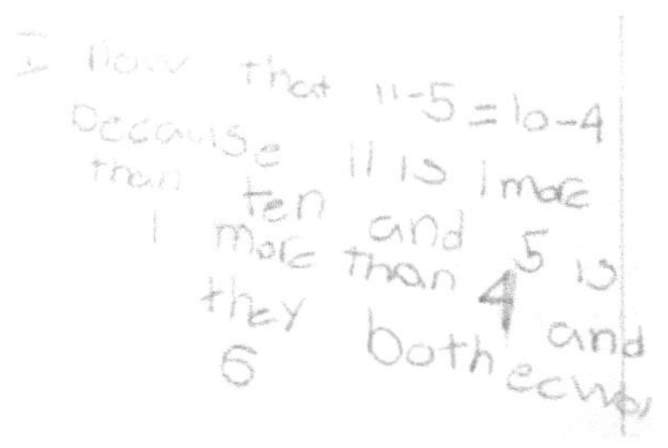

"I know 11 minus 5 is the same as 10 minus 4 because 11 is 1 more than 10, and 5 is 1 more than 4 and they both equal 6.

ematical idea that the left side of the number sentence (11 - 5) has two numbers that are each 1 larger than the numbers on the right side of the number sentence (10 - 4).

Takeaways from First Grade Task. All learners can benefit from opportunities to write about their mathematical ideas. In this task first grade students explored a task focused on the meaning of the equal sign and subtraction fluency, and then provided written and oral explanations. The way that the task was posed enacted equity-based mathematics teaching practices, especially the Power dimension of Gutiérrez (2012) Equity Framework (see Table 3). Additionally, the teachers' strong encouragement to show their work using sentences and then asking them to orally explain their thinking provided the teacher with insight into students' mathematical thinking.

Brainstorming and Graphic Organizers

Description. Empowering students with their own prior knowledge supports deeper understanding and connection. In this section, we share a few examples of how providing students with time to write and reflect about their own knowledge and using tools such as a graphic organizer builds a foundation for teachers to develop instruction. Rodriguez (2013) highlights that the theory of Funds of Knowledge (FoK) centers the focus on the knowledge students bring with them into the classroom from lived experiences and that this knowledge is valuable for learning. It moves to an asset view rather than a deficit view. Tapping into prior knowledge is essential when beginning any new concept (van De Walle, et al., 2019).

In this activity, fifth graders were asked to create a graphic organizer with the word "decimal" at the center and branches extending to record their prior knowledge of them, then write to explain their organizer. (For more information, see Martin and Polly [2019].) This activity allowed students to sit, reflect on their knowledge, and grapple with their own understanding. Table 4 describes how the use of graphic organizers supports Gutierrez' (2009; 2012) equity framework.

Examples of Student Work. In Figure 5, Student A shares their ideas, and it offers the teacher insight into how deeply the student grasps decimals. The student shares on their graphic organizer "less than 1" and "not a whole number." This indicates the student firmly grasps that decimals are less than one. In their first paragraph, they elaborate on this understanding and try to illustrate tenths, hundredths, and thousandths with their .9, 9, .900 examples. However, this sentence uses the terms "ten," "hundred," and "thousand." This raises a point of clarification for the teacher and opens discussion to the importance of precision of language. In the second paragraph, the student seems to be accessing prior knowledge that appears procedural, and the explanation is modest. The student states "all you do is when you have a dividing and a decimal you move the decimal to the left to make the number smaller and do the opposite for the multiplication." This statement could reveal limited understanding, or it could be referring to a strategy in which the student solves and uses estimation to place the decimal.

The teacher can probe the student for clarification. The graphic organizer can be revisited and revised. In the same way, Clarke, Waywood, and Stephens (1993) highlighted that students' journal entries advances from recount, to summary, and then to dialogue. This activity can also evolve in difficulty. It begins as a writing activity that reflects on prior knowledge, but as the class builds on understanding, this activity can be used to support revision.

Table 4. Alignment of the graphic organizers decimals example to Gutiérrez' (2012) equity framework

Dimension	Description of Alignment
Access	Students engaged in brainstorming and the creation of graphic organizers for concepts that aligned to grade-level content.
Achievement	The teacher chose the concept of decimals and operations with decimals that is foundational for future mathematics. While there were not a lot of scaffolds and supports, this was an introduction to the concept activity to give the teachers an understanding of what students already knew.
Identity	While there was no real-world context in this students' example, the open-ended activity helped students identify ideas about decimals from their previous experiences. In other students' responses discussion of where and when they interact with decimals was included.
Power	Students had power and agency about what they wrote as well as how they showed their current understanding of the concept of decimals.

Figure 5. Graphic organizer example

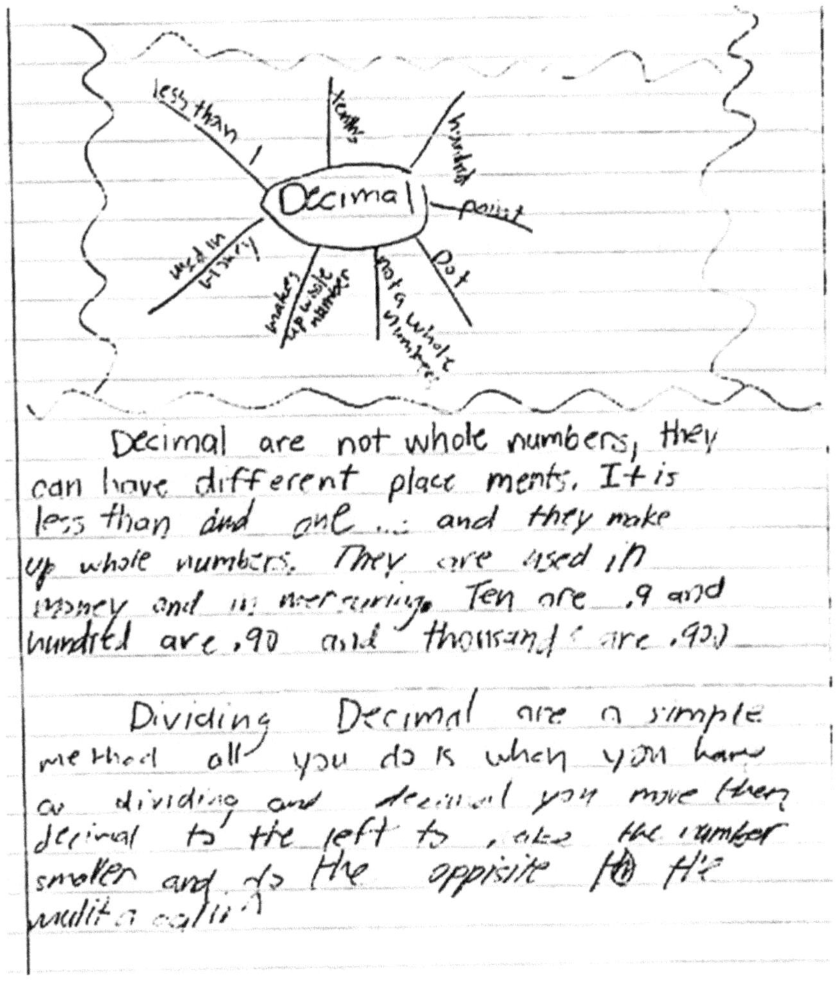

Takeaways From Graphic Organizers

This activity uses a graphic organizer as a tool to access prior knowledge of decimals. Starting with this activity is beneficial in several ways. Students receive a message that their prior knowledge and connection to the content is valuable and important. It encourages them to access their own ideas and work to express them in writing. Their written expression of their understanding is a formative assessment for the teacher. In the example shared, the teacher would have notes of what the student presents a solid understanding of and notes about further conversation and clarification that should follow. As noted, the activity should be a tool that is revisited and extended as greater understanding is built. This activity also opens the opportunity to build understanding by sharing their thinking in a whole group and small groups. The brainstorming activity began with an idea to reflect upon, in this next activity there is a task designed with instruction through a mathematics writing workshop.

Writing to Evaluate Mathematical Strategies

Description

Another way to integrate writing into the mathematics classroom is to provide students with strategies or examples of student work and have them reason about their accuracy and sense making. This example comes from a fifth-grade classroom, in which 22 of the 28 students were working on foundational concepts in multiplication and division from previous grades.

Nick and Tomika were looking at different strategies to multiply 37 and 26. Here are some ideas that they came up with. For each strategy solve the equation provided. Then determine if it is a strategy that is appropriate or not. Explain why.

Strategy A: $37\times26 = 30+7\times20+6$
Strategy B: $37\times26 = 30\times(20+6) + 7\times(20+6)$

Table 5 shows the alignment between activities where writing is used to evaluate strategies and Gutiérrez' (2012) Equity Framework. The teacher posed the task and had them work in their table groups to retell the task to each other.

Table 5. Alignment of writing to evaluate strategies to Gutiérrez' (2012) equity framework

Dimension	Description of Alignment
Access	Students were given a task with two strategies shown that aligned to Grade level content. Students were asked to determine if each strategy was correct.
Achievement	The task given to students focused on the distributive property of multiplication and the use of parentheses, which are both key concepts in Fifth Grade and foundational for later mathematics.
Identity	The task included two different strategies for students to analyze. The task included previously studied strategies which increased the likelihood that students would feel capable of successfully completing the task. There was no real-life context so there was no opportunity for students to see this relate to mathematics in their community or to identify with the context in the task.
Power	Students had power in the classroom as they were able to choose which strategy to analyze first. Students also had power as they were able to use whatever process or strategy to evaluate the two strategies.

Example of Student Work

Figure 6 shows student written work that was correct. It shows that students successfully analyzed the strategies. Based on the work shown the teacher asked the student to explain in writing why they know that each strategy worked or did not work.

Each of the students showed work, but the class did not write in detail about why each strategy worked or did not work. When the teacher asked students to write more about the strategies, the fifth-grade

Figure 6. Student work from 37 x 26 task

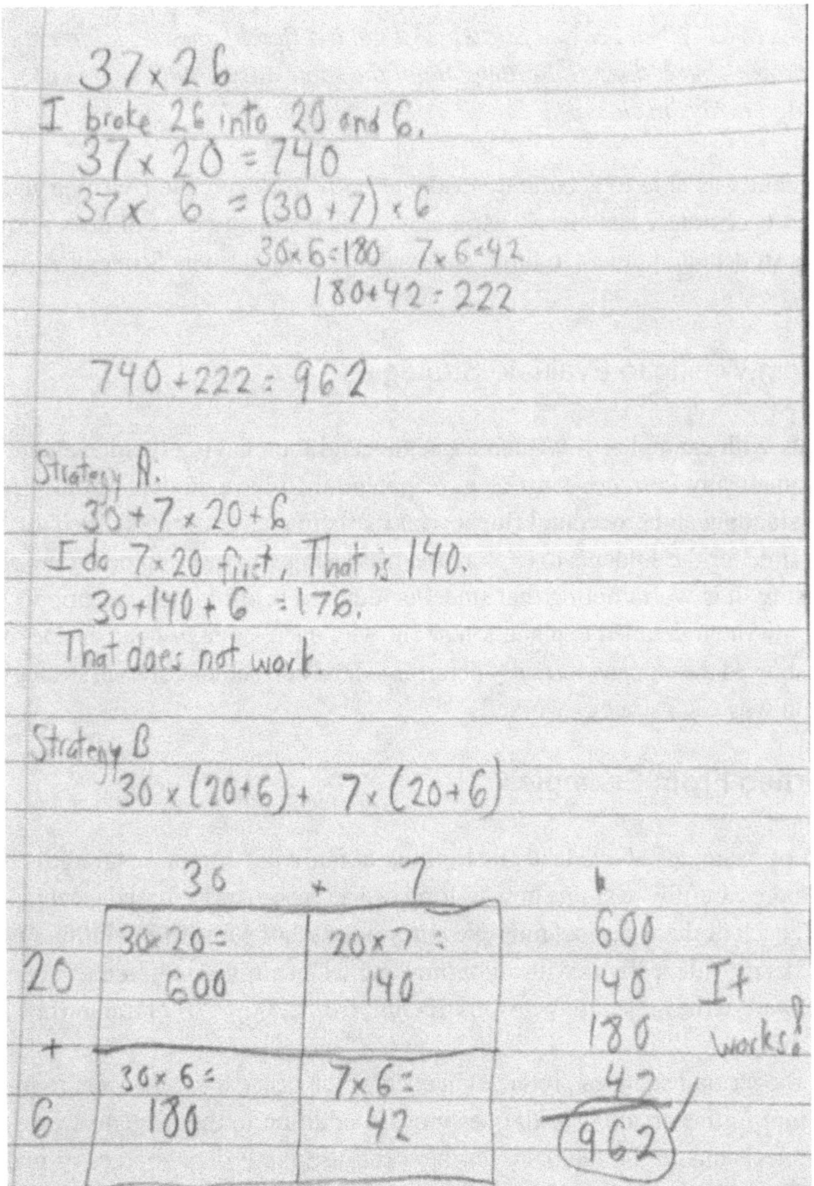

students responded and said, "We did. We showed our work." Based on students' responses, the teacher modified her plan and opted to have students meet with her in small groups and write their explanations while she was at the table with them. The use of whole class collaborative work at the beginning of the task allowed for the simultaneous exploration of the strategies in table groups to evaluate the strategies, while her decision to use small groups allowed the teacher to elicit students' mathematical ideas about the different strategies. Also, since students were showing their work, but not writing, the small groups provided a scaffolded and supported setting for students to write about their ideas.

The student whose work we looked at wrote the following response, which has been typed below for legibility:

I first found the answer to 37 x 26 by splitting 26 into 20 and 6. I then multiplied 37 x 20 and 37 x 6 and added them up to get 962. When I solved Strategy A I got a different answer. I don't know what they did and why it doesn't work but it doesn't. Strategy B got the same answer 962 so it works. I used the array model since we used arrays in class.

While the student was able to accurately evaluate both strategies their writing indicated that they were not yet able to accurately and clearly explain why each strategy does or does not work. When the teacher asked this student and others to think and write more about why Strategy A did not work, they were unable to.

Takeaways From Writing to Evaluate Strategies

Providing students with examples of written strategies and then having them evaluate if they work or if they are mathematically correct seems to have potential to have students think deeply about both conceptual understanding and procedural fluency for mathematics concepts (Liljedahl, 2023). Then the teacher provided time for the students to explore the task using whatever process they chose. While this activity is promising, it is worth noting that students required scaffolding and support when they were asked to provide a written detailed explanation about why each strategy does or does not work. In the example that we showed the student correctly determined the correctness of each strategy, but they could not further explain why each strategy worked.

Lessons Learned From Examples

The presentation of examples above leads us to think about what common strengths may exist across them in order to more carefully examine how writing can advance equity-based teaching in mathematics classrooms. First, each of the tasks had multiple entry points and paths to a solution (You Cubed, n.d.). These types of tasks provide learners with autonomy and when aligned to grade level expectations align to both the Power and Achievement dimensions of Gutierrez' (2009, 2012) framework for equity-based mathematics teaching.

Additionally, the writing samples produced in each of the tasks allow teachers to formatively assess students and prompt further thinking with questions. In addition to the formative assessment assisting teachers to make decisions in the moment, teachers can use these data sources to make decisions for future teaching. Equity is supported when students are provided opportunities in which they use their identity and prior knowledge to choose a solution path, the tasks provide access to rich mathematical content, their own writing places them in a position of power and provides the teacher with insights that are used to better instruction and this increase achievement.

IMPLICATIONS AND FUTURE DIRECTIONS

How Can We Better Support Teachers' Enactment of Mathematical Writing?

Colonnese (2020) proposed empirically aligned guidelines for mathematical writing. The three categories centered on are the purpose of the writing, techniques to implement writing, and aspects to attend to support a classroom community of writers. This chapter reflects the implementation suggested in the guidelines. The variety of writing activities presented in the chapter aligns with the guidelines in several ways. The cases shared show students learning and engaging with the different types and purposes of writing, having multiple opportunities to engage in mathematical writing, constructing and using graphic organizers, and the inclusion of writing frames in mini-lessons.

The examples included in this chapter offer examples of implementing writing in mathematics, the rationale for each practice, the connection to the equity framework, and ways to enact these activities. The guidelines for mathematical writing recommend creating a community of writers that grow in their mathematical understanding and communication. The effectiveness of the activities presented in this chapter depends on the environment teachers help establish and require a commitment to equity-based mathematics teaching. The guidelines suggest various techniques, such as establishing norms for mathematical writing, modeling writing, participating in the writing community, teaching specific strategies, gradually releasing responsibility, ensuring mathematical language is accessible through discourse, word walls, math dictionaries, and providing rubrics and feedback in multiple forms. These are several aspects of Colonnese's (2020) guidelines that teachers should have support and mentorship in creating for students to have mathematical writing experiences that are meaningful.

What Should Research Look Like Related to Mathematical Writing?

Mathematical writing aligns with and supports the development of mathematical literacy. The PISA assessment now includes mathematical literacy and how that is conceptualized is discussed within the framework; "OECD's Programme for the International Assessment of Adult Competencies (PIAAC) defines numeracy as the ability to access, use, interpret, and communicate mathematical information and ideas, in order to engage in and manage the mathematical demands of a range of situations in adult life" (p. 4). Research that examines students' writing in mathematics and how to best develop their ability is essential to support students in becoming mathematically literate.

REFERENCES

Bangert-Drowns, R. L., Hurley, M. M., & Wilkinson, B. (2004). The effects of school-based writing-to-learn interventions on academic achievement: A meta-analysis. *Review of Educational Research*, *74*(1), 29–58. doi:10.3102/00346543074001029

Bieda, K. N., & Staples, M. (2020). Justification as an equity practice. *Mathematics Teacher: Learning and Teaching PK-12, 113*(2), 102-108.

Brown, A. L., & Palincsar, A. (1982). Inducing strategic learning from texts by means of informed, self-control training. *Technical Report No.* 262.

Burns, M. (2004). Writing in math. *Educational Leadership*, 30–33.

Calkins, L. (1983). *Lessons from a child: On the teaching and learning of writing*. Heinemann.

Casa, T. M., Firmender, J. M., Cahill, J., Cardetti, F., Choppin, J. M., Cohen, J., & Zawodniak, R. (2016). Types of and purposes for elementary mathematical writing: Task force recommendations. University of Connecticut. http://mathwriting.education.uconn.edu

Clarke, D. J., Waywood, A., & Stephens, M. (1993). Probing the structure of mathematical writing. *Educational Studies in Mathematics, 25*(3), 235–250. doi:10.1007/BF01273863

Colonnese, M. W. (2020). The development of instructional guidelines for elementary mathematical writing. *School Science and Mathematics, 120*(3), 129–143. doi:10.1111sm.12391

Dacey, L. & Polly, D. (2012). The *Common Core State Standards for Mathematics*: The big picture. Invited article in *Teaching Children Mathematics, 18*(6), 378-383.

Ediger, M. (2006). Writing in the mathematics curriculum. *Journal of Instructional Psychology, 33*(2), 120–123.

Goldsby, D. S., & Cozza, B. (2002). Writing samples to understand mathematical thinking. *Mathematics Teaching in the Middle School, 7*(9), 517–520. doi:10.5951/MTMS.7.9.0517

Graham, S., Kiuhara, S. A., & MacKay, M. (2020). The effects of writing on learning in science, social studies, and mathematics: A meta-analysis. *Review of Educational Research, 90*(2), 179–226. doi:10.3102/0034654320914744

Graves, D. H. (1983). *Writing: Teachers and children at work.* Heineman.

Gutiérrez, R. (2009). Framing equity: Helping students "play the game" and "change the game. *Teaching for Excellence and Equity in Mathematics, 1*(1), 4-8.0.

Gutiérrez, R. (2012). Context matters: How should we conceptualize equity in mathematics education? In B. Herbel-Eisenmann, J. Chopin, D. Wagner, & D. Pimm (Eds.), *Equity in Discourse for Mathematics Education* (pp. 17–33). Springer. doi:10.1007/978-94-007-2813-4_2

Hattie, J. (2009). *Visible learning: A synthesis of over 800 meta-analyses relating to achievement*. Routledge.

Hattie, J., Fisher, D., Frey, N., Gojak, L. M., Moore, S. D., & Mellman, W. (2016). *Visible learning for mathematics, Grades K-12: What works best to optimize student learning (Corwin Mathematics Series).* Corwin.

Hillocks, G. (2005). At last: The focus on form vs. content in teaching writing. *Research in the Teaching of English, 40*, 238–248.

Karp, K. S., Bush, S. B., & Dougherty, B. J. (2022). 13 Rules That Expire. *Mathematics Teacher: Learning and Teaching PK-12*, *115*(2), 154-163.

Kline, S. L., & Ishii, D. K. (2008). Procedural explanations in mathematics writing: A framework for understanding college students' effective communication practices. *Written Communication*, *25*(4), 441–461. doi:10.1177/0741088308322343

Kosko, K. W., & Zimmerman, B. S. (2019). Emergence of argument in children's mathematical writing. *Journal of Early Childhood Literacy*, *19*(1), 82–106. doi:10.1177/1468798417712065

Liljedahl, P. (2023). *Building Thinking Classrooms, Grades K-12: Fourteen Teaching Practices for Enhancing Learning*. Corwin.

Martin, C. L. (2015). Writing as a tool to demonstrate mathematical understanding. *School Science and Mathematics*, *115*(6), 302–313. doi:10.1111sm.12131

Martin, C. S., & Polly, D. (2019). Examining the use of multiple writing and discourse tasks in 5th grade mathematics. *The Journal of Educational Research*, *112*(6), 663–675. doi:10.1080/00220671.2019.1678106

Murray, D. (1968). *A writer teaches writing: A practical method of teaching composition*. Houghton Mifflin.

Namkung, J. M., Hebert, M., Powell, S. R., Hoins, M., Bricko, N., & Torchia, M. (2020). Comparing and validating four methods for scoring mathematics writing. *Reading & Writing Quarterly*, *36*(2), 157–175. doi:10.1080/10573569.2019.1700858

National Council of Teachers of Mathematics. (2000). *Principles and standards for school mathematics*. NCTM.

National Governors Association Center for Best Practices & Council of Chief State School Officers. (2010). *Common Core State Standards for Mathematics*. NCTM.

National Research Council. (2001). *Adding It Up: Helping Children Learn Mathematics*. The National Academies Press.

Ntenza, S. (2006). Investigating forms of children's writing in grade 7 mathematics classrooms. *Educational Studies in Mathematics*, *61*(3), 321–345. doi:10.100710649-006-5891-0

Polly, D. (2021). Advancing equity-based mathematics teaching in the primary grades: The case of two clinical practice experiences. *International Journal of Teacher Education and Professional Development*, *4*(1), 68–88. doi:10.4018/IJTEPD.2021010105

Polly, D., & Martin, C. S. (in press). Considering the design and use of differentiated activities and fluency games to advance equity-based mathematics practices. *Educational Designer*.

Pugalee, D. K. (2004). A comparison of verbal and written descriptions of students' problem solving processes. *Educational Studies in Mathematics*, *55*(1), 27–47. doi:10.1023/B:EDUC.0000017666.11367.c7

Shanahan, C., & Shanahan, T. (2018). Disciplinary literacy. In D. Lapp & D. Fisher (Eds.), *Handbook of research on teaching the English language arts* (pp. 281–308). Routledge.

Shanahan, T., & Shanahan, C. (2008). Teaching disciplinary literacy to adolescents: Rethinking content-area literacy. *Harvard Educational Review*, *78*(1), 40–59. doi:10.17763/haer.78.1.v62444321p602101

Steen, L. A., Turner, R., & Burkhardt, H. (2007). Developing mathematical literacy. In *Modelling and applications in mathematics education* (pp. 285–294). Springer. doi:10.1007/978-0-387-29822-1_30

Van de Walle, J. A., Karp, K. S., & Bay-Williams, J. M. (2019). *Elementary and middle school mathematics: Teaching developmentally* (10th ed.). Pearson Education.

Chapter 10
A Mathematics Intervention to Engage Elementary Students in Mathematical Writing

Madelyn W. Colonnese
Reading and Elementary Education Department, Cato College of Education, University of North Carolina at Charlotte, USA

Fabiana Cardetti
https://orcid.org/0000-0002-3178-4659
Department of Mathematics, College of Liberal Arts and Sciences, University of Connecticut, USA

ABSTRACT

Mathematical writing is an important yet underutilized mode of discourse. Teachers play an important role in implementing and engaging their students' mathematical writing. However, research on mathematical writing does not offer suggestions for how teachers can engage students in mathematical writing. The purpose of this chapter is to contribute to this area of research by describing how teachers implemented a systematic mathematical writing intervention during a summer school program. The student writing resulting from the intervention demonstrated that students can successfully engage in mathematical writing. Further, students' written work showed key elements of mathematical writing and indicated an understanding of the subconstructs of fraction conceptualization that appeared to be absent prior to the intervention.

INTRODUCTION

Mathematical writing has many benefits for student learning (Graham et al., 2020; Bangert-Drowns et al., 2004). By writing students are able to clarify their mathematical ideas (NCTM, 2000), actively construct knowledge (Cross, 2009), and use mathematically precise vocabulary. Writing also provides students with an opportunity to make their thinking visible thus, allowing students to engage in the metacognitive

DOI: 10.4018/978-1-6684-6538-7.ch010

process (Cross, 2009; Pugalee, 2001; 2004). The metacognitive process contributes to students' ability to effectively problem solve, a key skill in mathematics (NCTM, 2014).

Teachers recognize the importance of having students write (Powell et al., 2021) however, writing has continued to be underutilized during mathematics class (Kosko, 2016). Curricular demands, the time needed for writing, and limited instructional support for teachers interested in implementing mathematical writing have contributed to the lack of mathematical writing (Fukawa-Connelly & Buck, 2010). Given the lack of mathematical writing and the importance of this form of discourse, the purpose of this chapter is to describe an example of a systematic mathematical writing intervention. In this chapter we will describe the professional learning that was facilitated to support teacher candidates with mathematical writing and its implementation, details of the intervention, and samples of the resulting student writing.

BACKGROUND

The National Council of Teachers of Mathematics ([NCTM] 2000; 2014) has long called for teachers to engage students in meaningful mathematical discourse that is designed to (a) advance students' understanding of mathematics and (b) engage students in mathematical reasoning. Writing is one way to engage students in such discourse. However; studies of mathematical writing have included a wide variety of definitions, types, and purposes with varied attention to advancing students' understanding of mathematics and engagement in reasoning (Colonnese et al., 2020). For this chapter, we drew from the Elementary Mathematical Writing Task Force to define the purpose of mathematical writing as writing to reason and communicate mathematically (Casa et al., 2016).

In studies of students writing to advance their understanding of mathematics (a), findings have demonstrated that mathematical writing can help to advance students' understanding of mathematics and improves students' mathematical communication skills (Graham et al., 2020; Bangert-Drowns et al., 2004). Further analysis of pre- and post- assessment of students' writing have demonstrated increases in students' mathematics achievement (Cohen et al., 2015; Tan & Graces-Bascal, 2013; Kostos & Shin, 2010; Cross, 2009).

In response to the need for discourse to engage students in mathematical reasoning (b), writing provides students with a way to justify and explain their reasoning (Kramarski & Mevarech, 2003). Students can use writing to both reason mathematically and share their reasoning with others (NCTM, 2000). Because of the permanent nature of writing, students are able to monitor their progress, reflect on, and evaluate their thinking (Graham et al., 2013) – all components of the metacognitive process (Cross, 2009; Pugalee, 2001; 2004).

The metacognitive process is essential in supporting students' development of problem-solving skills (Schoenfeld, 1987). However; researchers have found that explaining mathematical reasoning through writing is often challenging for students (Monoyiou et al., 2006; Kostos & Shin, 2010). For example, in a study of fifth- and sixth- grade students, Monoyiou et al. (2006) found that the students shared their solution but provided no justification or attempted to justify their thinking with numerical examples. This study also revealed that teachers valued reasoning-based justification similarly to justification with numerical examples. Such findings are in line with other research that suggests primary teachers' understanding of mathematical reasoning is often limited (e.g., Stylianides et al., 2006). The limited understanding of mathematical reasoning has the potential to impact how teachers can use student writing to engage in the formative assessment process.

A teacher's ability to effectively engage in the formative assessment process can significantly improve student achievement (Andersson & Palm, 2018; Hattie et al., 2016). However, currently the formative assessment process often happens in the moment during classroom discussions– making this a complex practice to enact (Sleep & Boerst, 2012). The complexity of this practice arises from the need for teachers to simultaneously ask questions, listen and make sense of a student's thinking to be able to provide feedback (Colonnese et al., 2022; Andersson & Palm, 2018; Sleep & Boerst, 2012). Further if teachers' understanding of mathematical reasoning is limited, this may impact their ability to respond in ways that further the students' thinking.

Writing offers another way for teachers to access their students' reasoning and engage in the formative assessment process. Unlike classroom discussions that are often held with a whole group of students, writing is typically an individual process. Teachers then have an opportunity to assess and respond to an individual student's in-the-moment thinking (Sadler, 1989). The permanent nature of writing also uniquely provides teachers with additional time to consider the students' writing before providing feedback or additional support to develop their mathematical reasoning skills. For teachers the additional time can potentially reduce the complexity of this practice since they would not have to read and understand students' writing while also providing in-the-moment feedback.

Effective feedback is needed given that researchers have found that while elementary students seem to understand the mathematics content, they tend to struggle in their ability to sufficiently communicate their reasoning in writing. High quality feedback from teachers through the formative assessment process is important (Bywater et al., 2022) to improve the quality of students' mathematical writing and further expand their reasoning skills.

Given the limited understanding teachers have of mathematical reasoning and the benefits of writing, it is worthwhile to consider how mathematics teacher educators can support teacher candidates in developing their understanding of mathematical writing. For example, Kenney et al., (2014) designed and implemented a writing-to-learn mathematics activity and reflection in their mathematics methods course. Danielson (2010) similarly had teacher candidates write a paper that focused on explaining the lattice algorithm in a mathematics content course for prospective teachers. The teacher candidates involved in studies of mathematical writing, like the aforementioned examples, have reported that they were likely to continue using writing in their mathematics instruction and felt positively towards this instructional practice (Kenney et al., 2014; McCarthy, 2008; Ward, 2005). Such findings support the likelihood that by providing teacher candidates with opportunities to engage with mathematical writing they will use writing in their future classrooms.

With this background in mind, our chapter offers the details of a mathematical writing intervention designed to engage students in mathematical reasoning and advance their understanding of fractions. In the following sections, we describe the preparation of teacher candidates prior to the mathematical writing intervention so that they could become familiar with mathematical writing and the instructional practices. We then share the structure of the intervention designed to support a purposeful launch that hooks students into the lessons, maintains the focus and extends the learning toward the lesson's goals. We follow by sharing the student writing that emerged before and during the mathematical writing intervention.

PREPARING TEACHERS TO IMPLEMENT A MATHEMATICAL WRITING INTERVENTION

The mathematical writing intervention took place during a summer school program designed to provide students with additional mathematics and reading instruction. All students in the school were invited to participate in the summer school program. All students were provided mathematics instruction that included mathematical writing. However; the rising fourth grade students were the only grade level to engage in the mathematical writing intervention described herein. The rising fourth grade students participated in forty-five minutes of mathematics instruction per day for 19 days of the program. The mathematics instruction included 30 minutes of the mathematical writing intervention and fifteen minutes of fluency work. The fluency work included number talks and fluency games drawn from the Kentucky Center for Mathematics. The lessons used during the mathematical writing intervention were informed by the mathematical writing process (Casa et al., in preparation).

Three teachers led the mathematics instruction for the students. The three teachers had different levels of teaching experience. The lead teacher in the classroom had recently completed an intensive clinical experience with the students in the summer school program. During the time of this program she was going into her junior year of a local teacher preparation program. The other two teachers provided general instructional support and worked one on one or with small groups of students. One of the support teachers was an instructional assistant at the elementary school and had worked with many of the students during the school year. The other support teacher was a recent graduate of a local teacher preparation program. All three teachers participated in professional learning before and during the summer school program.

The professional learning for the summer school program was facilitated by the first author, Madelyn. The teachers first reviewed the importance of fact fluency and routines such as number talks to support the development of students' fluency with addition, subtraction, multiplication and division. The focus on number talks also helped to review the importance of students' oral discourse in the classroom – an important part of the writing process. Teachers were then introduced to the four phases of the mathematical writing process: (a) shared mathematical experience; (b) connect mathematical talk and writing; (c) writing mathematically; and (d) reflect on mathematical writing (Casa et al., in preparation) by reviewing lessons from the mathematical writing intervention. Together we engaged in a shared mathematical experience, discussed the mathematical writing prompt, and wrote mathematically. We then reviewed student writing that was collected from another project and reflected on the characteristics of student writing such as a claim, evidence, and warrant as essential components of a mathematical argument. The purpose of reviewing student work was to help the teachers recognize some characteristics of a quality piece of mathematical writing and to develop a common understanding of such characteristics. We reasoned that this could help teachers when developing students' understanding of mathematical writing and to inform the kinds of feedback they provided students.

At the beginning of summer school, Madelyn co-taught the first mathematical writing lesson with the three teachers and then provided weekly in-the-moment mathematics coaching. She also met regularly with the teachers to review the lesson plans and student work. During these meetings she answered questions about the instructional techniques and discussed how to support the students in advancing their ability to write mathematically.

ENGAGING STUDENTS IN MATHEMATICAL WRITING

The content focus of the intervention was fractions, both unit fractions and non-unit fractions. This selection was informed by our team's observations and in collaboration with the students' third grade teacher to offer student participants additional support working with fractions and expressing in writing their fractional reasoning that helps them solve problems involving fractions.

The first part of the intervention was done through a series of lessons that focused on introducing students to mathematical writing, including the four types of mathematical writing: exploratory, argumentative, explanatory, and mathematically creative (Casa et al., 2016). These lessons built on students' number sense, rather than fractions, to allow students to focus on learning mathematical writing against a familiar content backdrop. The lessons exposed them to student writing samples that illustrated each of the different types of writing and allowed them to explore the characteristics for each one and recognize the differences between them.

The intervention then shifted to a series of lessons focused on developing students' understanding of unit fractions. During lesson 1, students engaged in a shared experience by playing a card game. In this game, players are dealt a number line with a unit fraction highlighted (see figure 1). A player then flips over a Spy Card (see figure 2), each representing a unit fraction using discrete (e.g., countable objects such as pipe cleaners) or continued models (e.g., measurable objects such as a cup filled to a certain level). Players see if they can match the Spy Card to the unit fraction highlighted on their number line board. The game ends when no more cards can be matched and the winner is the player who collected the most pairs of cards. During the lesson, the teachers walked around to support students, asked questions to probe thinking, and prompted students to justify had students justify why the card they selected matches their number line.

In Lesson 2 students were introduced to the argumentative prompt shown in Figure 3. As can be seen in the figure, students are asked for their agreement with Donnie's representation of ⅙, where the drawing actually represents 1/7.

The students first discussed their ideas related to the task. The teachers represented the mathematics discussion on the board using a talk frame (Casa, 2013). As seen in the example in Figure 4, the "Think" section is where the teacher records how the class interprets the writing prompt. The written representation of the discussion in the "Talk Ideas" section serves to model for students how to represent their thinking in writing. This section of the talk frame is intended to include both mathematically valid and developing ideas (i.e., Talk Ideas 1 & 3). Teachers continue to facilitate the discussion for the class to be able to work through developing ideas and transition them into valid ones. The notes in the "We Understand" section reflect the consensus of the class and only contain mathematically valid ideas.

Figure 1. Number line cards

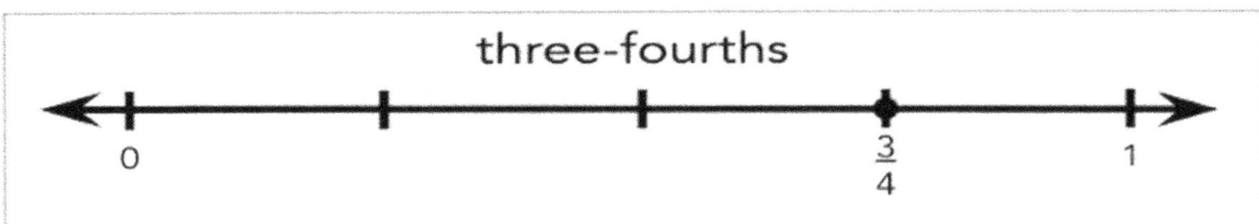

Figure 2. Spy card

Figure 3. Unit fractions argumentative task

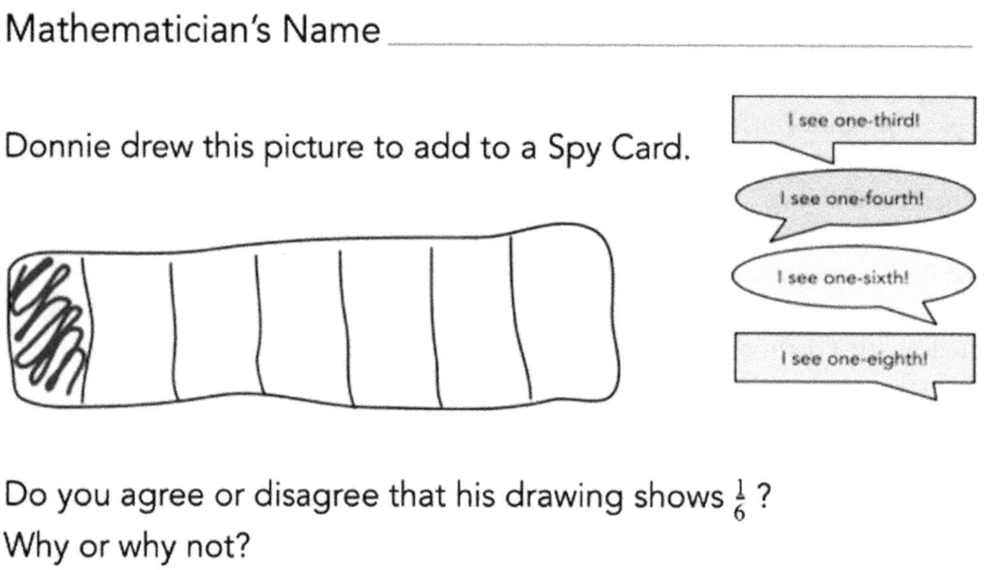

Lesson 3 provided students with an opportunity to write their own responses to the task in Figure 3. Then, in Lesson 4, students engaged in a discussion that helped them reflect on their own writing and the writing of their peers. The discussion focused on the characteristics of their writing, considering multiple aspects of mathematical writing. They examined the mathematics content used in the different responses. Students also reflected on the general writing and the writing characteristics specific to a mathematical argument.

Figure 4. An example of a talk frame used with the task in Figure 3

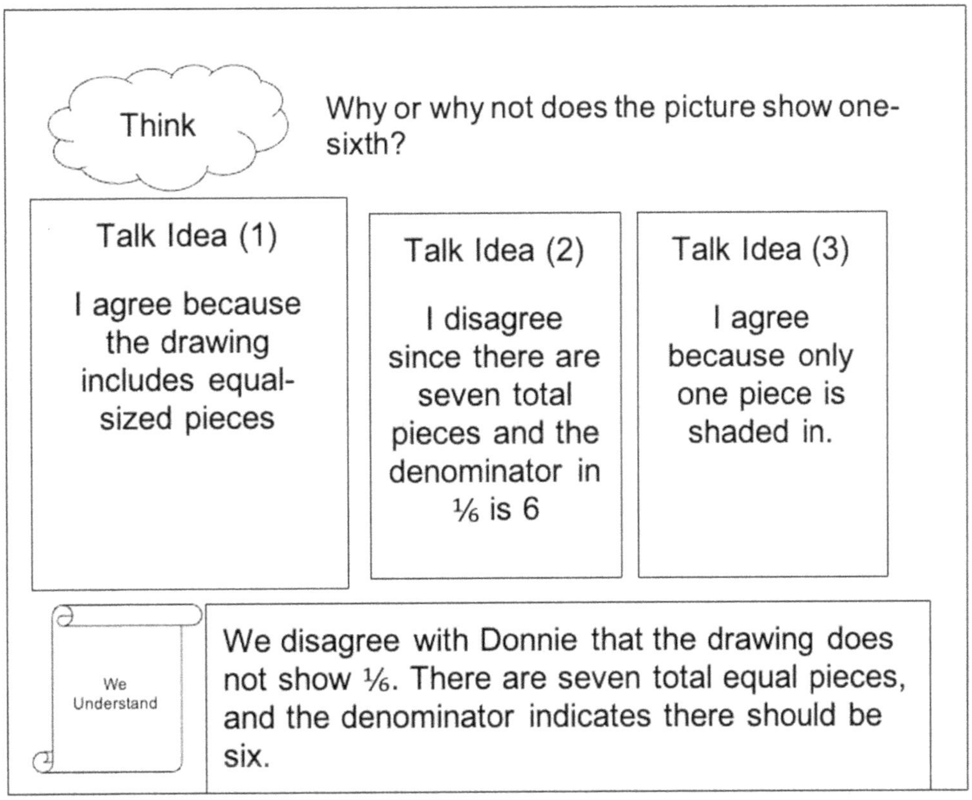

The last part of the intervention focused on non-unit fractions, following a similar sequence of lessons as the second part. For the shared learning experience, the students again played the same card game however; the Spy Cards had representations of non-unit fractions. This time the prompt presented to students focused on explanatory writing as shown in Figure 5. In this prompt, students are presented with the number line board and then asked to draw a card that would match the highlighted fraction, ⅚. Students discussed their ideas and the teacher recorded the class's ideas on the talk frame for how they could explain how their picture represented ⅚. The students then had a full lesson dedicated to drawing their card and writing their explanatory mathematical piece. The subsequent lesson focused on students reflecting on their own work and the work of their peers. In wrapping up the students' reflections, the teachers led the students in a discussion of the characteristics of a quality mathematical explanation.

The intervention used during the summer school program engaged students in a systematic approach to mathematical writing instruction: (a) a shared mathematics experience; (b) connecting talk to writing; (c) engaging in mathematical writing; and (d) reflecting on and sharing student writing. The students' mathematical arguments and explanations provide an opportunity to further understand the characteristics of students' writing after participating in such an intervention. Additionally, the writing prompts used in this intervention provide an opportunity to understand the kinds of reasoning students use when writing about unit and non-unit fractions. Such insight can help further the understanding of both mathematics researchers and educators of the ways students communicate about fractions in an explanation and an argument.

Figure 5. Non-unit fractions explanatory task

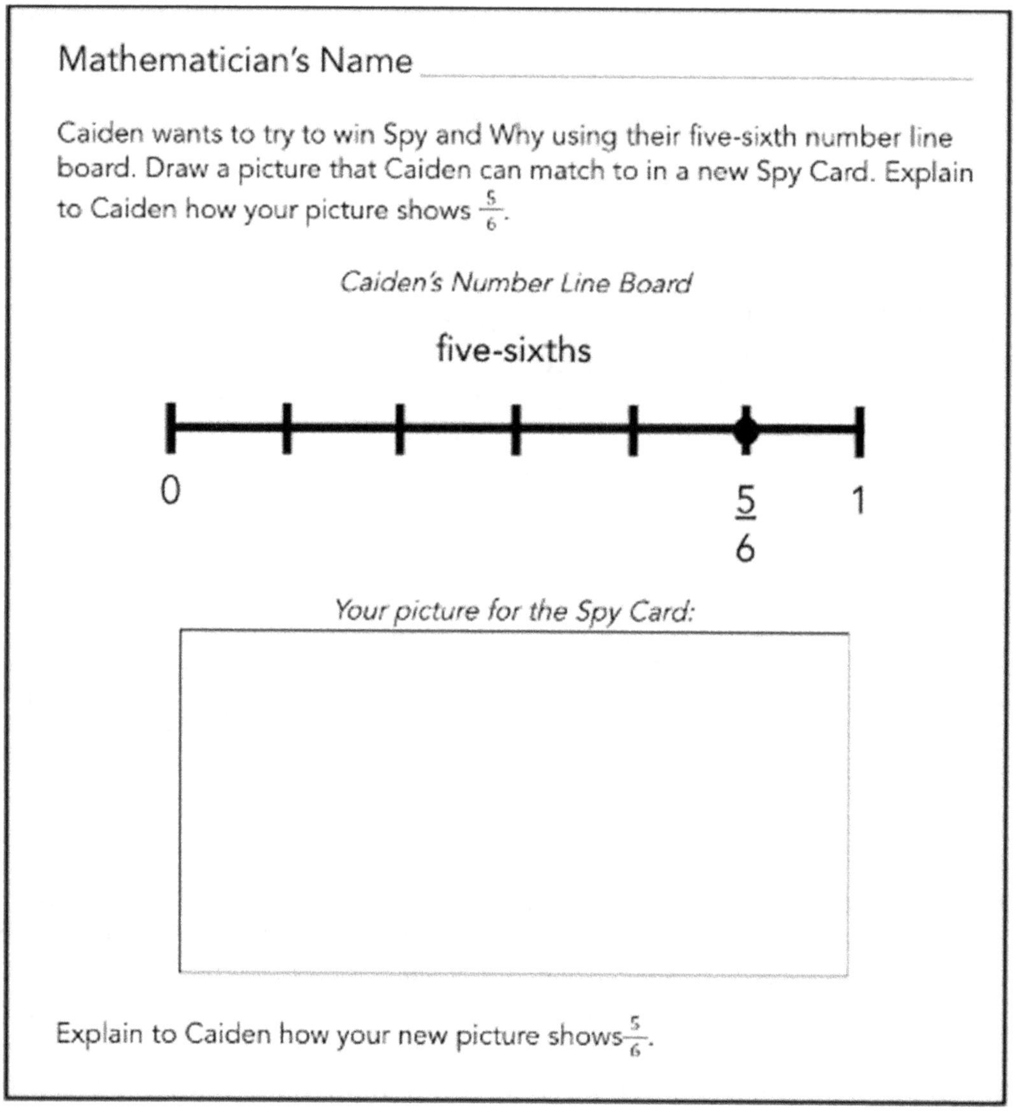

Mathematician's Name ______________________

Caiden wants to try to win Spy and Why using their five-sixth number line board. Draw a picture that Caiden can match to in a new Spy Card. Explain to Caiden how your picture shows $\frac{5}{6}$.

Caiden's Number Line Board

five-sixths

0 $\frac{5}{6}$ 1

Your picture for the Spy Card:

Explain to Caiden how your new picture shows $\frac{5}{6}$.

WHAT IS POSSIBLE

In this section, we present what is possible to achieve when we engage students in purposefully designed learning experiences as described in the previous section. The work samples presented here were created by students who participated in the summer school intervention. Many other samples collected prior to and during the intervention were also rich both in their mathematical content and writing, but due to scope and space within this chapter, we could not include them here. The ones presented here were selected specifically because they allow us to succinctly reveal the impact of the intervention by capturing differences between students' writing with and without instruction that purposefully and systematically targets the writing of mathematical explanations or arguments.

Figure 6. Pre-intervention student one work sample

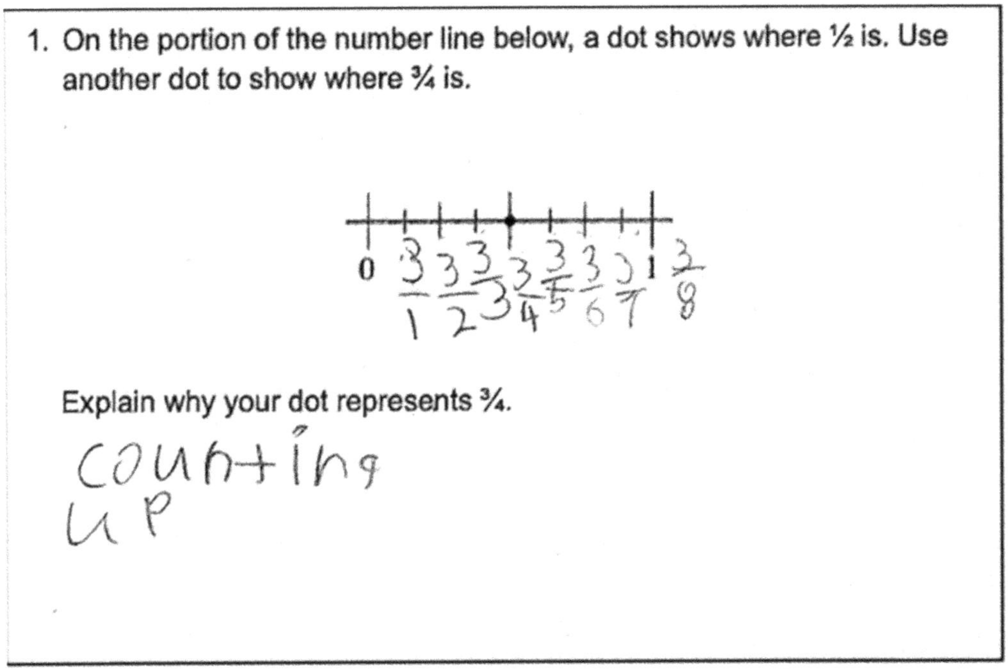

These examples are offered as an illustration, acknowledging that results of other implementations will depend on many different factors, some internal (e.g., preparation time, available support for PSTs) and some external (e.g., summer program infrastructure, attendance). For the case described in this chapter, lesson preparation and support for the teachers was as described previously. The intervention was implemented within a well-established summer program; and attendance within the intervention remained above 57% .

Figure 6 shows the response that one of the students, Student 1, provided to a prompt that was given to them prior to the start of the intervention. The prompt was selected from the NAEP Grade 4 (2003) sample items. As can be seen from the response, the student labeled the location for ¾ as the fourth tick mark from 0 on the number line provided. The student's response to support their choice consists simply of two words "counting up," referring to their labeling of the number line. The student's labeling provides evidence of their need for further experiences with the subconstructs of the conceptualization of fractions (Behr et al., 1983; Charambolos et al., 2007; Kieran, 1976) and with expressing their ideas in writing such that their reasoning is communicated in their writing.

The way the student expressed their thinking in Figure 6 has been similarly observed by others, including teachers and researchers (i.e., Monoyiou et al., 2006). The explanation provided by the student showcases their need to understand what "Explain why" means and/or what is expected in a response to such a prompt. In other words, what mathematical writing entails. It is important to note that the issue is not *quantity* or how much a student writes. As an example, Figure 7 shows the work of another student, Student 2, whose explanation is lengthier. The student chose the tick mark prior to ½ as the location for ¾. In their explanation they drew a zoomed-in version of that segment and supported the choice of location by changing the whole from the unit interval (0,1) to the interval (0,½) stating that in this new drawing the fraction ⅜ "now it is ¾." This response highlights not only that lengthier does not

Figure 7. Pre-intervention student two work sample

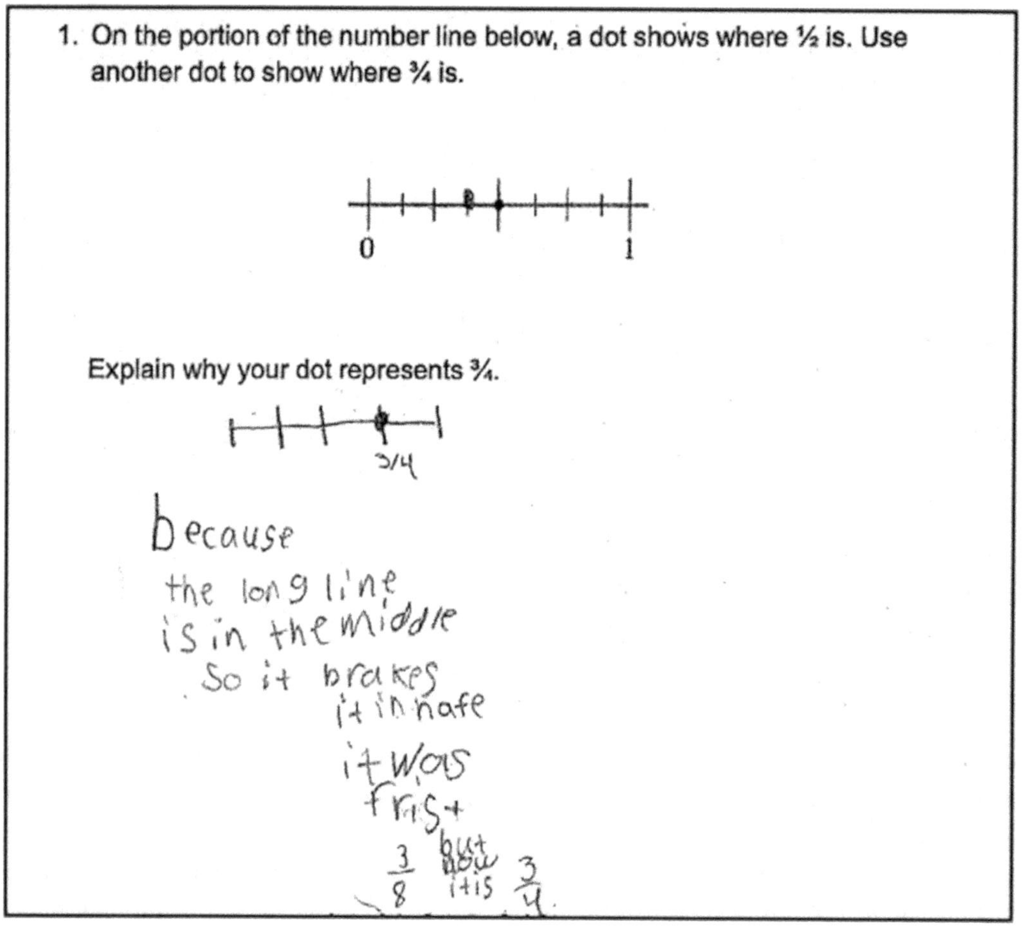

mean better, but also how valuable it is to get a glimpse at the student thinking about their response. The student's work shows a flawed understanding of the measure subconstruct of fraction conceptualization, which identifies fractions as numbers on the number line. While the student can correctly identify ⅜ within the unit interval, their work showcases a potential lack of understanding of equivalent fractions (6/8=¾) and their clear confusion about the whole associated with a fraction in the context of number line representations.

Compared to the pre-assessment, the work generated by all students during and after the intervention showcased a rich variety of responses that could be identified with every one of the components of fractional reasoning (Colonnese et al., submitted). Students' explanations showed concern for writing responses that made clear how they arrived at their answers, as well as revealed their understanding of the essential subconstructs of the conceptualization of fractions.

Figure 8 shows the response of Student 1 to the prompt in Figure 5, presented during the math intervention. Note that this is in stark contrast to this student's response in Figure 6 above, indicating a better understanding of the elements necessary to include in a mathematical explanation and a better understanding of fractions. While one cannot claim that the student understands the measure subconstruct of fraction, it is clear from the response that the student is able to connect the number line representation

Figure 8. During intervention student one work sample

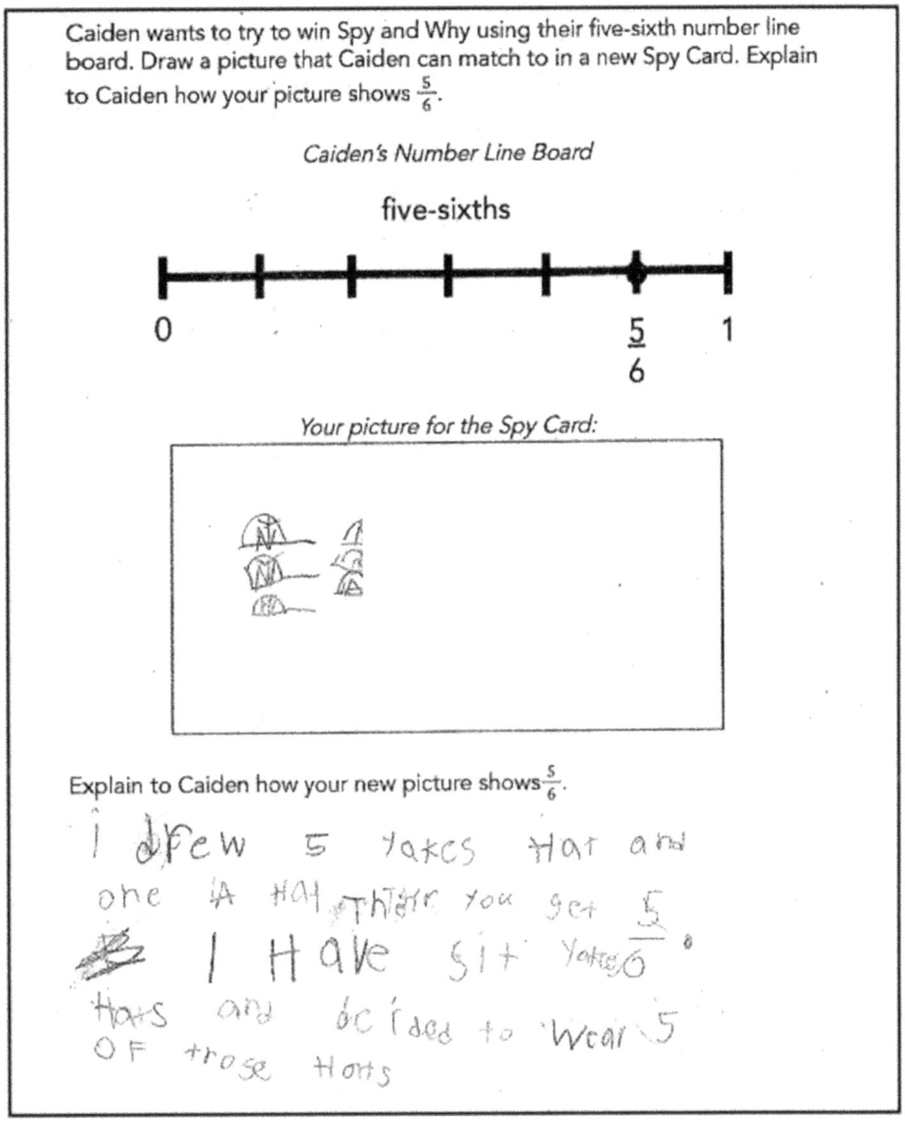

of the fraction with a collection of objects representing the same fraction and to provide a mathematical explanation for their representation. In their explanation, including the drawing, the student shows understanding of the part-whole subconstruct to conceptualize ⅚ as the comparison of the total number of hats, six, to the five he decided to wear. The response uses partitioning as the fractional reasoning (e.g., Steffe, 2002) to argue for why their drawing represents ⅚. The student ensures congruency of the parts, all parts are hats, and calls attention to the whole "I have six hats" to determine that the total number of parts, the denominator, is 6. The student further explains that only 5 of those need to be considered for this drawing to represent ⅚. Note that in the response prior to the intervention (Figure 6) the whole, 1, associated to ¾ was not addressed and was labeled ⅜.

Figure 9. During intervention student two work sample

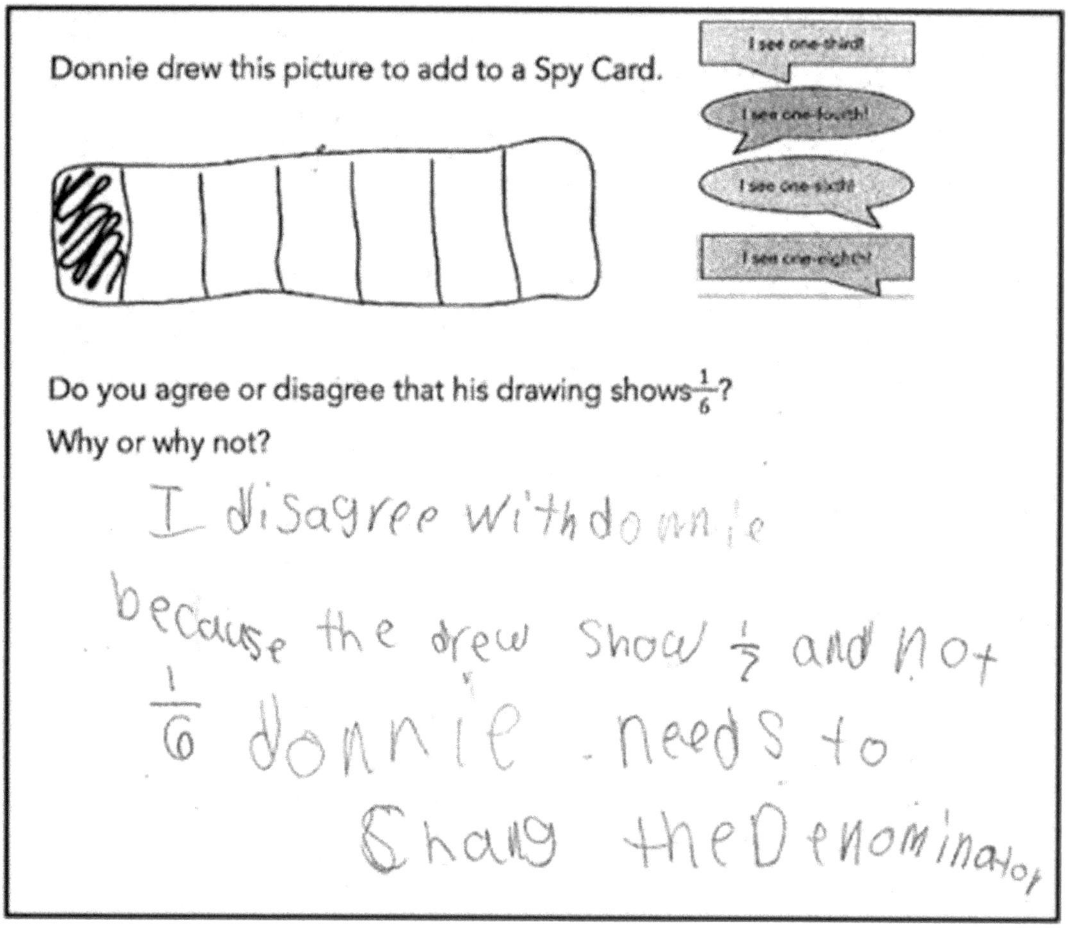

The student's writing also indicates growth in their understanding of what explanatory mathematical writing entails. It incorporates important characteristics such as presenting complete sentences that follow a logical flow; using precise, albeit informal mathematical vocabulary; and including valid mathematical ideas, all of which reflect their reasoning about fractions and connect tightly with their representation.

In Figure 9, one can observe the ability of Student 2 to distinguish between ⅙ and 1/7 of a given whole *without* changing the whole, something this student had done prior to the intervention (ref. Figure 7). This signals a better conceptual understanding of fractions. The differentiation between the two unit fractions within the same whole indicates the student used partitioning, disembedding, and iterating (Hackenberg, 2013; Colonnese et al., submitted) in their reasoning about this prompt. In addition, the student work displays a clear understanding of the representation of unit fractions (Donnie's does not show 1/6) which is the key issue with Donnie's card (it shows 1/7).

When focusing on the writing of Student 2, it is apparent that it is logically organized and the proper use of formal mathematical vocabulary, "denominator," is notable compared to what this student offered in their response prior to the intervention. The writing includes all necessary elements to support a mathematical argument (Toulmin, 1958/2003). Namely, the student includes a claim "I disagree with Donnie," and supports it with evidence "the [drawing] shows 1/7 not ⅙," that is based on the implicit warrant that the whole should be made of 6 equal parts. The writing extends further from the prompt by offering an idea for how Donnie's work could be fixed "Donnie needs to chang[e] the denominator."

FUTURE RESEARCH DIRECTIONS

The examples in Figures 6 and 7 (at the beginning of the section) pointing to areas of improvement in students' writing prior to the math intervention are no fault of the student or their educators. As mentioned in the introduction of this chapter, the mathematics education community is still working on gaining a better understanding of how to support both teachers and students to address the calls by NCTM, CC-SSM, and other large organizations - not an easy task. We offer these examples to spark motivation and ideas in others to engage students in mathematical writing and help them develop the skills necessary to create authentic explanations that allow us (and them) to "see" their thinking.

Additionally, teacher educators can draw some inspiration from the work presented here to create opportunities in their teacher preparation programs. The summer school program provided teachers with increased experiences working with elementary students and provided them with an opportunity to learn and practice instructional skills related to mathematical writing. Importantly, the first author was able to provide ongoing in-the-moment support to the teacher candidates since this was their first experience with mathematical writing. A next step is to design interventions such as this one that can be implemented during the school year. Further research is needed to understand the impact of a writing intervention for teacher candidates and elementary students.

The samples offered in this chapter can also be used to work with practicing teachers on professional development around mathematical writing. Research has demonstrated that teachers often have a limited understanding of reasoning (i.e., Monoyiou et al., 2006). The samples featured in this chapter can provide teachers with an opportunity to understand how students reason and the different kinds of reasoning that students may present when writing an argument or explanation. In addition to helping teachers identify the different kinds of reasoning, teachers can also begin to notice the characteristics of a mathematical argument and explanation. Additional research is needed to understand how understanding the characteristics may impact how teachers provide feedback to students.

CONCLUSION

In this chapter we shared a description of a mathematical writing intervention that was implemented during a summer school program by teacher candidates. The student samples included in the chapter provide examples of what is possible when students are engaged in a systematic mathematical writing intervention. The writing samples that emerged during and at the conclusion of the intervention demonstrate various components of fractional reasoning and reflect components of a mathematical argument and explanation. Further research is needed to understand the impact of such interventions on both teachers and students.

REFERENCES

Andersson, C., & Palm, T. (2018). Reasons for teachers' successful development of a formative assessment practice through professional development–a motivation perspective. *Assessment in Education: Principles, Policy & Practice*, *25*(6), 576–597. doi:10.1080/0969594X.2018.1430685

Bangert-Drowns, R. L., Hurley, M. M., & Wilkinson, B. (2004). The effects of school-based writing-to-learn interventions on academic achievement: A meta-analysis. *Review of Educational Research*, *74*(1), 29–58. doi:10.3102/00346543074001029

Bywater, J. P., Lilly, S., & Chiu, J. L. (2022). Examining technology-supported teacher responding and students' written mathematical explanations. *Journal of Mathematics Teacher Education*, 1–23.

Casa, T. M. (2013). Capturing thinking on the talk frame. *Teaching Children Mathematics*, *19*(8), 516–523. doi:10.5951/teacchilmath.19.8.0516

CasaT. M.CardettiF.ColonneseM. W.FirmenderJ. M. (in preparation)

Casa, T. M., Firmender, J. M., Cahill, J., Cardetti, F., Choppin, J. M., Cohen, J., & Zawodniak, R. (2016). *Types of and purposes for primary mathematical writing: Task force recommendations*. Math Writing. http://mathwriting.education.uconn.edu

Cohen, J. A., Casa, T. M., Miller, H. C., & Firmender, J. M. (2015). Characteristics of second graders' mathematical writing. *School Science and Mathematics*, *115*(7), 344–355. doi:10.1111sm.12138

Colonnese, M. W. (2020). The development of instructional guidelines for elementary mathematical writing. *School Science and Mathematics*, *120*(3), 129–143. doi:10.1111sm.12391

Colonnese, M. W., Cardetti, F., & Casa, T. M. (2023). *School Science and Mathematics*.

Colonnese, M. W., Reinke, L. T., & Polly, D. (2022). An analysis of the questions elementary education teacher candidates pose to elicit mathematical thinking. *Action in Teacher Education*, *44*(3), 196–211. doi:10.1080/01626620.2021.2020696

Cross, D. I. (2009). Creating optimal mathematics learning environments: Combining argumentation and writing to enhance achievement. *International Journal of Science and Mathematics Education*, *7*(5), 905–930. doi:10.100710763-008-9144-9

Danielson, C. (2010). Writing papers in math class: A tool for encouraging mathematical exploration by preservice primary teachers. *School Science and Mathematics*, *110*(8), 374–381. doi:10.1111/j.1949-8594.2010.00049.x

Fukawa-Connelly, T., & Buck, S. (2010). Using portfolio assignments to assess students' mathematical thinking. *Mathematics Teacher*, *103*(9), 649–654. doi:10.5951/MT.103.9.0649

Graham, S., Gillespie, A., & McKeown, D. (2013). Writing: Importance, development, and instruction. *Reading and Writing*, *26*(1), 1–15. doi:10.100711145-012-9395-2

Graham, S., Kiuhara, S. A., & MacKay, M. (2020). The effects of writing on learning in science, social studies, and mathematics: A meta-analysis. *Review of Educational Research*, *90*(2), 179–226. doi:10.3102/0034654320914744

Hackenberg, A. J. (2013). The fractional knowledge and algebraic reasoning of students with the first multiplicative concept. *The Journal of Mathematical Behavior*, *32*(3), 538–563. doi:10.1016/j.jmathb.2013.06.007

Hattie, J., Fisher, D., Frey, N., Gojak, L. M., Moore, S. D., & Mellman, W. (2016). *Visible learning for mathematics, Grades K-12: What works best to optimize student learning*. Corwin.

Kenney, R., Shoffner, M., & Norris, D. (2014). Reflecting on the use of writing to promote mathematical learning: An examination of preservice mathematics teachers' perspectives. *Teacher Educator*, *49*(1), 28–43. doi:10.1080/08878730.2013.848002

Kosko, K. W. (2016). Writing in mathematics: A survey of K–12 teachers' reported frequency in the classroom. *School Science and Mathematics*, *116*(5), 276–285. doi:10.1111sm.12177

Kostos, K., & Shin, E. (2010). Using Math Journals to Enhance Second Graders' Communication of Mathematical Thinking. *Early Childhood Education Journal*, *38*(3), 223–231. doi:10.100710643-010-0390-4

Kramarski, B., & Mevarech, Z. R. (2003). Enhancing mathematical reasoning in the classroom: The effects of cooperative learning and metacognitive training. *American Educational Research Journal*, *40*(1), 281–310. doi:10.3102/00028312040001281

McCarthy, D. S. (2008). Communication in mathematics: Preparing preservice teachers to include writing in mathematics teaching and learning. *School Science and Mathematics*, *108*(7), 334–340. doi:10.1111/j.1949-8594.2008.tb17846.x

Monoyiou, A., Xistouri, X., & Philippou, G. (2006). Primary students' reasoning in problem solving and teachers' evaluation of their arguments. *International Group for the Psychology of Mathematics Education*, *4*, 177.

Mudaly, V., & De Villiers, M. (2000). Learners' needs for conviction and explanation within the context of dynamic geometry. *Pythagoras*, *52*(8), 20–23.

National Council of Teachers of Mathematics. (2000). *Principles and standards for school mathematics*. NCTM.

National Council of Teachers of Mathematics. (2014). *Principles to actions: Ensuring mathematical success for all*. NCTM.

National Governors Association Centre for Best Practices, Council of Chief State School Officers. (2010). *Common core state standards*. CCSSI.

Powell, S. R., Hebert, M. A., & Hughes, E. M. (2021). How educators use mathematics writing in the classroom: A national survey of mathematics educators. *Reading and Writing*, *34*(2), 417–447. doi:10.100711145-020-10076-8

Pugalee, D. K. (2001). Writing, mathematics, and metacognition: Looking for connections Through students' work in mathematical problem solving. *School Science and Mathematics, 101*(5), 236–245. doi:10.1111/j.1949-8594.2001.tb18026.x

Pugalee, D. K. (2004). A comparison of verbal and written descriptions of students' problem solving processes. *Educational Studies in Mathematics, 55*(1-3), 27–47. doi:10.1023/B:EDUC.0000017666.11367.c7

Sadler, D. R. (1989). Formative assessment and the design of instructional systems. *Instructional Science, 18*(2), 119–144. doi:10.1007/BF00117714

Schoenfeld, A. H. (1987). *What's all the fuss about metacognition.* Cognitive Science and Mathematics Education.

Sleep, L., & Boerst, T. A. (2012). Preparing beginning teachers to elicit and interpret students' mathematical thinking. *Teaching and Teacher Education, 28*(7), 1038–1048. doi:10.1016/j.tate.2012.04.005

Steffe, L. P. (2012). On the construction of learning trajectories of children: The case of commensurate fractions. In *Hypothetical Learning Trajectories* (pp. 129–162). Routledge.

Steffe, L. P., Olive, J. (2010). Articulation of the reorganization hypothesis. *Children's fractional knowledge*, 49-74.

Stylianides, A. J., & Stylianides, G. J. (2006, July). Content knowledge for mathematics teaching: the case of reasoning and proving. In *Proceedings 30th conference of the international group for the psychology of mathematics education* (Vol. 5, pp. 201-208). IEEE.

Tan, T., & Garces-Bacsal, R. M. (2013). The effect of journal writing on mathematics achievement among high-ability students in Singapore. *Gifted and Talented International, 28*(1–2), 173–184. doi:10.1080/15332276.2013.11678412

Toulmin, S. E. (2003). The uses of argument (updated ed.). New York, NY: Cambridge University Press.

Ward, R. A. (2005). Using children's literature to inspire K–8 preservice teachers' future mathematics pedagogy. *The Reading Teacher, 59*(2), 132–143. doi:10.1598/RT.59.2.3

ADDITIONAL READING

Namkung, J. M., Hebert, M., Powell, S. R., Hoins, M., Bricko, N., & Torchia, M. (2020). Comparing and validating four methods for scoring mathematics writing. *Reading & Writing Quarterly, 36*(2), 157–175. doi:10.1080/10573569.2019.1700858

Powell, S. R., & Hebert, M. A. (2016). Influence of writing ability and computation skill on mathematics writing. *The Elementary School Journal, 117*(2), 310–335. doi:10.1086/688887

KEY TERMS AND DEFINITIONS

Mathematical Writing: Writing for the purpose of communicating and reasoning (Casa et al., 2016).

Chapter 11
A Framework for Assessing Students' Written Explanations of Numerical Reasoning

Sarah Quebec Fuentes
Texas Christian University, USA

Lindy Crawford
https://orcid.org/0009-0007-0911-5698
Lee Pesky Learning Center, USA

Ashley Rowe
Lee Pesky Learning Center, USA

Jacqueline Huscroft-D'Angelo
Oregon Research Institute, USA

ABSTRACT

Disciplinary writing in mathematics supports the use of words, symbols, and visual representations, allowing one to communicate more fully in writing than through speech alone. This chapter explores the disciplinary writing of 394 fourth-grade students who shared their numerical reasoning in written explanations to seven whole-number and fraction comparison tasks. Data were collected via whole group administration procedures and students' explanations were scored using a validated framework for evaluating numerical reasoning. Results include descriptive statistics and qualitative analyses of student responses using the framework's five categories. A difference was found between the types of reasoning students shared for whole-number tasks versus those shared for fraction comparison tasks, favoring the incorporation of more conceptual reasoning for whole-number tasks. The framework is a practical and effective tool that teachers can use to examine the depth of student reasoning and disciplinary writing to document such reasoning.

DOI: 10.4018/978-1-6684-6538-7.ch011

INTRODUCTION

Writing to learn mathematics significantly enhances the achievement of elementary school students (Graham et al., 2020). Writing to learn has been defined "as a vehicle for strengthening, extending, and deepening students' knowledge," (Graham et al., 2020, p. 180), and its benefits have been empirically supported in various meta-analyses (Bangert-Drowns et al., 2004; Graham et al., 2020; Graham & Perrin, 2007). Of similar value to writing to learn a discipline is learning to write in a discipline as emphasized through various standards documents (NGA & CCSSO, 2010a; National Council of Teachers of Mathematics [NCTM], 2000). Thus, disciplinary writing affords students the "dual benefit of communicating to learn mathematics and learning to communicate mathematically" (Pugalee, 2004, p. 27). As with any discipline, learning to write in mathematics requires an understanding of the purpose for writing (e.g., argument, justification, evaluation), the means available (e.g., words, symbols, visual representations), and the disciplinary norms for communicating (e.g., brevity, linearity).

Writing, as one form of disciplinary literacy, can provide benefits beyond those of disciplinary speaking, reading, or performing, particularly in the context of traditional classrooms. First, written work provides an immutable record of student reasoning at one time and thus can be used to measure growth in mathematical understanding over time. Second, writing allows for efficient large-group measurement of student reasoning sans the time and transcription demands of capturing verbal communication. Third, as Pugalee (2004) reported, disciplinary writing may encourage students to communicate their reasoning with more precision and accuracy.

In summary, empirical research on the benefits of writing to learn, professional standards emphasizing the importance of disciplinary writing, and the efficiency of measuring disciplinary literacy through writing provide three justifications for engaging students in mathematical writing. The intent of the present chapter is to share the exploration of the analysis of elementary-aged students' numerical reasoning expressed via writing, utilizing the *Framework for Evaluating Quantitative Reasoning Strategies*. Teachers, teacher educators, and researchers are provided with a range of examples that illustrate the complexity (or lack thereof) of student reasoning and their ability to communicate this reasoning through writing.

DISCIPLINARY LITERACY

Addressing disciplinary literacy in K-12 education has become a focus of discussions about learning and teaching in the various content areas (Fang, 2012a; Håland, 2016; Shanahan & Shanahan, 2008, 2012). Disciplinary literacy "is grounded in the beliefs that reading and writing are integral to disciplinary practices and that disciplines differ not only in content but also in the ways this content is produced, communicated, and critiqued" (Fang, 2012a, p. 20). In other words, discipline-specific literacy reflects the norms and values of experts in each field (Moschkovich, 2003). Research in disciplinary literacy has identified distinctions in how these experts read and write in the various disciplines (e.g., Johnson et al., 2011; Shanahan & Shanahan, 2008). Proponents of disciplinary literacy argue that literacy instruction should transcend general strategies that apply across content areas to include strategies that reflect disciplinary practices (Fang, 2012a; Shanahan & Shanahan, 2008, 2012) so that students develop "disciplinary habits of mind (e.g., reading-writing, viewing-representing, listening-speaking, thinking-reasoning, and problem-solving practices consistent with those of content experts)" (Fang, 2012a, p. 20).

Disciplinary Literacy in Mathematics

To identify the disciplinary distinctions between literacy in various content areas, researchers examine the activities of experts (e.g., Johnson et al., 2011; Shanahan & Shanahan, 2008). Mathematicians center their investigations on observing patterns, making conjectures, often times asking driving questions in the form of *What If* (Johnson et al., 2011). For example, some mathematicians conduct their work in a decontextualized, or abstract, system (Kleve & Penne, 2016) and may inquire: What if this condition is changed? This type of question leads to proofs, or justifications of conclusions that include a logical sequence of reasoning stemming from the implications of their explorations (Fang, 2012a; Shanahan & Shanahan, 2008). Some mathematicians generate models of real-world phenomenon and interrogate the accuracy of such models (e.g., What is the necessary information for this model to be true?) (Johnson et al., 2011). Communicating about such mathematical ideas (theoretical or applied) entails understanding the terms, grammar, and patterns of the language; following the logical flow of reasoning; and flexibly transitioning between various representations (e.g., symbolic, pictorial, and words; Fang, 2012b; Heuer, 2005; Johnson et al., 2011; Moschkovich, 2003; Shanahan & Shanahan, 2008, 2012).

Disciplinary Writing in Mathematics

Writing is one means of communicating mathematically (NCTM, 2000). Disciplinary writing in mathematics must extend beyond *recounting* the process for solving a problem (Marks & Mousley, 1990). Rather than documenting a *temporal sequence of events*, mathematical writing reflects a logical flow of reasoning (Solomon & O'Neill, 1998). The excerpts in Table 1 show the distinction between these two forms of writing. Additionally, disciplinary writing in mathematics has unique features as compared to writing in other content areas (Seo, 2016; Solomon & O'Neill, 1998). Mathematical writing involves precision and is often impersonal, expressed through short sentences, use of the present tense, and lack of a subject (Forman, 1996; Solomon & O'Neill, 1998).

The aforementioned precision in disciplinary writing in mathematics is realized through language as well as abstract symbols and visuals (Fang, 2012b; Seo, 2016; Sfard, 2007). The language of mathematics involves terms and grammar (Fang, 2012b; Johnson et al., 2011). Technical terms can be unique to mathematics (e.g., quotient) or colloquial terms with specific meanings in mathematics (e.g., difference, product) (Moschkovitch, 2003; Fang, 2012b). Single terms have abundant of meanings, described as "loaded vocabulary words" (Seo, 2016, p. 3). For example, a *square* is a parallelogram with four equal sides and four equal angles. Implied in the term is a hierarchical relationship. Since a square is a parallelogram, a square also has all the properties of a parallelogram (e.g., diagonals bisect each other). Additionally, other properties can be deduced (e.g., diagonals are congruent). Using technical terms in mathematical writing involves knowing the connections between and their logical implications of them (Moschkovitch, 2003; Fang, 2012b). The grammar of mathematics includes nominalizations, which are nouns that stem from adjectives or verbs. For instance, the transition from the verb *add* to the noun *addition* results in a change from an operation to an entire area of mathematics, altering cognitive expectations from performing a procedure to understanding a concept (Fang, 2012b). Mathematical writing also includes long noun phrases (Fang, 2012b). Table 2 demonstrates how multiple sentences can be reduced to a single noun phrase. Information-dense terms and grammatical structures allow for precision in mathematical writing.

Table 1. Distinction between recounting and logical reasoning expressed in writing

<table>
<tr><td colspan="2">Task
Adapted from Binary Arts Corporation (1995)</td></tr>
<tr><td colspan="2">Five miners charged out of the shaft as the timbers crashed behind them. Jeb was not the first out. Cain was neither first nor last. John Henry pushed out immediately behind Jeb. Preach was not second out. Zane was two men behind Preach.

a. In what order did the men race from the mine?
b. Explain how you determined the order.</td></tr>
<tr><td>Recounting</td><td>Logical Reasoning</td></tr>
<tr><td>
<table>
<tr><td></td><td>1</td><td>2</td><td>3</td><td>4</td><td>5</td></tr>
<tr><td>Jeb</td><td>X</td><td>X</td><td>X</td><td>√</td><td>X</td></tr>
<tr><td>Cain</td><td>X</td><td>√</td><td>X</td><td>X</td><td>X</td></tr>
<tr><td>John Henry</td><td>X</td><td>X</td><td>X</td><td>X</td><td>√</td></tr>
<tr><td>Preach</td><td>√</td><td>X</td><td>X</td><td>X</td><td>X</td></tr>
<tr><td>Zane</td><td>X</td><td>X</td><td>√</td><td>X</td><td>X</td></tr>
</table>
The order is Preach, Cain, Zane, Jeb, John Henry.
First, I made a table with the order across the top and the names down the side. I read the clues and eliminated positions by putting an X in the table. When I figured out the correct position, I put a checkmark. I continued with the process of elimination until I determined the order.
</td><td>Preach, Cain, Zane, Jeb, and then John Henry

Preach was the only minor who could be first since the clues indicate that Jeb nor Cain were first out, and John Henry and Zane came out behind others. Because Preach was first, Zane was two behind or third. The remaining available positions are 2, 4, and 5. Since John Henry pushed out immediately behind Jeb, Jeb must be fourth and John Henry fifth. Therefore, Cain is second, the only position left.</td></tr>
</table>

Language is not the only means of expressing mathematical ideas in writing. Other representations, such as abstract symbols and visuals, also hold meaning in a concise manner (Fang, 2012a, 2012b; Table 3). Abstract symbols achieve a level of precision that is difficult to achieve solely with language (Fang, 2012b), counter to others interpretation that writing needs to include words (Casa et al., 2023). For instance, the symbol < implies the words *is less than*. Visual representations appeal to the sensory experience of readers and include charts, tables, graphs, pictures, drawings, diagrams, and figures (Fang, 2012a, 2012b; Seo, 2016; Sfard, 2007). Although a single representation mode may suffice, writing in mathematics often involves the interplay between language, abstract symbols, and visuals (Fang, 2012b; Solomon & O'Neill, 1998).

Table 2. Example of how noun phrases condense mathematical writing

Deconstructed Noun Phrase	Noun Phrase
Sasha has a brother and sister. Her brother is 16 years old. Her sister is 9 years old. How much older is Sasha's brother?	Find *the difference in age between Sasha's 16-year-old brother and 9-year-old sister.*

Table 3. Three means for expressing a mathematical justification for a fraction magtude task

Which is greater $\frac{7}{12}$ or $\frac{2}{5}$?		
Language	**Symbols**	**Visual**
$\frac{6}{12}$ is one-half. $\frac{7}{12}$ is more than one-half. $\frac{2.5}{6}$ is one-half $\frac{2}{5}$ is less than one-half. Therefore, $\frac{7}{12}$ is greater than $\frac{2}{5}$.	$\frac{7}{12} > \frac{1}{2} > \frac{2}{5}$	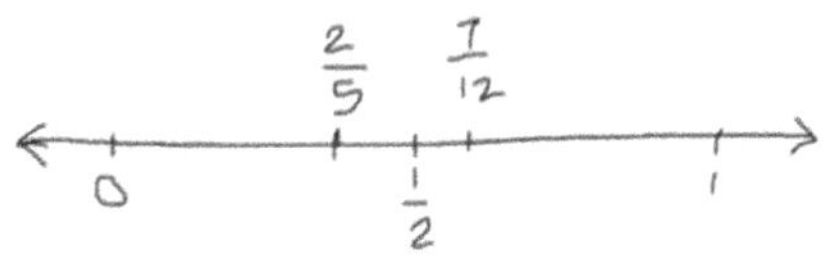

Standards Guidance for Disciplinary Writing in the K-12 Mathematics Classroom

How can students learn to write according to the aforementioned conventions in the field of mathematics? First, understanding how to write like a mathematician should start at a young age (Johnson et al., 2011). The *Common Core State Standards for English Language Arts & Literacy in History/Social Studies, Science, and Technical Subjects* (NGA & CCSSO, 2010b) calls for disciplinary literacy instruction to be a responsibility held across teachers of all content areas and all grade levels. Second, through purposefully designed instructional experiences, students must engage in an *apprenticeship* that reflects the disciplinary norms and activities in mathematics, including asking questions, investigating such questions, and communicating their reasoning via writing (Hillman, 2014; Johnson et al., 2011; Solomon & O'Neill, 1998). Standards documents provide guidance for the learning and teaching of disciplinary writing in mathematics (Brozo & Crain, 2018; Hillman, 2014; NCTM, 2000; NGA & CCSSO, 2010a). Three of the NCTM (2000) process standards (*Reasoning and Proof, Communication*, and *Representation*) speak to aspects of disciplinary writing, providing progressions of student learning. Likewise, critical components of disciplinary writing are embedded within the Common Core Standards for Mathematical Practice (NGA & CCSSO, 2010a), such as reason abstractly and quantitatively, construct viable arguments and critique the reasoning of others, use appropriate tools strategically, and attend to precision.

The activity of *reasoning* is critical to learning mathematics with understanding. Across the K-12 years, reasoning is based on mathematical explorations and evolves from informal ideas often grounded in examples and pattern identification (Johnson et al., 2011) to generalizations expressed through if-then statements (Temple & Doerr, 2018) to justifying those generalizations through the formal logical sequencing of proof (NCTM, 2000; NGA & CCSSO, 2010a). Students can *communicate* their reasoning both orally and through writing. In fact, in a study conducted by Pugalee (2014), students were more efficacious problem solvers when they shared their reasoning in writing rather than by thinking aloud. Writing in mathematics involves the incorporation of an array of mathematical *representations*. Initially, student writing includes unconventional pictorial representations with the eventual inclusion of some words. Students' writing then progresses to integrate the grammar of mathematics and more technical terms as well as other conventional mathematical representations such as abstract symbols and visuals (NCTM, 2000; NGA & CCSSO, 2010a). As students more seamlessly combine mathematical language, abstract symbols, and visuals, they attend to precision (Hillman, 2014; NGA & CCSSO, 2010a) and enhance their conceptual understanding (Moschkovich, 2003; NCTM, 2000).

Numerical Reasoning

Numerical reasoning is making sense of number and operations (Baek, 2017). Similarly, Ketterlin-Geller and Perry (2018) define early numeric relational reasoning as "the ability to mentally analyze relationships between numbers or expressions, often using knowledge of properties of operations, decomposition, and known facts" (para. 1). These definitions are grounded in seminal research (e.g., Carpenter & Moser, 1984; Fuson, 1988; Steffe, Cobb, & von Glasersfeld, 1988) and represented in components of the NCTM's (2000) Number and Operations standard, namely the understanding of numbers (number sense) and the meanings of operations (operation sense). The focus of the present chapter is to share the analysis of student reasoning (expressed in writing) about whole-number operations and the comparison of fractions. In the United States, the learning of operations with whole numbers spans grades Kindergarten through six (NGA & CCSSO, 2010a) with the relationship between addition and subtraction introduced in first grade (CCSS-M 1.OA.4) and foundations for the meaning of multiplication introduced in second grade (CCSS-M 2.OA.4). The learning of fractions starts in grade three with comparison of fractions featured in both grades three (CCSS-M 3.NF.3) and four (CCSS-M 4.NF.1 and CCSS-M 4.NF.2).

Number and Operation Sense for Whole Numbers

Number sense is described as "good intuition about numbers and their relationships. It develops gradually as a result of exploring numbers, visualizing them in a variety of contexts, and relating them in ways that are not limited by traditional algorithms" (Howden, 1989, p. 11). To move beyond counting, young children build number sense through understanding relationships between numbers. Three such relationships are one and two more and less, benchmarks of 5 and 10, and part-part-whole (i.e., decomposing a number into two or more component parts in a variety of ways) (Van de Walle et al., 2019). Developing this understanding of number within 10 can be extended to larger numbers. For instance, the relationship *two more than eight is 10* parallels the relationship between 18 and 20, 98 and 100, and 998 and 1000 (Van de Walle et al., 2019). Number sense has been found to predict general mathematics achievement in subsequent grades (Jordan et al., 2007; Jordan et al., 2009) as well as fluency with calculations (Locuniak, 2010). Further, it is foundational for the conceptual understanding of operating with numbers (Witzel et al., 2012).

Operation sense involves "understand[ing] the meaning of operations and how they relate to one another" (NCTM, 2000, p. 34). The four elementary operations are represented with different structures (e.g., multiplication of groups of equal size, Van de Walle et al., 2019) and can be interpreted using their inverse relationships (Caldwell et al., 2011; Otto et al., 2011). The combination of number and operation sense leads to fluency, which involves flexibility in the selection and appropriate implementation of efficient and accurate strategies (Heinze et al., 2018; Kling & Bay-Williams, 2014; NCTM, 2000; Parrish, 2011). Opportunities to generate, discuss, document, and evaluate an array of approaches of invented strategies and standard algorithms support the development of computational fluency (e.g., Heinze et al., 2018).

Fraction Magnitude

As students begin to work with fractions, they need to "recognize that fractions are numbers and that they expand the number system beyond whole numbers" (Siegler et al., 2010, p. 19). Initially, students may exhibit misconceptions by using whole-number reasoning in scenarios involving fractions (e.g., DeWolfe & Vosniadou, 2015; Siegler et al., 2011). For instance, students may perceive a fraction as two whole numbers instead of a single number (Cramer & Whitney, 2010; Malone & Fuchs, 2017). To counter such misconceptions, students must build their fraction sense. Initial engagement with concrete representations of fractions (e.g., area models, number lines) helps students visualize the various aspects of fraction sense, including the idea of the referent whole (Cramer & Whitney, 2010; Cramer et al., 2008); partitioning and subpartitioning (Johanning, 2011); unit fractions, inverse relationship between the size of unit fractions and the size of its denominator, and iteration of unit fractions (e.g., 4/5 consists of four parts of the unit fraction 1/5; Cramer et al., 2008; NGA & CCSSO, 2010a); and benchmark fractions (Fennell & Karp, 2017; Johanning, 2011).

Fraction sense supports the development of student understanding of fraction magnitude, or size (Cramer & Henry, 2002), through tasks such as the comparison of two fractions, one focus of the present chapter. Students, who demonstrate conceptual understanding, use their fraction sense to compare fractions. For instance, since $\frac{1}{8}$ is greater than $\frac{1}{12}$ (inverse relationship between the size of unit fractions and the size of its denominator), $\frac{5}{8}$ is greater than $\frac{5}{12}$ (iteration of unit fractions; Cramer et al., 2002). In contrast, algorithmic approaches, such as finding a common denominator and comparing the numerators or calculating and comparing the cross products, emphasize whole-number thinking (Cramer & Whitney, 2010). Students' orientations (conceptual or algorithmic) towards fraction magnitude mirrors their orientations as they transition to operating with fractions (Fennel & Karp, 2017) with algorithmic orientations often resulting in faulty or inappropriate application of procedures (Karp et al., 2015).

PURPOSE

The purpose of the present chapter is to illustrate the exploration of the use of the *Framework for Evaluating Quantitative Reasoning Strategies* (hereafter referred to as *Framework*) to analyze the reasoning, expressed in writing, of fourth-grade students in the area of number and operations. Specifically, the empirically developed *Framework* distinguishes five categories of numerical reasoning. The *Framework* was used to evaluate students' written responses to three operations with whole numbers and four comparison-of-fractions tasks. The findings demonstrate how use of the *Framework* is a practical and effective way to examine the depth of student reasoning and disciplinary writing to document such reasoning.

METHODS

Initial Development of the Framework

The *Framework* is the result of three-phases of studies. For the first phase, 523 students provided responses to whole number and fraction tasks from the Math Reasoning Inventory (MRI; Burns, 2012). Examiners administered the MRI with students in one-on-one sessions lasting approximately 30 minutes. During this session, the examiner read a question to the student, the student replied verbally (without paper, pencil, or calculator), and the examiner entered the student's answer onto the MRI website. Next, the examiner asked, "How did you figure that out?" Each explanation was entered verbatim into the MRI portal. The researchers analyzed responses to over 4000 tasks using a cluster analysis (Crawford et al., 2018), resulting in five distinct categories: (a) conceptual reasoning strategy, (b) partial reasoning strategy, (c) algorithmic reasoning strategy, (d) unfounded reasoning strategy, and (e) no reasoning strategy (Crawford et al., 2018).

A sample, consisting of 428 fifth-grade students, was obtained from the cluster analysis study for the second phase study. The researchers conducted a latent class analysis (LCA) only on the fraction magnitude tasks due to the implication that a lack of fraction sense, particularly fraction magnitude, has for learning operations with fractions with a strong procedural rather than conceptual orientation (Fennell & Karp, 2017). The LCA identified five relevant profiles very similar to those identified through the cluster analysis: (a) conceptual-fully developed (11.2%; $n = 48$), (b) conceptual-partially developed (29%, $n = 122$), (c) algorithmic (27%, $n = 115$), (d) faulty (18.7%; $n = 80$), and (e) absent (14.7%; $n = 63$) (Crawford et al., 2020). The reliance on algorithmic reasoning (e.g., comparing two fractions using the cross-products algorithm) could be due to the preponderance of procedurally-focused instruction in the United States or the MRI protocol, asking students *how* they arrived at a particular answer rather than *why* they arrived at a particular answer (Herringer, 2013; Tuttle, 2005).

Use of the Framework to Evaluate Numerical Reasoning Expressed via Writing

The first two phases established the evidence base for the reliability and validity of the *Framework* (Table 4) in a one-on-one setting with students sharing their reasoning verbally. Therefore, the *Framework* needs to be validated for use with written responses. Prior to such validation, the third phase, which is the primary focus of the present chapter, entails exploring the use the *Framework* generated from the LCA profiles to analyze students' *written* responses, explaining *why* they arrived at their answers. This phase also differs from previous phases in that data were collected on fourth-grade students in a whole-class setting with the understanding that, beginning at age nine, children have the capacity to develop disciplinary literacy by integrating technical language with everyday and educational language (Fang, 2012a). The tasks in all three phases remained similar.

Participants

Participants included 394 fourth-grade students, typically 9 to 10 years of age, all of whom completed the measures within their general education homerooms. Students attended 23 general education classrooms in six schools (four large public schools and two small private schools) in the Southwest United States. Eighty-one percent of students attended a public school and 19% attended a private school. Although not randomly selected, the sample was normative as it included all students in fourth grade at the participating school sites. Participants demonstrated a range of achievement levels from very low performing to very high performing based on the Wide Range Achievement Test – Fourth Edition mathematics computation subtest (WRAT-IV; Wilkinson & Robertson, 2006). Their standard scores ranged from 55 –120 with a mean score of M = 70.91(11.01). Forty-nine percent of the students were male and 51% female, 74.1% identified as White/Caucasian, 12.8% as multi-racial, 6.8% as Asian American, 5.2% as African American, and 1% American Indian/Alaska Native. With respect to ethnicity, 12.4% identified as Hispanic/Latino. Findings are reflective of those students for whom we had 100% complete data on whole-number tasks (n = 349) and fraction tasks (n = 385).

Procedures

During the test administration, students were provided with 15 minutes to complete the mathematics computation subtest of the WRAT-4 (Wilkinson & Robertson, 2006) and 28 minutes to complete three whole-number and four fraction magnitude tasks (Table 5). Students also shared written explanations of *why* each of their answers was appropriate. Prior to completing these tasks, the administrator modeled various explanations of reasoning, reading each example problem aloud, the hypothetical answer recorded, and the hypothetical response to the question of *Why?* The first example response used only words while the second example response consisted of words accompanied by an illustration. Neither example pertained to mathematics. The third example applied to mathematics ("Which segment is longer?") and was answered correctly in each response but justified using very different reasoning. The first explanation included a drawing accompanied by the explanation, "Segment A is about 4 paperclips long and Segment B is about two paperclips long. So, Segment A is longer." The second explanation included only words, "About two of Segment B equal one of Segment A," as did the third explanation, "If you used a ruler to measure each segment, Segment A would consist of more of the same size units (inches, centimeters, etc.), than Segment B."

After students completed the seven tasks and provided their written explanations, authors reviewed the *Framework* to discuss scoring procedures (only one author was new to the process). Two authors took responsibility for scoring all protocols with weekly meetings to discuss tasks for which agreement was not reached with the second author making the final decision. Interrater reliability was calculated as agreements divided by agreements plus disagreements and is reported at 77% on a total of 199 tasks during round one (85% for fraction tasks and 70% for whole number tasks). After additional training and discussion, interrater reliability is reported as 86% across 175 different tasks during round two (89% for fraction tasks and 83% for whole number tasks). Fraction tasks resulted in higher rates of agreement than whole-number tasks. Because the *Framework* is composed of five categories, more opportunities exist for non-agreements than if scoring for correct versus incorrect, thus 86% agreement was deemed to be acceptable.

FINDINGS

For students' written expressions of their reasoning across the three whole-number tasks, the percentages for the *algorithmic*, *emerging conceptual*, and *conceptual* categories were similar (33.1%, 29.6%, and 26.9%, respectively) as were those for the *absent* and *faulty* categories (6.4% and 4.0%, respectively;

Table 4. Framework for evaluating quantitative reasoning strategies

Category	Definition	Whole Number Explanations	Fraction Explanations
Fully Developed Conceptual Understanding	The student shares reasoning founded on the intrinsic mathematical properties of the components of the task with or without describing a specific procedure. Evidence of reasoning does not require written words and may be demonstrated through use of visual representations and/or abstract symbols.	Demonstration of number sense as well as operation sense. Depending on problem type, response may include evidence of one or two more or less,, benchmark, or part-part-whole (i.e., decomposing and composing numbers) number relationships, and/ or understanding of place value. If the most logical solution involves use of inverse operations, reasoning reflects this understanding beyond application of algorithm.	Demonstration of fraction sense (i.e., fully developed conceptual understanding of fractions as numbers). For example, illustration or model and/or explanation of concepts such as residual parts, use of benchmarks, or existence of the referent whole, with accurate partitioning of equal parts.
Emerging Conceptual Understanding	The student uses reasoning that is logical and heading in the right direction but is not fully developed.	Demonstration of initial number sense and/or operation sense. Depending on problem type, response may include partial understanding of whole-number relationships, place value, and/or inverse operations. If the latter, no additional evidence provided beyond algorithm.	Demonstration of initial fraction sense (i.e., conceptual understanding of fractions as numbers). For example, illustration of an area model with or without written explanation of equivalent parts, or application of a strategy such as benchmarking inaccurately or inefficiently.
Algorithmic	The student applies a set of rules that guarantees a correct solution will be reached; remaining reasoning parts are trivial for the reasoner.	No demonstration of number sense and/ or operation sense. For example, instead of counting on for simple subtraction problem, student uses algorithm. The algorithm may be shown numerically or with words describing the procedure without reference to place value.	No demonstration of fraction sense. For example, use of cross-multiplication (i.e., the "butterfly" method), that relies on a whole-number comparison to describe the size of two fractions.
Faulty	The student uses reasoning that is incorrect (a logical fallacy) or guesses at the solution. Response may be partially or fully developed.	No explicit and obvious demonstration of number sense and/or operation sense. For example, overgeneralization of a strategy that may be *effective* if used in a different context, or use of a novel strategy based on a misconception. In some cases, the difference between *faulty* and *emerging* resides in the level of inference that must be applied to response provided.	No explicit and obvious demonstration of fraction sense (i.e., conceptual understanding of fractions as numbers). For example, overgeneralization of a strategy that may be *effective* if used in a different context, or use of a novel strategy based on a misconception. In some cases, the difference between *faulty* and *emerging* resides in the level of inference that must be applied to response provided.
Absent	The student provides no illustration or explanation to support how he or she arrived at an answer.	No evidence of number sense and/or operation sense. For example, response may be shared but no explanation or justification beyond restatement of answer is provided.	No explanation of how student arrived at answer. For example, student shares "I don't know," or restates the answer to original problem.

Table 6). However, the type of reasoning students shared varied across task type. For example, 50.1% of responses for Task 1 were *algorithmic* with the use of a standard subtraction algorithm. The preponderance of responses for Task 2 were *emerging conceptual* (59.9%) due to a scoring decision that if students used a subtraction algorithm to find the difference 200 – 15 = 185, they demonstrated an understanding of the relationship between addition and subtraction. For Task 3, the percentage of responses was similar between *algorithmic* (33.8%) and *conceptual* (37.8%). The responses documented for the whole-number tasks displayed stronger *emerging* and *conceptual* reasoning than did the responses for the fraction tasks (56.5% versus 34.2%), which is developmentally appropriate for the fourth-grade participants. Table 7

Table 5. Whole-number and fraction magnitude tasks completed by students

Task Number	Question
1	What is 1000 minus 998?
2	Figure out the missing number: 15 + ____ = 200
3	The answer is 20 × 15 = 300. Use this information to figure out the answer to 21 × 15.
4	Which is greater $\frac{3}{8}$ or $\frac{5}{6}$?
5	Which is greater $\frac{5}{12}$ or $\frac{5}{8}$?
6	Which is greater $\frac{7}{12}$ or $\frac{2}{5}$?
7	Carolos lives $\frac{3}{4}$ of a mile from school. Terrell lives $\frac{6}{8}$ of a mile from school. Which of these is correct? Both boys live the same distance from school. One boy lives farther from school.

Table 6. Reasoning category percentages for whole-number tasks

Reasoning Category	Task 1	Task 2	Task 3	Total
Absent	8.0	5.7	5.4	6.4
Faulty	2.9	3.4	5.7	4.0
Algorithmic	50.1	15.5	33.8	33.1
Emerging Conceptual	11.5	59.9	17.2	29.6
Conceptual	27.5	15.5	37.8	26.9
Total	100.0	100.0	100.0	100.0

n = 349 responses per task

Table 7. Reasoning category percentages for fraction tasks

Reasoning Category	Task 4	Task 5	Task 6	Task 7	Total
Absent	2.3	4.9	5.5	10.6	5.8
Faulty	13.8	14.5	16.6	11.9	14.2
Algorithmic	43.4	36.9	42.1	60.5	45.8
Emerging Conceptual	28.6	32.2	24.7	8.1	23.4
Conceptual	11.9	11.4	11.2	8.8	10.8
Total	100.0	100.0	100.0	100.0	100.0

n = 385 responses per task

also shows that more reasoning was categorized as *faulty* or *algorithmic* for the fraction tasks in comparison to the whole-number tasks. In particular, Task 7 elicited *algorithmic* reasoning with students using the common denominator strategy quite often (60.5%). In comparison, the percentages for Tasks 4, 5, and 6 reflected the overall percentages for the fraction tasks.

The following sections include examples of students' written responses for the *algorithmic*, *faulty*, *emerging conceptual*, and *conceptual* categories across the whole-number and fraction tasks. The *absent* category represents no illustration or explanation to support how a student arrived at an answer.

Algorithmic

When students used a standardized procedure without any indication of number or operation sense, their response was coded as *algorithmic*. Figure 1 provides examples of algorithmic responses for the whole-number tasks. Student WN.A.1 uses a standard subtraction algorithm to get the correct answer. This student accompanies the symbolic representation of the algorithm with a detailed *recounting* of each step in the algorithm described in words. The explanation focuses on the digits rather than the numbers,

Figure 1. Examples of algorithmic reasoning for whole-number tasks

Student	Problem (Student Answer)	Reasoning
WN.A.1	What is 1000 minus 998? (2)	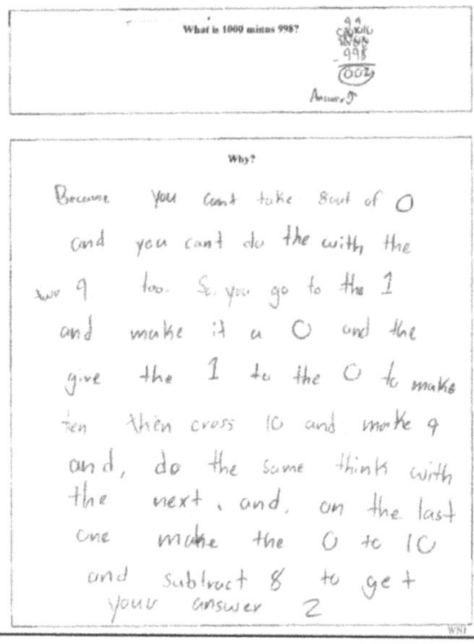
WN.A.2	The answer is to 20 × 15 = 300. Use this information to figure out the answer to 21 × 15. (315)	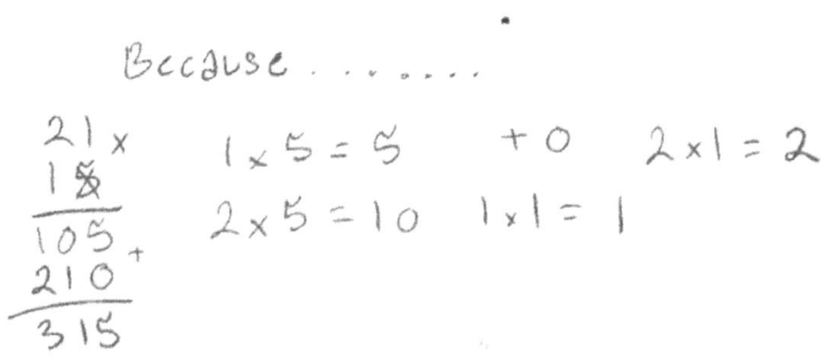

Figure 2. Example of algorithmic reasoning for fraction task

Student	Problem (Student Answer)	Reasoning
F.A.1	Which is greater $\frac{3}{8}$ or $\frac{5}{6}$? ($\frac{5}{6}$)	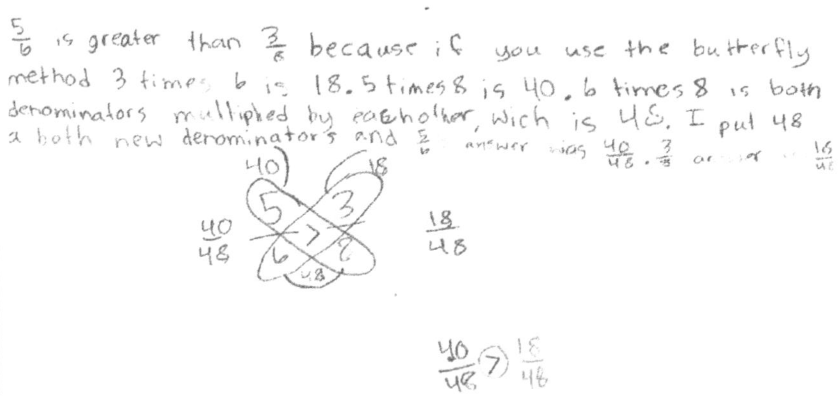

Figure 3. Examples of faulty reasoning for whole-number tasks

Student	Problem (Student Answer)	Reasoning
WN.F.1	What is 1000 minus 998? (1?)	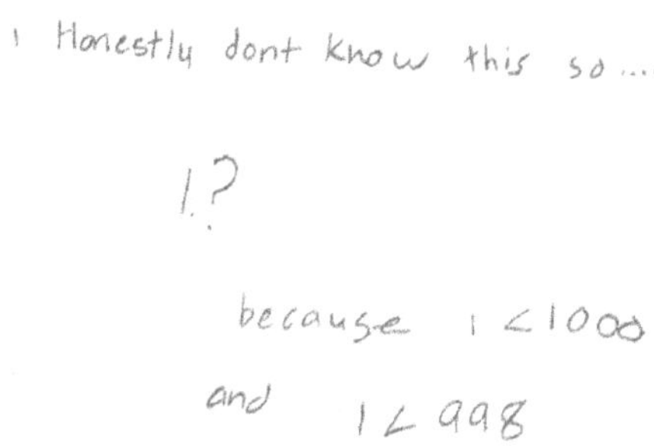
WN.F.2	The answer is to 20 × 15 = 300. Use this information to figure out the answer to 21 × 15. (25)	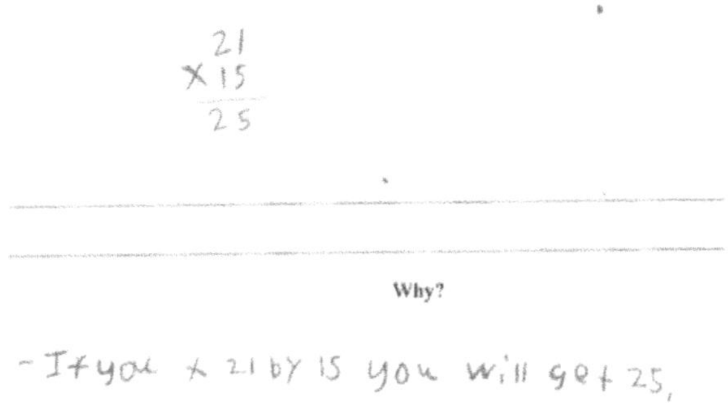

lacking any reference to place value or regrouping. For instance, the student treats the 1 in the thousands place (i.e., 1 thousand) simply as a 1, writing "*So, you go [sic] the 1 and make it a 0 and the [sic] give the 1 to the 0 to make ten.*" Similarly, Student WN.A.2 does not use the given information ($20 \times 15 = 300$) to solve the task; instead, the student uses a standard multiplication algorithm. This student justifies the algorithm symbolically by showing each of the partial products without considering the value of each number (e.g., $2 \times 5 = 10$ should be $20 \times 5 = 100$).

In Figure 2, Student F.A.1 shows two common algorithmic approaches for comparing fractions (common denominator and cross products, referenced as the butterfly method). The student accompanies both procedures, represented symbolically, with words describing each step. Rather than demonstrating fraction sense, the student relies on whole-number thinking by comparing the cross products 40 and 18.

Faulty

A *faulty* response is characterized by reasoning that is incorrect (a logical fallacy) or a guess at an answer. The examples of *faulty* reasoning for whole-number tasks in Figure 3 each demonstrate one of the types of *faulty* reasoning. Student WN.F.1 indicates both in words, writing "*Honestly don't* [sic] *know this*" and symbolically, writing "1?" that they guessed the incorrect answer of 1. Further, the student attempted to justify the answer, again with a mix of symbols and words, with two true statements that do not make sense in relation to the task itself, writing '*because* $1 < 1000$ *and* $1 < 998$". Student WN.F.2 used an incorrect procedure for determining the product. The student multiplied the digits in the ones place of each factor to get the digit in the ones place of the product ($1 \times 5 = 5$) and did the same for the tens place ($2 \times 1 = 2$). The words accompanying the symbolic representation simply restate the problem and answer in *if-then* form. Further, the student does not recognize that the product of 25 does not make sense; that is, if $20 \times 15 = 300$, then 21×15 should be greater than 300.

The three *faulty* examples for the fraction comparison tasks in Figure 4 all include words to document the reasoning. Student F.F.1 uses whole-number reasoning. Instead of considering $\frac{3}{4}$ and $\frac{6}{8}$ as numbers, this student just compares the numerators to arrive at an incorrect answer. In the written explanation, the student demonstrates a lack of understanding of the role of the denominator in determining the size of a fraction by stating that "*its the 2 numbers on top that gives you the answer.*" Students F.F.2 and F.F.3 both only attend to one fraction when deciding which fraction is greater. Rather than considering how the two fractions relate to each other, they justify their choice claiming it is closer to one. With this *faulty* strategy, Student F.F.2 got the correct answer and Student F.F.3 got the incorrect answer. The written explanation is necessary to see that both students were using the same *faulty* reasoning despite the difference in the correctness of their answers.

Emerging Conceptual

Emerging conceptual understanding is characterized by logical reasoning that is headed in the right direction but not yet fully developed. Figure 5 has three examples of *emerging conceptual* reasoning for the whole-number tasks. Student WN.EC.1 makes an error using the standard algorithm. Although the student gets an incorrect answer of 1, the student justifies their answer by making the statement that *998 is almost the same as 1000*, an observation grounded in number sense with respect to magnitude. Student WN.EC.2 uses two different approaches to determine the answer to the task. For the strategy on

Figure 4. Examples of faulty reasoning for fraction tasks

Student	Problem (Student Answer)	Reasoning
F.F.1	Carlos lives $\frac{3}{4}$ of a mile from school. Terrell lives $\frac{6}{8}$ of a mile from school. Which of these is correct? Both boys live the same distance from school. (One boy lives farther from school.)	Terrell lives farther because 6 is grater than 3 and its the 2 numbers on top that gives you the answar
F.F.2	Which is greater $\frac{5}{12}$ or $\frac{5}{8}$? ($\frac{5}{8}$)	$\frac{5}{8}$ is greater because it is almost a whole
F.F.3	Carlos lives $\frac{3}{4}$ of a mile from school. Terrell lives $\frac{6}{8}$ of a mile from school. Which of these is correct? Both boys live the same distance from school. (One boy lives farther from school.)	I think one boy Lives farther from school wich is $\frac{6}{8}$ because he is almost a hole away from school.

the right, the student uses an area model for a visual representation of the distributive property, (20 + 1) × (10 + 5), to calculate and add the four partial products to get the overall product (200 + 10 + 100 + 5 = 315). In their written explanation, the student uses the descriptor *window pane* (i.e., the visual looks like a window pane) rather than the appropriate technical term, *distributive property*. For the second strategy on the left, Student WN.EC.2 demonstrates an understanding of the meaning of multiplication in words (*you could count 21, 15 times*) and by making a table with 21 listed 15 times. For both strategies, the student determines the product of 21 × 15 rather than using the given information that 20 × 15 = 300. Student WN.EC.3 creates a story problem that represents 185 + 15 = , using benchmarks of

Figure 5. Examples of emerging conceptual reasoning for whole-number tasks

Student	Problem (Student Answer)	Reasoning
WN.EC.1	What is 1000 minus 998? (1)	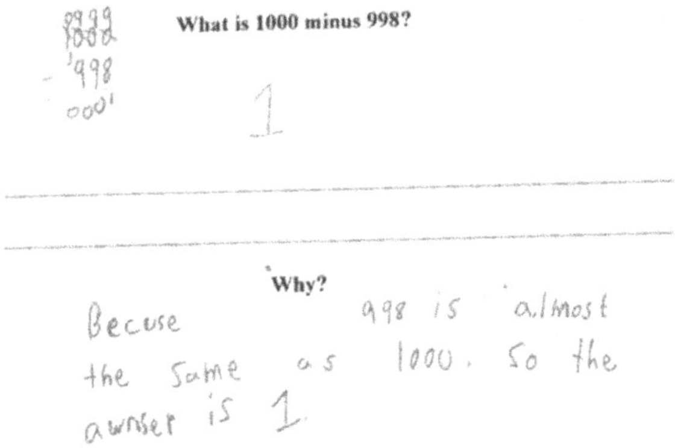
WN.EC.2	The answer is to 20 × 15 = 300. Use this information to figure out the answer to 21 × 15. (315)	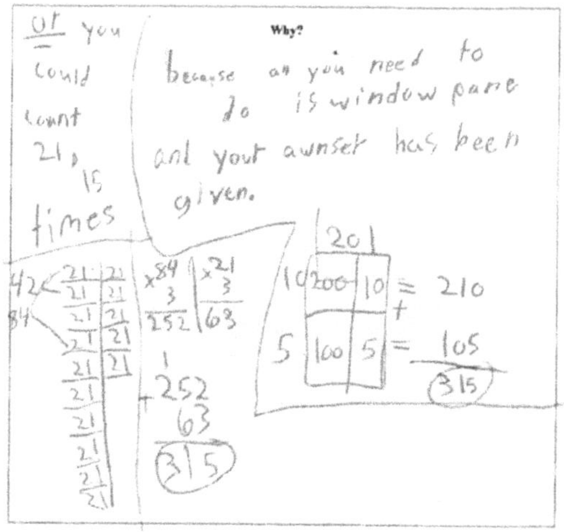
WN.EC.3	Figure out the missing number: 15 + ___ = 200 (185)	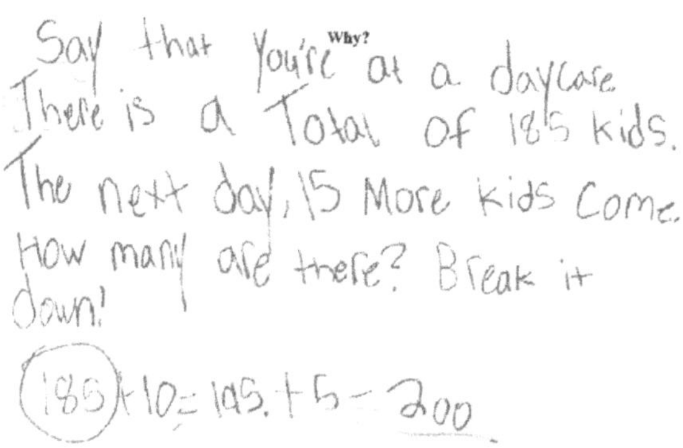

10 and 5 to add on from 185 to verify that 185 + 15 = 200; however, the student does not provide any explanation for how they determined that the missing addend is 185. Additionally, the use of the equal sign in their work is incorrect since 185 + 10 does not equal 200.

Figure 6 provides examples of emerging conceptual reasoning written by students for the fraction comparison tasks. Student F.EC.1 tries to use a residual strategy by comparing each fraction to the benchmark of one. However, in their explanation, the student transitions to whole-number thinking, expressing the residuals as one and five rather than $\frac{1}{6}$ and $\frac{5}{8}$, respectively. Student F.EC.2 also attempts to use benchmarks but, in their written explanation, compares each fraction to different benchmarks, $\frac{1}{2}$ and 1, and omits to compare the benchmarks to support their conclusion. In contrast, Student F.EC.3 employs the visual representation of a number line to compare each fraction to what appears to be 0, $\frac{1}{2}$, and 1; however, none of the benchmarks are explicitly labeled and the associated words do not explain why "$\frac{5}{6}$ *is further to the right*."

Figure 6. Examples of emerging conceptual reasoning for fraction tasks

Student	Problem (Student Answer)	Reasoning
F.EC.1	Which is greater $\frac{3}{8}$ or $\frac{5}{6}$? ($\frac{5}{6}$)	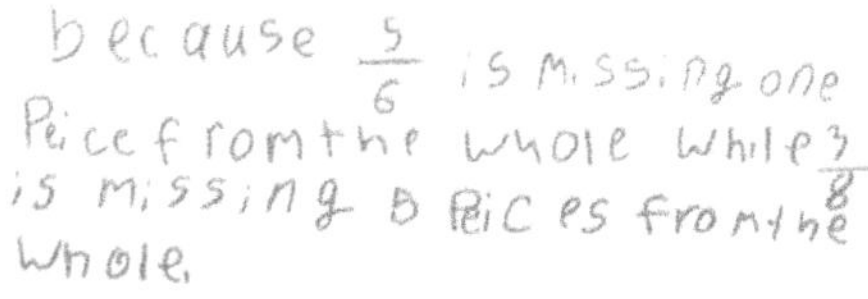
F.EC.2	Which is greater $\frac{3}{8}$ or $\frac{5}{6}$? ($\frac{5}{6}$)	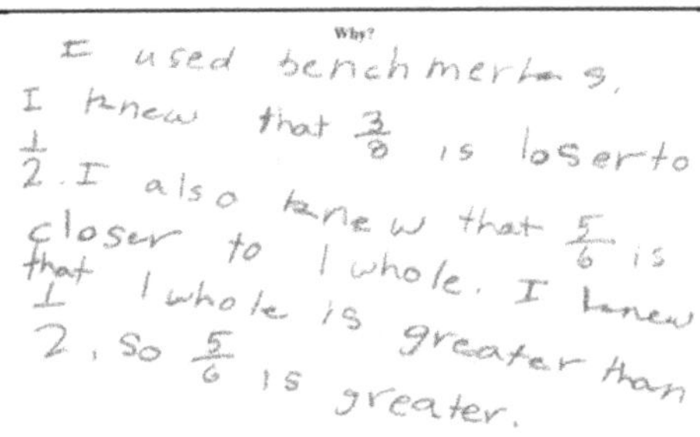
F.EC.3	Which is greater $\frac{3}{8}$ or $\frac{5}{6}$? ($\frac{5}{6}$)	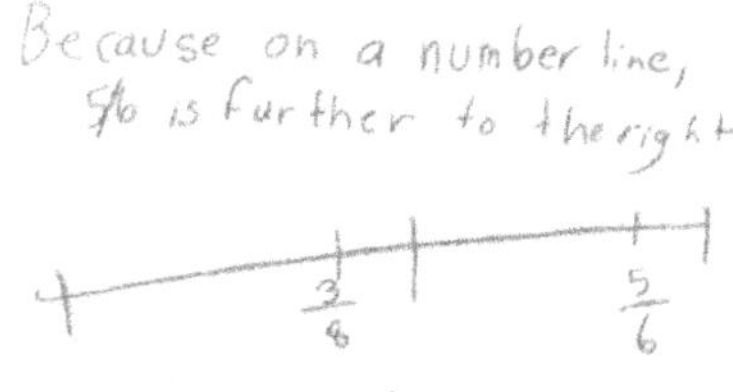

Conceptual

Fully developed *conceptual* understanding reflects reasoning founded on the intrinsic mathematical properties of the components of the task with or without describing a specific procedure. Figures 7 and 8 provides examples of *conceptual* reasoning across the three whole-number tasks. Student WN.C.1 uses both operation sense (i.e., relationship between addition and subtraction) and number sense (benchmark of 10, two more than, and part-part-whole number relationships) to determine that "*you only need two more to get from 998 to 1000.*" Student WN.C.2 adds on to 15 to make 200 using benchmarks of five and 10. Although their use of the equal sign is incorrect, $15 + 5$ does not equal 100, the student uses visual markers to show how they arrived at the answer of 185. Student WN.C.3 extends their knowledge of the fact that $85 + 15 = 100$ by adding on 100 to make 200, explaining their reasoning in a combination of words and abstract symbols. For the last three examples, the students combine their understanding of multiplication (i.e., 20×15 is 20 equal groups of size 15) and the given information $20 \times 15 = 300$) to explain that they need one additional group of 15 to determine the answer to 21×15. The documentation of their reasoning demonstrates different mediums for communicating in writing (visual, words, and symbols) and an increasing level of precision. Student WN.C.4 uses circles enclosing the number 15 to represent the equal groups, Student WN.C.5 employs a mix of words and symbols (e.g., 20×15 *is* $15 + 15 +15 + 15$ … 20 *times*), and Student WN.C.6 symbolically expresses the additional group of 15 ($20 \times 15 + 15$).

For comparison of fractions tasks (Figures 9 and 10), the implicit assumption is that the referent whole is the same size. Student F.C.1 does not make this assumption and presents area models to demonstrate that $\frac{2}{5}$ could be larger than $\frac{7}{12}$ and show that the answer depends on the size of the referent whole. Students F.C.2 and F.C.3 both explain their answer by relying on equivalent fractions, $\frac{1}{4} = \frac{2}{8}$, using different representations in their explanations, area model and table, respectively. With a combination of area models and words, Student F.C.4 represents and explains several fraction sense concepts (e.g., same size referent whole with equal size partitions, inverse relationship between the size of unit fractions and the size of its denominator, and iteration of unit fractions) to show that since $\frac{1}{8} > \frac{1}{12}$, and given that both have the same numerators then $\frac{5}{8} > \frac{5}{12}$. The last three students used the benchmark of $\frac{1}{2}$ to compare the two fractions. Student F.C.5 used a number line with three benchmarks 0, $\frac{1}{2}$, and 1 labeled, and used words to name the representation employed. In contrast, the words of Student F.C.6 provide a mathematical support to the answer that is also represented with abstract symbols and area models. Any one of these means of written communication (words, abstract symbols, or visuals) could be used to explain the reasoning, as demonstrated with the precise symbolic representation of Student F.C.7.

DISCUSSION

The National Academy of Education (1999) states: "Carefully designed tools that educators find useful in their practice can … become a powerful means of changing educational practice" (p. 35). The *Framework* presented in this chapter can serve as a means to enhance teachers' content knowledge as well as knowledge of the components of disciplinary writing in mathematics. To teach mathematics with understanding and recognize conceptual reasoning of their students, teachers must have a strong

Figure 7. Examples of conceptual reasoning for whole-number tasks

Student	Problem (Student Answer)	Reasoning
WN.C.1	What is 1000 minus 998? (2)	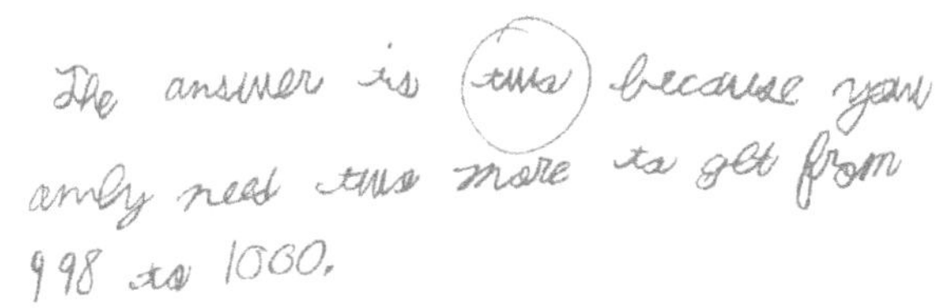
WN.C.2	Figure out the missing number: 15 + ___ = 200 (185)	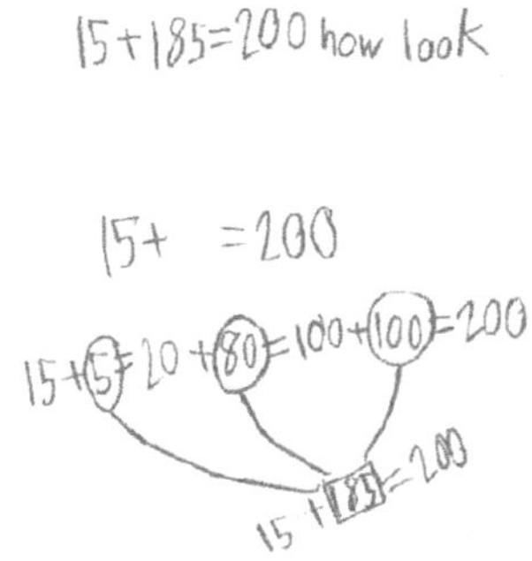
WN.C.3	Figure out the missing number: 15 + ___ = 200 (185)	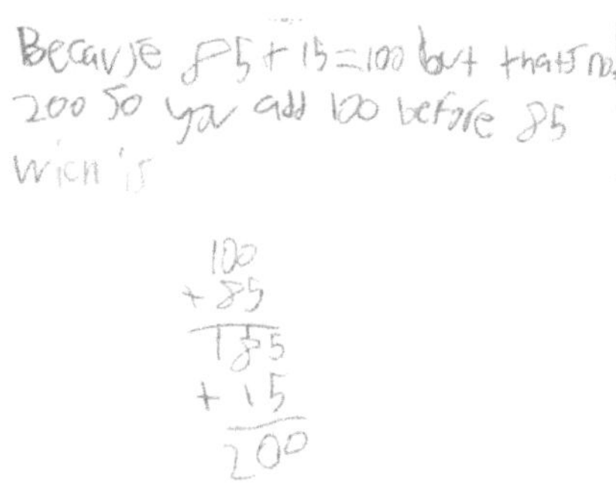
WN.C.4	The answer is to 20 × 15 = 300. Use this information to figure out the answer to 21 × 15. (315)	

Figure 8. Examples of conceptual reasoning for whole-number tasks

WN.C.5	The answer is to 20 × 15 = 300. Use this information to figure out the answer to 21 × 15. (315)	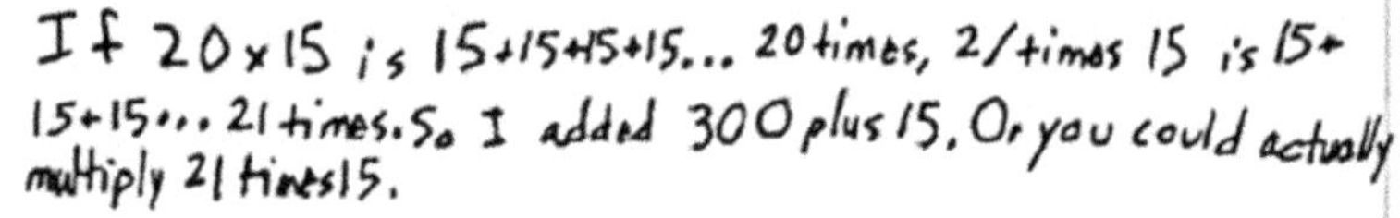
WN.C.6	The answer is to 20 × 15 = 300. Use this information to figure out the answer to 21 × 15. (315)	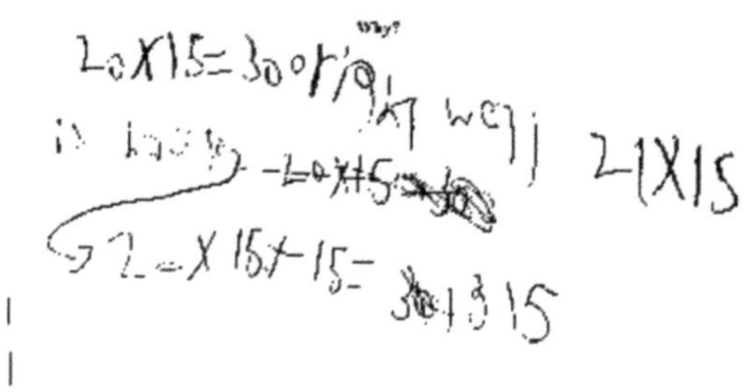

understanding of concepts themselves (Temple & Doerr, 2018). The *conceptual* responses presented provide insight into the nature of number and operation sense for whole numbers and fraction sense. The analysis of student responses also highlights the various aspects of disciplinary writing in mathematics. The intersection of the two areas of knowledge, content and disciplinary writing, opens the opportunity to apprentice students in the disciplinary practices of mathematics and to develop students' conceptual knowledge (Hillman, 2014; Johnson et al., 2011; Solomon & O'Neill, 1998; Temple & Doerr, 2018).

Students need more than the chance to write in the mathematics classroom; they need targeted opportunities to write, assess, and revise their writing so that it reflects the disciplinary norms of the field of mathematics (Fang, 2012a). The *Framework* helps teachers identify the unique features of disciplinary writing in mathematics. In particular, the *algorithmic* and *conceptual* categories make a distinction between recounting a process (e.g., Students WN.A.1 and F.A.1) and providing a mathematical justification (e.g., Students WN.C.2 and F.C.6; Marks & Mousley, 1990; Solomon & O'Neill, 1998). The *conceptual* responses also demonstrate the integration of words, abstract symbols, and visual representations (e.g., Students WN.C.2 and F.C.4; NCTM, 2000; Seo, 2016; Sfard, 2007). In addition, comparing *conceptual* responses provides insight into how to achieve precision (e.g., Students WN.C .4 – 7 and Students F.C.4 – 7; NGA & CCSSO, 2010a; Solomon & O'Neill, 1998).

The *Framework* can also serve as a structure around which teachers can collaborate. NCTM (2014) calls for teachers to "work collaboratively with colleagues to plan instruction, solve common challenges, and provide mutual support as they take collective responsibility for student learning" (p. 5). As demonstrated, the *Framework* was used to assess student reasoning for different areas of number and operations that are focal points across several elementary grades (NGA & CCSSO, 2010a). The *Framework* has the potential to be a common formative assessment to evaluate instruction and student progress across elementary grade levels, supporting instructional alignment focused on effective teaching practices (e.g.,

Figure 9. Examples of conceptual reasoning for fraction tasks

<table>
<tr><th>Student</th><th>Problem
(Student Answer)</th><th>Reasoning</th></tr>
<tr><td>F.C.1</td><td>Which is greater $\frac{7}{12}$ or $\frac{2}{5}$?

(Both could be)</td><td>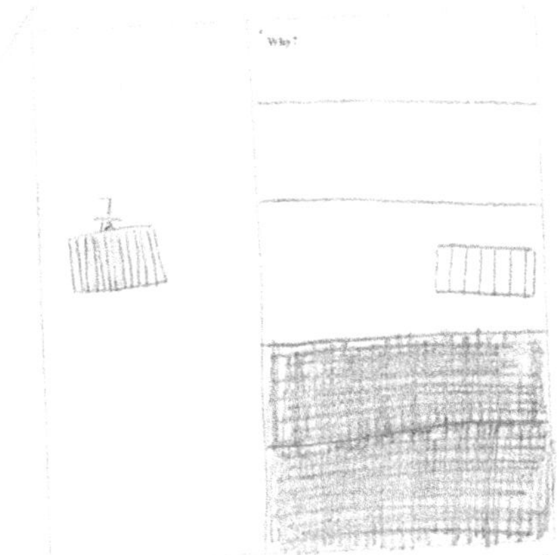</td></tr>
<tr><td>F.C.2</td><td>Carlos lives $\frac{3}{4}$ of a mile from school. Terrell lives $\frac{6}{8}$ of a mile from school. Which of these is correct?

(Both boys live the same distance from school.)

One boy lives farther from school.</td><td>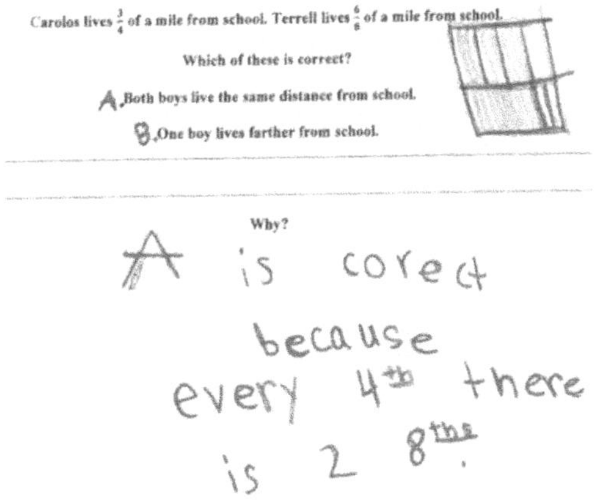
</td></tr>
<tr><td>F.C.3</td><td>Carlos lives $\frac{3}{4}$ of a mile from school. Terrell lives $\frac{6}{8}$ of a mile from school. Which of these is correct?

(Both boys live the same distance from school.)

One boy lives farther from school.</td><td>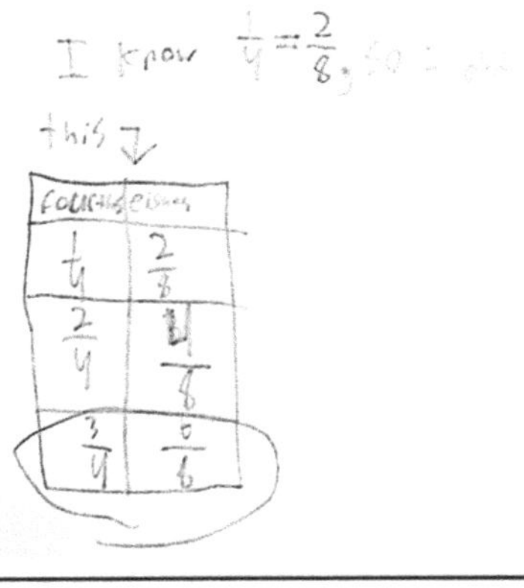
</td></tr>
</table>

Use and connect mathematical representations; Build procedural fluency from conceptual understanding; and Elicit and use evidence of student thinking [NCTM, 2014, p. 3]).

Figure 10. Examples of conceptual reasoning for fraction tasks

F.C.4	Which is greater $\frac{5}{12}$ or $\frac{5}{8}$? ($\frac{5}{8}$)	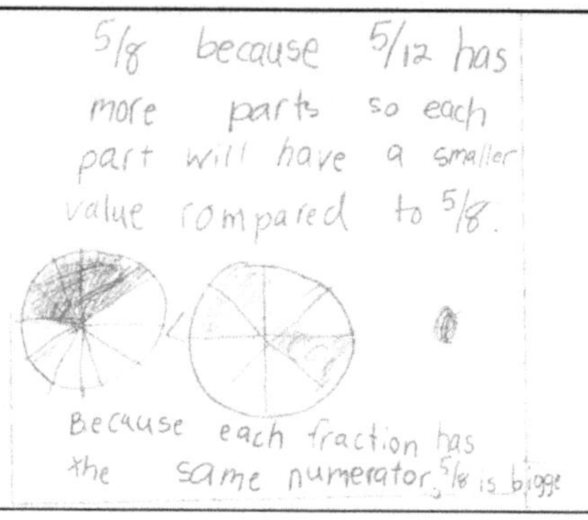
F.C.5	Which is greater $\frac{5}{12}$ or $\frac{5}{8}$? ($\frac{5}{8}$)	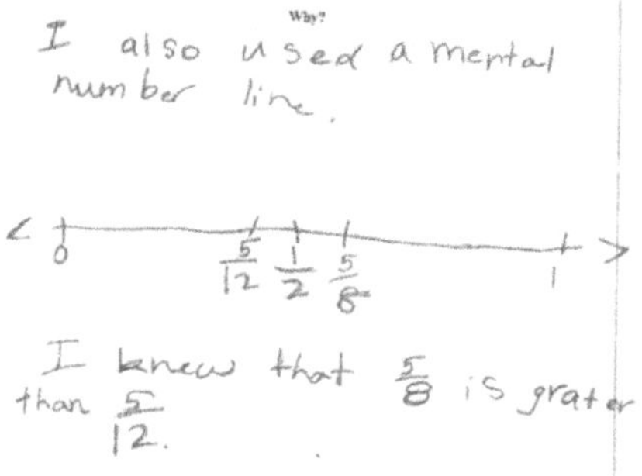
F.C.6	Which is greater $\frac{3}{8}$ or $\frac{5}{6}$? ($\frac{5}{6}$)	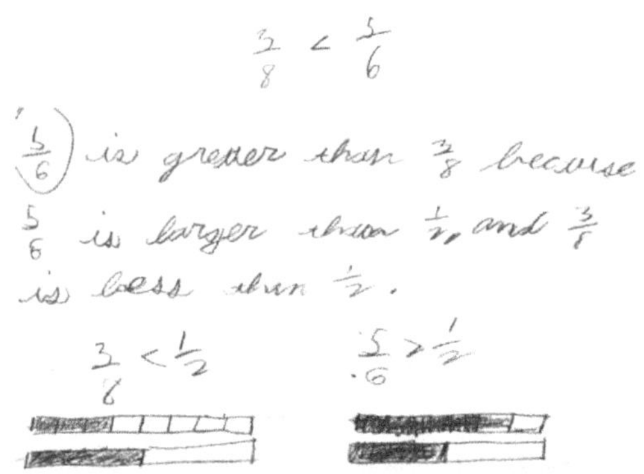
F.C.7	Which is greater $\frac{7}{12}$ or $\frac{2}{5}$? ($\frac{7}{12}$)	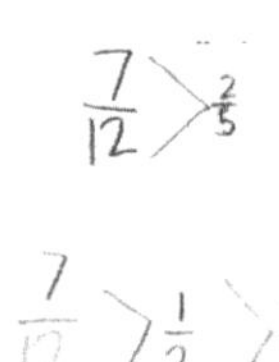

Since the *Framework* is a screener rather than a diagnostic tool, its use allows for rich discussions about the learning and teaching of mathematics within grade-level or vertical planning teams. First, with the *Framework*, teachers move beyond assessing students based on correct or incorrect answers to analyzing the nature of student reasoning expressed via writing. Students can get correct answers but have *faulty* or *emerging conceptual* reasoning; that is, underdeveloped number and operation sense (e.g., Students F.F.3 and F.EC.4; Geller et al., 2017). Second, teachers also need to consider the current expectations for students. For instance, standards related to comparison of fractions fall in third and fourth grade (NGA & CCSSO, 2010a). Many fourth-grade students' reasoning may be *emerging conceptual* rather than *fully conceptual* (as seen for Tasks 1, 2, and 3 in Table 7) since they are still developing and applying their fraction sense in this context. Third, teachers may have difficulty distinguishing between two of the reasoning categories (e.g., *emerging conceptual* and *conceptual*). One approach to using the *Framework* in such situations is to consider if assumptions about student thinking need to be made to justify a category. For instance, the response of Student WN.EC.3 could be categorized as *conceptual* only if assumptions were made about how the answer of 185 was determined. Another approach to avoid making assumptions is to evaluate responses from a colleague's classroom. In all cases, the challenge of using the *Framework* can form the basis for discussion and collaboration between teachers. As incorporating writing in the mathematics class is not fully exploited despite its benefits (Powell et al., 2021), conversations about student-written responses enhance teacher development and student learning. In other words, teachers can use the evaluation of student reasoning to brainstorm ways to work with students within a vertically aligned curriculum, whether they need to clarify the student reasoning to make a more accurate assessment, address misconceptions, or move students towards a more developed number and operation sense.

Task choice is another important consideration. The *Framework* itself can be used to assess students' written responses to tasks in the area of number and operations; it does not include the tasks themselves. First, the tasks should require more than just an answer. The prompt to elicit student reasoning should be *why* rather than *how*. *How* prompts tend to result in a recounting of what students did whereas *why* prompts encourage mathematical justification (Herringer, 2013; Tuttle, 2005). To *build procedural fluency from conceptual understanding* (NCTM, 2014, p. 3), tasks should allow for consideration of an efficient and appropriate strategy grounded in flexible application of number and operation sense (Kling & Bay-Williams, 2014; NCTM, 2000; Parrish, 2011). For example, students can reason about the fraction task by comparing $\frac{5}{8}$ and $\frac{5}{12}$ in different ways (e.g., Students F.C.4 and F.C.5). The whole-number multiplication task encourages students to use operation sense to efficiently add one group (e.g., Students WN.C.4 – 7) rather than use an algorithm (e.g., Student WN.A.2). Opportunities to document such thinking, in writing, for purposefully designed tasks enhances both students conceptual understanding and disciplinary writing (e.g., Graham et al., 2020; Johnson et al., 2011; NCTM, 2000).

CONCLUSION

Two of the recommendations related to research on the role of language in mathematics made by Moschkovich (2010) were "use frameworks for recognizing the mathematical reasoning learners construct in, through, and with language [and] design studies that build on the findings and instruments used in previous relevant research literature" (p. 155). The *Framework* presented in this chapter is grounded in a set of previously conducted studies. The findings presented herein demonstrate how the *Framework* can be used to assess student reasoning expressed in writing (in the area of number and operations).

Further research with the *Framework* can continue to advance this work. First, the psychometric properties of the *Framework* for evaluating students' written responses need to be investigated. Second, potential expanse of use of the *Framework* to assess student written reasoning in other areas of number and operations should be explored. Third, future studies can investigate task choice and its implications on student reasoning and associated disciplinary writing. Fourth, one of the limitations of the current work is the lack of consideration of the nature of instruction or teacher knowledge. Investigating the interplay of teacher knowledge, modes of instruction, and students' reasoning and disciplinary writing will address this limitation. Finally, the researchers found that using the *Framework* to code student-written responses and discussing their coding was a powerful activity. This anecdotal experience can be the impetus for a formal examination of the impact of the use of the *Framework* in collaborative settings on the development of teacher content knowledge and knowledge of disciplinary writing in mathematics.

REFERENCES

Baek, J. M. (2017). Numerical reasoning: Making sense of numbers and operations through multiplication in grades 3-5. In M. T. Battista (Ed.), *Reasoning and sense making in the mathematics classroom: Grades 3-5* (pp. 23–42). National Council of Teachers of Mathematics.

Bangert-Drowns, R. L., Hurley, M. M., & Wilkinson, B. (2004). The effects of school-based writing-to-learn interventions on academic achievement: A meta-analysis. *Review of Educational Research, 74*(1), 29–58. doi: /. doi:10.3102/00346543074001029

Binary Arts Corporation. (1995). *Visual brainstorms: The smart thinking game*. BAC.

Brozo, W. G., & Crain, S. (2018). Writing in math: A disciplinary literacy approach. *The Clearing House: A Journal of Educational Strategies. The Clearing House: A Journal of Educational Strategies, Issues and Ideas, 91*(1), 7–13. doi:10.1080/00098655.2017.1342435

Burns, M. (2012). *Math reasoning inventory. Overview: About the assessments.*

Caldwell, J. H., Karp, K., & Bay-Williams, J. M. (2011). *Developing essential understanding of addition and subtraction for teaching mathematics in prekindergarten-grade 2*. National Council of Teachers of Mathematics.

Carpenter, T. P., & Moser, J. M. (1984). The acquisition of addition and subtraction concepts in grades one through three. *Journal for Research in Mathematics Education, 15*(3), 179–202. doi:10.2307/748348

Casa, T. M., Arsenault, T. L., Powell, S. R., & Hebert, M. A. (2023). Assessing and scoring elementary mathematical writing: Research and practice considerations. In K. L. Wright & T. S. Hodges (Eds.), *Assessing disciplinary writing in both research and practice* (pp. 151–187). IGI Global. doi:10.4018/978-1-6684-8262-9.ch006

Cramer, K., & Henry, A. (2002). Using manipulative models to build number sense for addition and fractions. In B. Litwiller & G. Bright (Eds.), *Making sense of fractions, ratios, and proportions* (pp. 41–48). National Council of Teachers of Mathematics.

Cramer, K., & Whitney, S. (2010). Learning rational number concepts and skills in elementary school classrooms. In D. V. Lambdin & F. K. Lester (Eds.), *Teaching and learning mathematics: Translating research for elementary school teachers* (pp. 15–22). National Council of Teachers of Mathematics.

Cramer, K., Wyberg, T., & Leavitt, S. (2008). The role of representations in fraction addition and subtraction. *Mathematics Teaching in the Middle School, 13*(8), 490–496. doi:10.5951/MTMS.13.8.0490

Crawford, L., Quebec-Fuentes, S., Huscroft-D'Angelo, J., & Higgins, K. N. (2018). Evaluating quantitative reasoning strategies for comparing fractions: A tool for teachers. *Intervention in School and Clinic, 54*(1), 225–234.

Crawford, L., Quebec-Fuentes, S., Huscroft-D'Angelo, J., & Lambert, M. (2020, April). *Using latent class analysis to validate a classroom based numerical reasoning tool* [Panel presentation]. American Educational Research Association (AERA), San Francisco, CA, United States.

DeWolfe, M., & Vosniadou, S. (2015). The representation of fraction magnitudes and the whole number bias reconsidered. *Learning and Instruction, 37*, 39–49. doi:10.1016/j.learninstruc.2014.07.002

Fang, Z. (2012a). Language correlates of disciplinary literacy. *Topics in Language Disorders, 32*(1), 19–34. doi:10.1097/TLD.0b013e31824501de

Fang, Z. (2012b). The challenges of reading disciplinary texts. In T. L. Jetton & C. Shanahan (Eds.), *Adolescent literacy in the academic disciplines: General principles and practical strategies* (pp. 34–68). Guilford Publications.

Fennell, F., & Karp, K. (2017). Fraction sense: Foundational understandings. *Journal of Learning Disabilities, 50*(6), 648–650. doi:10.1177/0022219416662030 PMID:27506552

Forman, E. (1996). Learning mathematics as participation in classroom practice: Implications of sociocultural theory for educational reform. In L. Steffe, P. Nesher, P. Cobb, G. Goldin, & B. Greer (Eds.), *Theories of mathematical learning* (pp. 115–130). Lawrence Erlbaum Associates.

Fuson, K. C. (1988). *Children's counting and concepts of number*. Springer-Verlag. doi:10.1007/978-1-4612-3754-9

Geller, E. H., Son, J. Y., & Stigler, J. W. (2017). Conceptual explanations and understanding fraction comparisons. *Learning and Instruction, 52*, 122–129. doi:10.1016/j.learninstruc.2017.05.006

Graham, S., Kiuhara, S. A., & MacKay, M. (2020). The effects of writing on learning in science, social studies, and mathematics: A meta-analysis. *Review of Educational Research, 90*(2), 179–226. doi:10.3102/0034654320914744

Graham, S., & Perin, D. (2007). *Writing next: Effective strategies to improve writing of adolescent middle and high school—A report to Carnegie Corporation of New York. Alliance for Excellent Education.* Carnegie Corporation of New York. https://production-carnegie.s3.amazonaws.com/filer_public/3c/f5/3cf58727-34f4-4140-a014-723a00ac56f7/ccny_report_2007_writing.pdf

Håland, A. (2016). Disciplinary literacy in elementary school: How a struggling student positions herself as a writer. *The Reading Teacher, 70*(4), 457–468. doi:10.1002/trtr.1541

Heinze, J. A., Gruessing, M., & Lipowsky, F. (2018). Instructional approaches to foster third graders' adaptive use of strategies: An experimental study on the effects of two learning environments on multi-digit addition and subtraction. *Instructional Science, 46*(6), 869–891. doi:10.100711251-018-9457-1

Herringer, N. (2013). *Making sense of mathematics through reasoning*. 3planning. https://www.3plearning.com/blog/making-sense-mathematics-reasoning-2/

Heuer, L. (2005). Graphic representation in the mathematics classroom. In J. M. Kenney (Ed.), *Literacy strategies for improving mathematics instruction* (pp. 51–71). Association for Supervision and Curriculum Development.

Hillman, A. M. (2014). A literature review on disciplinary literacy: How do secondary teachers apprentice students into mathematical literacy? *Journal of Adolescent & Adult Literacy, 57*(5), 397–406. doi:10.1002/jaal.256

Howden, H. (1989). Teaching number sense. *The Arithmetic Teacher, 36*(6), 6–11. doi:10.5951/AT.36.6.0006

Johanning, D. I. (2011). Estimation's role in calculations with fractions. *Mathematics Teaching in the Middle School, 17*(2), 96–102. doi:10.5951/mathteacmiddscho.17.2.0096

Johnson, H., Watson, P. A., Delahunty, T., McSwiggen, P., & Smith, T. (2011). What is it they do? Differentiating knowledge and literacy practices across content disciplines. *Journal of Adolescent & Adult Literacy, 55*(2), 100–109. doi:10.1002/JAAL.00013

Jordan, N. C., Kaplan, D., Locuniak, M. N., & Ramineni, C. (2007). Predicting first grade math achievement from developmental number sense trajectories. *Learning Disabilities Research & Practice, 22*(1), 36–46. doi:10.1111/j.1540-5826.2007.00229.x

Jordan, N. C., Kaplan, D., Ramineni, C., & Locuniak, M. N. (2009). Early math matters: Kindergarten number competence and later mathematics outcomes. *Developmental Psychology, 45*(3), 850–867. doi:10.1037/a0014939 PMID:19413436

Karp, K. S., Bush, S. B., & Dougherty, B. J. (2015). 12 math rules that expire in the middle grades. *Mathematics Teaching in the Middle School, 21*(4), 208–215. doi:10.5951/mathteacmiddscho.21.4.0208

Ketterlin-Geller, L. R., & Perry, L. (2018). Measuring early mathematical reasoning skills: Developing tests of numeric relational reasoning and spatial reasoning. *Poster presented at the 2018 National Science Foundation Principal Investigators Meeting*. CADREK. https://cadrek12.org/posters/measuring-early-mathematical-reasoning-skills-developing-tests-numeric-relational-reasoning

Kleve, B., & Penne, S. (2016). Learning subjects in school – being outsiders or insiders in the disciplinary discourses on mathematics and Language 1. *International Journal of Educational Research, 78*, 41–49. doi:10.1016/j.ijer.2016.05.014

Kling, G., & Bay-Williams, J. B. (2014). Assessing basic fact fluency. *Teaching Children Mathematics, 20*(8), 488–497. doi:10.5951/teacchilmath.20.8.0488

Locuniak, M. N. (2010). *Early predictors of calculation fluency in second grade* (Publication No. 3397013) [Doctoral dissertation, University of Delaware]. ProQuest Dissertations Publishing.

Malone, A. S., & Fuchs, L. S. (2017). Error patterns in ordering fractions among at-risk fourth-grade students. *Journal of Learning Disabilities*, *50*(3), 337–352. doi:10.1177/0022219416629647 PMID:26966153

Marks, G., & Mousley, J. (1990). Mathematics education and genre: Dare we make the process writing mistake again? *Language and Education*, *4*(2), 117–135. https://doi-org.ezproxy.tcu.edu/10.1080/09500789009541278. doi:10.1080/09500789009541278

Moschkovich, J. N. (2003). What counts as mathematical discourse? *International Group for the Psychology of Mathematics Education*, *3*, 325–332.

Moschkovich, J. N. (2010). Recommendations for research on language and mathematics education. In J. N. Moschkovich (Ed.), *Language and mathematics education: Multiple perspectives and directions for research* (pp. 151–170). Information Age Publishing, Inc.

National Council of Teachers of Mathematics. (NCTM). (2000). *Principles and standards for school mathematics*. NCTM. https://www.nctm.org/Standards-and-Positions/Principles-and-Standards/

National Council of Teachers of Mathematics. (NCTM). (2014). *Principles to action: Executive summary*. NCTM. https://www.nctm.org/uploadedFiles/Standards_and_Positions/PtAExecutiveSummary.pdf

National Governors Association & Council of Chief State School Officers (NGA & CCSSO). (2010a). *Common Core State Standards for mathematics*. CCSSO. https://learning.ccsso.org/wp-content/uploads/2022/11/Math_Standards1.pdf

National Governors Association & Council of Chief State School Officers (NGA & CCSSO). (2010b). *Common Core State Standards for English language arts & literacy in history/social studies, science, and technical subjects*. CCSSO. https://learning.ccsso.org/wp-content/uploads/2022/11/ELA_Standards1.pdf

Otto, A. D., Caldwell, J. H., Lubinski, C. A., & Hancock, S. W. (2011). *Developing essential understanding of multiplication and division for teaching mathematics in grades 3-5*. National Council of Teachers of Mathematics.

Parrish, S. D. (2011). Number talks build numerical reasoning. *Teaching Children Mathematics*, *18*(3), 198–206. doi:10.5951/teacchilmath.18.3.0198

Powell, S. R., Hebert, M. A., & Hughes, E. M. (2021). How educators use mathematics writing in the classroom: A survey of mathematics educators. *Reading and Writing*, *34*(2), 417–447. doi:10.100711145-020-10076-8

Pugalee, D. (2004). A comparison of verbal and written descriptions to students' problem solving processes. *Educational Studies in Mathematics*, *55*(1-3), 27–47. doi:10.1023/B:EDUC.0000017666.11367.c7

Seo, B. (2016). Written discourse analysis: Looking at adolescent mathematical writing. *SAGE Research Methods Cases*, (Part 2), 1–21.

Sfard, A. (2007). When the rules of discourse change, but nobody tells you: Making sense of mathematics learning from a commognitive standpoint. *Journal of the Learning Sciences*, *16*(4), 565–613. doi:10.1080/10508400701525253

Shanahan, T., & Shanahan, C. (2008). Teaching disciplinary literacy to adolescents: Rethinking content-area literacy. *Harvard Educational Review*, *78*(1), 40–59. doi:10.17763/haer.78.1.v62444321p602101

Shanahan, T., & Shanahan, C. (2012). What is disciplinary literacy and why does it matter? *Topics in Language Disorders*, *32*(1), 7–18. doi:10.1097/TLD.0b013e318244557a

Siegler, R. S., Carpenter, T., Fennel, F., Geary, D., Lewis, J., Okamoto, Y., & Wray, J. (2010). *IES practice guide: Developing effective fractions instruction for kindergarten through 8th grade* (NCEE Report 2010-4039). National Center for Education Evaluation.

Siegler, R. S., Thompson, C. A., & Schneider, M. (2011). An integrated theory of whole number and fractions development. *Cognitive Psychology*, *62*(4), 273–296. doi:10.1016/j.cogpsych.2011.03.001 PMID:21569877

Solomon, Y., & O'Neill, J. (1998). Mathematics and narrative. *Language and Education*, *12*(3), 210–221. doi:10.1080/09500789808666749

Steffe, L. P., Cobb, P., & von Glasersfeld, E. (1988). *Construction of arithmetical meanings and strategies*. Springer-Verlag. doi:10.1007/978-1-4612-3844-7

Temple, C., & Doerr, H. M. (2018). How do teachers develop and enact a disciplinary view of literacy in mathematics? *Journal of Adolescent & Adult Literacy*, *61*(5), 483–488. doi:10.1002/jaal.664

Tuttle, C. L. (2005). Writing in the mathematics classroom. In J. M. Kenney (Ed.), *Literacy strategies for improving mathematics instruction* (pp. 24–50). Association for Supervision and Curriculum Development.

Van de Walle, J. A., Karp, K. S., & Bay-Williams, J. M. (2019). *Elementary and middle school mathematics: Teaching developmentally*. Pearson.

Wilkinson, G. S., & Robertson, G. J. (2006). *Wide Range Achievement Test 4 professional manual*. Psychological Assessment Resources.

Witzel, B. S., Ferguson, C. J., & Mink, D. V. (2012). Number sense: Strategies for helping preschool through grade 3 children develop math skills. *Young Children*, *67*(3), 89–94.

ADDITIONAL READING

Craig, T. S. (2011). Categorisation and analysis of explanatory writing in mathematics. *International Journal of Mathematical Education in Science and Technology*, *42*(7), 867–878. doi:10.1080/002073 9X.2011.611909

Cross, D. I. (2009). Creating optimal mathematics learning environments: Combining argumentation and writing to enhance achievement. *International Journal of Science and Mathematics Education*, *7*(5), 905–930. doi:10.100710763-008-9144-9

Gillis, V. (2014). Disciplinary literacy: Adapt not adopt. *Journal of Adolescent & Adult Literacy*, *57*(8), 614–623. doi:10.1002/jaal.301

Porter, M. K., & Masingila, J. O. (2000). Examining the effects of writing on conceptual and procedural knowledge in calculus. *Educational Studies in Mathematics*, *42*(2), 165–177. doi:10.1023/A:1004166811047

Shanahan, C., Shanahan, T., & Misischia, C. (2011). Analysis of expert readers in three disciplines: History, mathematics, and chemistry. *Journal of Literacy Research*, *43*(4), 393–429. doi:10.1177/1086296X11424071

Siebert, D., & Draper, R. J. (2008). Why content-area literacy messages do not speak to mathematics teachers: A critical content analysis. *Literacy Research and Instruction*, *47*(4), 229–245. doi:10.1080/19388070802300314

Wilcox, B., & Monroe, E. E. (2011). Integrating writing and mathematics. *The Reading Teacher*, *64*(7), 521–529. doi:10.1598/RT.64.7.6

Zygouris-Coe, V. I. (2012). Disciplinary literacy and the Common Core State Standards. *Topics in Language Disorders*, *32*(1), 35–50. doi:10.1097/TLD.0b013e31824561a2

KEY TERMS AND DEFINITIONS

Algorithmic Reasoning Strategy: Application of a set of rules that guarantees a correct solution will be reached; remaining reasoning parts are trivial for the reasoner.

Conceptual Understanding: Reasoning founded on the intrinsic mathematical properties of the components of the task with or without describing a specific procedure. Evidence of reasoning does not require written words and may be demonstrated through use of visual representations or abstract symbols.

Disciplinary Literacy: Reading, writing, listening, and speaking that reflect the norms and values of experts in the field.

Disciplinary Writing in Mathematics: Precise means (words, symbols, or abstract symbols) of conveying a logical flow of reasoning.

Fraction Sense: A conceptual understanding of fractions as numbers.

Framework: Rubric consisting of multiple, non-continuous categories.

Numerical Reasoning: Making sense of number and operations.

Chapter 12
What to Assess for Mathematical Writing:
The Dimensionality of Written Expression for Mathematical Problem Solving and Reasoning

Joo-Young Lee
https://orcid.org/0000-0002-4483-6331
The University of Maine, USA

Elizabeth M. Hughes
The Pennsylvania State University, USA

Paul J. Riccomini
The Pennsylvania State University, USA

ABSTRACT

This chapter investigates the internal structure of informative/explanatory mathematical writing (MW) as a latent variable. The study defines it as written communication in mathematics to explain and describe one's problem solving and reasoning by using mathematical language and representation. To assess the dimensions of MW, ten scales were developed based on literature and expert reviews to quantify students' MW performance. The dataset utilized for statistical analyses consisted of MW samples from 153 4th-grade students attending a public school in a Mid-Atlantic state. Through exploratory factor analysis (EFA) conducted on the three MW prompts, two factors emerged: mathematics problem solving and reasoning and writing. Subsequently, a confirmatory factor analysis (CFA) was conducted, resulting in a cross-factor loading of reasoning onto both factors. The role of representations in MW remains uncertain. This finding contributes to an understanding of the underlying dimensions, as well as offers a validated approach for assessing MW.

DOI: 10.4018/978-1-6684-6538-7.ch012

INTRODUCTION

Written communication for mathematical thinking and reasoning is one of the key standards for mathematics education (National Council of Teachers of Mathematics, 2000; National Governors Association Center for Best Practices & Council of Chief State School Officers, 2010). According to the National Council of Teachers of Mathematics (NCTM), students should learn to communicate their mathematical problem solving, justify mathematical conclusions, and evaluate mathematical arguments (NCTM, 2000). The Common Core State Standards for Mathematics (CCSSM; National Governors Association Center for Best Practices & Council of Chief State School Officers, 2010) requires students to "construct viable arguments and critique the reasoning of others" (p. 6). In line with the curriculum standards, many state assessments in mathematics (e.g., Connecticut, Illinois, Massachusetts, Michigan, New York, Pennsylvania, Texas) include one or two open-ended items asking students to (a) explain their *mathematical problem solving* and (b) justify their *reasoning* in a logical and coherent sequence with written language (Hughes et al., 2019).

In response to the increasing demands to incorporate mathematical writing (MW) into the mathematics curriculum, numerous researchers have sought to clarify its definitions and dimensions. According to Baxter et al. (2005), MW is a writing activity "to explain and justify a (mathematical) problem solution" (p. 120). Similarly, Cross (2009) and Cohen et al. (2015) defined MW as writing in mathematics to explain how to *solve* a mathematical problem and *justify* reasoning and procedures. Kiuhara and her colleagues (2020) defined MW as a learning activity (Klein, 1999) to use *mathematical language* to justify the mathematical conclusions that students made. Lepak et al. (2018) placed substantial emphasis on the significance of *pictorial* and *symbolic representations* in written mathematical reasoning. These representations serve as a means for mathematical communication, visualizing mathematical relationships between salient quantities within the mathematical context of the given problem statement.

Similar to other forms of content writing, MW employs various types of writing to serve different purposes. Casa and colleagues (Casa et al., 2016; Colonesse, et al., 2018) described four genres of MW: (a) exploratory, (b) informative/explanatory, (c) argumentative, and (d) creative writing. The most frequent type of MW prompts in students' textbook and the state assessments is informative/explanatory writing (Casa et al., 2016; Colonesse et al., 2018; Hughes et al., 2019). Informative/explanatory writing in the context of problem solving describes the reasoning used to find a solution to a problem, highlighting the understanding of mathematical concepts and representations used. From these perspectives, the current study focuses on informative/explanatory writing in mathematics, specifically defining it as written communication in mathematics wherein individuals explain their own problem solving procedures and reasoning by using mathematical language and representation (Bossé & Faulconer, 2008; Hughes & Lee, 2020; Namkung et al., 2020).

Based on our definition of informative/explanatory MW, the purpose of this chapter is to explore the dimensionality of MW as a latent variable. Over the past two decades, there has been growing interest to explore *how to effectively teach* MW for students at various grade levels and verify their effectiveness. However, relatively less attention has been devoted to understanding *what to assess* in students' MW performance (Powell et al., 2017). To begin to address this question, this chapter focuses on (a) building and validating a hypothesized model to explain the internal structure of MW and (b) discussing what components should be assessed based on the quantitative results.

LATENT STRUCTURE OF MATHEMATICAL WRITING

MW serves as an academic mode of written communication that delves into the depths of mathematical thinking and reasoning. In a general sense, it comprises two academic domains: mathematics and writing (Bossé & Faulconer, 2008; Hebert & Powell, 2016). To ensure high-quality MW, students should be equipped with a comprehensive skill set for both *mathematics problem solving and reasoning* and *writing. Mathematics problem solving and reasoning* constructs (See Table 1) may include students' abilities (a) to identify critical information (quantities) from the problem statement (e.g., Lepak et al., 2018), (b) to visually represent the mathematical relationship between the quantities (e.g., Santos & Semana, 2015), (c) to create an abstract representation (i.e., equation or expression) using accurate numerals, symbols, and operations (e.g., Hebert & Powell, 2016), (d) to verbalize a logical sequence of mathematical problem solving and reasoning (e.g., Kiuhara et al., 2020), and (e) to precisely execute computations and procedures (e.g., Arsenault et al., 2022).

Table 1. Hypothesized constructs for mathematical problem solving and reasoning

Constructs	Definition
Identifying *salient quantities*	Ability to identify critical quantities and information necessary to solve the problem from the problem statement
Creating *a visual representation*	Ability to create a visual representation to depict a mathematical relationship among quantities
Building *an equation*	Ability to select operations and write a mathematical expression (equation) using numerals and symbols.
Explaining problem solving and *reasoning* procedures	Ability to describe problem solving and reasoning procedures in a logical sequence
Executing *computations*	Ability to execute calculations and elicit a correct conclusion

Note. Adapted from Lepak et al. (2018).

Table 2. Hypothesized constructs for writing

Constructs	Definition
Focusing on a single main topic	Ability to maintain a focused central point, demonstrating an understanding of the task and its relevance to a particular topic
Generating the content of the topic	Ability to elaborate ideas that are supported by factual information, examples, reasons, and compelling explanations
Organizing the content	Ability to organize the content in a logical order, using transitional words to ensure coherence, and effectively structuring the introduction, body, and conclusion sections
Using mathematics vocabulary	Ability to use appropriate mathematics vocabulary to explain problem solving and reasoning
Following conventions	Ability to adhere to grammatical rules, mechanics, usage, and sentence formation conventions

Note. Adapted from Deatline-Buchman & Jitendra (2006).

Writing constructs (See Table 2) may embrace students' competencies (a) to maintain a focus on a specific topic (e.g., Heyes, 2012), (b) to develop and elaborate content to serve a purpose of writing (e.g., Mo & Troia, 2017), (c) to organize the ideas using transition words to effectively connect the introduction and conclusion sections (e.g., Hebert & Powell, 2016), (d) to select and use appropriate and accurate mathematics vocabulary words to deliver clear and concise meanings (e.g., Hughes et al., 2016), and (e) to form a structure of words and sentences minimizing spelling and grammatical errors (e.g., Namkung et al., 2020).

What to Assess for Mathematical Problem Solving and Reasoning

Identifying Salient Quantities

Prior to executing problem solving, students should be able to identify what quantities are essential to use from the problem statement and discern what information is irrelevant to the context of the problem (Lepak et al., 2018). According to Polya (1957), the first step of mathematics problem solving is to thoroughly read and comprehend the problem statement, making sense of the mathematical context. Meyer's model (1985) conceptualizes this initial phase of problem solving as a *problem translation*, which requires finding the critical numbers to translate them into a mathematical representation (Kingsdorf & Krawec, 2014). In this phase, it is crucial to develop students' ability to process the information collected from the problem statement, subsequently evaluating which information to select and which to disregard based on its relative significance (Moran et al., 2014).

Creating a Visual Representation

After identifying salient quantities from the problem statement, students should be able to represent mathematical relationships among the quantities (Kingdorf & Krawec, 2014; Lepak et al., 2018). Mathematical representations, as another means for mathematical communication (NCTM, 2000), refers to any configurations of representing mathematical ideas, relationships, and concepts, including concrete objects, pictures, diagrams, charts, graphs, symbols, and characters (Bruner & Kenney, 1965; Jitendra et al., 2016; Mainali, 2021). To understand and visualize abstract, underlying mathematical structures between the quantities, skilled problem solvers construct different modes of representations such as Venn diagrams, tables, number lines, or bar charts (Van Garderen & Montague, 2003). These visual representations facilitate students' understanding of the mathematical concepts and numerical relationships and the reasoning of selecting operations (Zahner & Corter, 2010). They also reduce the cognitive load placed on students, as they no longer need to remember all the story elements from the problem statement (Jitendra et al., 2016).

Building an Equation

In addition to creating visual representations, students should also possess the capability to represent a mathematical relationship using symbolic language and expressions (Namkung & Bricko, 2021). The CCSSM (2010) explicitly documents the significance of students' proficiency in formulating and interpreting mathematical expressions and equations across various mathematics domains. Transitioning from visual representations (i.e., semi-concrete representations) to this abstract phase (i.e., equation) (Witzel, 2005), students should learn to determine an accurate operation(s) (Lepak et al., 2018) and use numerals, symbols, and notations to represent quantitative relations.

Explaining Problem Solving and Reasoning Procedures

Mathematical reasoning involves students' ability to justify how they arrived at the mathematical conclusion and why they used a particular method to solve the problem in a logical solution (Hughes & Lee, 2020; Lithner, 2000). Students should acquire the ability to discern the underlying reasons behind their selection of specific problem-solving methods and operations (Lepak et al., 2018). It is important to note that mathematical reasoning goes beyond the accuracy of mathematical solutions and justifications; it also requires students' effective mathematical communication skills (NCTM, 2000) and appropriate use of mathematical language (Hebert & Powell, 2016; Hughes et al., 2016; Arsenault et al., 2022). To enhance communication for mathematical problem solving and reasoning, students can use visual representations and equations to effectively convey mathematical relationships among quantities and facilitate a clearer understanding of the problem-solving process they have chosen (Hebert & Powell, 2016; Lepak et al., 2018). Additionally, by comparing alternative solutions, students can bolster the mathematical argumentation and reasoning provided in their writing (Kiuhara et al., 2020). This counterargument enhances the strengths and robustness of their justification for why the selected solutions are optimal.

Executing Calculations

Returning to the Polya's approach (1957) and Meyer's model (1985), the final stage of math problem solving entails executing calculations with precision. To successfully perform calculations, students should master fundamental arithmetic operations and acquire procedural knowledge of algorithms (Fuchs et al., 2006; Villeneuve et al., 2019). For example, complex calculations (e.g., 308×73) require retrieving basic arithmetic facts (e.g., 8×3, 8×7) and using procedural knowledge (e.g., 308×3, 308×70, then $(308 \times 3) + (308 \times 70)$) to complete the problem-solving sequence (McCloskey et al., 1985). In this final phase, the problem solver can focus on providing a precise mathematical conclusion from the accurate calculation procedures.

What to Assess for Writing

Focusing on a Single Main Topic

Across diverse genres of writing, good writers establish and maintain a single main topic throughout the composition, demonstrating a profound understanding of the writing task or prompt (Deatline-Buchman & Jitendra, 2006). The purposeful writing in mathematics (Casa et al., 2016) requires students' attention to what the math problem asks and what should be addressed in writing, while adhering to a single focal point. From the initial to concluding sentences, students should construct their mathematics text with careful attention to coherence and consistency in all mathematical ideas (Hughes et al., 2019; Hughes & Lee, 2020), including visual representations and numerical statements (i.e., equations), all of which should align with the designated topic.

Generating the Content of the Topic

When confronted with a specific topic, students should not only generate main ideas but also clearly elaborate on these ideas by providing compelling reasons to support them (Engelhard, 1992). Generally, informative/explanatory writing requires students to provide detailed explanations on the rationale behind their written statement, elucidating the reasons or justifications for its reasonable accuracy (Jefferey, 2009). In the context of formal assessments that demand informative/explanatory MW (Hughes et al., 2019), students are expected to describe a particular method they have chosen, and present persuasive rationales to support their solutions in a logical sequence (Colonnese et al., 2018; Kiuhara et al., 2020). To achieve this, students should learn to provide details pertaining to mathematical concepts, algorithms, strategies, and analysis, effectively elaborating their reasoning to the intended audience (Casa et al., 2016.).

Organizing the Content

Organization encompasses the skillful manifestation of introduction, body, and conclusion sections while ensuring a seamless and cohesive flow between sentences and ideas (Correnti et al., 2013). This competency includes the use of appropriate cohesive devises as well as organization of ideas in a logical sequence, considering the specific genres of writing (Hebert & Powell, 2016). For informative/explanatory MW, for example, it may entail a problem restatement as an introductory sentence, representations, problem solutions, reasons for the choice of problem solving methods, and a conclusion that ends with the correct answer to the problem (Hughes & Lee, 2020). Argumentative MW may require a claim on a mathematical statement (e.g., solution), compelling reasons (evidence) to justify the claim, a counterclaim, and a concluding statement to wrap up the writing (Kiuhara et al., 2020). To effectively demonstrate a cohesive sequence of these genres' elements, students should be proficient in utilizing various types of transition words and phrases across the genres (Kiuhara et al., 2020; Powell & Hebert, 2016).

Using Mathematics Vocabulary

Mathematics vocabulary refers to academic vocabulary terms that convey meanings of mathematical concepts and knowledge (Hughes et al., 2016), which is critical to effectively communicate mathematical thinking and reasoning (Kiuhara et al., 2020; Powell & Driver, 2015; Riccomini et al., 2015). Recent empirical studies have verified the relationships between students' mathematics vocabulary proficiency and various aspects of mathematics performance. These include mathematics computational skills (e.g., Purpura et al., 2017), conceptual understanding (e.g., Fuchs et al., 2021), word problem solving (e.g., Chan & Kwan, 2021), reasoning (e.g., Peng et al., 2020), and MW (e.g., Forsyth & Powell, 2017) that requires all the aforementioned skills. Notably, Peng and Lin (2019) highlighted the significant contribution of mathematics vocabulary to word problem solving, after controlling for other cognitive abilities such as IQ, working memory, general vocabulary, and processing speed. Their findings also indicate that the effects of mathematics vocabulary may vary across different mathematics domains and skills, underscoring the need for explicit instruction in mathematics vocabulary words when introducing new skills and concepts.

Following Conventions

Writing quality involves students' application of grammatical and syntactic knowledge, mechanics, spelling, punctuations, and capitalization in constructing sentences (Ritchey & Cocker, 2013). In MW, consistent with other content areas, students' adherence to conventions such as proper grammar, accurate spelling of mathematics vocabulary, and sentential structure is essential (Arsenault et al., 2022). However, MW differs from other content area writing in that it requires students' precise and accurate use of symbols and numerals in the text. For example, students may combine numerals and mathematical symbols within equations or expressions (e.g., I did 10 + 5 to get 15), but the symbols should not be used as substitutes for words outside of equations and expressions (e.g., I + 10 and 5) (Namkung et al., 2020). These grammatical rules should be explicitly taught in a mathematics classroom to facilitate students' understanding of unique conventions in MW.

EMPIRICAL EVIDENCE FOR THE HYPOTHESIZED MODEL

Based on the ten constructs for MW, a two-factor hypothesized model was created as depicted in Figure 1. To statistically verify the hypothesized model, quantitative data analyses were conducted using the data collected from the participating students. The subsequent section outlines the data collection and analysis employed in this study.

Participants and Settings

The participants consisted of 153 4th-grade students attending at a public school in a Mid-Atlantic state, all of whom consented to take part in the research. Among these students, 22 had one or more Individualized Education Program (IEP) goals related to mathematics, reading, speech, and/or behavior, while 2 students were identified as gifted. In terms of ethnic composition, the student body comprised 88.2% European Americans, 3.1% African Americans, 5.0% Hispanic and Latino Americans, and 3.9% from other ethnic backgrounds. The previous year's state assessment results for mathematics indicated that 41.7% of the students achieved an advanced level, 26.4% performed at a proficient level, 18.4% were classified as basic, and 6.7% fell below the basic level. In the state assessment for language arts, it was found that 14.7% of the students demonstrated an advanced proficiency, 50.9% were classified as proficient, 20.2% fell within the basic range, and 7.4% scored below the basic level.

Mathematical Writing Prompts and Analytic Rubric

For MW prompts, three algebraic word problems were selected and adapted from the Pennsylvania System of School Assessment for Mathematics (PSSA-M), a state standardized test. To enhance the original word problems, specific directions such as *show your work and explain your reasoning* were added. Each MW prompt was designed to target different mathematical skills and concepts. The first prompt focused on determining a subtractive relation between a three-digit number and a two-digit number, requiring students to perform the subtraction operation (e.g., 146 – 79). In the second prompt, students were asked to execute a two-step computation involving multiplication and subtraction (e.g., 90 – (20 × 4)). To successfully solve the problem, students needed to grasp the correct sequence of operations.

Figure 1. Hypothesized model for mathematical writing as a latent variable
Note. qt = identifying salient quantities; rp = creating a visual representation; eq = building an equation; rs = explaining problem solving and reasoning; cp = executing computations; fc = focusing on a single main topic; ct = generating the content of the topic; og = organizing the content; vc = using mathematics vocabulary; cv = following conventions

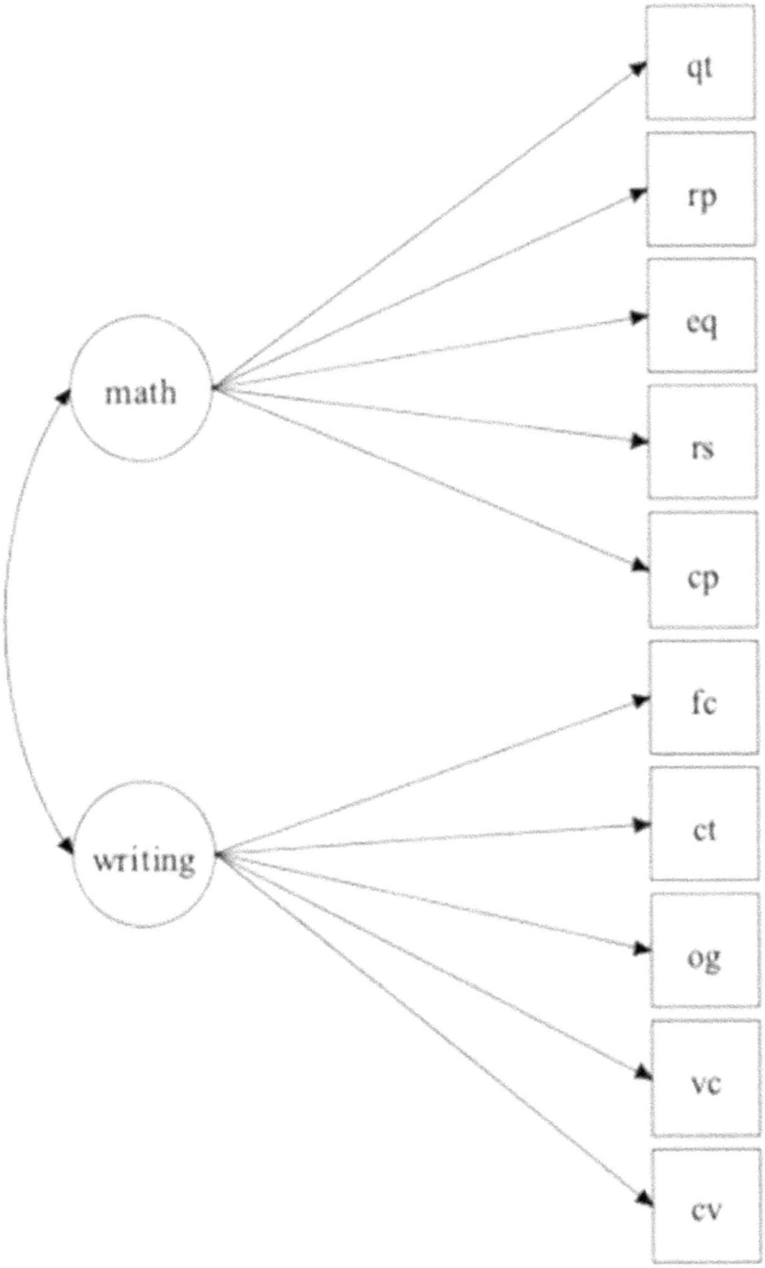

The third prompt challenged students to perform division with a three-digit number including '0' in the ones place, divided by a two-digit number with '0' in the ones place as well (e.g., 120 ÷ 10). All three prompts were selected and modified to align with the mathematical knowledge and skills expected at the

Figure 2. A sample of a student's written response

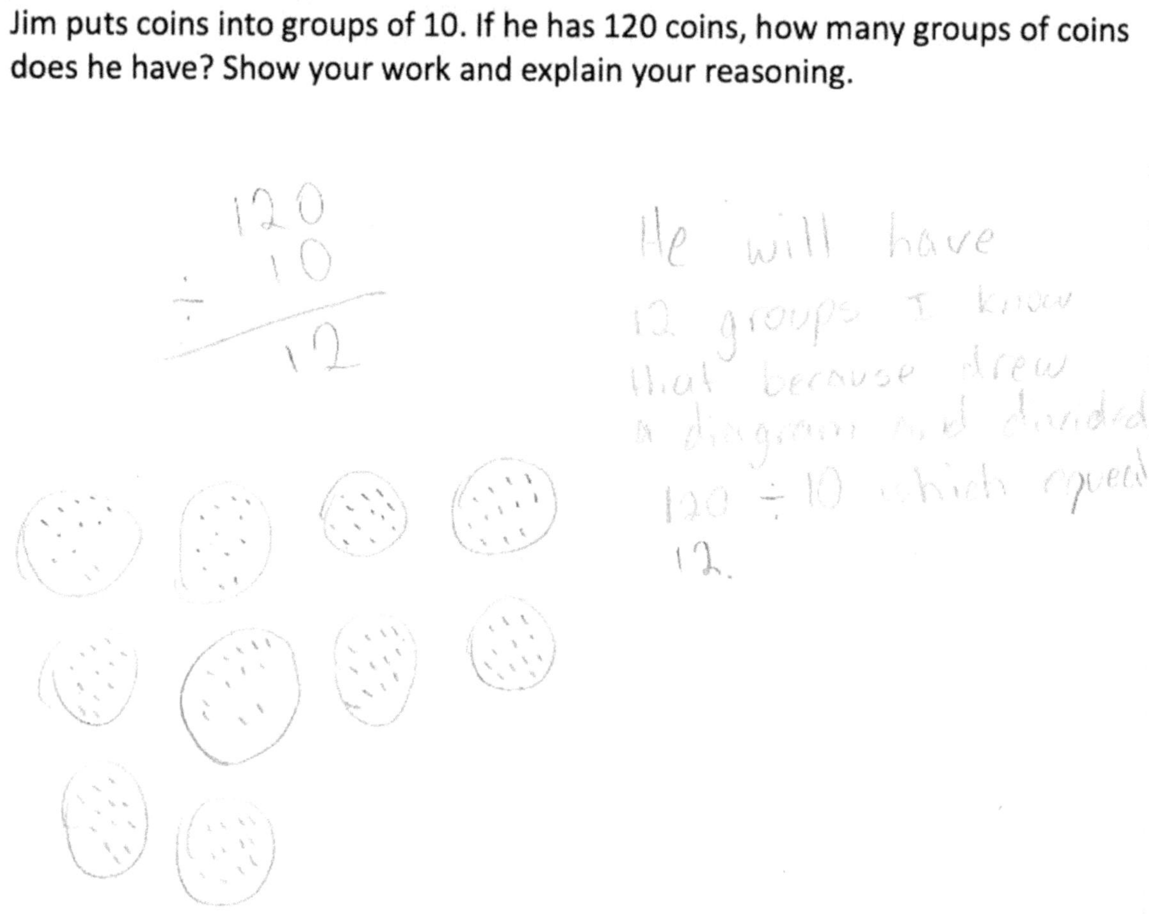

third-grade level, ensuring that the assessed content had been covered within the curriculum. Nine teachers administered the assessment under the guidance of Paul and distributed the assessment packet to the students and read the directions. Students were provided 45 minutes to complete the three MW prompts.

To score students' written responses, instead of using the holistic 5-point scale of PSSA-M, which categorizes students' MW performance as a *thorough* understanding, a *general* understanding, a *partial* understanding, a *minimal* understanding, *no correct answer and insufficient evidence for any* understanding of the mathematical concepts and procedures required by the task (Pennsylvania Department of Education, 2023), we used an analytic rubric consisting of 10 scales presented in Table 3, aligning with the constructs of the hypothesized MW model. Unlike the holistic rubric, which uses a single scale to evaluate multiple MW components in one score (Hughes et al., 2018), the analytic rubric was designed to provide more precise information about students' strengths and weaknesses across the constructs of mathematics and writing (Namkung et al., 2020).

Each scale represented a specific construct and was assigned a score ranging from 0 to 2. A score of 2 indicated clear evidence of demonstrating the competency associated with the construct, while a score of 1 indicated partial evidence. No points were awarded when no evidence of demonstrating the compe-

Table 3. Analytic rubric for MW

Factors	Constructs	2	1	0
Mathematics Problem Solving and Reasoning	Identifying *salient quantities*	Clear evidence of finding all the salient quantities required to solve the problem from the problem statement	Evidence of partially finding the salient quantities required to solve the problem from the problem statement	No evidence of finding any quantities required to solve the problem from the problem statement
	Creating *a visual representation*	Clear evidence of creating a pictorial representation to visualize the mathematical relationship among quantities accurately	Evidence of creating a partially accurate pictorial representation to visualize the mathematical relationship among quantities	No evidence of creating a pictorial representation to visualize the mathematical relationship among quantities accurately
	Building an *equation*	Clear evidence of building equations with accurate operations and symbols	Evidence of building equations with partially accurate operations and symbols	No evidence of building equations with accurate operations and symbols
	Explaining problem solving and *reasoning*	Clear evidence of providing accurate statements regarding why the operation(s) were chosen	Evidence of providing partially accurate statements regarding why the operation(s) were chosen	No evidence of providing accurate statements regarding why the operation(s) were chosen
	Executing *computations*	Clear evidence of accurate procedures of computations	Evidence of partially accurate procedures of computations	No evidence of accurate procedures of computations
Writing	*Focusing* on a single main topic	Clear evidence of making an apparent point about a single topic with sufficient awareness of task	Evidence of making a point about a single topic based on the task, but including some irrelevant statements to the topic	No evidence of making an apparent point about a single topic nor being aware of the task
	Generating the *content* of the topic	Clear evidence of developing sufficient content with adequate elaboration or explanation required in the task	Evidence of developing content with partially adequate elaboration or explanation required in the task	No evidence of developing any content with adequate elaboration or explanation required in the task
	Organizing the content	Clear evidence of functionally arranging content in a way that maintains a logical order and uses appropriate transition words	Evidence of arranging the content in a partially illogical sequence, whether or not an attempt was made to use transition words	No evidence of arranging the content in a logical sequence and not using appropriate transition words
	using *mathematics vocabulary*	Clear evidence of using appropriate mathematical vocabulary words precisely to elaborate on the content	Evidence of demonstrating limited word choice and somewhat inaccurate use of math vocabulary to elaborate on the content	No evidence of using any appropriate and precise mathematical vocabulary words to elaborate on the content
	Following *conventions*	Clear evidence of controlling grammar, mechanics, spelling, usage, and sentence formation	Evidence of using partially accurate grammar, mechanics, spelling, usage, and sentence formation with minor errors, but clearly conveying meanings	No evidence of using grammar, mechanics, spelling, usage, and sentence formation accurately enough to convey meanings

tency was observed. Figure 2 includes a sample of a student's written response to the MW prompt. To score this student's writing for the first scale, *salient quantities*, for example, this student demonstrates clear evidence of finding all the salient quantities (i.e., *120* and *10*) in the equation and the written text (2 points). The student's *visual representation* illustrates a division relationship between the quantities;

however, it represents '10 groups of 12 coins,' instead of '12 groups of 10 coins,' which demonstrates partial understanding of mathematical representation (1 point). Next, the student establishes an *equation* by choosing an accurate operation (2 points). The student knows division should be used to solve the problem but does not provide any expiation for the choice (e.g., *I chose division because he puts [distributes] 10 coins in each group equally*), thus failing to earn any point on it. However, the computation of the division is correct, so the student earns a full credit for it (2 points). On the writing scales, the student's response maintains a single *focus* without any irrelevant information (2 points). However, the response describes what the student did to solve the problem (i.e., *I drew a diagram and divided 120 ÷ 10*) but does not include any reasoning procedures on why the student chose the method to solve the problem, as the prompt requires. Thus, the student obtains a partial credit for the *content* (1 point). Next, in terms of *organization*, the student uses transition and linking words such as 'because', 'and', and 'which.' However, the response is organized in a partially logical sequence (e.g., *We will have 12 groups. I know that because drew a diagram*), thus earning a partial credit (1 point). Regarding *mathematics vocabulary*, the student appropriately uses words such as 'groups,' 'to divide,' and 'to equal,' all related to division (2 points). Lastly, the student's response includes some errors in punctuation (e.g., *He will have 12 groups*), preposition usage (e.g., *divided 120 ÷ 10*, not *divided 120 into 10*), and subject-verb agreements (e.g., *which equal 12*), but the sentences are all comprehensible (1 point).

An expert review was conducted with two math teachers, an English language arts teacher, and a reading and writing specialist with special education license, and a researcher in mathematics teacher education to ensure content validity. The content validity ratios (CVR) were calculated for each scale, ranging from 0.60 and 1.00. Interrater reliability was examined among the three raters including the first author and two doctoral students in special education who were not involved in this study. The interrater reliability, Fleiss' kappa coefficients, ranged from .344 to 1.000 across the three MW prompts. The internal consistency, Cronbach's alpha coefficients, among the scales resulted in $a = .838$ for the first prompt, $a = .853$ for the second prompt, and $a = .784$ for the third MW prompt.

Data Analyses

The exploratory factor analysis (EFA) and confirmatory factor analysis (CFA) were conducted to explore and evaluate the underlying dimensionality of the MW as a latent variable using Mplus version 8.7. Through the EFA using the principal component analysis and oblique promax rotation, the number of factors was determined and a hypothesized model for MW was created. The CFA using robust weighted least squares estimation (WLSMV) confirmed the number of factors and the pattern of relations among the categorical, ordinal variables under the factors. It also investigated whether the hypothesized model in Figure 1 would account for students' MW scores by examining fit indices of the model and suggested alternative models to better explain students' MW performance from the data.

Results

Exploratory Factor Analysis

Kaiser-Meyer-Olkin (KMO) Measure of Sampling Adequacy and Bartlett's Test of Sphericity indicated that students' data for the three MW prompts were adequate for the factor analysis (See Table 4). The EFAs resulted in a cumulative percentage of variance of over 50% for the three MW prompts (55.912%

Table 4. Suitability of the data for EFA

MW prompts	Kaiser-Meyer-Olkin (KMO) Measure of Sampling Adequacy	Bartlett's Test of Sphericity		
		x^2	*df*	*p*
Subtraction	.854	541.545	45	<. 001
Two-step computation	.795	676.760	45	<. 001
Division	.740	520.703	45	<. 001

Table 5. Factor loadings for individual scales from EFA

Scales	Subtraction		Two-Step Computation		Division	
	Factor 1	Factor 2	Factor 1	Factor 2	Factor 1	Factor 2
Salient Quantities	**.844**	-.096	**.727**	-.054	**.605**	-.026
Representation	**.271**	-.192	**.739**	-.362	**.790**	-.095
Equation	**.823**	.022	**.921**	-.270	**.699**	-.486
Reasoning	**.536**	.343	**.480**	.323	**.697**	.239
Computation	**.789**	.133	**.733**	-.136	**.541**	.306
Focus	.117	**.775**	.125	**.671**	.152	**.777**
Content	-.073	**.826**	.399	**.637**	.173	**.728**
Organization	-.111	**.883**	.064	**.778**	-.017	**.747**
Mathematics Vocabulary	-.124	**.783**	.091	**.728**	-.283	**.554**
Convention	.153	**.701**	.046	**.756**	.138	**.717**

for the subtraction prompt; 57.616% for the two-step computation prompt; 55.384% for the division prompt) explained by two factors having an eigenvalue > 1 (Hair et al., 1995; Williams et al., 2010). Table 5 provides factor loadings for the 10 scales. Across the three problems, the pattern matrix indicated that the first five scales (i.e., *salient quantities, representation, equation, reasoning, computations*) were associated with Factor 1, and the next five scales (i.e., *focus, content, organization, mathematics vocabulary, convention*) were associated with Factor 2. Based on the hypothesized model for MW in Figure 1, Factor 1 indicates the scales of *mathematics problem solving and reasoning* (hereafter referred to as '*mathematics*' within the model), and Factor 2 indicates the scales of *writing*.

Confirmatory Factor Analysis

A CFA was conducted to evaluate the hypothesized two-factor measurement model for MW using WLSMV estimation (Muthén, et al., 1997). As recommended by Hu and Bentler (1999), a good model fit between the hypothesized model and the observed data is obtained when (a) the chi-square statistic (x^2) is not significant, (b) the comparative fit index (CFI) and Tucker-Lewis index (TLI) is close to .95 or greater, (c) the root mean square error of approximation (RMSEA) is close to .08 or below, and (d) the standardized root mean square residual (SRMR) is close to .06 or below (Brown, 2015). When the goodness-of-fit indices for the hypothesized model were not adequate for CFA, an alternative model(s) would be established as necessary.

Table 6. Goodness-of-fit indices for model one and model two

Prompt		Fit Indices							Model 2 – Model 1		
	Model	x^2	*df*	*p*	CFI	TLI	RMSEA	SRMR	Δx^2	Δ *df*	Δ CFI
Subtraction	Model 1	48.603	34	.050	.972	.963	.053	.063	-	-	-
	Model 2	35.184	33	.365	.996	.994	.021	.037	-13.419	-1	.024
Two-step computation	Model 1	93.639	34	<.001	.976	.968	.107	.092	-	-	-
	Model 2	66.413	33	<.001	.986	.981	.081	.066	-27.226	-1	.010
Division	Model 1	78.076	34	<.001	.994	.959	.066	.053	-	-	-
	Model 2	75.633	33	<.001	.995	.957	.068	.050	-2.443	-1	.001

Note. CFI = comparative fit index; Model 1 = hypothesized model; Model 2 = alternative model; RMSEA = root mean square error of approximation; SRMR = standardized root mean square residual; TLI = Tucker-Lewis index

Figure 3. Hypothesized model and alternatively specified model for the subtraction MW prompt
Note. qt = identifying salient quantities; rp = creating a visual representation; eq = building an equation; rs = explaining problem solving and reasoning; cp = executing computations; fc = focusing on a single main topic; ct = generating the content of the topic; og = organizing the content; vc = using mathematics vocabulary; cv = following conventions

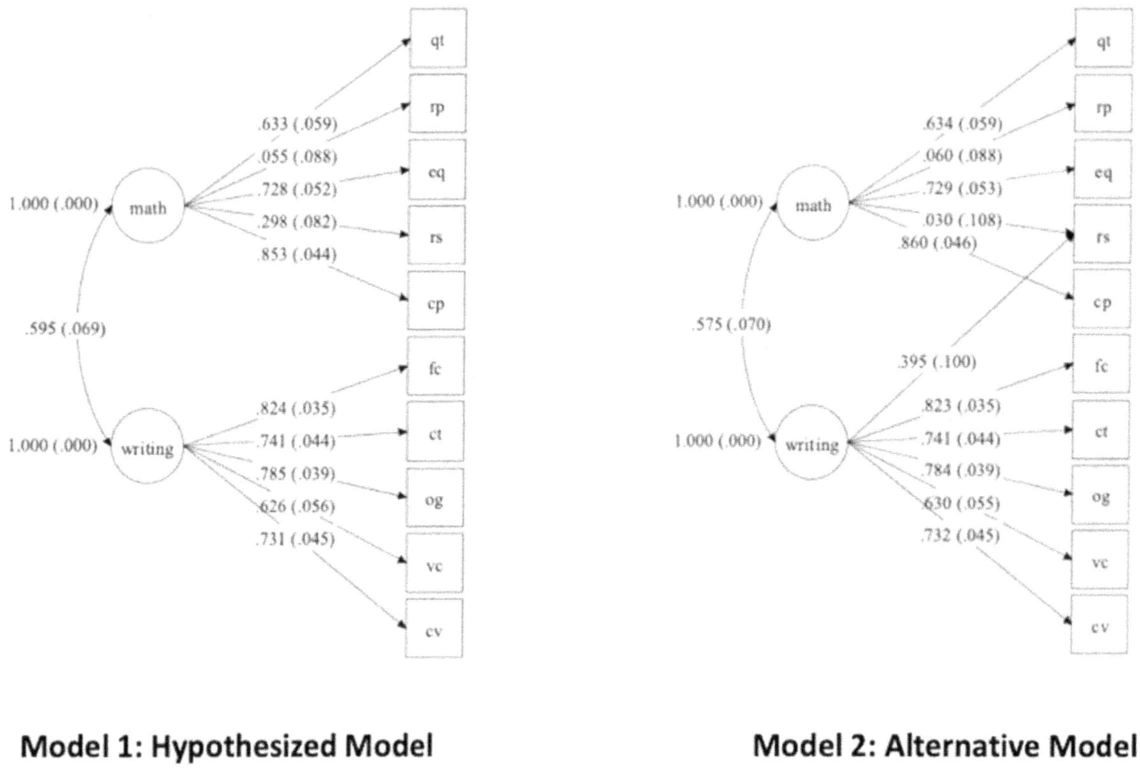

Across the three MW prompts, goodness-of-fit indices suggested that the observed student data were adequate for the CFA for the two-factor hypothesized model. Because the chi-square index was statistically significant in all three prompts, an alternative model (i.e., Model 2 in Figure 3 to 5) was specified, where the observed variable, *reasoning* was to load additionally onto the second factor, *writing*. This cross-loading resulted in the changes in other fit indices. Table 6 compares the goodness-of-fit indices for the first hypothesized model (Model 1) with the alternatively specified model (Model 2).

Figure 3 displays the subtraction prompt's standardized factor loadings for Model 1 and Model 2. In both Model 1 and Model 2, the factor loading, *mathematics* by *representation*, was not significant (standardized estimate = 0.055, SE = 0.088, p = .530 in Model 1; standardized estimate =0.060, SE = 0.088, p = .493 in Model 2). All other standardized factor loadings in Model 1 were statistically significant, which ranged from 0.298 to 0.853. The factor covariance, *mathematics* with *writing*, was also significant in both Model 1 (standardized estimate = 0.595, SE = 0.069, $p < .001$) and Model 2 (standardized estimate = 0.575, SE = 0.070, $p < .001$). In Model 2, the cross-factor loading for *writing* by *reasoning* was additionally estimated. The factor loading for *writing* by *reasoning* was statistically significant (standardized estimate =0.395, SE = 0.100, $p < .001$), while the factor loading for *mathematics* by *reasoning* was not significant (standardized estimate =0.030, SE = 0.108, p = .780). The statistically significant standardized factor loadings in Model 2 ranged from 0.395 to 0.860.

The proportions of the variance in each observed variable explained by its related factor, R^2values, were estimated. In Model 1, R^2 values of *representation* (R^2= .003, SE = .010, p = .754) and *reasoning* (R^2= .089, SE = .010, p = .071) were not statistically significant, and the other variables were significant at a = .05. The significant R^2 values ranged from .392 to .727. In this model, 40.1%, 52.9%, and 72.7% of variance in *salient quantities*, *representation*, and *equation*, respectively, were explained by *mathematics*, respectively. Also, 68.0%, 54.9%, 61.5%, 39.2%, and 53.5% of variances in *focus*, *content*, *organization*, *mathematics vocabulary*, and *convention*, respectively, were explained by *writing*. In Model 2, R^2 value of *representation* was not statistically significant (R^2 = .004, SE = .011, p = .732), and the other variables were significant at a = .05. The significant R^2 values ranged from .171 to .740. In this model, 40.2%, 53.2%, and 74.0% of variances in *salient quantities*, *equation*, and *computation*, respectively, were explained by *mathematics*, while 67.7%, 54.9%, 61.4%, 39.7%, and 53.6% of variances in *focus*, *content*, *organization*, *mathematics vocabulary*, and *convention*, respectively, were accounted for by *writing*. Different from Model 1, 17.1% of variance in *reasoning* was explained by the two factors, *mathematics* and *writing* in Model 2.

Figure 4 exhibits standardized factor loadings in Models 1 and 2 for the two-step computation MW prompt. In Model 1 and Model 2, all factor loadings were significant at a = .05, while the factor loading of *mathematics* by *representation* was not significant (standardized estimate = 0.016, SE = 0.049, p = .747 in Model 1; standardized estimate = 0.012. SE = 0.049. p = .809 in Model 2). Except for this nonsignificant factor loading, all the other standardized factor loadings ranged from 0.570 to 0.957 in Model 1 and 0.570 to 0.969 in Model 2. The factor covariance, *mathematics* with *writing* was also significant in both Model 1 (standardized estimate = 0.801, SE = 0.043, $p < .001$) and Model 2 (standardized estimate = 0.733, SE = 0.050, $p < .001$). Model 2 included the cross-factor loading for *writing* by *reasoning,* which was statistically significant (standardized estimate = 0.899, SE = 0.158, $p < .001$). The factor loading for *mathematics* by *reasoning* was not significant (standardized estimate = -0.089, SE = 0.187, p = .635).

Consistent with the subtraction prompt, R^2 values were calculated. In both Model 1 and Model 2, R^2 values for *representation* were not significant (R^2 = 0, SE = .002, p = .872 in Model 1; R^2 = 0, SE = .001, p = .904 in Model 2). Except for *representation*, the significant R^2 values for the other variables in Model 1 ranged from .325 to .916. In this model, 79.5% of variance in *salient quantities*, 86.7% of variance in *equation*, 75.8% of variance in *reasoning*, and 72.9% of variance in *computation* were accounted for by *mathematics*. Also, 84.9% of variance in *focus*, 91.6% of variance in *content*, 77.4% of variance

Figure 4. Hypothesized model and alternatively specified model for the two-step computation MW prompt
Note. qt = identifying salient quantities; rp = creating a visual representation; eq = building an equation; rs = explaining problem solving and reasoning; cp = executing computations; fc = focusing on a single main topic; ct = generating the content of the topic; og = organizing the content; vc = using mathematics vocabulary; cv = following conventions

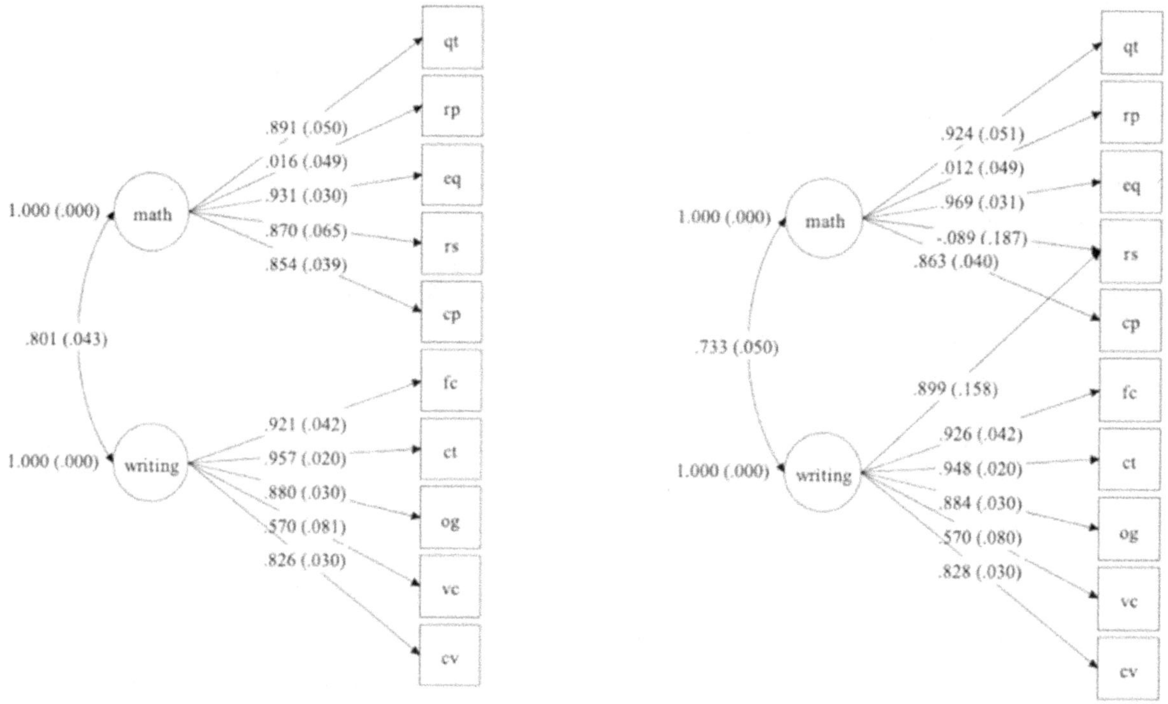

Model 1: Hypothesized Model **Model 2: Alternative Model**

in *organization*, 32.5% of variance in *mathematics vocabulary*, and 68.3% of variance in *convention* were explained by the factor, *writing*. In Model 2, the significant R^2 values ranged from .325 to .939. In this model, 85.3%, 93.9%, and 74.5% of variances in *salient quantities*, *equation*, and *computation*, respectively, were explained by *mathematics*. Also, 85.7%, 89.9%, 78.2%, 32.5%, and 68.6% of variances in *focus*, *content*, *organization*, *mathematics vocabulary*, and *convention,* respectively, were explained by the *writing* factor. For the cross-factor loadings, 69.9% of variance in *reasoning* was explained by *mathematics* and *writing*.

Figure 5 shows standardized factor loadings of Model 1 and Model 2 for the division prompt. The statistical results of Model 1 indicated that the factor loading, *mathematics* by *representation*, was not significant (standardized estimate = 0.070, *SE* = 0.096, $p = .469$). The other standardized factor loadings were statistically significant, ranging from 0.542 to 0.989. The factor covariance, *mathematics* with *writing*, was significant in Model 1 (standardized estimate = 0.855, *SE* = 0.054, $p < .001$). In Model 2, similar to Model 1, the factor loading for *mathematics* by *representation* was not statistically significant (standardized estimate = 0.050, *SE* = 0.098, $p = .611$). The cross-factor loading for *writing* by *reasoning* was additionally estimated, which was statistically significant (standardized estimate = 0.711, *SE* = 0.291, $p = .015$). In this model, the factor loading for *mathematics* by *reasoning* was not statistically

Figure 5. Hypothesized model and alternatively specified model for the division MW prompt
Note. qt = identifying salient quantities; rp = creating a visual representation; eq = building an equation; rs = explaining problem solving and reasoning; cp = executing computations; fc = focusing on a single main topic; ct = generating the content of the topic; og = organizing the content; vc = using mathematics vocabulary; cv = following conventions

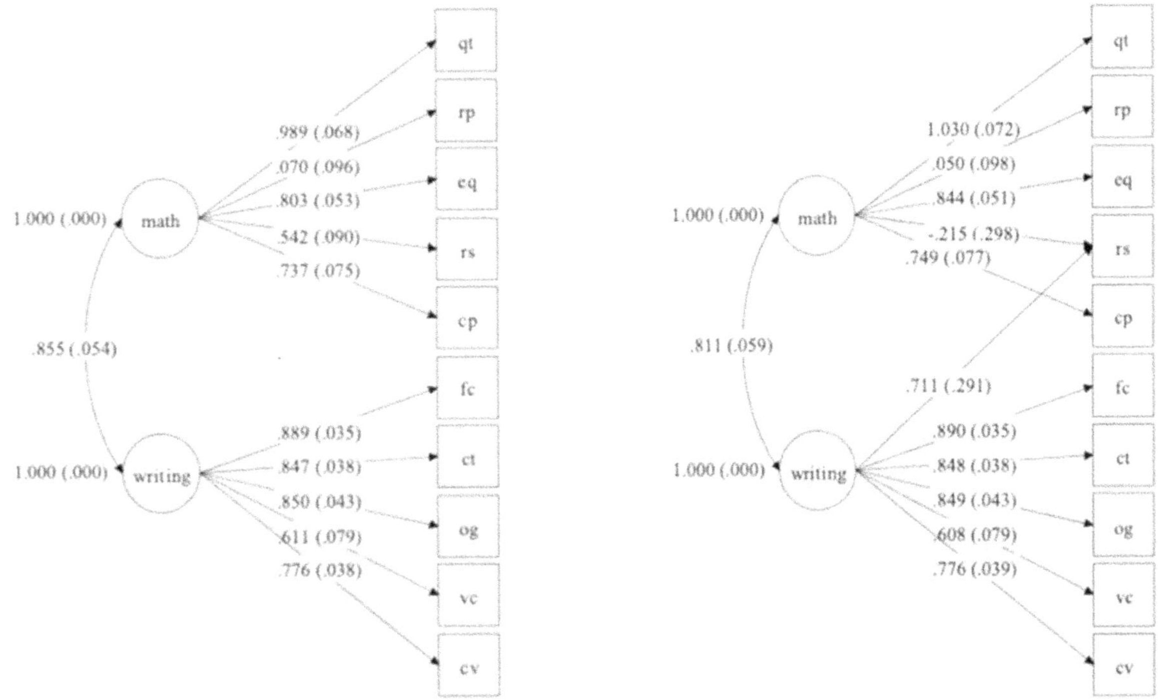

Model 1: Hypothesized Model **Model 2: Alternative Model**

significant (standardized estimate = -0.215, *SE* = 0.298, $p = .470$). The other statistically significant standardized factor loadings in Model 2 ranged from 0.608 to 1.030. The factor covariance, *mathematics* with *writing*, was significant in Model 2 (standardized estimate = 0.811, *SE* = 0.059, $p < .001$).

For the division prompt, R^2 statistics were estimated to calculate the proportions of the variances in individual variables explained by the two factors, *mathematics* and *writing*. In Model 1, the R^2 statistic for *representation* was not significant ($R^2 = .005$, $SE = .013$, $p = .718$). The other R^2 values were significant at $a = .05$, ranging from .294 to .978. In this model, 97.8%, 64.5%, 29.4%, and 54.3% of variances in *salient quantities*, *equation*, *reasoning*, and *computation*, respectively, were explained by *mathematics*, and 79.0%, 71.7%, 72.2%, 37.3%, and 60.2% of variances in *focus*, *content*, *organization*, *mathematics vocabulary*, and *convention*, respectively, were explained by *writing*. In Model 2, the R^2 statistic for *representation* was not significant ($R^2 = .002$, $SE = .010$, $p = .799$). The R^2 statistics for the other variables were significant at $a = .05$, ranging from .304 to .999. In this model, 99.9%, 71.2%, 56.1% of variances in *salient quantities*, *equation*, and *computation*, respectively, were explained by the factor, *mathematics*, and 79.2%, 71.8%, 72.0%, 37.0%, and 60.2% of variances in *focus*, *content*, *organization*, *mathematics vocabulary*, and *convention*, respectively, were explained by the factor, *writing*. The Model 2 included the cross-factor loading for *reasoning*, and 30.4% of variance in *reasoning* was explained by the two factors, *mathematics* and *writing*.

DISCUSSION

The primary objective of this study was to establish a solid theoretical foundation by identifying the underlying constructs that contribute to students' MW performance. Different from previous studies, which primarily centered around *how to teach* MW, this study shifts the focus towards identifying *what to assess* for students' MW performance. Focusing on informative/explanatory MW frequently required in mathematics formal assessments and textbooks (Casa et al., 2016; Hughes et al., 2019), the definition of MW was established, and the underlying constructs were determined based on a body of literature related to MW (e.g., Bossé & Faulconer, 2008; Cohen et al., 2015; Cross, 2009; Hughes & Lee, 2020; Lepak et al., 2018; Nakung et al., 2020). By delineating these constructs, the study aimed to enhance our understanding of the intricate interplay between *mathematics* and *writing* constructs in the context of MW.

Building upon the literature, the two-factor hypothesized model in Figure 1 was developed. This model conceptualizes MW as comprising two key factors: *mathematics problem solving and reasoning* and *writing*. Each factor was further dissected into five distinct constructs, as presented in Tables 1 and 2. The validity of the hypothesized model was explored and confirmed through EFA and CFA conducted using the data collected from 153 fourth grade students. Across all the three MW prompts, the results obtained from the EFA indicated that the latent structure of MW was best characterized by a two-factor model, which was represented by *mathematics* and *writing*. The results align with the theoretical frameworks of MW proposed in previous studies (e.g., Hughes & Lee, 2020; Powell & Hebert, 2016). The pattern matrix revealed that the first five scales under *mathematics* and the other five scales under *writing* were associated with their respective factors. Collectively, these preliminary findings provide compelling evidence supporting the fit between the two-factor hypothesized model and the data, which reasonably explains the ten constructs that impact students' MW proficiency.

The subsequent CFA was conducted to provide a more comprehensive evaluation of the model fit indices, offering substantial evidence regarding the adequacy of the MW constructs to assess students' MW proficiency. While the chi-square fit indices did not demonstrate optimal fit between the data and the model, the other fit indices such as CFI, TLI, RMSEA, and SRMR provided substantial support for the presence of two distinct latent variables characterizing students' MW scores. To improve the fit indices, particularly addressing the significant chi-square fit indices observed in the three MW prompts, the initial hypothesized model (Model 1) was modified to the alternatively specified model (Model 2). This alternative model involved cross-factor loading of *reasoning* onto both *mathematics* and *writing,* based on the factor loadings obtained from the EFA and modification indices derived from the CFA. This alternative model resulted in non-significant chi-square fit index for the subtraction prompt, while in the other two prompts, the chi-square fit indices remained significant. However, it is important to note that the chi-square index is not widely utilized in applied research as the sole indicator for evaluating the model fit, as it can be inflated by sample size (Brown, 2015). Considering that the other fit indices fell within the acceptable ranges, these results provide support for a good model fit between the proposed models and the observed data.

Taking into account the aforementioned considerations, the cross-factor loading of *reasoning*, which reflects the ability to provide clear and accurate statements for how and why the operation(s) was selected, significantly improved the overall model fit indices of the model. Mathematical reasoning inherently requires the utilization of mathematical inference, encompassing mathematical argumentation to justify the chosen operations and problem solving method and providing a plausible explanation for them (Jonsson et al., 2014; Kiuhara et al., 2020). This process necessitates the use of mathematics vocabulary and syntactic knowledge to formulate coherent and well-structured sentences, requiring both mathematics and writing competencies (Hebert & Powell, 2016; Hughes et al., 2016; Namkung et al., 2020).

This interdependence between mathematical reasoning and language have been empirically investigated through many studies (e.g., Chow & Ekholm, 2019; Espinas & Fuchs, 2022; Lin et al., 2021; Peng et al., 2020). Meta-analyses conducted by Lin et al. (2021) and Peng et al. (2020) have provided support for the significant correlations between mathematical reasoning and language. Their findings indicate that mathematical language serves as a critical medium for facilitating mathematical reasoning, even after controlling for other linguistic and cognitive factors such as listening comprehension, phonological processing, working memory, and intelligence (Espinas & Fuchs, 2022). Students use linguistic abilities to retrieve mathematical knowledge and to comprehend the lexically dense word problem statements to establish a coherent and logical sequence of mathematical reasoning (Chow & Ekholm, 2019; Peng et al., 2020). By understanding the relationship between mathematical reasoning and language, researchers and practitioners can design targeted interventions and instructional strategies that explicitly teach mathematics vocabulary and expressions (Powell & Driver, 2015).

While the alternative model with the cross-factor loading of *reasoning* improved overall fit indices of the initial hypothesized model, the impact of *visual representations* on the MW performance remains uncertain within the model. One of the possible reasons for this is that the MW prompts used in this study did not explicitly require students to include pictorial representations in their written responses, resulting in a limited number of students who included such representations (n = 11 for the subtraction prompt; n = 24 for the two-step computation prompt; n = 72 for the division prompt). Consequently, it was challenging to precisely examine whether the ability to visualize mathematical relations between the quantities represents a distinct latent variable that consists of MW proficiency (Hebert & Powell, 2017).

However, many studies have established a significant connection between students' ability to visually represent mathematical relations and their mathematical problem solving and reasoning (e,g., Campbell et al., 1995; Van Garderen & Montague, 2003; Lepak et al., 2018). According to Brenner et al. (1997), during the transition from problem representation to solution planning in the cognitive problem solution process, a visual representation plays an important role in guiding students to select arithmetic and/or algebraic steps for problem solving. These visual aids help students comprehend how the quantities are related within the problem context, reducing their cognitive load (Lepak et al., 2018). Students strategically utilize these visual representations to plan how to solve the problem and then translate them to an abstract representations, such as equations, using symbolic languages (Brenner et al., 1997; Riccomini et al., 2022). The NCTM (2000) also emphasizes the importance of students' mastering representation skills as one of the process standards to articulate and organize mathematical concepts and relationships in problem solving.

In the context of MW, Hebert and Powell (2016) observed students' employing visual representations in their written responses to elaborate their approach to solving fraction word problems. These students utilized visual representations to elaborate on their written explanations, integrating them within the written paragraph. As suggested by Bossé & Faulconer (2008) and Cohen et al. (2015), it is important to explicitly teach students to include pictorial representations in their MW illustrating conceptual understanding and problem solving. It would be helpful for students to organize their mathematical ideas in a more logical sequence (Lepak et al., 2018) as well as more strategically plan their problem solutions (Brenner et al., 1997). Further empirical efforts are needed to explore how the visual representations impact students' MW performance, while considering their ability to use abstract symbols for equation building.

Limitations

This study conceptualizes MW as a latent variable, considering it as an essential means for mathematical communication within the context of the mathematics curriculum and standards (National Council of Teachers of Mathematics, 2000; National Governors Association Center for Best Practices & Council of Chief State School Officers, 2010). The authors also delve into key constructs associated with students' MW proficiency conducting statistical data analyses. The results validated the adequacy of the collected student data for the hypothesized model comprising 10 variables. These results have implications for researchers, (a) offering insights into what competencies should be fostered to improve the quality of MW through interventions and (b) identifying what variables should be measured when profiling academic and cognitive characteristics reflected in students' written responses for mathematics problem solving and reasoning.

Despite the potential benefits of this preliminary study, we noted some limitations when interpreting the findings of the study and inferring students' performance based on the statistical results. Firstly, the data collected from 153 4th-grade students at an elementary school in a rural school district may not be fully representative of the broader student population. Although the study took into account the sample size in statistical analyses, the results should be cautiously interpreted when generalizing to other grade levels and diverse ability groups, especially students with disabilities. Further investigation with a larger and more diverse sample is necessary to determine if the scales used in this study can effectively identify patterns of academic strengths or weaknesses of the students. This will enable researchers to draw informed conclusions about the students' performance based on the data.

Secondly, the study focused on three MW prompts related to subtraction, two-step computations involving multiplication and subtraction, and divisions, all of which were limited to arithmetic computations. However, it is important to evaluate the model in other areas of the mathematics curriculum such as measurement, geometry, and statistics (National Governors Association Center for Best Practices & Council of Chief State School Officers, 2010). Future research might explore whether the identified constructs also influence students' MW performance across different mathematical domains.

Thirdly, the construct of *visual representations* requires further empirical investigation to validate the hypothesized model of MW. Previous studies have emphasized the significance of students' pictorial representations in MW (e.g., Hebert & Powell, 2016; Kosko & Zimmerman, 2019; Nakung et al., 2020). However, the results from the CFA indicate the necessity for additional research to thoroughly scrutinize the effects of visual representations on students' MW performance, particularly to determine whether it is crucial to explicitly ask students to visualize the mathematical relationship between the quantities presented in the problem statements.

Fourthly, the current study has shed light on the underlying dimensions to be assessed for MW, but it does not provide specific guidance on the content that should be taught to enhance MW skills. To bridge this gap, future research could focus on empirical analyses of MW teaching resources and materials (Siebert & Draper, 2008) and/or the mathematical discourse between students and mathematics teachers (Moschkovich, 2015) to offer concrete evidence on which strands are important in MW instruction.

Lastly, this study specifically focused on students' informative/explanatory writing in mathematics. Other lines of research should investigate whether the 10 identified constructs can be applied to other genres of writing in mathematics such as exploratory, argumentative, and creative writing (Casa et al., 2016), to gain a more comprehensive understanding of students' MW competencies.

Implications for Practice

As evident from the results, MW requires proficiency in both mathematics and writing skills. This implies that teachers should consider students' current performance levels in both aspects. Even in a mathematics classroom, for students who lack mastery of prerequisite writing skills, we recommend that teachers provide scaffolded instruction to teach fundamental writing skills required for MW, such as paraphrasing (Moran et al., 2014), transition words (Hebert & Powell, 2016), and sentence formation rules (Namkung et al., 2020). We also suggest incorporating explicit instruction on mathematics vocabulary that aligns with the specific areas of mathematics content being taught (Peng & Lin, 2019; Powell & Driver, 2015).

In addition, this study provides valuable insights on what aspects of students' MW proficiency should be assessed. Rather than relying solely on a single holistic scale like state assessments (e.g., PSSA-M), analyzing students' written responses across various constructs allows teachers and practitioners to identify distinct patterns of strengths and weaknesses represented in specific areas. This information can be utilized to make cognizant instructional decisions. Thus, this book chapter suggests the evaluation of 10 constructs represented in students' written responses to the open-ended questions and prompts in students' textbooks and formal assessments. By doing this, educators can assess whether students require additional support in particular domains (Namkung et al., 2020).

CONCLUSION

This book chapter explores what constructs impact students' MW proficiency and offers educational implications for assessing the quality of informative/explanatory writing in mathematics. Prior to investigating the empirical evidence, a two-factor hypothesized model comprising 10 constructs that influence students' MW proficiency was established, assuming the integration of both mathematics and writing aspects. Across the three MW prompts that required different arithmetic skills, the results of EFA and CFA consistently supported the adequacy of the hypothesized model in explaining the underlying dimensionality of MW proficiency based on student data. Notably, the cross-factor loading of *mathematical reasoning* on *mathematics problem solving and reasoning* and *writing* improved the overall fit for the model, indicating that students' ability to provide valid explanations on the method and operation they chose requires proficiency in both domains (Lepak et al., 2018; Powell & Hebert, 2016).

However, the impact of *visual representations* on students' MW proficiency remains inconclusive within this model. In light of the importance highlighted by previous studies regrading visual representation in MW, it is recommended that future research thoroughly examines their influence on students' performance. This can be achieved by directly prompting students to incorporate visual representations in their written responses, demonstrating their understanding of the mathematical context and relationships between quantities presented in the problem statement.

REFERENCES

Arsenault, T. L., Powell, S. R., Hebert, M. A., King, S. G., Lin, X., & Lang, D. (2022). Mathematics-writing profiles for students with mathematics difficulty. *Reading and Writing*. doi:10.100711145-022-10375-2

Baxter, J. A., Woodward, J., & Olson, D. (2005). Writing in mathematics: An alternative form of communication for academically low-achieving students. *Learning Disabilities Research & Practice*, *20*(2), 119–135. doi:10.1111/j.1540-5826.2005.00127.x

Bossé, M. J., & Faulconer, J. (2008). Learning and assessing mathematics through reading and writing. *School Science and Mathematics*, *108*(1), 8–9. doi:10.1111/j.1949-8594.2008.tb17935.x

Brenner, M. E., Mayer, R. E., Moseley, B., Brar, T., Duran, R., Reed, B. S., & Webb, D. (1997). Learning by understanding: The role of multiple representations in learning algebra. *American Educational Research Journal*, *34*(4), 663–689. doi:10.3102/00028312034004663

Brown, T. A. (2015). *Confirmatory factor analysis for applied research* (2nd ed.). The Guilford Press. doi:10.4324/9781315757421

Bruner, J. S., & Kenney, H. J. (1965). Representation and mathematics learning. *Monographs of the Society for Research in Child Development*, *30*(1), 50–59. doi:10.2307/1165708

Campbell, K. J., Collis, K. F., & Watson, J. M. (1995). Visual processing during mathematical problem solving. *Educational Studies in Mathematics*, *28*(2), 177–194. doi:10.1007/BF01295792

Casa, T. M., Firmender, J. M., Cahill, J., Cardetti, F., Choppin, J. M., Cohen, J., & Zawodniak, R. (2016). *Types of and purposes for elementary mathematical writing: Task force recommendations*. Math Writing. http://mathwriting.eduation.uconn.edu

Chan, W. W. L., & Kwan, J. L. Y. (2021). Pathways to word problem solving: The mediating roles of schema construction and mathematical vocabulary. *Contemporary Educational Psychology*, *65*, 1–12. doi:10.1016/j.cedpsych.2021.101963

Chow, J. C., & Ekholm, E. (2019). Language domains differentially predict mathematics performance in young children. *Early Childhood Research Quarterly*, *46*, 179–186. doi:10.1016/j.ecresq.2018.02.011

Cohen, J. A., Casa, T. M., Miller, H. C., & Firmender, J. M. (2015). Characteristics of second grader's mathematical writing. *School Science and Mathematics*, *115*(7), 344–355. doi:10.1111sm.12138

Colonesse, M. W., Amspaugh, C. M., LeMay, S., Evans, K., & Field, K. (2018). Writing in the disciplines: How math fits into the equation. *The Reading Teacher*, *72*(3), 379–387. doi:10.1002/trtr.1733

Correnti, R., Matsumura, L. C., Hamilton, L., & Wang, E. (2013). Assessing students' skills at writing analytically in response to texts. *The Elementary School Journal*, *114*(2), 142–177. doi:10.1086/671936

Cross, D. I. (2009). Alignment, cohesion, and change: Examining mathematics teachers' belief structures and their influence on instructional practices. *Journal of Mathematics Teacher Education*, *12*(5), 324–346. doi:10.100710857-009-9120-5

Deatline-Buchman, A., & Jitendra, A. K. (2006). Enhancing argumentative essay writing of fourth-grade students with learning disabilities. *Learning Disability Quarterly*, *29*(1), 39–54. doi:10.2307/30035531

Engelhard, G. Jr. (1992). The measurement of writing ability with a Many-Faceted Rasch Model. *Applied Measurement in Education*, *5*(3), 171–191. doi:10.120715324818ame0503_1

Espinas, D. R., & Fuchs, L. S. (2022). The effects of language instruction on math development. *Child Development Perspectives*, *16*(2), 69–75. doi:10.1111/cdep.12444 PMID:35493789

Forsyth, S. R., & Powell, S. R. (2017). Differences in the mathematics-vocabulary knowledge of fifth-grade students with and without learning difficulties. *Learning Disabilities Research & Practice*, *32*(4), 231–245. doi:10.1111/ldrp.12144

Fuchs, L. S., Fuchs, D., Compton, D. L., Powell, S. R., Seethaler, P. M., Capizzi, A. M., Schatschneider, C., & Fletcher, J. M. (2006). The cognitive correlates of third-grade skill in arithmetic, algorithmic computation, and arithmetic word problems. *Journal of Educational Psychology*, *98*(1), 29–43. doi:10.1037/0022-0663.98.1.29

Fuchs, L. S., Seethaler, P. M., Sterba, S. K., Craddock, C., Fuchs, D., Compton, D. L., Geary, D. C., & Changas, P. (2021). Closing the word-problem achievement gap in first grade: Schema-based word-problem intervention with embedded language comprehension instruction. *Journal of Educational Psychology*, *113*(1), 86–103. doi:10.1037/edu0000467 PMID:33776137

Hair, J., Anderson, R. E., Tatham, R. L., & Black, W. C. (1995). *Multivariate data analysis* (4th ed.). Prentice-Hall Inc.

Hebert, M. A., & Powell, S. R. (2016). Examining fourth-grade mathematics writing: Features of organization, mathematics vocabulary and mathematical representations. *Reading and Writing*, *29*(7), 1511–1537. doi:10.100711145-016-9649-5

Heyes, J. R. (2012). Modeling and remodeling writing. *Written Communication*, *29*(3), 369–388. doi:10.1177/0741088312451260

Hu, L., & Bentler, P. M. (1999). Cutoff criteria for fit indexes incovariance structure analysis: Conventional criteria versus new alternatives. *Structural Equation Modeling*, *6*(1), 1–55. doi:10.1080/10705519909540118

Hughes, E. M., & Lee, J. (2020). Effects of a mathematical writing intervention on middle school students' performance. *Reading & Writing Quarterly*, *36*(2), 176–192. doi:10.1080/10573569.2019.1677537

Hughes, E. M., Marketz, A. M., & Cozad, L. E. (2019). Evaluating various undergraduate perspectives of elementary-level mathematical writing. *International Journal of Science and Mathematics Education*, *17*(5), 1031–1048. doi:10.100710763-018-9903-1

Hughes, E. M., Powell, S. R., & Stevens, E. A. (2016). Supporting clear and concise mathematics language: Say this, not that. *Teaching Exceptional Children*, *49*(1), 7–17. doi:10.1177/0040059916654901

Jeffery, J. V. (2009). Constructs of writing proficiency in US state and national writing assessments: Exploring variability. *Assessing Writing*, *14*(1), 3–24. doi:10.1016/j.asw.2008.12.002

Jitendra, A. K., Nelson, G., Pulles, S. M., Kiss, A. J., & Houseworth, J. (2016). Is mathematical representation of problems an evidence-based strategy for students with mathematics difficulties? *Exceptional Children, 83*(1), 8–25. doi:10.1177/0014402915625062

Jonsson, B., Norqvist, M., Liljekvist, Y., & Lithner, J. (2014). Learning mathematics through algorithmic and creative reasoning. *The Journal of Mathematical Behavior, 36*, 20–32. doi:10.1016/j.jmathb.2014.08.003

Kingsdorf, S., & Krawec, J. (2014). Error analysis of mathematical word problem solving across students with and without learning disabilities. *Learning Disabilities Research & Practice, 29*(2), 66–74. doi:10.1111/ldrp.12029

Kiuhara, S. A., Gillespie Rouse, A., Dai, T., Witzel, B. S., Morphy, P., & Unker, B. (2020). Constructing written arguments to develop fraction knowledge. *Journal of Educational Psychology, 112*(3), 584–607. doi:10.1037/edu0000391

Klein, P. D. (1999). Reopening inquiry into cognitive processes in writing-to-learn. *Educational Psychology Review, 11*(3), 203–270. doi:10.1023/A:1021913217147

Kosko, K. W., & Zimmerman, B. S. (2019). Emergence of argument in children's mathematical writing. *Journal of Early Childhood Literacy, 19*(1), 82–106. doi:10.1177/1468798417712065

Lepak, J. R., Wernet, J. L., & Ayieko, R. A. (2018). Capturing and characterizing students' strategic algebraic reasoning through cognitively demanding tasks with focus on representations. *The Journal of Mathematical Behavior, 50*, 57–73. doi:10.1016/j.jmathb.2018.01.003

Lin, X., Peng, P., & Zeng, J. (2021). Understanding the relation between mathematics vocabulary and mathematics performance: A meta-analysis. *The Elementary School Journal, 121*(3), 357–540. doi:10.1086/712504

Lithner, J. (2000). Mathematical reasoning in task solving. *Educational Studies in Mathematics, 41*(2), 165–190. https://www.jstor.org/stable/3483188

Mainali, B. (2021). Representation in teaching and learning mathematics. [IJEMST]. *International Journal of Education in Mathematics, Science, and Technology, 9*(1), 1–21. doi:10.46328/ijemst.1111

Mayer, R. E. (1985). Implications of cognitive psychology for instruction in mathematical problem solving. In E. A. Silver (Ed.), *Teaching and learning mathematical problem solving: Multiple research perspectives* (pp. 123–138). Lawrence Erlbaum Associates.

McCloskey, M., Caramazza, A., & Basili, A. (1985). Cognitive mechanisms in number processing and calculation: Evidence form dyscalculia. *Brain and Cognition, 4*(2), 171–196. doi:10.1016/0278-2626(85)90069-7 PMID:2409994

Mo, Y., & Troia, G. A. (2017). Similarities and differences in constructs represented by U. S. States' middle school writing tests and the 2007 National Assessment of Educational Progress writing assessment. *Assessing Writing, 33*, 48–67. doi:10.1016/j.asw.2017.06.001

Moran, A. S., Swanson, H. L., Gerber, M. M., & Fung, W. (2014). The effects of paraphrasing interventions on problem-solving accuracy for children at risk for math disabilities. *Learning Disabilities Research & Practice, 29*(3), 97–105. doi:10.1111/ldrp.12035

Moschkovich, J. N. (2015). Academic literacy in mathematics for English learners. *The Journal of Mathematical Behavior, 40*, 43–62. doi:10.1016/j.jmathb.2015.01.005

Muthén, B., du Toit, S. H. C., & Spisic, D. (1997). Robust inference using weighted least squares and quadratic estimating equations in latent variable modelling with categorical and continuous outcomes (Technical Report). University of California, Los Angeles.

Namkung, J. M., & Bricko, N. (2021). The effects of algebraic equation solving intervention for students with mathematics learning difficulties. *Journal of Learning Disabilities, 54*(2), 111–123. doi:10.1177/0022219420930814 PMID:32508218

Namkung, J. M., Hebert, M. H., Powell, S. R., Hoins, M., Bricko, N., & Torchia, M. (2020). Comparing and validating four methods for scoring mathematics writing. *Reading & Writing Quarterly, 36*(2), 157–175. doi:10.1080/10573569.2019.1700858

National Council of Teachers of Mathematics. (2000). *Principles and standards for school mathematics*. NCTM.

National Governors Association Center for Best Practices & Council of Chief State School Officers. (2010). *Common Core State Standards for Mathematics*. NCTM. http://www.corestandards.org/wp-content/uploads/Math_Standards1.pdf

Peng, P., & Lin, X. (2019). The relation between mathematics vocabulary and mathematics performance among fourth graders. *Learning and Individual Differences, 69*, 11–21. doi:10.1016/j.lindif.2018.11.006

Peng, P., Lin, X., Ünal, Z. E., Lee, K., Namkung, J., Chow, J., & Sales, A. (2020). Examining the mutual relations between language and mathematics: A meta-analysis. *Psychological Bulletin, 146*(7), 595–634. doi:10.1037/bul0000231 PMID:32297751

Pennsylvania Department of Education. (2023). *The Pennsylvania System of School Assessment Grade 3 Mathematics Items and Scoring Sampler*. PA DoE. https://www.education.pa.gov/Documents/K-12/Assessment%20and%20Accountability/PSSA/Item%20and%20Scoring%20Samples/2023%20PSSA%20ISS%20Math%20Grade%203.pdf

Polya, G. (1957). *How to solve it: A new aspect of mathematical method* (2nd ed.). Doubleday.

Powell, S. R., & Driver, M. K. (2015). The influence of mathematics vocabulary instruction embedded within addition tutoring for first-grade students with mathematics difficulty. *Learning Disability Quarterly, 38*(4), 221–233. doi:10.1177/0731948714564574

Powell, S. R., & Hebert, M. A. (2016). Influence of writing ability and computational skill on mathematics writing. *The Elementary School Journal, 117*(2), 310–335. doi:10.1086/688887

Powell, S. R., Hebert, M. A., Cohen, J. A., Casa, T. M., & Firmender, J. M. (2017). A synthesis of mathematics writing: Assessment, interventions, and surveys. *Journal of Writing Research, 8*(3), 493–526. doi:10.17239/jowr-2017.08.03.04

Purpura, D. J., Logan, J. A. R., Hassinger-Das, B., & Napoli, A. R. (2017). Why do early mathematics skills predict later reading? The role of mathematical language. *Developmental Psychology, 53*(9), 1633–1642. doi:10.1037/dev0000375 PMID:28758785

Riccomini, P. J., Smith, G. W., Hughes, E. M., & Fries, K. M. (2015). The language of mathematics: The importance of teaching and learning mathematical vocabulary. *Reading & Writing Quarterly*, *31*(3), 235–252. doi:10.1080/10573569.2015.1030995

Riccomini, P. J., Witzel, B. S., & Deshpande, D. S. (2022). Combining visual representations and a powerful retention strategy with peer-mediated strategies to improve mathematical outcomes for students with EBD. *Beyond Behavior*, *31*(1), 42–52. doi:10.1177/10742956211072555

Ritchey, K. D., & Coker, D. L. Jr. (2013). An investigation of the validity and utility of two curriculum-based measurement writing tasks. *Reading & Writing Quarterly*, *29*(1), 89–119. doi:10.1080/10573569.2013.741957

Santos, L., & Semana, S. (2015). Developing mathematics written communication through expository writing supported by assessment strategies. *Educational Studies in Mathematics*, *88*(1), 65–87. doi:10.100710649-014-9557-z

Siebert, D., & Draper, R. J. (2008). Why content-area literacy messages do not speak to mathematics teachers: A critical content analysis. *Literacy Research and Instruction*, *47*(4), 229–245. doi:10.1080/19388070802300314

Van Garderen, D., & Montague, M. (2003). Visual-spatial representation, mathematical problem solving, and students of varying abilities. *Learning Disabilities Research & Practice*, *18*(4), 246–254. doi:10.1111/1540-5826.00079

Villeneuve, E. F., Hajovsky, D. B., Mason, B. A., & Lewno, B. M. (2019). Cognitive ability and math computation developmental relations with math problem solving: An integrated, multigroup approach. *School Psychology*, *34*(1), 96–108. doi:10.1037pq0000267 PMID:29985022

Williams, B., Onsmans, A., & Brown, T. (2010). Exploratory factor analysis: A five-step guide for novices. *Australasian Journal of Paramedicine*, *8*(3), 1–13. doi:10.33151/ajp.8.3.93

Witzel, B. S. (2005). Using CRA to teach algebra to students with math difficulties in inclusive settings. *Learning Disabilities (Weston, Mass.)*, *3*(2), 49–60.

Zahner, D., & Corter, J. E. (2010). The process of probability problem solving: Use of external visual representations. *Mathematical Thinking and Learning*, *12*(2), 177–204. doi:10.1080/10986061003654240

Section 4

Engaging All Learners in Mathematical Writing

Chapter 13
Academic Literacy in Mathematics to Frame Mathematical Writing Research and Practice:
Writing and Revising for Mathematical Practices

Judit N. Moschkovich
https://orcid.org/0000-0003-1522-9337
University of California, Santa Cruz, USA

ABSTRACT

Oral student contributions are important for learning mathematics, but without a written record of students' reasoning, revising is difficult or impossible. While writing makes revising possible, research needs to theoretically frame what to revise, how to revise, and how to support revisions. The chapter addresses the question: How can the academic literacy in mathematics framework inform research and practice on mathematical writing? The examples illustrate how that framework provides a complex view of language and a focus on mathematical practices to frame research and practice on mathematical writing and revising. The chapter considers why student writing and revising are important and describes how both research and instruction can emphasize mathematical practices. The examples focus on explanatory and argumentative writing and several mathematical practices central to revising explanations or arguments: generating conjectures, connecting representations, justifying claims, constructing arguments, and attending to precision.

DOI: 10.4018/978-1-6684-6538-7.ch013

INTRODUCTION

This chapter describes how the *academic literacy in mathematics* framework provides a complex view of language and an emphasis on mathematical practices to frame research on K-12 students' mathematical writing and revising. The examples illustrate how research can explore mathematical writing by emphasizing mathematical practices and how instruction can support revising by focusing on mathematical practices.

This chapter considers two components of how to theoretically frame research on mathematical writing: views of the nature of language and mathematics, because both inform views of mathematical writing. Researchers and practitioners alike bring multiple assumptions about what language and mathematics are, and these assumptions inform how we imagine supporting mathematics learning through mathematical writing.

Writing reveals and records one's thinking and reasoning, thus providing opportunities for reflection, revision, and discussion in more enduring (and perhaps also more concrete) ways than talk. A research agenda focused on writing to reason mathematically first requires clarifying what actually constitutes *academic literacy in mathematics*, so that we can theoretically frame what kinds of writing (and revising) matter for learning important mathematics.

Contemporary views on communication in mathematics education research provide a view of language as a resource for making meaning. These views also assume meanings are situated, contextual, and dynamic (not static or provided by textbooks or definitions) and emphasize the multimodal nature of discourse as oral, written, pictorial, and gestural. This chapter uses a complex view of mathematical language to define writing in mathematics classrooms. Language in general, and talking or writing in particular, are not forms or rules (e.g., vocabulary, grammar) but tools and processes for creating meaning. Researchers from the perspective of language as a meaning-making resource use phrases such as "writing as meaning-making" (Bunch & Willett, 2013, p. 142) and describe how such approaches "shift the focus away from writing 'skills' and native-speaker norms, toward expanding students' semiotic resources through apprenticeship into academic and literacy practices" (p. 142).

These complex views of language frame it not as something to learn but, instead, see using language as a process to construct meaning. This assumption is now central to work on oral aspects of discourse in mathematics classrooms and needs to be incorporated into further research on mathematical writing. Importantly, narrow views of language can result in deficit views of learners and may lead to impoverished approaches to instruction focusing on low level language skills (Moschkovich, 2015). An emphasis on the precise use of formal disciplinary vocabulary and the exclusive use of the official language of instruction (e.g., standard English in the US) over the use of other languages or informal/everyday language enacts deficit views of students whose first language is not the language of instruction or who use everyday language when they first encounter new ideas and as a resource to make sense of ideas. Instead, an expanded view of language for meaning, sense making, and reasoning as it relates to instruction can support learners as they refine both their thinking and the language they use to express that thinking.

A focus on writing to reason mathematically must include a contemporary perspective on language, including the definition of language as a tool for making meaning, as well the assumptions that meanings are situated and dynamic. A complex view of mathematical language that emphasizes its multimodal nature means that lessons need to include multiple modes, not only listening, reading and talking, but also writing. Mathematical writing needs to include multiple representations (gestures, objects, drawings, tables, graphs, symbols, etc.), because connecting representations has been documented as supporting student understanding (Hiebert & Carpenter, 1992). Lastly, but just as importantly, a complex view of language as a tool for making meaning also assumes that there are multiple types of language[1] (formal school mathematical language, home languages, everyday[2] language) and ways to use those discourses to support meaning making. A central issue for instruction is noticing the mathematical content in students' writing when they use everyday language in their mathematical writing, building on that everyday language, and supporting more formal ways of writing.

Narrow views of mathematical language in the classroom as merely words severely limit the linguistic resources that teachers can use to teach mathematics and that students can use to learn mathematics. Mathematical writing that uses a narrow view of language and focuses on words, vocabulary, or definitions, limits learners' access to the five strands of mathematical proficiency (Kilpatrick, Swafford, & Findell, 2001) and curtails students' opportunities to participate in valued activities for learning mathematics documented in research, such as the mathematical practices described in national or state standards (e.g. CCSSM, 2010) or mathematical processes (e.g. NCTM 2000). Not allowing students to use informal language when writing also limits the resources they can use to learn mathematics. Lastly, focusing on correct vocabulary curtails the opportunities for students to express themselves mathematically in what are likely to be imperfect ways, especially initially. In contrast, the view of *academic literacy in mathematics* described here provides a complex and expanded view of mathematical language that is connected to the five strands of mathematical proficiency, emphasizes mathematical practices, and includes informal language as a resource.

Research and practice on mathematical writing needs to shift from simplified views of mathematical language as words, vocabulary, or definitions. Researchers and teachers need to carefully consider when and how it is appropriate and timely to emphasize correct vocabulary and formal language, and when it is pedagogically appropriate and useful for students to use informal language or imperfect ways of communicating, to express emergent mathematical thinking through writing. This shift from a narrow to an expanded view of mathematical language is important for all students, and is also an issue of equity and access.

Supporting mathematical writing focused on reasoning may have important implications for student participation. Emphasizing writing that attends to reasoning may expand the ways that students can participate, and thus has the potential to make classrooms more equitable, and ideally allow students to draw on the full range of their cultural, linguistic, and social resources. Moschkovich (2013a) defines equitable teaching practices for students from non-dominant communities in mathematics classrooms as "those practices that a) support mathematical reasoning, conceptual understanding, and discourse—because we know such practices lead to learning important mathematics, and b) broaden participation for students from non-dominant communities—because we know that participation is connected to opportunities to learn" (pp. 27-28). Equitable classroom practices focus on honoring student resources, in particular the "repertoires of practice" (Gutiérrez & Rogoff, 2003 p. 22) that students from non-dominant communities bring to classrooms. These repertoires are defined as "ways of engaging in activities stemming from observing and otherwise participating in cultural practices" (p. 22).

Rogoff et al (2003) provide an example of how opening up classroom communication to multiple modes would lead to more equitable opportunities for students to draw on their cultural practices to participate in classrooms. They explain that a participation structure or tradition common in some home communities, what they call intent participation, involves children actively observing, watching, and listening but producing little talk. They describe intent participation as "actively observing and 'listening-in' on ongoing activities as they participate in shared endeavors" (p. 175) and clarify that intent participation is not just observing or listening, but refers to "keenly observing and listening in anticipation of or in the process of engaging in an endeavor" (p. 178). In contrast, school instruction involves large amounts of talk. This research suggests that silence and intent observation may be an important student resource to include when designing instruction. If this is the case, then mathematical writing may provide more equitable participation opportunities than whole-class discussions, which privilege producing talk as the central, and sometimes only, mode of participation. Attending to writing in both research and practice is thus an important goal for more equitable instruction.

Incorporating mathematical writing and revising into research and practice is central to equitable mathematics instruction. This is an equity issue for students who talk less (regardless of the reasons why) and for students who may not produce a perfect argument the first time and need to revise it. It is also an issue for students who are learning how to construct mathematical arguments and notice mathematical structure. For instruction to support equitable participation in mathematical discussions, it is crucial to include the possibility that some students may choose to speak less, for whatever reasons, and to consider how students participate besides producing or contributing talk. Actively observing and listening-in while participating in shared work, or intent participation as defined above (Rogoff et al, 2003) seems important for making sense of mathematical discussions (especially those that involve symbols, figures, representations, inscriptions, etc.), and may turn out to be a positive side of some silences. If that is the case, those silences should not be interpreted as the absence of participation in a mathematical discussion. Equitable teaching practices that broaden student participation would need to address student silences in multiple ways that are productive for as many learners as possible. Concerns with a few talkative students monopolizing discussions can be balanced with different ways to include engaged silence as participation. For example, evidence that quiet students who were engaged still learned has been provided by one study that explored vocal and silent student participation (O'Connor et al, 2016). Writing and revising can become a powerful tool to support those quiet students, increase their participation in mathematical practices, and further their overall understanding of the mathematics.

BACKGROUND

More research needs to focus on mathematical writing for many reasons, including that writing supports revising, revising supports learning (Jansen, 2020) and revising can support conceptual understanding (Hiebert & Carpenter, 1992). Although oral student contributions are important for learning mathematics with understanding, if they are not recorded those contributions (and a record of students' reasoning) will disappear, making revising difficult or impossible.

Views of language matter for mathematical writing. If we imagine that language is text (oral or written) and that text is produced in perfect form, then we will imagine mathematical writing as text that students may produce only once. If, instead, we use a view of language as necessarily involving the production of imperfect text (oral or written), that second view allows for revising text to be a part of learning to write mathematically. The first view is often what prevents mathematical writing from being included in lessons for students who are learning the language of instruction. Surely, those students cannot possibly be asked to include mathematical writing in their math classrooms, that would be too heavy a burden or increase the language demands. That opinion uses a view of mathematical writing that does not include revising. In contrast, writing instruction outside of math classrooms typically includes revising and feedback (Huixilopotchli, 2022).

Once we include revising in our definition of mathematical writing, we then need to decide how to support revisions: what to revise, how to revise, how to provide feedback. Decisions about revising are also informed by our views of language, because views of language inform how teachers and students understand the goals for revising and decisions on what, how, and when to revise. A simplified view of language leads to a focus on revising that focuses on form, formal vocabulary, or grammar. In contrast, a complex view of language implies that revisions should not focus on form, formal vocabulary, or grammar but instead on the *function* of the writing: the mathematical purpose or goal, the mathematical ideas or strategies used, the explanations used, or the arguments presented. In particular, revisions should not emphasize precision at the word level, but instead focus on the logic of an argument and whether a claim is precise or the argument includes precise constraints. Thus, revising should focus on precision not at the word level but at the claim or argument level.

This chapter uses a definition of *academic literacy in mathematics* as three integrated components: mathematical proficiency· mathematical practices, and mathematical discourse. This definition incorporates the complex views of language described above. Furthermore, the definition also incorporates contemporary views of what constitutes mathematical proficiency and emphasizes the central role of participation mathematical practices in learning mathematics with understanding. These three components of *academic literacy in mathematics* are assumed to work together, not separately, and this interrelatedness needs to be considered when designing research that examines mathematical writing or lessons that support mathematical writing. This is especially important when considering equity issues, because "isolating mathematical proficiency from mathematical practices or discourse limits access to an impoverished version of mathematical activity that does not parallel mathematical expertise" (Moschkovich, 2015, p. 15).

Mathematical proficiency (Kilpatrick, Swafford, & Findell, 2001) includes five intertwined strands: 1) Conceptual understanding, or comprehension of mathematical concepts, operations, and relations; 2) Procedural fluency, or skill in carrying out procedures flexibly, accurately, efficiently, and appropriately; 3) Strategic competence, or competence in formulating, representing, and solving mathematical problems (novel problems, not routine exercises); 4) Adaptive reasoning, or logical thought, reflection, explanation, and justification; and 5) Productive disposition, a habitual inclination to see mathematics as sensible, useful, and worthwhile, coupled with a belief in diligence and one's own efficacy. The *academic literacy in mathematics* framework emphasizes the first strand, conceptual understanding, not because it is more important than the other four, but because it is more difficult to clarify exactly how to identify conceptual understanding, support it, assess it, and recognize it in student contributions. When contrasted with the other four strands, conceptual understanding is murkier and more elusive. It can be easily misunderstood as memorizing definitions or using correct vocabulary, and is thus directly connected to views of language and goals for mathematical writing. Conceptual understanding is defined in the *academic literacy in mathematics* framework as the connections students make Hiebert & Carpenter (1992) and the meanings they construct (Hiebert, 1990). Beyond this focus on conceptual understanding, the *academic literacy in mathematics* framework also emphasizes mathematical practices, defined to include not only those mentioned in national standards (e.g. CCSS, 2010) but, more broadly, those practices identified by researchers as central to learning mathematics such as generalizing, justifying, and connecting representations (Moschkovich, 2013b).

Views of mathematical proficiency matter for mathematical writing and revising because they inform decisions on what, how, and when to revise. The *academic literacy in mathematics* framework used here suggests that revising not focus primarily on revising mathematical procedures, calculations, or the steps taken to reach a solution to a procedural mathematical exercise. Although that may sometimes be an appropriate goal for writing and revising, how to support students in revising mathematical calculations and supporting procedural fluency has been addressed with great depth and breadth in both research and recommendations for practice. Instead, this chapter focuses on the goals of and the processes for revising explanations or arguments through an emphasis on a broad set of mathematical practices: connecting representations, generating conjectures, explaining solutions, constructing arguments. Following Casa et al (2016) [3], the chapter focuses on writing with the primary purpose to reason mathematically and on two types of writing closely related to mathematical practices: *explanatory writing* where students describe and explain and *argumentative writing* where students construct or critique an argument.

What is Academic Literacy in Mathematics?

This chapter uses a definition of *academic literacy in mathematics* as including three integrated components: mathematical proficiency, mathematical practices, and mathematical discourse. Although these three are described as components, they should not be seen as separate; instead they are intertwined and should not be separated.

The view of academic literacy in mathematics presented here is different from previous approaches to academic language in several ways. First, the definition includes not only cognitive aspects of mathematical activity—activities that happen in one's mind such as mathematical reasoning, thinking, concepts, and metacognition—but also social and cultural aspects—activities that happen with other people, such as participation in mathematical practices—and discourse aspects—activities that happen when using mathematical language, reading, listening, talking, and in this case, writing about mathematics. Most importantly, this framework provides an integrated view of these three components working in unison, rather than isolating mathematical language from mathematical proficiency or mathematical practices.

The definition used in this chapter expands *academic literacy in mathematics* beyond narrow views of mathematical language as words. Narrow views of academic language limit learners' access to a high quality curriculum. A focus on single words or vocabulary limits access to both complex texts and high level mathematical ideas, as well as opportunities for students to understand and make sense of those texts. The assumption that meanings in mathematical texts (textbooks, written explanations, written arguments, etc.) are static and given by definitions limits students' opportunities to make sense of mathematics for themselves. The assumption that mathematical ideas should always and only be communicated using formal mathematical language (formal mathematical vocabulary, terms, or phrases) limits the resources (including informal language) that students can use to communicate mathematically. In contrast, the view of mathematical language used here assumes that meanings for academic language are situated and grounded in the mathematical activity that students are actively engaged in.

For example, the meanings for the words in a word problem come not from a definition in a word list provided by the teacher before a lesson using a word problem; instead students develop these meanings as they work on the problem, communicate about the word problem with their peers, and develop their solutions. A complex view of mathematical language also means that lessons need to include multiple modes (not only reading and talking but also listening and writing), multiple representations (not only symbols but also gestures, objects, drawings, illustrations, sketches, tables, graphs, etc.), and multiple ways of using language, not only formal school mathematical language but also everyday language and students' home languages).

In addition, the definition used in this chapter expands *academic literacy in mathematics* beyond simplified views of mathematics as computation or arithmetic. First, this definition includes the full spectrum of mathematical proficiency, balancing procedural fluency, conceptual understanding, and reasoning. Conceptual understanding involves the connections, reasoning, and meaning that learners construct. Conceptual understanding is more than performing a procedure accurately and quickly (or memorizing a definition or theorem). It involves understanding why a particular result is the correct answer, what that results means, i.e. what the number, solution, or result represents. For example, explaining (or showing using a picture) why the result of multiplying a number by 1/2 will result in an answer that is less than the original number. Another aspect of conceptual understanding involves connecting representations (such as words, drawings, symbols, diagrams, tables, graphs, equation, etc.), procedures, and concepts (Hiebert & Carpenter, 1992). For example, if a student understands the relationship between addition and multiplication, we would say they have learned to make connections between these two operations, and expect that they would be able to explain how multiplication and addition are related. If they understand the procedures for multiplying and dividing negative numbers, we would say they have learned to make connections between these two operations, and expect that they would be able to explain how the procedures for multiplication and division are similar and different, and explain why.

Most importantly, the definition of *academic literacy in mathematics* emphasizes mathematical practices as a quintessential aspect of mathematical activity and an important emphasis for both research and instruction.

What Can the *Academic Literacy in Mathematics* Framework Provide for Research and Practice on Mathematical Writing?

First, the *academic literacy in mathematics* framework provides a complex view of mathematical activity focused on mathematical practices. Writing should not focus on students recounting procedures, but instead focus on core mathematical practices: explaining, justifying, connecting representations. Revising should not focus on revising procedures, but instead on making explanations and arguments "stronger and clearer[4]" (Zwiers et al, 2017, p. 5). Research can examine how mathematical writing and revising can support students in learning to use and connect representations[5], construct arguments[6], and notice mathematical structure[7]. Students can learn to make their explanations or arguments clearer and stronger by providing more representations, connecting representations to text, making claims more precise, and noticing mathematical structure. Teachers can provide feedback and comments for revising that focus on those mathematical practices (see examples in subsequent sections).

Second, the framework provides a complex view of mathematical language that makes the following assumptions:

1. Writing needs to include and connect multiple representations, text needs to be used to describe, explain and make meaning for symbols, drawings, tables, graphs, etc. and the relationships between those.
2. Revising can focus on increasing the clarity of a representation and/or the text that explains that representation,
3. There are multiple ways of using language in mathematical writing: formal school mathematical language, everyday language, and students' home languages;

4. Researchers and teachers need to learn to notice mathematical practices when students express them in informal language.

The first step in supporting learners in writing and revising to learn mathematics is to shift from focusing on single words to using a broader definition of academic literacy, not just learning words but learning how to communicate mathematically. This shift from a simplified view of academic language as single words to an expanded view of *academic literacy in mathematics* (Moschkovich, 2015) that integrates mathematical proficiency and practices is crucial for issues of equity and access (Moschkovich, 2013a). Research and policy have repeatedly, clearly, and strongly called for mathematics instruction to maintain high-cognitive demand (American Educational Research Association 2006). In order to accomplish that goal, mathematics instruction needs to shift from defining academic literacy in mathematics as low-level language skills (i.e., vocabulary or single words) or mathematical skills (i.e., arithmetic computation) and use an expanded definition of *academic literacy in mathematics* to design lessons that support writing to reason mathematically.

In general, lessons that support writing to reason mathematically need to:

(a) Include the full spectrum of mathematical proficiency (Kilpatrick et al, 2001), balancing computational fluency with high-cognitive-demand tasks (Stein et al, 1996) that require conceptual understanding and reasoning (Hiebert, & Carpenter, 1992; Hiebert, & Grouws, 2007);
(b) Provide students opportunities to participate in mathematical practices;
(c) Allow students to use multiple resources for mathematical reasoning such as modes of communication, symbol systems, registers, and languages (Moschkovich, 2014), and
(d) Support students in negotiating meanings for mathematical language that are grounded in student mathematical work, instead of merely giving students definitions separate from mathematical activity.

These recommendations run counter to some interpretations of typical approaches to teaching vocabulary, such as pre-teaching a list of math words using definitions. The question is not whether students should learn vocabulary but rather when and how instruction can best support students as they learn not only the meanings of words and phrases but also how to participate in mathematical practices. One important contribution that a complex view of language provides is a research based understanding of how humans learn vocabulary. Vocabulary drill, practice, definitions, or lists that are provided to students before a lesson are not the most effective way to learn to communicate mathematically. Instead, research has documented how vocabulary acquisition (whether it is in a first or second language) occurs most successfully in instructional contexts that are language-rich, actively involve students in using language, require both receptive and expressive understanding, and require students to use words in multiple ways over extended periods of time (Blachowicz & Fisher, 2000; Pressley, 2000). To develop oral and written communication, students need to participate in negotiating meanings (Savignon, 1991) and in tasks that require student output (Swain, 2001). Therefore, instruction must provide opportunities for students to actively use mathematical language to communicate about, make meaning, and negotiate collective meanings for mathematical situations.

Focusing on Mathematical Practices

For defining mathematical practices, in my work I have expanded beyond the list of eight Common Core practices listed below by including a broader set of other practices identified by researchers as central to learning mathematics such as generalizing, justifying, and connecting representations (Moschkovich, 2013b). I also assume that mathematical practices are social and cultural in origin and that children develop proficiency in these practices through opportunities to participate in them. (Moschkovich, 2013b). Without that assumption, some learners may be viewed as "deficient" and others "talented" or "gifted" because they (supposedly) developed these practices individually.

In addition to research, two other sources for descriptions of mathematical practices are the National Council of Teachers of Mathematics (NCTM) Standards (NCTM, 1989) and the Common Core State Standards (CCSS) for Mathematical Practices (CCSS, 2010). The CCSS standards for mathematical practice overlap in important ways with the NCTM standards and the definition of mathematical proficiency (Kilpatrick et al, 2001). Although the NCTM Standards did not explicitly label these practices, they focused on problem solving, sense-making, reasoning, modeling, and looking for patterns, structure, or regularity (Koestler et al, 2013). The CCSS introduced a list of eight "Standards for Mathematical Practice": MP1-Make sense of problems and persevere in solving them; MP2-Reason abstractly and quantitatively; MP3-Construct viable arguments and critique the reasoning of others; MP4-Model with mathematics; MP5-Use appropriate tools strategically; MP6-Attend to precision; MP7-Look for and make use of structure; MP8-Look for and express regularity in repeated reasoning.

This chapter focuses on two kinds of mathematical writing, *explanatory writing* and *argumentative writing* because these lend themselves to writing and revising by focusing on mathematical practices. The examples illustrate how writing and revising can support three mathematical practices:

- Connecting representations: Explanations and arguments are usually supported by using multiple representations; they can be revised by refining drawings, tables, or diagrams and by clarifying the connections between a visual image and the written text.
- Constructing arguments: Writing is essential to learning to construct and improve one's arguments, oral arguments may disappear but written arguments can be revised and improved.
- Attending to precision: Writing and revising can support students in attending to the most useful kind of precision for mathematical communication, not in using the exact perfect word, but in making precise and logical arguments.

Shifting to a Complex View of Precision in Mathematical Writing and Revising

Work on mathematical writing and revising needs to shift to a complex view of precision that goes beyond using precise vocabulary. Precise mathematical language is highly valued by communities of people who use mathematics, such as mathematicians, scientists, and engineers. However, it is important to note that, for students learning mathematics, informal language is also important, especially when students are exploring a mathematical concept or first learning a new concept, or discussing a mathematics problem in small groups. Informal language can be used by students and teachers during exploratory talk (Barnes, 1992) or when working in a small group communication context (Herbel-Eisenmann et al, 2013). Such informal language can reflect important student mathematical thinking (for examples see Moschkovich

1996, 1998). In other situations, for example, when making a presentation or developing a written account of a solution, using more formal academic mathematical language may be more important.

Academic mathematical discourse is not principally about formal or technical vocabulary (Moschkovich, 2007). The mathematics register is a complex construct that includes styles of meaning, modes of argument, and mathematical practices and has several dimensions such as the concepts involved, how mathematical discourse positions students, and how mathematics texts are organized. When describing mathematical language we should not confuse "mathematical" with "formal." Formal definitions are only one aspect of school mathematical discourse. In classrooms students use multiple resources, including everyday registers and experiences, to make sense of mathematics. It is also important to avoid construing everyday and academic registers as opposites (Moschkovich, 2010).

The *academic literacy in mathematics* framework provides a complex view of mathematical writing. A simplified view of mathematical writing can lead to the assumption that precision lies primarily in individual word meaning. This misguided assumption can have serious consequences for research and practice on mathematical writing, as students are likely to use imperfect language to describe their mathematical thinking (for analyses of accurate mathematical thinking expressed through imperfect utterances see Moschkovich, 2002). One example of a consequence for practice is that, if teachers are focusing only on precision at the individual word level, they would miss students' use of this mathematical practice when students use informal individual words but make a claim that is precise in when it applies (see example about claims about multiplication in later section). An example of a consequence for research is that, if researchers are focusing only on precision at the word level, they would miss students attention to precision when that attention occurs through a whole utterance or through a set of connected utterances (see example about comparing two graphs in later section).

Views of mathematical writing and revising need to shift from a simplified view of precision as existing only at the word level. Such views separate language from mathematical proficiency and practices, and do not provide students with access to the full spectrum of *academic literacy in mathematics*. These views also severely limit the linguistic resources teachers can use to teach mathematics and students can use to learn mathematics by requiring that only formal ways of talking or formal mathematical vocabulary be used for mathematical writing. Focusing mathematical writing on using precise words limits students' access to the five strands of mathematical proficiency[8] and curtails students' opportunities to participate in mathematical practices. In contrast, the view of *academic literacy in mathematics* used here provides a complex view of mathematical writing as fundamentally connected to mathematical practices.

The ambiguity and multiplicity of meanings in everyday or informal language should be recognized and treated not as a failure to be mathematically precise but as fundamental to making sense of mathematical meanings and to learning mathematics[9]. It is even possible that mathematical writing may not always involve the most precise words. Although many of us who are mathematically educated may be deeply attached to the precision we imagine mathematics provides, ambiguity and vagueness have been reported as common in mathematical conversations (even among mathematicians) and have been documented as resources in teaching and learning mathematics (e.g., Barwell, 2005; Barwell, Leung, Morgan, & Street, 2005; O'Halloran, 2000; Rowland, 1999). Even definitions are not a monolithic mathematical practice, because they were documented as presented differently in some lower grade textbooks--as static and absolute facts to be accepted--while in mathematics journal articles they were presented as dynamic, evolving, and open to revisions by the mathematician (Morgan, 2004). Neither should textbooks be seen as homogeneous. Higher grade textbooks have been documented to be more like journal articles in allowing for more uncertainty and evolving meaning than early grade textbooks

(Morgan, 2004), providing evidence that there are multiple approaches to the issue of precision, even in mathematical texts.

USING *ACADEMIC LITERACY IN MATHEMATICS*

Examples of Mathematical Writing and Revising Focused on Mathematical Practices

This section will use three examples to illustrate how to use the *academic literacy in mathematics framework* for research and instruction. The examples illustrate how research and practice can use a view of language as a resource for making meaning and focus on mathematical practices. First and foremost, the examples emphasize how writing and revising can support students in using language to make meaning. The examples also problematize the practices of "attending to precision" and "looking for mathematical structure" and show how writing and revising can support students in learning to participate in those crucial mathematical practices. The *academic literacy in mathematics* framework provides both researchers and practitioners with principled ways to design research and instructional tasks, so that those tasks frame language as a resource for making meaning and support revising focused on mathematical practices.

Using a Complex View of Language for Mathematical Writing and Revising

The mathematical practice "Attending to precision" (CCSS 2010) should not be interpreted as using the perfect word or making the perfect measurement. Attending to precision can also refer to deciding when and what kind of precision is necessary during a computation, including when an exact answer is or is not necessary, a mathematical practice that does not require a precise word. Attending to precision is also involved in making precise claims, a practice that occurs not at the word level but at the claim level. The belief that precision lies primarily in individual word meaning reflects a simplified view of mathematical language. We could imagine that *attending to precision* means using two different words for the set of symbols "$x + 3$" and the set of symbols "$x + 3 = 10$." For example, an explanatory writing prompt could ask students to compare and contrast these three mathematical objects:

Writing prompt #1: Comparing and contrasting
Describe any differences or similarities between

a) $x + 3$
b) $x + 3 = 10$
c) $x + 3 = y$

If we are being precise at the level of individual word meaning, a) is an expression, b) is an equation, and c) is an equation and a linear function. One possible student response might be to simply provide those three labels:

The first one is an expression, the second one is an equation, and the third one is a linear function.

A researcher or practitioner using a simplified view of mathematical language would be satisfied with a response that uses those precise words. A researcher exploring mathematical writing who is using a view of language as single precise words or vocabulary would assess that response as correct, high level, or a multitude of other positive labels. In contrast, a researcher using the complex view of language provided by the academic literacy in mathematics perspective, would label this response as "vocabulary" and either revise that initial prompt to provide more depth or provide follow up questions that require learners to describe the meaning of the objects or the words. However, a researcher cannot simply ask a student "What does "$x + 3$" mean to you?" to elicit descriptions of meaning. One way to elicit such meanings is to ask students to use and connect representations to show meaning, for example using the following revised prompt:

Revised-Writing prompt #1: Comparing and contrasting
Describe any differences or similarities between, $x + 3$, $x + 3 = 10$, and $x + 3 = y$.
First write down any ideas that come to you.

After you have written down some ideas, try describing how each one looks (or what each one shows you). Try using a number line or a graph to show how each one can be represented. Make sure to write an explanation in words that describes whether and how the x or the 3 appear, in the number line or graph.

The revised version accomplishes several goals:

1. Overall, the *academic literacy in mathematics* framework was used to frame mathematical language and writing as a resource for making meaning, not simply learning precise vocabulary.
2. It used the *academic literacy in mathematics* framework to provide an outline for how a student might produce a response showing evidence of *academic literacy in mathematics*: first by brainstorming ideas, then using mathematical representations, and then making sure the initial ideas and text are connected to those representations.
3. It requires that the learner focus on meaning and show their conceptual understanding, because describing what an object means is one way to show conceptual understanding and connecting representations is also a way to show meaning.

Using the *academic literacy in mathematics* framework for revisions means that the teacher would focus on how students are showing conceptual understanding or using any mathematical practices. A practitioner using the *academic literacy in mathematics* framework to review, provide feedback, and suggest revisions for responses to that original task, would not focus on whether a student used precise words, but instead on the meaning the student provided for the objects, because meaning is a fundamental (if not defining) aspect of conceptual understanding. The *academic literacy in mathematics* framework, by requiring a focus beyond single words and an emphasis on conceptual understanding, leads to ways to review and provide feedback for that student response that would emphasize meaning, suggest using more representations, and ask questions that lead students to connect representations, for example:

Good start! Those are the formal terms and phrases we use in math class. Could you revise this by:

1. Describing how each one looks (or what each one shows you, or what each one represents)? try using a number line or a graph to show what each one means. Make sure to write an explanation in words that describes whether and how the x or the 3 appear, in the number line or graph.
2. Thinking and writing about the numbers that work in each one: How many numbers can work for x in each example? Only one number, a few numbers, many numbers, an infinite quantity of numbers?

The first comment and suggestion for revising focus the student on the mathematical practice "using and connecting representations." The second comment focuses the student on the mathematical practice "looking for mathematical structure." Neither of the comments are for a revising focus on precision at the word level. Instead, the comments that focus on mathematical practices provide students with much needed support in learning how to use those practices.

The example above shows that the mathematical practice "attending to precision" is not only about using the perfect word. Furthermore, an important mathematical practice is making *precise claims about precise situations*. To further illustrate how that mathematical practice "attend to precision" might work beyond the word level in mathematical writing, I use the claim "Multiplication makes bigger," to describe how this claim can be used to support mathematical writing and revising focused on making a claim more precise.

Writing Prompt Two: Critiquing a Claim

A student in my class last year was convinced that this was a true statement (or claim or conjecture): "Multiplication makes bigger." What do you think about this statement?

If we are being precise at the level of individual word meaning, one possible student response might be: "The statement is not precise, the result of multiplying is called a product. A possible revised statement would be "Multiplication makes the product of two numbers bigger[10]."

Another student might respond: "Multiplication makes bigger, but only when you multiply by a positive number greater than 1."

First, if we contrast the claim "Multiplication makes bigger," with the claim "Multiplication makes bigger, but only when you multiply by a positive number greater than 1," it is clear that the first claim is not precise. When contrasting the two claims, using a complex view of language as more than single vocabulary words helps us to notice two things. First, precision does not lie in the individual words used. Second, the words used in the second claim are not more formal mathematical words. Instead, the precision of the second claim lies in specifying when the claim is true[11].

In a classroom, if a teachers' response to the first claim focused on precision at the word level, feedback for revising that claim might be to ask a student to use a formal word such as "product" or a more formal word for "bigger." In contrast, if a teacher was focusing on precision at the claim level, feedback for revising that claim might be to ask "When does multiplication make a result bigger?"

For research, the prompt could be revised to provide an outline for how a response showing academic literacy in mathematics might be produced, first by brainstorming ideas, then using math representations, and then making sure the initial ideas and text are connected to those representations.

Revised Writing Prompt Two: Critiquing a Claim

A student in my class last year was convinced that this was a true statement (or claim or conjecture): "Multiplication makes bigger." What do you think about this statement?

First write down any ideas that come to you.

After you have written down some ideas, consider these questions: Is this true always? Sometimes? Never?

Follow up Questions

-Think and write about different kinds of numbers that would make the statement true or not: Whole numbers? Fractions? Negative numbers?
-Try drawing a picture to show why a statement is true. For example, you can use arrays or area models or a number line to show multiplication. Make sure to write an explanation in words that describes where you see whether the result is bigger or not.

If those follow up questions do not generate more writing, a researcher or teacher can always provide students with examples of conjectures or claims: "Another student said that multiplication makes bigger is only true for whole numbers, but not for fractions. What do you think about that claim?"

The follow up questions provide crucial support for showing whether a learner can learn to construct a clear and strong argument, rather than simply expecting a learner to provide a clear and strong argument on the first try. Overall, this revision frames mathematical language and writing as a resource for making meaning, not simply using precise vocabulary. The follow up questions provide the mathematical structure for considering a conjecture's truth (always, never, sometimes). Next, the follow-up questions also emphasize mathematical structure by suggesting the sets of numbers to consider. The follow up questions also require using drawings or pictures to show meaning. The prompt to provide an explanation requires that a learner connect the written text to representations, not simply create a drawing. Instead, the prompt asks students to describe what the drawing shows, what assumptions they made in creating a drawing, and where in the drawing they see support for their claim. For example, if using a drawing of an area model to show that 1/2 x 1/3 is less than either of those two fractions, one assumption is that the rectangle shows the unit or the whole. Written text can also describe where they actually "see" that the product of 1/6 is less than 1/2.

Revised Writing Prompt #2 explicitly asks students to carefully consider when a claim is true, for example asking "always, sometimes, never?" This request is a start in showing students what mathematical structure is and how we look for it, so they can start to notice and identify it themselves. Writing Prompt #2 illustrates one aspect of mathematical structure, the sets of numbers we consider: whole numbers, fractions, positive or negative, etc. Other follow up questions can ask students to consider special cases such as 0 and 1, illuminating another aspect of mathematical structure in the properties of the operations. In other situations, such as exploring the sum of consecutive numbers in the next example, mathematical structure might be about patterns.

Writing Prompt Three: Evaluating and Revising Statements[12]

The number 18 can be made by adding three consecutive whole numbers:

$5 + 6 + 7 = 18$

Which other numbers can be made by adding three consecutive whole numbers?

1. Experiment with some numbers and try to state a conjecture about which numbers can be made.
2. Try to prove your conjecture.

The lesson provides students with examples of conjectures to discuss, such as the one below:

Conjecture: The sum of four consecutive whole numbers is always even.
$5 + 6 + 7 + 8 = 26$
Or
Conjecture: The sum of three consecutive numbers is 3 times the middle number.

The materials also provide sample responses from imaginary students to discuss and prompts for how to discuss those sample responses:

In your groups:

1. Try to understand what the student has done. You may want to annotate the work/do some math on your mini-whiteboard.
2. Explain the students' reasoning in your own words for each piece of work.
3. Explain why each piece of work does or does not convince you.
4. Decide which of the approaches you have looked at you prefer. **Be prepared to explain your choice.**

The lesson also provides students with the structure for claims and arguments:

1. Decide whether each conjecture is Always, Sometimes or Never True.
    - If the conjecture is Always or Never true, write down how you can be sure.
    - If the conjecture is Sometimes true, write down when it is true and when it is not.
2. Use ideas from the sample work you have looked at and make notes on your mini-whiteboards.
3. Agree about each conjecture before moving on.
4. Explain your thinking on your poster.

A writing prompt at any of those different stages in that lesson would support students in recording their explanations and revising them after a small group or whole class discussion. Students could learn to revise their explanations making a claim more precise, a mathematical practice already supported by the questions in the lesson. Furthermore, students could also learn to revise by using and/or connecting representations, because the three sample responses use different representations of algebra symbols, written text, and a drawing to show the relationship between a number (n) and the numbers preceding (n - 1) or following it (n + 1).

Noticing How Informal Language is a Resource for Mathematical Writing

Elsewhere, I have described how students used informal language to describe mathematical ideas. Here I use one of those examples to suggest that informal language can also support mathematical writing. The example summarized here comes from an eighth grade classroom discussion between two students and a teacher as they compared two graphs[13]. The discussion involved a hybrid mix of everyday and academic language (Moschkovich, 2008). The example shows how participants engaged in *academic literacy in mathematics* by using hybrid language resources, especially formal and informal registers (everyday

Figure 1. David's graph

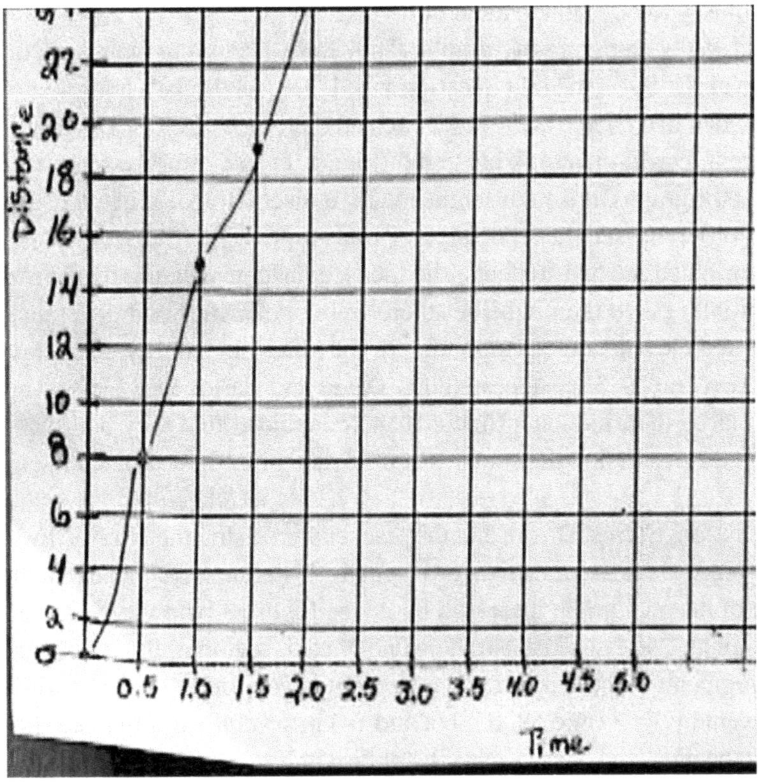

Figure 2. Carlos's graph

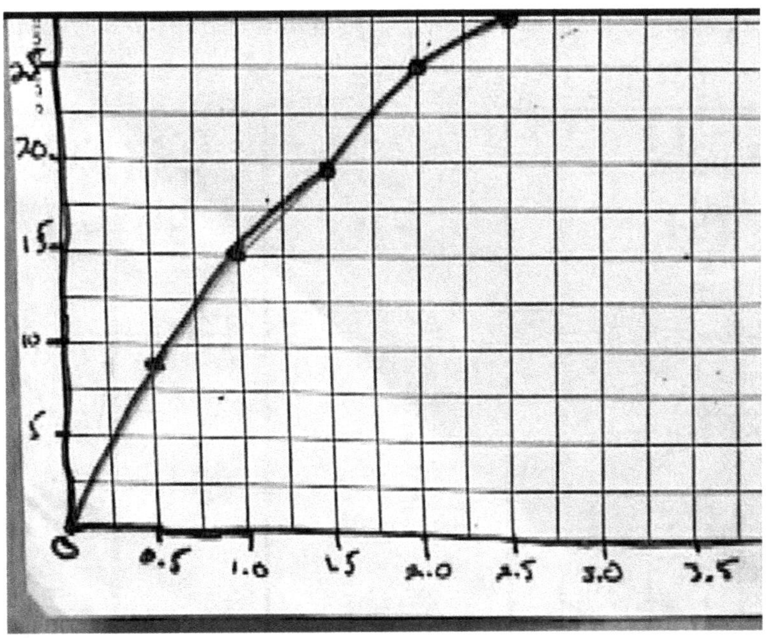

and academic). The example illustrates that mathematical text, in this case oral, need not involve formal mathematical vocabulary to be mathematical or precise.

Carlos and David were comparing two graphs. They were discussing their work on a problem about a bicycle trip (Lappan et al, 1998); the first question asked them to "Make a coordinate graph of the (time, distance) data" provided in a table. As the students read their answers, Carlos remarked that the written answers were different. David then took his graph (Figure 1) and compared it to Carlos' graph (Figure 2). They noticed that the graphs looked different and their discussion originated in their questioning why their graphs looked different (see the Appendix for transcript).

When the teacher joined the two students, she focused the conversation on the meaning of the scales on the axes. She initially asked them whether there were similarities and differences between the two graphs. Next, she asked the students to compare how they had labeled their axes and consider what was different about the two graphs. She suggested that David and Carlos look at their numbers on the axes and how they had placed their numbers (line 40) and concluded that they both had put time on the x-axis. At this point in the discussion the students started using multiple situated meanings for phrases of the form "I went by. . . . "

During this discussion, Carlos, David, and the teacher were using the phrase "I went by" with at least three different meanings. David seemed to use "I went by 2" in line 56 to mean that the value of each tick mark on the y-axis of his own graph increased by 2 (see Figure 3 below). In David's graph tick marks corresponded to segments, so "two" is also the value of each segment. In Carlos' graph, tick marks did not correspond to segments since Carlos had labeled only every other segment with a tick mark.

Carlos used "I went by. . . ." (line 42, 63, 66, and 74) in several ways that are different than David's or the teacher's. On the one hand, Carlos used the phrase in line 63 to refer to how many segments there were between tick marks on his graph, in this case 2 segments (see Figure 4).

Figure 3. David describes his scale as "I went by twos."

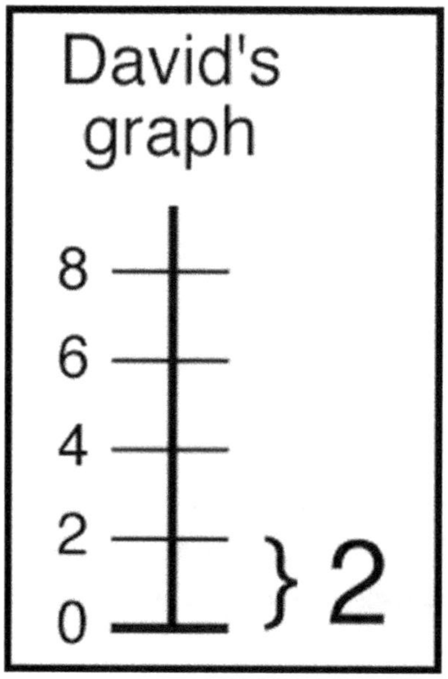

Figure 4. Carlos describes his scale as "I went by twos."

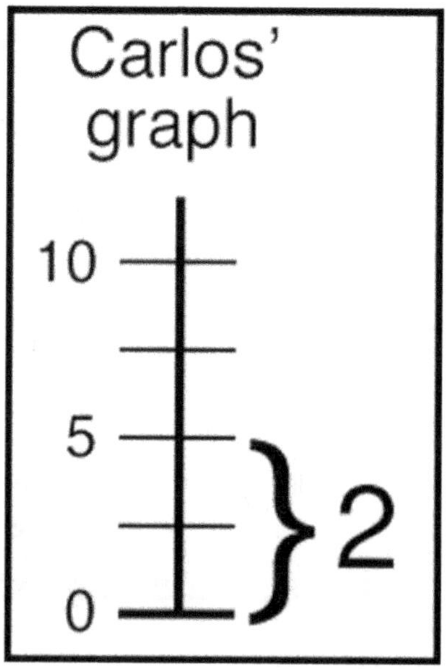

Figure 5. Carlos describes his scale as "I went by fives."

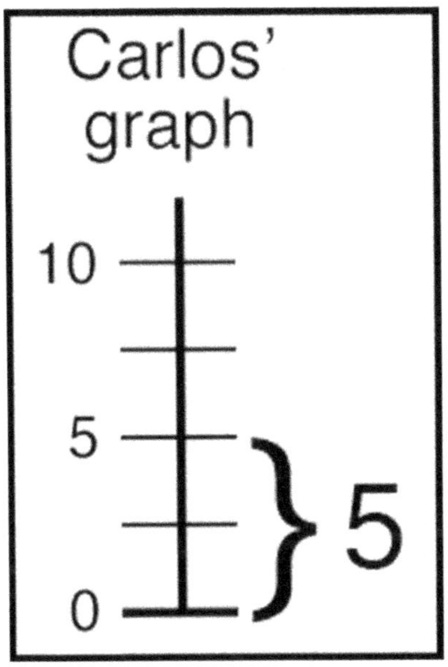

Figure 6. Teacher describes Carlos' scale as "You went by two and a halves" and "each one is two and a half"

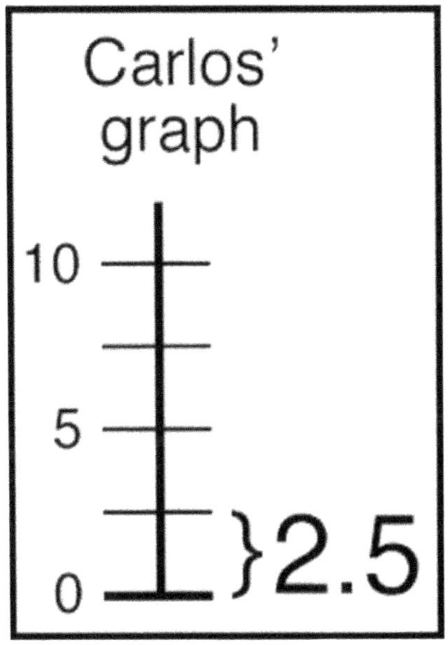

On the other hand, in lines 66 and 74, Carlos used the phrase to refer to how much the value increased for each tick mark on his graph, in this case by 5 (see Figure 5).

In contrast, the teacher was referring to the value of one segment in Carlos's graph (lines 67 and 86, see Figure 6), first saying "No, actually you didn't go by fives. You actually went by two and a halves because you did every two spaces was five" and later, "The numbers you wrote are by fives but since you skipped a line in between, each one is two and a half. " When saying "every two spaces was five" and "each one is two and a half" the teacher connected the meaning of the informal phrase to a mathematical concept, unitizing, to describe the scales.

This example illustrates that mathematical discussions need not involve formal mathematical vocabulary to be mathematical or precise. The phrase "I went by. . ." is not particularly mathematical and can be labeled as everyday or informal, rather than formal. The example illustrates how, although the phrase is informal or vernacular, the teacher and the students used this phrase repeatedly during this discussion (and at other times in this classroom), this phrase grounded the discussion on the graphs that students had created and had in front of them, and the students used this phrase to communicate their mathematical ideas and compare competing claims in an argument about the graphs. The phrase "I went by…" is not in itself mathematical, but rather how the phrase was used in this classroom. The ways participants used this phrase is an example of informal language used by the participants as a resource to participate in important mathematical practices: making mathematical claims, supporting an argument, constructing an argument, critiquing others' reasoning, and expressing regularity. What made student contributions mathematical was not the use of formal mathematical words. Instead, this discussion was mathematical because of the focus on sharing, co-constructing, and clarifying the underlying mathematical concepts. In general, mathematical concepts can be expressed using informal words and phrases, and mathematical practices, such as justifying a claim, which is not evident in a single word but in a series of connected student utterances.

One mathematical practice evident in this discussion is "attending to precision[14]," even though that attention to precision was accomplished through the use of informal language. What makes a claim mathematical is, in part, the attention to describing in detail when the claim applies and when it does not. In this discussion it was the teacher who connected the meaning of an informal phrase to a mathematical concept, unitizing, to describe a mathematical situation more precisely. As a follow up, a writing prompt following such discussions could be used to support students revising the meanings for the phrase "I went by . . . " Such revising could occur with or without providing formal mathematical vocabulary, such as the word "unit" or "scale," depending on the focus of the revisions.

When using the *academic literacy in mathematics* framework what matters most about writing and revising is when and how students might use a mathematical practice, not necessarily that teachers provide them the most precise or most formal mathematical word. In this case, students' writing would necessarily have to refer to the graph, thus also using the mathematical practice of connecting representations. A prompt to compare and contrast the three claims in the discussion "I went by 2s", "You went by 5s", and "No, you went by 2 and a halves" would certainly frame mathematical language and writing as a resource for making meaning, not simply learning precise vocabulary. Such a prompt would also help students see that, while informal language can provide resources for making meaning, it can also be revised to be clearer and more formal.

RECOMMENDATIONS FOR INSTRUCTION

Writing reveals and records one's thinking and reasoning, thus providing opportunities for reflection, revision, and discussion in more enduring ways than talk alone. Vocabulary should not be the main focus of mathematical writing. When vocabulary is the focus, it should be after students have had opportunities for engaging in meaningful mathematical activity to develop meaning for math ideas and procedures. *Academic literacy in mathematics* is much more than vocabulary; it includes mathematical proficiency, mathematical practices, and mathematical discourse. A complex view of language and an emphasis on mathematical practices can guide instruction in mathematical writing and revising.

Teachers can learn from research on writing to improve general approaches to mathematical writing and revising, such as conferencing, giving feedback, and encouraging peer editing. Beyond general ways to support student writing, there are specific ways to support students in mathematical writing and revising, a few include:

- Providing students with examples of conjectures or claims, both true and false
- Asking students to carefully consider when a claim is true
- Providing support and prompts that focus on mathematical structure (which set of numbers, patterns, regularity, etc.)
- Following oral discussions with a prompt to write an explanation or argument
- Emphasizing mathematical practices when providing feedback, especially using and connecting representations
- Providing feedback on what makes mathematical writing stronger and clearer. This need not be individual feedback, it can be provided in a small group or whole class setting.
- Asking students to revise peers' mathematical writing

CONCLUSION

Research needs to focus on mathematical writing because writing supports revising, and revising supports conceptual understanding. Although oral student contributions are important for learning mathematics with understanding, if they are not recorded those contributions (and the record of students' reasoning) will disappear, making revising difficult or impossible. A research agenda focused on writing to reason mathematically first requires clarifying our assumptions about language. The *academic literacy in mathematics* framework provides an expanded view of language and a focus on mathematical practices that theoretically frame what kinds of writing and revising matter for learning important mathematics. Future research needs to use a complex view of language. A complex view of language assumes that writing supports meaning making and that revisions are part of writing. Therefore, research should consider not only mathematical writing, but also revising, and examine how to support revising, the details of what to revise, how to revise, and when to revise. An emphasis on mathematical practices is also important for future research to avoid mathematical writing or revising that over emphasizes procedural fluency. Instead, mathematical writing and revising with an emphasis on mathematical practices maintains the important balance between procedural fluency and conceptual understanding.

Research and practice need to recognize the complexity of language in mathematics classrooms and the need to explore language in all its complexity. We need to move away from simplified views of language as vocabulary, embrace the multimodal and multi-semiotic nature of mathematical activity, shift from monolithic views of mathematical talk or dichotomized views of the everyday and academic mathematics registers. Research and practice need to leave behind a sole emphasis on correct vocabulary and formal language because that limits the linguistic resources teachers and students can use in the classroom to learn mathematics with understanding.

Overall, research and practice need to avoid dichotomies such as everyday/academic or formal/informal. Classroom mathematical writing can draw on hybrid resources from both academic and everyday contexts and multiple registers co-exist in mathematics classrooms. Most importantly for supporting the success of students in classrooms, mathematical writing needs to build on and link with the language students bring from their home communities. Therefore, everyday language should not be seen as an obstacle to participation in mathematical writing, but as a resource teachers can build on to support students in developing more formal mathematical writing as they learn mathematics. Revising mathematical writing is an opportunity to build on students everyday language, support their meaning making, and develop more formal mathematical writing that is connected to that meaning.

Research and practice on mathematical writing need to address the issue of precision in much more detail. Mathematical writing should address precision at many levels, not only at the word level, but also consider when students may be making precise claims by including the details of constraints or questioning the meaning of written text. It is important to consider what we mean by *precision*, because students are likely to need time and support for moving from expressing their reasoning and arguments in imperfect form[15]. Although students' use of imperfect language is likely to interact with teachers' own multiple interpretations of precision, we should not confuse the two. In particular, we should remember that precise claims can be expressed in imperfect language and that attending to precision at the individual word meaning level will get in the way of students' expressing their emerging mathematical ideas in writing. And finally, much more work is certainly needed to clarify how to guide practitioners in helping students become more precise in their written language over time.

REFERENCES

American Educational Research Association (2006). Do the math: Cognitive demand makes a difference. *Research Points, 4*(2).

Barnes, D. (1992). *From communication to curriculum.* Boynton/Cook.

Barwell, R. (2005). Ambiguity in the mathematics classroom. *Language and Education, 19*(2), 117–125. doi:10.1080/09500780508668667

Barwell, R., Leung, C., Morgan, C., & Street, B. (2005). Applied linguistics and mathematics education: More than words and numbers. *Language and Education, 19*(2), 141–146. doi:10.1080/09500780508668670

Blachowicz, C., & Fisher, P. (2000). Vocabulary instruction. In M. L. Kamil, P. B. Mosenthal, P. D. Pearson, & R. Barr (Eds.), *Handbook of Reading Research* (Vol. 3, pp. 545–561). Erlbaum.

Bunch, G. C., & Willett, K. (2013). Writing to mean in middle school: Understanding how second language writers negotiate textually-rich content-area instruction. *Journal of Second Language Writing, 22*(2), 141–160. doi:10.1016/j.jslw.2013.03.007

Casa, T. M., Firmender, J. M., Cahill, J., Cardetti, F., Choppin, J. M., Cohen, J., Cole, S., Colonnesse, M., Copley, J., DiCicco, M., & Dieckmann, J. (2016). *Types of and purposes for elementary mathematical writing: Task force recommendations.* published by the University of Connecticut. http://mathwriting.education.uconn.edu

Common Core State Standards. (2010). *Common Core State Standards for mathematical practice.* CCSS. http://www.corestandards.org/Math/Practice

Gutierrez, K., & Rogoff, B. (2003). Cultural ways of learning: Individual traits or repertoires of practice. *Educational Researcher, 32*(5), 19–25. doi:10.3102/0013189X032005019

Herbel-Eisenmann, B., Steele, M., & Cirillo, M. (2013). (Developing) teacher discourse moves: A framework for professional development. *Mathematics Teacher Educator, 1*(2), 181–196. doi:10.5951/mathteaceduc.1.2.0181

Hiebert, J. (1990). The role of routine procedures in the development of mathematical competence. In T. Cooney & C. Hirsch (Eds.), *Teaching and learning mathematics in the 1990's* (pp. 31–40). National Council of Teachers of Mathematics.

Hiebert, J., & Carpenter, T. P. (1992). Learning and teaching with understanding. In D. A. Grouws (Ed.), *Handbook of research on mathematics teaching and learning: A project of the National Council of Teachers of Mathematics* (pp. 65–97). Macmillan Publishing.

Hiebert, J., & Grouws, D. (2007). The effects of classroom mathematics teaching on students' learning. In F. Lester (Ed.), *Second handbook of research on mathematics teaching and learning* (pp. 371–404). National Council of Teachers of Mathematics.

Huixilopotchli, S. (2022). *Multilingual students' arguments in early algebra: Writing, refining, and revising through conferencing* [Unpublished doctoral dissertation, University of California, Santa Cruz].

Jansen, A. (2020). *Rough draft math: Revising to learn.* Stenhouse Publishers.

Kilpatrick, J., Swafford, J., & Findell, B. (2001). Adding it up: Helping children learn mathematics. J. Kilpatrick, J. Swafford, & B. Findell (Eds.). National Research Council. Washington, DC: National Academy Press.

Koestler, C., Felton, M., Bieda, K., & Otten, S. (2013). *Connecting the NCTM process standards and the CCSSM practices.* National Council of Teachers of Mathematics.

Lappan, G., Fey, J. T., Fitzgerald, W. M., Friel, S. N., & Phillips, E. D. (1998). *Connected Mathematics.* Dale Seymour Publications.

Mathematics Assessment Resource Service. (2015). *Mathematics Assessment Project classroom challenges: Formative assessment lessons.* MARS. http://map.mathshell.org

Morgan, C. (2004). Word, definitions and concepts in discourses of mathematics, teaching and learning. *Language and Education, 18*, 1–15.

Moschkovich, J. N. (1996). Moving up and getting steeper: Negotiating shared descriptions of linear graphs. *Journal of the Learning Sciences, 5*(3), 239–277. doi:10.120715327809jls0503_4

Moschkovich, J. N. (2002). A situated and sociocultural perspective on bilingual mathematics learners. *Mathematical Thinking and Learning, 4*(2 & 3), 189–212. doi:10.1207/S15327833MTL04023_5

Moschkovich, J. N. (2007). Examining mathematical discourse practices. *For the Learning of Mathematics, 27*(1), 24–30.

Moschkovich, J. N. (2008). "I went by twos, he went by one:" Multiple interpretations of inscriptions as resources for mathematical discussions. *Journal of the Learning Sciences, 17*(4), 551–587. doi:10.1080/10508400802395077

Moschkovich, J. N. (2010). Language(s) and learning mathematics: Resources, challenges, and issues for research. In J. N. Moschkovich (Ed.), *Language and mathematics education: Multiple perspectives and directions for research* (pp. 1–28). Information Age Publishing.

Moschkovich, J. N. (2013a). Equitable practices in mathematics classrooms: Research based recommendations. *Teaching for Excellence and Equity in Mathematics, 5*(1), 26–34.

Moschkovich, J. N. (2013b). Issues regarding the concept of mathematical practices. In Y. Li & J. N. Moschkovich (Eds.), *Proficiency and Beliefs in Learning and Teaching Mathematics: Learning from Alan Schoenfeld and Günter Toerner* (pp. 257–275). Sense Publishers. doi:10.1007/978-94-6209-299-0_15

Moschkovich, J. N. (2015). Academic literacy in mathematics for English Learners. *The Journal of Mathematical Behavior, 40*, 43–62. doi:10.1016/j.jmathb.2015.01.005

Moschkovich, J. N. (2015b). A sociocultural approach to academic literacy in mathematics for adolescent English Learners: Integrating mathematical proficiency, practices, and discourse. In D. Molle, E. Sato, T. Boals, & C. Hedgspeth (Eds.), *Multilingual Learners and Academic Literacies: Sociocultural Contexts of Literacy Development in Adolescents* (pp. 75–104). Routledge.

National Council of Teachers of Mathematics. (1989). *Curriculum and evaluation standards for school mathematics*. NCTM.

O'Connor, C., Michaels, S., Chapin, S., & Harbaugh, A. G. (2016). The silent and the vocal: Participation and learning in whole-class discussion. *Learning and Instruction*.

O'Halloran, K. (2000). Classroom discourse in mathematics: A multisemiotic analysis. *Linguistics and Education*, *10*(3), 359–388. doi:10.1016/S0898-5898(99)00013-3

Pressley, M. (2000). What should comprehension instruction be the instruction of? In M. L. Kamil, P. B. Mosenthal, P. D. Pearson, & R. Barr (Eds.), *Handbook of reading research* (Vol. 3, pp. 545–561). Erlbaum.

Rogoff, B., Paradise, R., Mejía Arauz, R., Correa-Chávez, M., & Angelillo, C. (2003). Firsthand learning through intent participation. *Annual Review of Psychology*, *54*(1), 175–203. doi:10.1146/annurev.psych.54.101601.145118 PMID:12499516

Rowland, T. (1999). *The pragmatics of mathematics education: Vagueness in mathematical discourse*. Routledge, Taylor & Francis Group.

Savignon, S. (1991). Communicative language teaching. *TESOL Quarterly*, *25*(2), 261–277. doi:10.2307/3587463

Stein, M. K., Grover, B., & Henningsen, M. (1996). Building student capacity for mathematical thinking and reasoning: An analysis of mathematical tasks used in reform classrooms. *American Educational Research Journal*, *33*(2), 455–488. doi:10.3102/00028312033002455

Swain, M. (2001). Integrating language and content teaching through collaborative tasks. *Canadian Modern Language Review*, *58*(1), 44–63. doi:10.3138/cmlr.58.1.44

Zwiers, J., Dieckmann, J., Rutherford-Quach, S., Daro, V., Skarin, R., Weiss, S., & Malamut, J. (2017). *Principles for the design of mathematics curricula: Promoting language and content development*. Stanford University, UL/SCALE. https://ell.stanford.edu/content/mathematics-resources-additional-resources

ADDITIONAL READING

Jansen, A. (2020). *Rough draft math: Revising to learn*. Stenhouse Publishers.

Moschkovich, J. N. (2016). Hearing mathematical competence expressed in emergent language. In D. White, S. Crespo & M. Civil (Eds.), Cases for mathematics teacher educators: Facilitating conversations about inequities in mathematics classrooms (pp. 161–170). The Association of Mathematics Teacher Educator (AMTE) Professional Book Series, Information Age Publishing.

Smith, M. S., & Stein, M. K. (2011). *Five practices for orchestrating productive mathematics discussions*. National Council of Teachers of Mathematics.

KEY TERMS AND DEFINITIONS

Argumentative Writing: Mathematical writing where students construct or critique an argument.

Explanatory Writing: Mathematical writing where students describe and explain.

Math Practices (MPs) in "Common Core": Mathematical practices can also include some of the 8 practices listed in the Common Core (CCSS, 2010), such as MP#1-Make sense of problems and persevere in solving them, MP#3-Construct viable arguments and critique the reasoning of others, MP#6-Attend to precision, MP#7-Look for and make use of structure, and MP#8-Look for and express regularity in repeated reasoning.

Mathematical Discourse: Mathematical discourse is not only talk, it includes multiple modes (oral, written, and pictorial). Mathematical discourse is also much more than vocabulary or textbook definitions, because meanings are situated, contextual, and dynamic (not static or provided by definitions).

Mathematical Practices: practices identified by researchers as central to learning mathematics such as generalizing, justifying, and connecting representations; mathematical practices are social and cultural in origin and that children develop proficiency in these practices through opportunities to participate in them.

Mathematical Proficiency: Mathematical proficiency (Kilpatrick, Swafford, & Findell, 2001) includes five intertwined strands: 1) Conceptual understanding, 2) Procedural fluency, 3) Strategic competence, 4) Adaptive reasoning, and 5) Productive disposition.

ENDNOTES

1 Types of language are typically labeled registers or discourses.

2 Many different labels can be used for everyday language, colloquial, vernacular, conversational, etc. I will use everyday and informal language to refer to all of those registers and discourses not associated with school, professions, or academic settings.

3 That document identified four types: exploratory writing, mathematically creative writing, explanatory writing, and argumentative writing.

4 "Stronger and Clearer Each Time" is one of eight mathematical language routines suggested by Zwiers et al (2017) for supporting mathematical language use and development .

5 Although this was not included in the 8 Math Practices in the Common Core Standards, it was documented as a core mathematical practice before the Common Core Standards appeared.

6 This general mathematical practice documented in research appears in the Common Core Standards as MP3: Construct viable arguments and critique the reasoning of others. In the rest of this chapter, I will use MP# to refer to any of the 8 Common Core list of mathematical practices.

7 This general mathematical practice documented in research appears in the Common Core Standards as MP7: Look for and make use of structure.

8 The strands are conceptual understanding, procedural fluency, strategic competence, adaptive reasoning, and productive disposition (Kilpatrick, Swafford, & Findell, 2001).

9 Note that this ambiguity and multiplicity of meanings is not exclusive to everyday or informal language. For example, the word "base" varies in meanings across plane geometry, solid geometry, exponential functions or logarithms.

10 Note that this is only a hypothetical example for the sake of illustration, the words "bigger" and "makes" can also be issues.

11 Note that the second claim still not completely or perfectly precise. If one multiplies a negative number by a positive number greater than 1, the result is not bigger—depending on the meaning by 'bigger." The claim would be true if 'bigger' refers only to the absolute value or the magnitude, irrespective of direction.

12 The materials for this lesson (MARS, 2015) provide a set of classroom activities based on this exploration that focuses on student thinking. Here I will only add some thoughts on how to integrate writing and revising into that lesson. Many of these ideas arose from working with students in my MA course, who use this task for an interview, and with Dr. Sal Huixilopotchli on his dissertation, which used this lesson to examine mathematical writing and revising and support revisions through feedback and/or conferencing (see Huixilopotchli, 2022).

13 Different analyses of this discussion appeared in Moschkovich (2008) and Moschkovich (2015a).

14 An important mathematical practice described in research and labeled as MP6 in Common Core Standards.

15 It may also be important to distinguish between precision and specificity.

APPENDIX: TRANSCRIPT

(When reading the transcript, it is important to look at the graphs in the figures)

37. Teacher: Did you have the same, did you put both of the same things on the x-axis and on the y-axis?
38. Teacher's gesture: ((Looking through students' graphs.))
39. Carlos: No, ((0.2 sec)) yes, actually.
40. Teacher: You both put time here (referring to the x-axes). OK, so what's different about these? I don't think it's, I don't think it's the positioning of them. **Look at your numbers, the way you placed your numbers.**
41. Teacher's gesture: ((Points at the x-axis on both papers, back and forth. Then turns David's paper in the same direction as Carlos' paper.))
42. **Carlos: Oh, that's true, 'cause I went by twos, I went 1, 2** ((0.2 sec)) and then I put that one (---) **he went by one**.
43. Carlos' gesture: ((Begins counting with his right finger following the numbers on his paper.))
44. Teacher: Aha, You skipped one. So how does that change how it looks?
45. Carlos: 'Cause it doesn't go up as far, it only goes, it's more steeper. It looks more steeper.
46. Carlos' gesture: ((Moving his right hand outward. Then moving his right hand straight up.))
47. Teacher: Remem-. Similar to the difference between this one and ((0.1 sec.))and this one here. Right?
48. Teacher's gesture: ((Makes a sign with right thumb and index of her hand to show interval differences on their papers. Then she points to a graph on the blackboard. Next she points to a second graph on the blackboard. The first and second graphs have different scales on the x-axis so that the second graph is compressed along the x direction.))
49. David: That one.
50. Carlos: Yeah.
51. Teacher: Here the numbers are closer together so it looks looks steeper. Other than that, are they the same graph?
52. Teacher's gesture: ((Makes a sign with thumb and index finger. Then she gestures upwards with her right hand.)) ((0.1 sec.))
53. Carlos: No, also here in the x-axis.
54. Carlos' gesture: ((Carlos point to the x-axis on his paper.))
55. David: (the distance) ((Points to the axis on his paper.))
56. **David: I went by twos.**
57. Carlos: This is the x-axis. Right?
58. ((Carlos points to the axis.))
59. Teacher: This is the y-axis, ((0.1 sec.)) this is the x-axis.
60. Teacher's gesture: ((Sweeps her pencil vertically to represent the y-axis and then horizontally to represent the x-axis.))
61. David: I went by twos.
62. Teacher: You went by twos and you went b:y- ((0.2 sec.))
63. **Carlos: I went by twos.** You (didn't) **you went by ones**! What are you talking about.

64. Teacher: No, here on the y-axis.
65. Teacher's gesture: ((Points to the axis.))
66. Carlos: Oh, I went by fives.
67. Teacher: You went by fives. ((0.2 sec.)) No, actually you didn't go by fives. You actually went by two and a halves because you'd, you did every two spaces was five.
68. Teacher's gesture: ((She points to Carlos' paper while she explains.)) ((0.3 sec.))
69. Carlos: Then he only went by one.
70. Carlos' gesture: ((Carlos points to David's paper.)) ((0.2 sec))
71. Teacher: Every one space was two of his. You see, they're almost the same. If you look at the next two-((Puts down her notebook and points to the graphs.))
72. Carlos: Wait! But I don't get what you're saying.
73. Teacher: OK.
74. Carlos: 'Cause I went by fives. ((David stands up.))
75. **Teacher: OK, your numbers, right, the numbers you have are by five ((0.2 sec.))** OK ((0.1 sec.)). If you look at one line here, what number is he at?
76. Teacher's gesture: ((Takes David's paper and places it next to Carlos' paper then points to David's graph.))
77. Carlos: Two.
78. Teacher: What number would you be at if you had a number here?
79. Teacher's gesture: ((Points to Carlos' graph.))
80. Carlos: Three.
81. Teacher: Almost, two and a half.
82. Carlos: Yeah.
83. Teacher: Because that'd be half way to five. OK. ((0.1 sec.)) At this point, after 1, 2, 3, he's got 6. For you after three, 1, 2, 3, you'd be at 7 and a half.
84. Teacher's gesture: ((She counts the squares with her pencil.))
85. Carlos: O:K:.
86. Teacher: See what I mean? So it's actually two and a half. The numbers you wrote are by fives but since you skipped a line in between, each one is two and a half.
87. Teacher's gesture: ((Raises her hand and in the air uses thumb and index to show interval.))

FIGURES

Note: Figures 1 through 6 are reprinted by permission of Taylor & Francis LLC from Moschkovich, J. N. (2008). "I went by twos, he went by one:" Multiple interpretations of inscriptions as resources for mathematical discussions. *The Journal of the Learning Sciences*, 17(4), 551-587.

Chapter 14
Mathematical Writing and English Language Learners:
Strengths-Based Instructional Strategies

Salvador Huitzilopochtli
University of California, Santa Cruz, USA

ABSTRACT

The purpose of this chapter is to examine the role of mathematical writing to enhance mathematics instruction for students bureaucratically labeled as English Language Learners (ELLs) and multilingual students. The chapter addresses the specific needs of these students and focuses on two major claims: first, mathematical writing is useful for ELLs learning mathematics, and second, mathematics instruction for ELLs should support them to generate mathematical claims and provide opportunities for them to revise their work. The claims are illustrated using data from a dissertation study that examined how multilingual students learned mathematical arguments in a classroom that used mathematical writing and conferencing. This chapter per the author frames mathematical arguments using proof schemes and Toulmin's model and takes a strengths-based approach to instruction. The final section suggests directions for future research and focuses on expanding beyond writing as an individual exercise.

The product of mathematical activity might be justified with a deductive proof, but the product does not represent the process of coming to know. (Lampert, 1990, p.30)

It is worth noting that the roots of "prove" come from "to probe," meaning "to test the validity of." Thus the maxim "the exception proves the rule" is not an oxymoron, but a correct statement: "The exception probes (the truth of) the rule" and finds it wanting. (Schoenfeld, 2009, p.xvi)

DOI: 10.4018/978-1-6684-6538-7.ch014

INTRODUCTION

Teachers are increasingly called on to incorporate writing in mathematics instruction (Common Core State Standards Initiative, 2017; Leinwand et al., 2014). Unfortunately, many teachers of mathematics feel ill-prepared to teach writing and assign writing less than teachers in other disciplines (Applebee & Langer, 2011; Powell et al., 2017). This is a problem because mathematical writing is known to support metacognition and learning (Bangert-Drowns et al., 2004; Pugalee, 2004). For students who are multilingual, bilingual, or learning English, the problem is worse. Teachers often hold deficit views of these students (Faltis & Valdés, 2016; Flores & Rosa, 2015), feel unprepared to teach them mathematics (de Araujo et al., 2018), and curricula do not, generally, provide the kinds of supports necessary for their success (de Araujo & Smith, 2022). Multilingual students and English Language Learners (ELLs) represent populations of students who would benefit from access to mathematical writing (NASEM, 2017) but are not likely to have opportunities to engage in the practice.

The purpose of this chapter is to examine the use of mathematical writing to enhance mathematics instruction for multilingual students and ELLs. Throughout this chapter, I will use the term "ELL" even in the presence of abundant critique for this term. Terms such as "ELL," "Limited English Proficient," "English as a Second Language," and "English Learner" are inherently deficit oriented as they focus on what students "lack" (de Araujo et al, 2018; Walqui & Bunch, 2019), privilege the language of instruction (Barwell et al., 2017), and reflect social hierarchies (Faltis & Valdés, 2016; Flores & Rosa, 2015). At the same time, "ELL" is a term that is commonly used by states, districts, and practitioners, so I will use it with the full recognition of its limitations and critiques. Unfortunately, different states and localities define and redesignate students labeled "ELL" inconsistently. So, an "ELL" in one locality might not be designated an "ELL" in another. Consequently, I will use the term "ELL" to include any student who ever held that label, students who have been screened and deemed "Initially Fluent," and other multilingual students (including bilinguals) who might benefit from additional language support.

Here, I discuss some of the educational context that surrounds ELLs, including demographic information, theoretical assumptions about learning language and mathematics that foreground the strengths ELLs bring to the classroom, and how to use argumentative and informative/explanatory writing (Casa et al., 2016) to support ELLs to generate mathematical claims and engage in reasoning when writing arguments. The chapter is intended to support teachers, teacher leaders, administrators, researchers, and other stakeholders who seek to improve educational opportunities in mathematics classrooms for K-12 students bureaucratically labeled as "English Language Learners" (ELLs).

This chapter presents research results and data from the author's dissertation study (Huitzilopochtli, 2022). The study took place in a bilingual charter school in an agricultural community in California's Central Valley. Many of the students attended the bilingual school where instruction took place in Spanish and English since kindergarten. While language designations of individual students might have shifted during their years of schooling (e.g., some students were redesignated from ELL to "Fluent English Proficient"), these students can be considered "ELLs" for the purposes of this chapter.

This chapter is organized into four major parts. First, I describe some characteristics and background information about ELLs. Secondly, I discuss the role of mathematical writing and the recommendations for instructional practice. In the third section, I describe and illustrate how to use mathematical writing to support ELLs to:

1. Generate mathematical claims

2. Engage and cultivate their empirical reasoning

In the fourth section, I discuss implications for future research. I focus on upper elementary and middle school mathematics. Specifically, I focus on facilitating students' engagement with the Common Core State Standards (National Governors Association Center for Best Practices & Council of Chief State School Officers, 2010) Mathematics Practices (MPs), such as students' *constructing viable arguments and critiquing the reasoning of others* (MP 3), *attending to precision* (MP 6), and *looking for and making use of structure* (MP 7). I emphasize *justifying* (Stephens et al., 2017; Smith, 2008) as an additional mathematics practice related to constructing arguments.

BACKGROUND

Who are ELLs?

The National Academies of Sciences, Engineering, and Medicine's (NASEM; 2017) report entitled *Promising Futures* details demographic information and promising instructional strategies to support ELLs. In summary, ELLs are a diverse group of students. They vary across a range of demographic categories, including home language, age, immigration status and circumstance, generational status in the United States, geographic location, socioeconomic status and parent characteristics, achievement, disability, and more. Mexico is the most common origin country for foreign-born ELLs (41.3%), followed by China (4.5%). More importantly, most ELLs—at every age—were *born in the United States*, and in some cases, both of their parents were also US-born.

How do these students experience education, broadly? Most ELLs are segregated in low-performing schools (NASEM, 2017). Students might also be segregated into linguistically "sheltered" classes that purport to offer additional language support. Unfortunately, teachers report feelings of under preparedness to teach mathematics to ELLs (de Araujo et al., 2018). Language segregation, whether within or across schools, can contribute to a variety of negative side effects, including stigmatizing students, diminishing access to grade-level content, prolonging the English language development process, and sapping motivation. The school context and teachers' backgrounds may explain why some ELLs are deemed to be "unmotivated." For example, less than 5% of pre-service and in-service teachers come from multilingual homes (Faltis & Valdés, 2016), so they might not understand experiences that are common in immigrant communities. Moreover, many pre-service teachers have had minimal experience with people from outside of their own primary cultural and language groups and have rarely encountered spoken English as an obstacle to their social and academic progress and purposes (Faltis & Valdés, 2016). This can have profound implications for the interactions between teachers and students. Therefore, how can we ensure that teachers, educators, and other stakeholders have what they need for teaching, advocating for, and ensuring educational opportunity for ELLs—who embody an experience that is largely unfamiliar to teachers?

Furthermore, ELLs often have fewer opportunities to engage with high-quality mathematics instruction because of how deficit views are imbued into educational policies and practices (NCSM & TODOS, 2020), and even curriculum (de Araujo & Smith, 2022). The demographics and experiences of ELLs are important because teachers and curriculum writers should take the diversity of ELLs into consideration when designing activities and supports. Secondly, teachers and curriculum writers should promote mathematical writing with ELLs, even though it will involve struggle for those teachers who feel under prepared to teach ELLs (de Araujo et al., 2018) or use writing in a mathematics classroom (Powell et al., 2017). Whether at the individual level (where we examine our own beliefs) or at the structural level, educators, curriculum designers, and advocates should consider these aspects of students' experiences to make informed decisions about the instruction of ELLs.

Strengths-Based Assumptions

In this brief section, I summarize three assumptions about the use of writing to support ELLs learning mathematics. Naming a few strengths-based assumptions is an important step to recognizing and addressing persistent, unproductive beliefs. First, *academic language* (including *academic language in mathematics*; Moschkovich, 2015) is understood as a set of discipline-connected language practices that *all* students can learn (NASEM, 2018). Secondly, students are presumed to have linguistic and mathematical competencies that are sufficient for participation in academic and mathematical practices. Moreover, it is not necessary to master language *before* engaging with disciplinary content, nor is language learned separately or sequentially with disciplinary content (Walqui & Bunch, 2019). Language proficiency is assumed to be an *outcome of*, not a prerequisite to, effective content instruction (NASEM, 2018; Walqui & Bunch, 2019). Third, under the right conditions, ELLs learn mathematics at least as well as their monolingual counterparts (Barwell et al., 2017). To foster this learning, instruction should focus on mathematical reasoning and practices, build on student thinking, invite student strengths, facilitate student interactions, and include everyday ways of communicating.

These assumptions work together to support a view that writing is an important practice for ELLs as students make sense of the mathematics at play. The concept of "academic language," which includes written forms of language, can create a "symbolic border" that impedes access to rigorous curricula (Valdés, 2017), especially when academic language is thought of as "decontextualized," positioned as normative, or narrowly understood as "vocabulary" (p.336). Therefore, writing should be used to support students as they make sense of the mathematics and by teachers as they formatively assess student thinking. Students' everyday language practices, which might include the use of non-privileged dialects of English (e.g., African American Vernacular English or Chicano English), should be embraced as valuable resources for constructing new knowledge (Bunch, 2014; Wang et al., 2021), including through writing.

What is Mathematical Writing?

In this chapter, mathematical writing is defined to include "any representation… that students record on paper (including prose, symbols, and other visuals) resulting from prompts that press to include one's mathematical reasoning" (Cohen et al., 2015, p.345). There are two important aspects to this definition. First, that mathematical writing includes the kind of prose, alphabetic writing, that is commonly used to explain concepts or procedures, but it also includes symbols (e.g., those used in equations or expressions) and a variety of visuals (e.g., graphs, geometric visuals, informal visual representations that students might generate). Each modality has the unique potential to contribute meaning when students communicate mathematically (Moschkovich, 2015; O'Halloran, 2015). This is especially important for ELLs who need access to their entire repertoire of sense-making tools.

The second aspect is the intimate connection to mathematical reasoning. Mathematical writing is generated in response to a press for reasoning, so writing prompts are assumed to require or elicit some form of student reasoning. This is important not only for language development but also for focusing on mathematical concepts, which is critical for ELLs (Moschkovich, 2013). Most importantly, both aspects include the use of students' informal spoken and written language, as well as more formal means, to communicate mathematically.

In this chapter, I will focus on two of the purposes suggested by Casa and colleagues' (2016) mathematical writing framework:

1. Informative/explanatory writing: Writing to describe, represent, connect, and define
2. Argumentative writing: constructing, justifying, and critiquing arguments

These purposes can be used strategically to support different kinds of mathematical activity and assessment with various writing activities.

Questions remain about how mathematical writing can be used to support ELLs in the classroom. In the next section, I describe how to support ELLs to 1) generate mathematical claims and 2) engage and cultivate their empirical reasoning when arguing mathematically. First, I will discuss how mathematical arguments are typically described and evaluated using Toulmin's (2003) structure and Harel and Sowder's (1998) notion of *proof schemes.* Then, I describe how mathematical writing can be used with ELLs to connect their everyday mathematical arguments to canonical ways of arguing mathematically. I pay particular attention to the mathematical strengths that ELLs bring to the classroom and highlight the key role that those strengths can play in the process of learning about mathematical argument.

SUPPORTING ELLS WITH MATHEMATICAL WRITING

How instruction leverages the strengths that students bring depends on how those student contributions are framed. Part of the work to support ELLs in learning mathematical argument involves eschewing deficit views of their linguistic and mathematical capacities. In doing so, the utility of mathematical writing becomes possible (because one no longer assumes that they "can't do it"), and the strengths they present can be made useful to facilitate learning. Historically, "traditional" instruction in algebra emphasizes a symbolic orientation, simplifying expressions, and solving equations and inequalities (Kieran, 2007). However, more contemporary approaches to algebra education emphasize four essential practices: generalizing, representing, justifying, and reasoning with mathematical structure and relationships (Blanton et al., 2018; Kaput, 2008). This aligns with calls to engage ELLs with mathematical practices (Goffney et al., 2018; Moschkovich, 2013). As Lampert (1990) once stated, "The product of mathematical activity might be justified with a deductive proof, but the product does not represent the process of coming to know" and students need time and support to engage in that process (p.37). Time and support to explore examples and engage in the proof process are crucial for ELLs because more modes of communication can support greater participation in mathematical practices and activity. In this way, educators can support students, including ELLs, to "probe" mathematical truths in the way that Schoenfeld (2009) describes:

It is worth noting that the roots of "prove" come from "to probe," meaning "to test the validity of." Thus the maxim "the exception proves the rule" is not an oxymoron, but a correct statement: "The exception probes (the truth of) the rule" and finds it wanting. (p.xvi)

Next, I describe what is meant by "mathematical argument" and illustrate some ways that mathematical writing can support ELLs to engage in this practice and take time to probe the truth of mathematical claims.

What's a Mathematical Argument?

Mathematical argument is difficult for any student of mathematics to learn (Blanton et al., 2009; Healy & Hoyles, 2000; Huitzilopochtli, 2022; Stylianides & Stylianides, 2009). Part of the difficulty lies in having to learn a new "basis of belief" that is different from all other domains of knowledge (including everyday experience or even science; ECEMS, 2011).

In terms of structure, mathematical arguments can be thought of as consisting of these basic parts: a claim (or conclusion), data, warrants, and backing (Toulmin, 2003). While mathematical claims can be referred to in a variety of ways (e.g., "conclusions," "conjectures," or "assertions"), I will primarily use "claim" to refer to mathematical assertions. In general, warrants refer to the mathematical knowledge that ratifies steps from data (or evidence) to claim. Backing includes additional mathematical knowledge that justifies the warrant and "data" can refer to facts and calculations upon which claims are based. A diagram of a generic argument structure is shown in Figure 1. In the case of ELLs, argumentative writing can be used to prompt students to generate and refine mathematical arguments as they are asked to record mathematical representations, different types of examples, or generate narrative text to connect data to warrants (e.g., mathematical theorems, properties, definitions, or even claims previously established in class).

In addition, mathematical arguments can be holistically characterized by the *proof scheme* that a person might employ to justify their claims (Harel & Sowder, 1998). Typically, students make use of one of three kinds of proof schemes: external conviction, analytical, and empirical (Harel & Sowder, 1998). Unless students have had instruction on mathematical argument, they are most likely to make use of empirical proof schemes (Harel & Sowder, 1998; Healy & Hoyles, 2000; Stylianides & Stylianides, 2009). External conviction proof schemes include appeals to authorities (like teachers or textbooks) or

Figure 1. Toulmin model of a mathematical argument
(Toulmin, 2003)

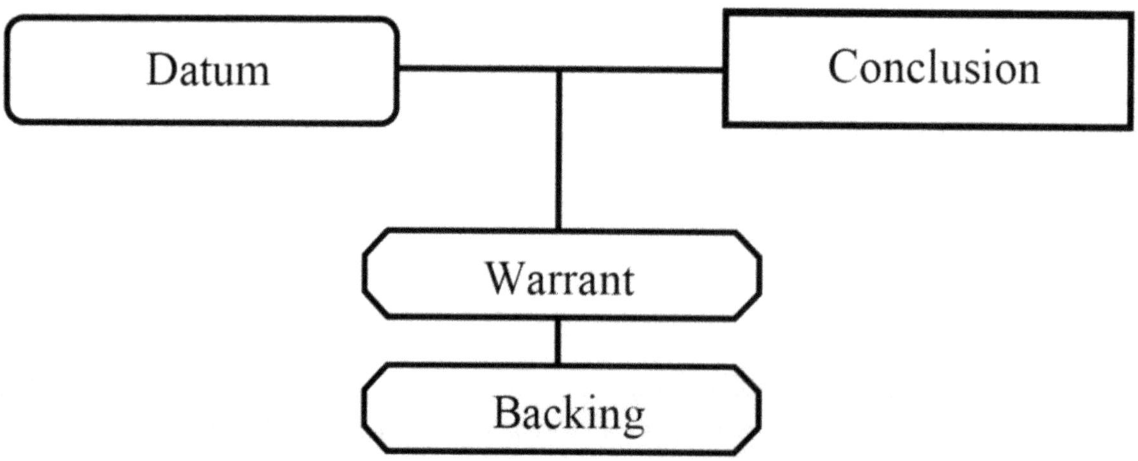

Figure 2. Example of a mathematical claim (left) and inductive empirical argument (right)
(Huitzilopochtli, 2022)

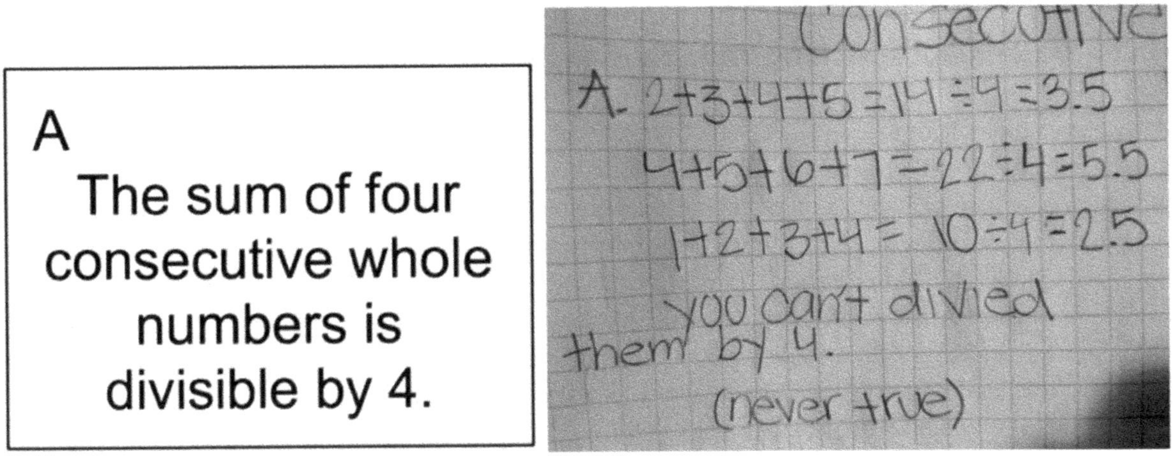

other ritual or symbolic arguments. For example, doubts are removed by the mere *appearance* of a valid proof or doubts are not removed because the proof did not look "proof-y" enough (e.g., it did not use mathematical symbols). Analytical proof schemes seek to settle a claim in general. These proof schemes reflect a student's use of logic to connect data and claims and their view of mathematics as an organized body of knowledge. Empirical proof schemes, the most common among middle schoolers, rely on facts or observations. Harel and Sowder (1998) describe two classes of empirical proof: perceptual and inductive. Perceptual empirical proof schemes rely on "rudimentary mental images," without consideration for the implications of what is observed, while inductive empirical proof schemes rely on one (or more) specific cases to test the truth of a claim. An example of a student using an inductive empirical proof scheme is shown in Figure 2.

In this chapter, I focus on *justifying* as a core algebraic practice that involves seeking mathematical certainty with informal mathematical methods (Carpenter et al., 2003; Smith, 2008; Stephens et al., 2017) and prepares students to engage formal proof and proving in higher mathematics (Leinwand et al., 2014). Justifying is a core algebraic practice (Stephens et al., 2017) that involves using informal arguments to construct certainty (Smith, 2008). Justifying has also been described as an equity practice because it creates opportunities for students to reason mathematically, learn deeply, and foster agency (Bieda & Staples, 2020).

As much as educators might be tempted to rush students "past" an empirical approach to justifying toward more analytic ones, supporting students to learn mathematical argument requires meeting students where they are and accompanying them as they engage "cross country" mathematics—zigzagging through "jagged and uncertain" territories—to apprentice them into a new way of believing (Lakatos, 1976; Lampert, 1990). Examples can be calculated for various purposes at different stages of the proof process. Some purposes include "first-thought-of examples," "understanding," "checking," "conjecture breaking," "generic," or "supporting a general argument" (Ellis et al., 2012). In any case, examples are never totally absent from the proof process. Hence, efforts to frame students as "remedial" for calculating examples to enter a task is disingenuous and unproductive. Supporting engagement with examples and mathematics practices, like justifying, is especially important for ELLs who benefit from instruction that focuses on mathematical reasoning, mathematics practices, and invites informal language as a resource for learning (Moschkovich, 2013).

Using Mathematical Writing With ELLs to Support Instruction and Rigor

ELLs have unique learning needs that can be met at the intersections provided by mathematical writing and the introduction of challenging tasks. Good mathematics instruction for ELLs should include tasks with high cognitive demand, center mathematical concepts, foster peer interactions and co-construction of knowledge, and offer language support for understanding and engaging disciplinary practices (Hamburger & Chu, 2018; Moschkovich, 2013). Informative/explanatory and argumentative writing can be used in tandem with peer-to-peer and whole-group discussions to provide opportunities for multimodal communication about mathematical concepts and language support as students engage in mathematical practices like justifying. Rich tasks can invite and sustain this kind of activity.

It is important to choose tasks with high cognitive demand because the nature of the task can determine what students learn (NCTM, 1991; Stylianou & Blanton, 2018). Tasks can be thought of in one of four ways (Smith & Stein, 1998):

1. Memorization tasks—involving recall of facts
2. Procedures-without-connection tasks—involving procedures (or algorithms) without connecting to concepts or the meanings underlying the procedures
3. Procedures-with-connections tasks—involving procedures but also inviting students to make meaningful connections to concepts
4. "Doing mathematics" tasks—requiring complex, nonalgorithmic thinking and almost always requiring justifications or explanations

The first two categories involve low cognitive demand (Smith & Stein, 1998). These types of tasks involve little (if any) connection to mathematics concepts and seek to product "right answers". Higher cognitive demand items (such as those described in the two latter categories) emphasize concepts, the use of multiple representations, exploration of processes and relationships, and complex thinking. In the case of ELLs, task selection is critical as they are more likely to experience low-quality mathematics instruction (NASEM, 2018), which typically includes tasks with memorization and completing procedures-without-connections. Mathematics instruction for ELLs should include higher cognitive demand tasks as it aligns with recommendations for good instruction (Hamburger & Chu, 2018; Moschkovich, 2013) and can be supported by Informative/Explanatory and Argumentative mathematical writing.

In my dissertation study (Huitzilopochtli, 2022), I used the following tasks to introduce mathematical argument to a group of 8th-grade students (over 80% of whom were designated as ELLs—primarily Spanish speaking): *Fibonacci Rectangles* (Dietiker et al., 2013), *Happy Numbers* (Dietiker et al., 2013), and *Evaluating Statements: Consecutive Sums* (or "Consecutive Sums"; MARS, 2015). These tasks prompt students to generate either Informative/explanatory writing (e.g., "describe patterns" or explain their "thinking") or Argumentative writing (e.g., generate and justify predictions or prove a claim that is chosen from a list). The tasks involve high cognitive demand and provide opportunities for ELLs to write mathematically. At this point, I briefly describe the two tasks before showing samples of student work to show what kind of learning resulted.

Task 1. In the Fibonacci Rectangles task, students were asked to predict sums of areas for the rectangles produced using terms in the Fibonacci sequence, and then justify their predictions. Original questions and instructions given to students:

Step 1: How is the sequence growing? [1, 1, 2, 3, 5, 8, 13, …]

Step 2: If the measures of the sides of a rectangle are consecutive Fibonacci numbers, it is called a "Fibonacci Rectangle." Here are the first four [illustration]. What is the area of each rectangle? What is the sum of the first two? The first three?

[Insert rectangle]

Step 3: Copy/Complete the table (up to 8)

Number of Rectangles	Sum of Areas
2	3
3	9
4	
5	
6	
7	
8	

Step 4: Use your table. Look for any special numbers. Can you find any patterns?

Step 5: How can you predict the sum of the areas of Fibonacci rectangles? Use your patterns to predict the sum of the areas of the first 15 Fibonacci rectangles.

Task 2. In the Happy Numbers task, students were given a definition for "Happy" number and asked to find all the Happy numbers between 1 and 100. Based on what they learned, they were also asked to conjecture and justify which numbers "must be Happy" (given 478 is a Happy number). The definition of Happy number and original instructions are shown in Figure 2:

Task 3. The Consecutive Sums task begins with a card-sorting activity. Students sorted six mathematical claims as either *always, sometimes,* or *never* true. Then, students were asked to justify their responses for two of the claims they made. The original instructions are shown below, and the card sort activity is illustrated in Figure 3.

Instructions: [Consecutive Sums Task]

In your groups, decide whether each conjecture is Always, Sometimes, or Never True. If the conjecture is Always or Never true, write down how you can be sure.

If the conjecture is Sometimes true, write down **when** it is true and **when** it is not.

Use ideas from the sample work you have looked at and make notes on your math notebooks—we will submit photos.

Agree about each conjecture before moving on. Explain your thinking on your slides (and upload photos there).

After placing the cards, pick two conjectures to prove!

For each one, use symbols, words, visuals, examples, and other ideas to prove your conjecture is Always, Sometimes, or Never true.

Figure 3. Happy Numbers task
(Dietiker et al., 2013; Huitzilopochtli, 2022)

Definition: Happy Numbers are found by separating the digits of a number, squaring each digit, and then adding the resulting products. If the sum of the products is "1", then the number is Happy. If not, repeat the process until it results in "1" or a sum repeats, in which case the number is not Happy. Here are two examples:

Example 1: 23

$2^2 + 3^2 = 4 + 9 = 13$

$1^2 + 3^2 = 1 + 9 = 10$

$1^2 + 0^2 = 1$

Happy

Example 2: 34

$3^2 + 4^2 = 9 + 16 = 25$

$2^2 + 5^2 = 4 + 25 = 29$

$2^2 + 9^2 = 4 + 81 = 85$

$8^2 + 5^2 = 64 + 25 = 89$

...

$5^2 + 8^2 = 25 + 64 = 89$ (repeat)

Not Happy

A. *There are 17 Happy Numbers (between 1 and 100). Find as many as you can.*
B. *Find five 3-digit Happy Numbers.*
C. *Eva found out that 478 is a Happy Number.*
    a. *What other numbers must be Happy?*
    b. *How do you know?*
    c. *Find at least 10 new Happy Numbers.*

Using Mathematical Writing With ELLs: Generating Claims

In this section I discuss the role of mathematical writing in supporting ELLs to generate claims. Explanations can be an important way for ELLs to bring mathematical language to life (Nemirovsky, Barros, Noble, Schnepp, & Solomon, 2005). I discuss different kinds of student writing elicited by prompts for mathematical writing and how to leverage that writing to support learning. Then, I discuss the use of non-neutral qualifiers to scaffold student efforts to generate a claim and create points of entry into argumentative writing as they begin to justify their claims.

All students need support when learning to make mathematical claims. It is not always clear to students what is meant by a "claim." Mathematical writing can play a supportive role in developing the kind of awareness that students need for making general mathematical claims (i.e., referring to a class of numbers, for example, rather than a single calculation or example). When asked for a proof for a mathematical statement, it is typical for middle school students to write a "procedural explanation" or description (Hebert & Powell, 2016). Procedural explanations focus on and describe the steps or *proce-*

Figure 4. Conjectures for Consecutive Sums task
(MARS, 2015)

Sample Conjectures to Discuss

A	B
The sum of four consecutive whole numbers is divisible by 4.	The sum of five consecutive whole numbers is divisible by 5.
C	**D**
To find the sum of consecutive whole numbers, find the middle number and then multiply it by how many numbers there are.	The sum of five consecutive whole numbers is divisible by the sum of the first and last numbers.
E	**F**
The sum of two consecutive whole numbers is an odd number.	The sum of six consecutive whole numbers is divisible by the sum of the middle two numbers.

dures a student employs, rather than mathematical concepts. This kind of Informative/explanatory writing can be used to formatively assess whether students are understanding the task and the mathematics at play, as well as their use of mathematical language.

An example of Informative/explanatory writing is given below (see Figure 5). As part of the Fibonacci Rectangles task, a student named "Nathan" (designated as "English Only," but a full member of

Figure 5. Nathan's description in Fibonacci rectangle task (right), including table (left)
(Huitzilopochtli, 2022)

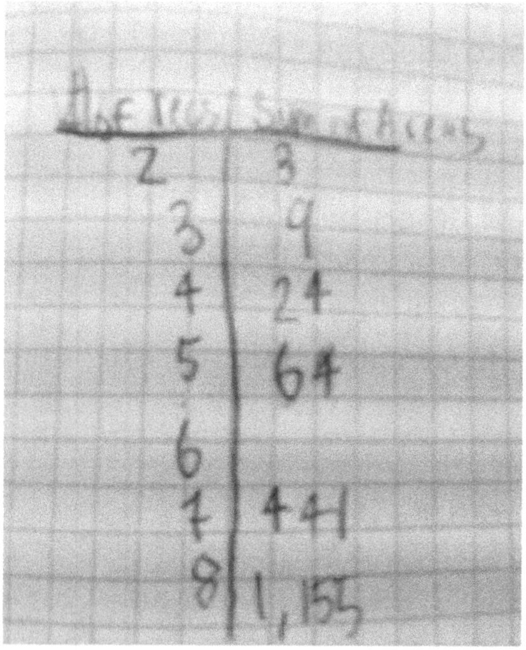

"I started by finding the size of the rectangles then I continued on by putting them onto a chart and adding them all together. The first + the second + the third etc."

the classroom bilingual community) responded to the prompt to "upload a final draft of his argument" by describing how he completed a table of values. Note that the student's description includes a table for reference (Figure 5, left) and a recursive description of how Nathan generated each new term (Figure 5, right). The writing describes how Nathan predicted each new term. Unfortunately, the recursive description only vaguely describes how each new term could be calculated—i.e., if the previous term is already known, then he could determine the next one. The task is to predict a remote term (in this case, the 15th term) with a *value* and to present and justify a means to predict additional terms.

How can Nathan be encouraged or guided to "see" beyond the examples he is calculating? There are possibilities for supporting Nathan with writing. For example, the teacher could invite him to generate a visual representation, provide an opportunity to *select* a claim and then justify the selection, or prompt him to think of ways to predict that don't require the previous term.

Mathematical writing might help students engage the proof process by targeting the reasoning it demands—that is, by fostering different kinds of reasoning. The proof process of mathematical inquiry can be thought of as involving three kinds of reasoning: abduction, induction, and deduction (Minnameier, 2004; Rivera & Becker, 2007). Inquiry begins with abduction (which considers what "may be" possible), followed by induction (which tests what is "actually operative"), and deduction (which states what "must be"; Rivera & Becker, 2007). In order to target abductive reasoning, for example, students could answer the following prompt: "Is there a way to predict the 15th term *without* calculating each term in between?" In this way, students are encouraged to reconsider the examples they have in-hand and continue the creative work of generating a general claim. Alternatively, instructors could provide students like Nathan a set of non-neutral qualifiers to support generating claims.

Figure 6. Nathan's examples as evidence
(Huitzilopochtli, 2022)

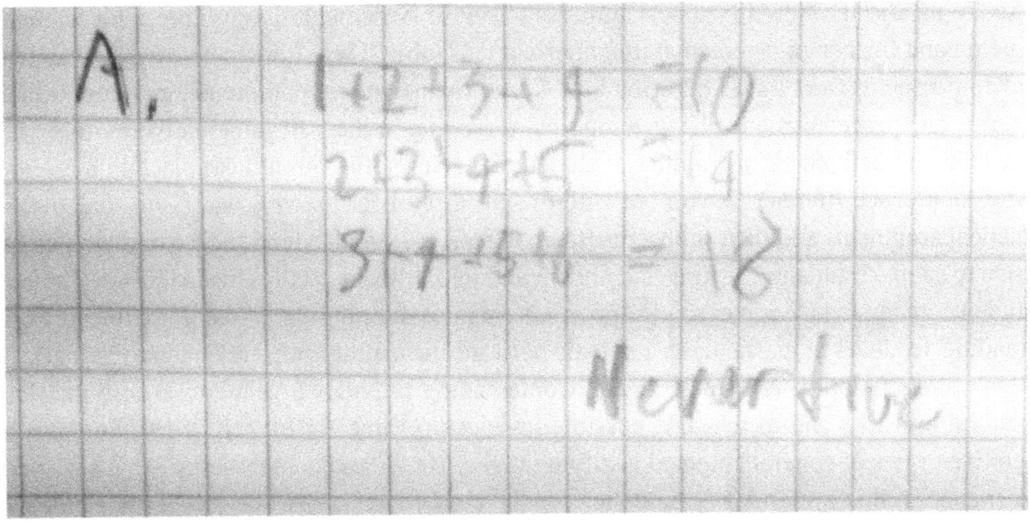

Non-neutral qualifiers scaffold student efforts to generate a claim by providing a list of possible claims and modeling the language practices involved when generating claims. This kind of language support is important for all students but crucial for ELLs. For example, consider the conjectures in Figure 4, above. In the Consecutive Sums task, students were asked to decide whether each claim is *always, sometimes,* or *never* true. The use of a selected-claim task clarifies mathematical expectations for students when they are just beginning to understand what a mathematical claim is and might even scaffold language requirements for entering the task. In my dissertation study, I observed that many more students made claims and attempted to justify them when provided with non-neutral qualifiers.

With this additional support, Nathan generated a claim by selecting a qualifier and justified his claim with a general argument. Nathan's argument was based on Claim A:

Claim A: *The sum of four consecutive whole numbers is divisible by 4. Always, Sometimes, or Never true?*

"I think that a is *never true* because it keeps adding 4 to 10, and 10 + any multiple of 4 will not be divisible by 4. For example, 1 + 2 + 3 + 4 = 10 and 10, is NOT divisible by 4." (Nathan)

In addition, Nathan provided three examples (Figure 6). Nathan's asserted that Claim A is "never true". As evidence, Nathan used three examples (as shown in Figure 6) but also reasoned that every sum of four consecutive numbers can be thought of as the sum of 10 and "any multiple of 4": i.e., "10 + 4n" is not divisible by 4. While Nathan stops just short of using variables, the strength of his argument lies in his *understanding* of the concept of variable and his use of example for the purpose of supporting a general argument.

In the following section, I described how to support students to generate mathematical claims by pressing students beyond Informative/explanatory writing and scaffolding them to make argumentative claims by providing options to select from.

Focus on Strengths: Engaging Empirical Reasoning

In this section, I discuss how to support students to revise Argumentative writing with mathematical conferencing and further engagement with empirical reasoning. First, I describe what I mean by mathematical conferencing, and then I describe how I used mathematical conferencing to support ELLs to clarify terms and procedures and engage empirical reasoning in their Argumentative writing.

Mathematical conferences are similar to student interviews that might take place in a research setting (for example, see Brenner, 2006), but with the purpose of formatively assessing students' written mathematical arguments and then supporting them to reflect on and extend their work. Conferencing is productive to support students' writing and understanding (Hillocks, 2006; Swain, 2009). Conferencing helps students understand feedback and supports them in developing the nascent argument. For ELLs, understanding feedback is particularly difficult because they might not understand feedback offered in only one mode (Aguirre & Bunch, 2012). Conferencing provides a venue to ensure that feedback is understood by ELLs and to rehearse possible revisions (Ortmeier-Hooper, 2013). Students benefit from iterative cycles of generalizing and justifying (Ellis, 2007), and conferencing provides a venue for them to engage in this cycle. A key guideline for the conference is that the teacher does their best to understand and follow the *student's reasoning*. This is difficult as educators might hold a predetermined mathematical "destination" in mind. Conferencing works to affirm student contributions and accompany their reasoning with students leading the way.

Some might protest that individual student conferences are too time-consuming to be done in a math class, but I assert that students and the outcomes are worth the time and effort. Learning to write mathematical arguments is difficult for all students, but ELLs can benefit from conferencing and revising. First, students and teachers can clarify feedback to ensure understanding. This oral check-in, based on targeted written feedback, is important for ELLs that might not understand feedback offered in only one mode. Moreover, teachers have an opportunity to focus on concepts, rather than "imperfect" language. Secondly, additional exploration can provide much-needed time for students to "see" beyond examples toward mathematical structure. Finally, conferencing provides a venue for language support where students can orally rehearse revisions to written arguments or relevant vocabulary as it arises *in context*. Conferencing is a worthwhile activity that can be used as part of a larger repertoire of teaching strategies and formative assessment. As with writing conferences, mathematical conferences do not need to be conducted frequently. They could be done twice a year in support of larger assignments or projects. Moreover, they provide teachers a unique opportunity to address ELLs' individual needs that might not be possible to address in other instructional settings (such as a whole-class discussion).

Student Conferences

I conducted conferences that involved reviewing students' work in a dyad for 10-12 minutes, and then asking reflective questions to refine their understandings of concepts or procedures, (re)organize data and ideas to seek insight or extend their work. I used conferencing as a follow-up activity to support the initial written work students submitted. After reviewing their work (which might have included Informative/explanatory or Argumentative writing), I prepared 1-3 questions to elicit further reflection and press for additional reasoning.

When considering my follow-up questions, I recognized that students' incipient proofs schemes are important springboards toward a goal that includes mastering deductive and generalized proof schemes (i.e., analytical; Ellis, 2007), but they need opportunities to reflect on and revise their arguments. Now I describe two examples of the kinds of revisions ELLs could make while conferencing. In the first example, "Yolanda" and I work on refining her use of relevant vocabulary to enter the task. In the second example, "Nina" and I reorganize her data to see how the examples are related and generate a claim.

Yolanda and Nina are both ELLs but benefitted from conferencing in different ways. Yolanda's conference focused on clarification of vocabulary and cases. These clarifications supported Yolanda to extend her thinking and begin to define cases where the claim was true and where it was false. Nina's conference focused on generating a claim. While Nina seemed to master the procedure for finding Happy Numbers, it was still difficult for her to articulate a general claim about how she could find Happy Numbers without the need for calculation.

Example 1—Yolanda. Until the conference, Yolanda participated minimally in classroom activities. For example, she would contribute to classroom discussions, but inconsistently submitted written work. In the Consecutive Sums task, Yolanda selected *always, sometimes,* or *never* for each of the claims, but did not submit an argument to justify her selections. We started the conference by deciding to discuss Claim C:

To find the sum of consecutive whole numbers, find the middle number and then multiply it by how many numbers there are.

Yolanda made the claim that statement C was "sometimes" true but offered no argument beyond the claim. We tested the case of "3 + 4 + 5" together and found that the example was true (i.e., 4 × 3 = 12; as the product of the middle number, "4," and the number of numbers, 3). When I asked her to come up with her own example, she suggested "2 + 4 + 6," a sum of consecutive *even* numbers. I asked her if 2, 4, and 6 are consecutive numbers and she responded, "No… How do you know what a consecutive number is?" At this point we clarified the definition together and she went on to test a few more examples.

The conference was a useful space for us to refine how relevant vocabulary could be used. In fact, definitions that we had already previewed in class required *additional clarification* in the conference and were supported by the context. Other terms that required clarifications during student conferences included "average," "squaring," and the use of the circumflex ("^") to denote exponentiation. The conference was a useful space precisely because the vocabulary was no longer decontextualized but a necessary part of engaging the activity. By the end of the conference, Yolanda was beginning to define cases where her claim that statement C was sometimes true (3 and 5 addends) and where it might be false (2 or 4 addends), stating that, "it works when there is [sic] 3 and 5+ numbers but it won't work if it's 4 or 2 numbers."

Example 2—Nina. Nina was a very quiet student in class. She rarely participated out loud, although she frequently contributed to shared documents and listened attentively. After working on her argument for the Happy Numbers task, she submitted a page full of calculations and a list of numbers she found that were Happy (see Figure 7). Per the instructions, Nina found all the Happy Numbers between 1 and 100.

The conference began with a review of the calculations she submitted. Since she had made no claim about the examples she calculated, I asked her if she noticed anything now. Nina said that she "noticed" that 68 and 86 were both Happy and that they have the same digits. I asked her if any other numbers had the same digits and she correctly observed that "…94 and 49… and 97, 79" (Turn 16, Transcript). I asked her why she thought this might be happening (Turn 17).

Figure 7. Nina's calculations for Happy Numbers task
(Huitzilopochtli, 2022)

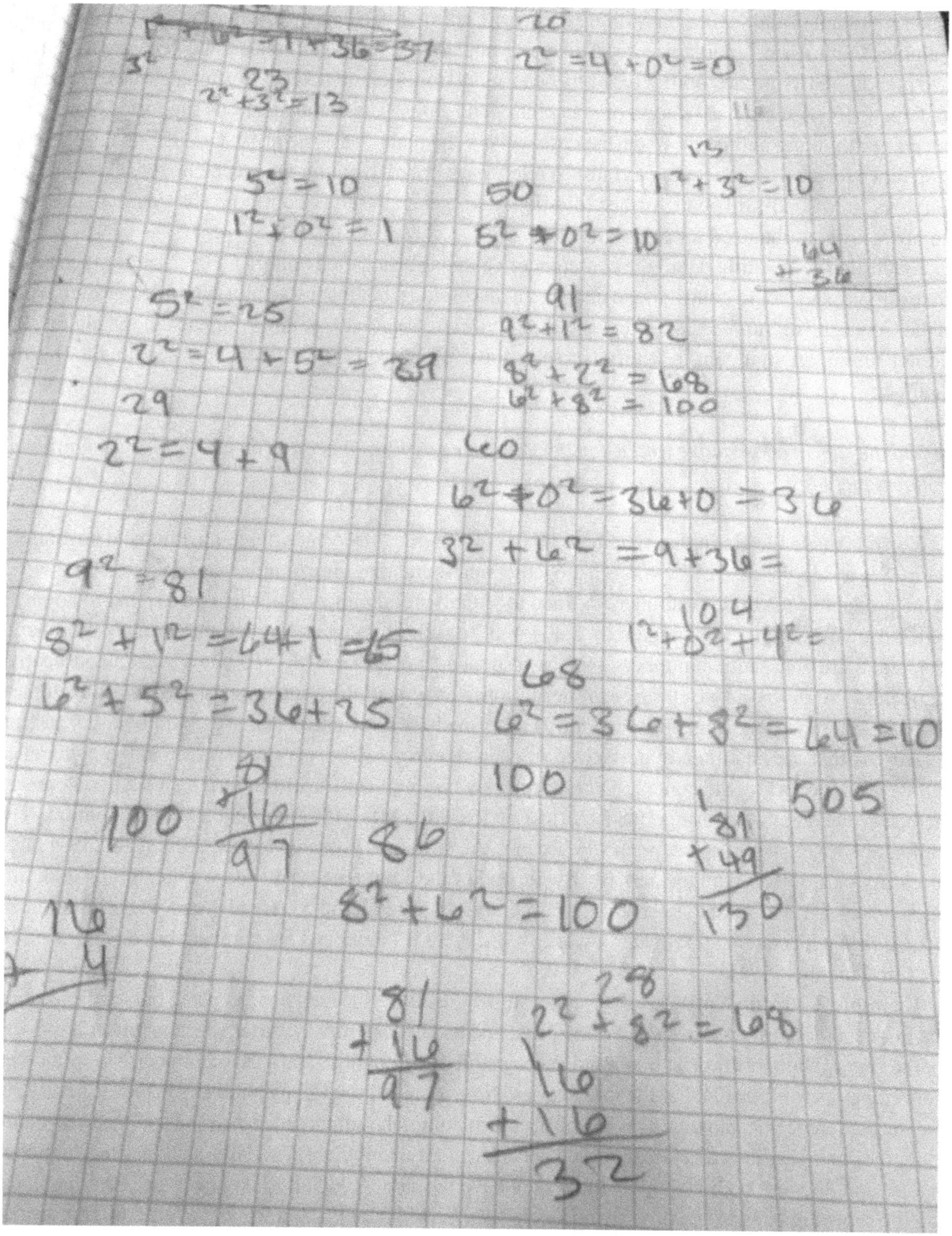

13 **Salvador:** Well, you said they have the same digits. Do any of the other numbers have the same digits?
14 **Nina:** Yes.
15 **Salvador:** Which ones?
16 **Nina:** Like, 94 and 49. and 97, 79.
17 **Salvador:** Good. Good. Good. So. So… Now if I ask you… How about this then, now if I ask you, why is that? Do have any idea why that's happening? Like if 68 is a happy number why does that mean 86 will be a happy number? How is that working?
18 **Nina:** Because you're... Uh... Square the individual numbers and … yeah.
19 **Salvador:** Cause you square the individual numbers? And then what?
20 **Nina:** And then you add them?
21 **Salvador:** Uh huh.
22 **Nina:** So, you're gonna get the same, like, sequence for the … yeah.

Nina explained that Happy Numbers can have reversed digits because squaring the individual digits ("numbers"; Turn 18), and then adding them (Turn 20), will yield the same sum ("sequence"; Turn 22). She implicitly relies on the commutative property of addition but is using this property as part of a larger argument that Happy Numbers can be combinations of digits.

Nina still had difficulty articulating a claim about Happy Numbers. So, I asked her if she could think of any three-digit Happy Numbers. She asserted that "you can get the same digits but put a zero in between or at the end" (Turn 28). When I asked her why that would work, she said "Because when you square zero, you're gonna get zero. So, it doesn't make any change in the final answer" (Turn 34). At that point I continued with the task (Happy Numbers, Part C) and asked if 478 is a Happy Number, then what other numbers *must* be Happy?

37 **Salvador:** …and use your pattern. I'm asking you to use your pattern. So, maybe not to calculate too much but if your pattern… if your ideas are right, then what other numbers could you make with that 4-7-8?
38 **Nina:** You can add a zero in between.
39 **Salvador:** Ok. So, like... And get what? What number would work, then?
40 **Nina:** Like, 4,780?
41 **Salvador:** Ok. That's good. What other numbers, maybe?
42 **Nina:** Like 4,078 or eight hun-- eight thousand seven hundred and forty… and forty. Yeah.
43 **Salvador:** Oh, you changed the order, too, that time. 8,740?
44 **Nina:** Yeah.
45 **Salvador:** Would it still work if you just changed the order of the 478?
46 **Nina:** Yeah. Because they're the same numbers… you use the same numbers.

By generating novel Happy Numbers, without a need for calculation, Nina demonstrated some implicit knowledge of the mathematical structure at play. She used a zero-digit and combinations of digits to generate novel Happy Numbers. Still, articulating a claim was difficult, but Nina has rehearsed important parts of an argument that could emerge in the revision. Nina made an important shift between empirical and analytical arguments.

Mathematical conferencing could be an important tool for supporting ELLs to reflect, gain clarity, and rehearse revisions. Conferencing is not an end, but an invitation for ELLs to extend their reasoning and an opportunity to revise written work.

IMPLICATIONS FOR FUTURE RESEARCH

In the future, research on mathematical writing should expand to include collaborative forms of writing (such as with shared documents or annotation) so that culturally aligned forms of participation can be invited into classroom practice. One important implication of using mathematical writing to support ELLs, specifically, emerges from what seems to be a focus on writing as an individual activity. Future research should consider how to organize *collaborative* writing activities that allow students to leverage each other's ideas and do so in a way that aligns with their cultural practices and values. This could take the form of students contributing to a larger, shared endeavor or annotating each other's work. Opportunities for collaborative writing align with the tenets for quality learning for ELLs (Walqui & Bunch, 2019) in three ways. First, social interaction supports the development of mathematical understanding and language. Secondly, collaborative writing can focus on disciplinary practices, such as justifying and proving. Finally, collaborative writing can provide opportunities to focus on language forms in "contextual, contingent, and supportive ways" (Walqui & Bunch, 2019, p.23), as was demonstrated in Yolanda's case.

A second implication is that teachers and researchers need learning theories and classroom organizational schemes that can account for and support flexible collaboration. One way to organize collaborative activity could be informed by the Learning by Observing and Pitching In to family and community endeavors (LOPI) framework (Rogoff, 2014). Such a framework might provide insight and a theoretical foundation for organizing activities in which students could contribute to classroom activities in accordance with their own initiative and in a variety of ways to a larger, shared endeavor. Typically, "participation" in mathematics classrooms has been narrowly defined as talk contributed by individuals (Martínez Hinestroza, 2022; O'Connor et al., 2017). Consequently, equity-oriented teachers might view their role as distributing talk turns equally among students on the presumption that more talk will result in more learning (Martínez Hinestroza, 2022). Unfortunately, these beliefs are not supported by research. O'Connor and colleagues (2017) conducted a study that explored vocal versus silent student participation in mathematics discussions and found "no appreciable relationship" between what students learned and what they contributed verbally (p.11). Interestingly, they found that while the presence of mathematical discussions supported better learning in general (when compared to classes with no discussion or classes with initiate-respond-evaluate or I-R-E patterns of talk), the number of words spoken did not improve outcomes at the individual level! O'Connor and colleagues (2017) also mentioned the importance of a "culture of active participation" that echoes the kind of silent participation discussed in Moschkovich (2018). LOPI reflects how a culture of participation is organized. What are some additional ways that students can participate in mathematics classrooms? These might include the use of shared documents; opportunities to write, review, and revise arguments (Moschkovich, 2018); draw, observe, read, write, or work with objects (Martínez Hinestroza, 2022); or participate in ways that are culturally aligned (Civil & Hunter, 2015; Goffney et al., 2018).

Research should also consider *annotation* as a form of written interaction (Kalir & Garcia, 2021). Annotation—a "note added to a text"—is a form of self-expression where students can "create, document, and curate new knowledge" (Kalir & Garcia, 2021, p.7). Kalir and Garcia (2021) present annotation as an everyday activity with five affordances: providing information, sharing commentary, sparking conversation, expressing and contesting power, and aiding learning. These affordances appear to align with some purposes for mathematical writing, such as exploratory writing (when sparking conversation), explanatory writing (when providing information), and possibly argumentative writing (when sharing commentary, especially on a mathematical argument). As an everyday practice that is similar to "comments" found in social media, annotating each other's work is an activity that ELLs might enter with less difficulty.

CONCLUSION

Mathematical writing is an important tool that is underutilized with ELLs. Unproductive beliefs about students that are deficit-based do not allow educators to see the strengths that students bring and deny students, especially ELLs, access to environments with rigorous mathematics. In the case of ELLs, these beliefs can contribute to a "symbolic language border" (Valdés, 2017) that inhibits, or even obstructs, their access to rigorous mathematical activity and teachers' access to the strengths students bring to the classroom.

Mathematical writing can be used effectively to support ELLs. In this chapter, I described how to support students to generate mathematical claims and how to embrace the strengths that they bring to the classroom vis-à-vis their empirical reasoning to revise arguments through mathematical conferencing. Writing is an important modality that complements mathematical discussion. The presence of multiple modalities for participation is crucial for ELLs, who might have difficulty communicating mathematically in a single modality. Moreover, the use of writing can preserve and record the mathematical ideas at play so that students have additional opportunities to examine and refine them.

ACKNOWLEDGMENT

This material is partially based upon work supported by the National Science Foundation (under Award No. DGE-2222234) and the National Academy of Education (NAEd)/Spencer Research Development Award. Any opinions, findings and conclusions or recommendations expressed in this material are those of the author and do not necessarily reflect the views of the National Science Foundation or NAEd/Spencer.

REFERENCES

Aguirre, J. M., & Bunch, G. C. (2012). What's language got to do with it?: Identifying language demands in mathematics instruction for English language learners. In S. Celedón-Pattichis & N. G. Ramirez's (Eds.) Beyond good teaching: Advancing mathematics education for ELLs (pp. 183-194). The National Council of Teachers of Mathematics.

Applebee, A. N., & Langer, J. A. (2011). "EJ" Extra: A Snapshot of Writing Instruction in Middle Schools and High Schools. *English Journal, 100*(6), 14–27.

Bangert-Drowns, R. L., Hurley, M. M., & Wilkinson, B. (2004). The effects of school-based writing-to-learn interventions on academic achievement: A meta-analysis. *Review of Educational Research, 74*(1), 29–58. doi:10.3102/00346543074001029

Barwell, R., Moschkovich, J., & Setati Phakeng, M. (2017). *Language diversity and mathematics: Second language, bilingual and multilingual learners. Compendium for research in mathematics education.* National Council of Teachers of Mathematics.

Bieda, K. N., & Staples, M. (2020). Justification as an equity practice. *Mathematics Teacher: Learning and Teaching PK-12, 113*(2), 102-108.

Blanton, M. L., Brizuela, B. M., Stephens, A., Knuth, E., Isler, I., Gardiner, A. M., & Stylianou, D. (2018). Implementing a framework for early algebra. In C. Kieran, C. Kieran, & Ohmer's (Eds.) Teaching and learning algebraic thinking with 5-to 12-year-olds: The global evolution of an emerging field of research and practice (pp.27-49). Springer. doi:10.1007/978-3-319-68351-5_2

Blanton, M. L., Stylianou, D. A., & David, M. (2009). Understanding instructional scaffolding in classroom discourse on proof. In D.A. Stylianou, M.L. Blanton, & E.J. Knuth's (Eds.) Teaching and learning proof across the grades: A k-16 perspective (pp.290-306). Routledge.

Brenner, M. E. (2006). Interviewing in Educational Research. In J. Green, G. Camilli, & P. Elmore (Eds.), *Complementary Methods for Research in Education* (pp. 357–370). American Educational Research Association/Erlbaum.

Bunch, G. C. (2014). The language of ideas and the language of display: Reconceptualizing "academic language" in linguistically diverse classrooms. *International Multilingual Research Journal, 8*(1), 70–86. doi:10.1080/19313152.2014.852431

Carpenter, T. P., Franke, M. L., & Levi, L. (2003). *Thinking mathematically.* Heinemann.

Casa, T. M., Firmender, J. M., Cahill, J., Cardetti, F., Choppin, J. M., Cohen, J., Cole, S., Colonnese, M. W., Copley, J., DiCicco, M., Dieckmann, J., Dorl, J., Gavin, M. K., Hebert, M. A., Karp, K. S., LaBella, E., & Moschkovich, J. N., Moylan, K. Olinghouse, N. G., & Zawodniak, R. (2016). *Types of and purposes for elementary mathematical writing: Task force recommendations.* Elementary Mathematical Writing Task Force. http://mathwriting.education.uconn.edu

Civil, M., & Hunter, R. (2015). Participation of non-dominant students in argumentation in the mathematics classroom. *Intercultural Education, 26*(4), 296–312. doi:10.1080/14675986.2015.1071755

Cohen, J. A., Miller, H. C., Casa, T. M., & Firmender, J. M. (2015). Characteristics of second graders' mathematical writing. *School Science and Mathematics, 115*(7), 344–355. doi:10.1111sm.12138

de Araujo, Z., Roberts, S. A., Willey, C., & Zahner, W. (2018). English learners in K–12 mathematics education: A review of the literature. *Review of Educational Research, 88*(6), 879–919. doi:10.3102/0034654318798093

de Araujo, Z., & Smith, E. (2022). Examining English language learners' learning needs through the lens of algebra curriculum materials. *Educational Studies in Mathematics, 109*(1), 65–87. doi:10.100710649-021-10081-w PMID:34934250

Dietiker, L., Baldinger, E., Kassarjian, M. and Shreve, B. (2013). *Core Connections 3*. V(5). Elk Grove, CA: CPM Educational Program.

Education Committee of the European Mathematical Society (ECEMS). (2011). Do theorems admit exceptions? Solid findings in mathematics education on empirical proof schemes. *EMS Newsletter*, *82*, 50–53.

Ellis, A. B. (2007). Connections between generalizing and justifying: Students' reasoning with linear relationships. *Journal for Research in Mathematics Education*, *38*(3), 194–229.

Ellis, A. E., Lockwood, E., Williams, C. C. W., Dogan, M. F., & Knuth, E. (2012). Middle school students' example use in conjecture exploration and justification. In L. R. Van Zoest, J. J. Lo, & J. L. Kratky (Eds.), *Proceedings of the 34th Annual Meeting of the North American Chapter of the Psychology of Mathematics Education* (pp. 135–142). Western Michigan University.

Faltis, C. J., & Valdés, G. (2016). Preparing teachers for teaching in and advocating for linguistically diverse classrooms: A vade mecum for teacher educators. In D. Gitomer & C. Bell's (Eds.) Handbook of Research on Teaching (pp. 549-592). American Educational Research Association. doi:10.3102/978-0-935302-48-6_8

Flores, N., & Rosa, J. (2015). Undoing appropriateness: Raciolinguistic ideologies and language diversity in education. *Harvard Educational Review*, *85*(2), 149–171. doi:10.17763/0017-8055.85.2.149

Goffney, I., Gutiérrez, R., & Boston, M. (2018). *Rehumanizing mathematics for Black, Indigenous, and Latinx students*. National Council of Teachers of Mathematics.

Hamburger, L., & Chu, H. (2019). Making slope a less slippery concept for English Learners. In A. Walqui & G. Bunch's (Eds.) Amplifying the curriculum: Designing quality learning opportunities for English Learners (115-137), Teachers College Press.

Harel, G., & Sowder, L. (1998). Students' proof schemes: Results from exploratory studies. American Mathematical Society, 7, 234-283.

Healy, L., & Hoyles, C. (2000). A study of proof conceptions in algebra. *Journal for Research in Mathematics Education*, *31*(4), 396–428. doi:10.2307/749651

Hebert, M. A., & Powell, S. R. (2016). Examining fourth-grade mathematics writing: Features of organization, mathematics vocabulary, and mathematical representations. *Reading and Writing*, *29*(7), 1511–1537. doi:10.100711145-016-9649-5

Hillocks, G. (2006). Middle and high school composition. In P. Smagorinsky's (Ed.). Research on composition: Multiple perspectives on two decades of change (pp.48-77). Teachers College Press.

Huitzilopochtli, S. (2022). *Multilingual Students' Arguments in Early Algebra: Writing, Refining, and Revising through Conferencing*. University of California.

Kalir, R. H., & Garcia, A. (2021). *Annotation*. MIT Press. doi:10.7551/mitpress/12444.001.0001

Kaput, J. J. (2008). What is algebra? What is algebraic reasoning. In J.J. Kaput, D.W. Carraher, & M.L. Blanton's (Eds.) Algebra in the early grades (pp. 5-17), Routledge.

Kieran, C. (2007). Learning and teaching algebra at the middle school through college levels: Building meaning for symbols and their manipulation. In F.K. Lester's (Ed.) Second handbook of research on mathematics teaching and learning (Vol. 2, pp.707-762). NCTM.

Lakatos, I. (1976). *Proofs and refutations: the logic of mathematical discovery*. Cambridge University Press. doi:10.1017/CBO9781139171472

Lampert, M. (1990). When the problem is not the question and the solution is not the answer: Mathematical knowing and teaching. *American Educational Research Journal, 27*(1), 29–63. doi:10.3102/00028312027001029

Leinwand, S., Brahier, D. J., & Huinker, D. (2014). *Principles to actions: Ensuring mathematical success for all*. National Council of Teachers of Mathematics.

Martínez Hinestroza, J. (2022, June). (Re)Learning what it means to participate: Bringing student and teacher perspectives into dialogue. *The Elementary School Journal, 122*(4), 616–641. doi:10.1086/719465

Mathematics Assessment Resource Service (MARS). (2015) *Evaluating Statements: Consecutive Sums*. Mathematics Assessment Project Classroom Challenges: A formative assessment lesson. http://map.mathshell.org

Minnameier, G. (2004). Peirce-suit of truth–why inference to the best explanation and abduction ought not to be confused. *Erkenntnis, 60*(1), 75–105. doi:10.1023/B:ERKE.0000005162.52052.7f

Moschkovich, J. (2013). Principles and guidelines for equitable mathematics teaching practices and materials for English language learners. *Journal of Urban Mathematics Education, 6*(1), 45–57. doi:10.21423/jume-v6i1a204

Moschkovich, J. (2015). Academic literacy in mathematics for English Learners. *The Journal of Mathematical Behavior, 40*, 43–62. doi:10.1016/j.jmathb.2015.01.005

Moschkovich, J. (2018). Considering silence and precision: Communication practices that support equitable mathematical discussions for marginalized students. In R. Hunter's (Ed.) Mathematical discourse that breaks barriers and creates space for marginalized learners (pp. 61-83). Brill.

National Academies of Sciences, Engineering, and Medicine (2017). *Promoting the educational success of children and youth learning English: Promising futures*. National Academies Press.

National Academies of Sciences, Engineering, and Medicine (2018). *English Learners in STEM Subjects: Transforming Classrooms, Schools, and Lives*. The National Academies Press. . doi:10.17226/25182

National Council of Supervisors of Mathematics. TODOS: Mathematics for all (2016). *Mathematics Education Through the Lens of Social Justice: Acknowledgment, Actions, and Accountability*. NCSM. https://www.todos-math.org/socialjustice

National Council of Teachers of Mathematics (NCTM). (1991). *Professional standards for teaching mathematics*. NCTM.

National Governors Association Center for Best Practices & Council of Chief State School Officers. (2010). *Common Core State Standards for Mathematics*. NGACBP.

Nemirovsky, R., Barros, A., Noble, T., Schnepp, M., & Solomon, J. (2005). Learning Mathematics in High School: Symbolic Places and Family Resemblances. In T.A. Romberg, T.P. Carpenter, & F. Dremock's (Eds) Understanding Mathematics and Science Matters (pp.185-206). Routledge.

O'Connor, C., Michaels, S., Chapin, S., & Harbaugh, A. G. (2017). The silent and the vocal: Participation and learning in whole-class discussion. *Learning and Instruction*, *48*, 5–13. doi:10.1016/j.learninstruc.2016.11.003

O'Halloran, K. L. (2015). The language of learning mathematics: A multimodal perspective. *The Journal of Mathematical Behavior*, *40*, 63–74. doi:10.1016/j.jmathb.2014.09.002

Ortmeier-Hooper, C. (2013). *The ELL writer: Moving beyond basics in the secondary classroom*. Teachers College Press.

Powell, S. R., Hebert, M. A., Cohen, J. A., Casa, T. M., & Firmender, J. M. (2017). A synthesis of mathematics writing: Assessments, interventions, and surveys. *Journal of Writing Research*, *8*(3), 493–526. doi:10.17239/jowr-2017.08.03.04

Pugalee, D. K. (2004). A comparison of verbal and writing descriptions of students' problem solving processes. *Educational Studies in Mathematics*, *55*(1), 27–47. doi:10.1023/B:EDUC.0000017666.11367.c7

Rivera, F. D., & Becker, J. R. (2007). Abduction–induction (generalization) processes of elementary majors on figural patterns in algebra. *The Journal of Mathematical Behavior*, *26*(2), 140–155. doi:10.1016/j.jmathb.2007.05.001

Rogoff, B. (2014). Learning by observing and pitching in to family and community endeavors: An orientation. *Human Development*, *57*(2-3), 69–81. doi:10.1159/000356757

Schoenfeld, A. (2009). The soul of mathematics (Series editor's forward). In D.A. Stylianou, M.L. Blanton, & E.J. Knuth's (Eds.) Teaching and learning proof across the grades: A K-16 perspective (pp. xii-xvi). Routledge.

Smith, E. (2008). Representational thinking as a framework for introducing functions in the elementary curriculum (133-160). In J.J. Kaput, D.W. Carraher, & M.L. Blanton's (Eds.) Algebra in the early grades. Routledge.

Smith, M. S., & Stein, M. K. (1998). Reflections on practice: Selecting and creating mathematical tasks: From research to practice. *Mathematics Teaching in the Middle School*, *3*(5), 344–350. doi:10.5951/MTMS.3.5.0344

Stephens, A. C., Ellis, A. B., Blanton, M., & Brizuela, B. M. (2017). Algebraic thinking in the elementary and middle grades. In J. Cai's (Ed.) Compendium for research in mathematics education. National Council of Teachers of Mathematics.

Stylianides, G. J., & Stylianides, A. J. (2009). Facilitating the transition from empirical arguments to proof. *Journal for Research in Mathematics Education*, *40*(3), 314–352. doi:10.5951/jresematheduc.40.3.0314

Stylianou, D., & Blanton, M. (2018). *Teaching with mathematical argument: Strategies for supporting everyday instruction*. Heinemann.

Swain, M. (2009). Languaging, agency and collaboration in advanced second language proficiency. In H. Byrnes' (Ed.), *Advanced language learning: The contribution of Halliday and Vygotsky* (pp. 95–108). Continuum.

Toulmin, S. (2003). *The uses of argument*. Cambridge University Press. doi:10.1017/CBO9780511840005

Valdés, G. (2017). Entry visa denied: The construction of symbolic language borders in educational settings. In O. García, N. Flores, & M. Spotti's. (Eds.) The Oxford handbook of language and society (pp. 321-348). Oxford University Press.

Walqui, A., & Bunch, G. C. (Eds.). (2019). *Amplifying the curriculum: Designing quality learning opportunities for English learners*. Teachers College Press.

Wang, S. H., Lang, N., Bunch, G. C., Basch, S., McHugh, S. R., Huitzilopochtli, S., & Callanan, M. (July 2021). Dismantling Persistent Deficit Narratives About the Language and Literacy of Culturally and Linguistically Minoritized Children and Youth: Counter-Possibilities. *Frontiers in Education* (*Vol. 6*). Frontiers Media SA. doi:10.3389/feduc.2021.641796

ADDITIONAL READING

Goodrich, J. M., Thayer, L., & Leiva, S. (2021). Evaluating Achievement Gaps Between Monolingual and Multilingual Students. *Educational Researcher*, *50*(7), 429–441. doi:10.3102/0013189X21999043

Huitzilopochtli, S., Foxworthy-Gonzalez, J., Moschkovich, J. N., McHugh, S. R., & Callanan, M. A. (2021). Noticing multilingual and non-dominant students' strengths for learning mathematics and science. In A. Essien & A. Msimanga's (Eds.) Multilingual education yearbook 2021: Policy and practice in STEM multilingual contexts (pp.155-173). Springer.

Küchemann, D., & Hoyles, C. (2009). From empirical to structural reasoning in mathematics: Tracking changes over time. In D.A. Stylianou, M.L. Blanton, & E.J. Knuth's (Eds.) Teaching and learning proof across the grades: A k-16 perspective (pp.171-190). Routledge.

Louie, N. L. (2018). Culture and ideology in mathematics teacher noticing. *Educational Studies in Mathematics*, *97*(1), 55–69. doi:10.100710649-017-9775-2

Mejía-Arauz, R., Rogoff, B., & Dexter, A. (2022). Advances in understanding Learning by Observing and Pitching In to family and community endeavours (LOPI). Introduction to the special issue (Avances en la comprensión de Aprender por medio de Observar y Acomedirse en las actividades de la familia y la comunidad (LOPI). Introducción al número especial). *Journal for the study of education and development (infancia y apredizaje)*, *45*(3).

Schifter, D. (2009). Representation-based proof in the elementary grades. In D.A. Stylianou, M.L. Blanton, & E.J. Knuth's (Eds.) Teaching and learning proof across the grades (pp. 71-86). Routledge.

Sepúlveda, E. III. (2011). Toward a pedagogy of acompañamiento: Mexican migrant youth writing from the underside of modernity. *Harvard Educational Review*, *81*(3), 550–573. doi:10.17763/haer.81.3.088mv5t704828u67

Valencia, R. R. (2010). *Dismantling contemporary deficit thinking: Educational thought and practice*. Routledge. doi:10.4324/9780203853214

Watson, A., & Mason, J. (2005). *Mathematics as a constructive activity: Learners generating examples*. Lawrence Earlbaum Associates.

KEY TERMS AND DEFINITIONS

Abductive Reasoning: In the early stages of mathematical inquiry, this form of creative reasoning infers possible claims from available data (Rivera & Becker, 2007; Minnameier, 2004).

Deductive Reasoning: The elaboration of a logically sequenced argument that includes data, warrants, and claims (Rivera & Becker, 2007; Minnameier, 2004).

Inductive Reasoning: Part of mathematical inquiry where one tests the validity of claims (Rivera & Becker, 2007; Minnameier, 2004).

Justifying: A mathematical practice involving the use of informal arguments to construct mathematical certainty about generalizations (Stephens et al, 2017).

Symbolic Language Border: Valdés (2017) defines the symbolic language border as beliefs, practices, and policies that make characteristics of language salient and then uses those characteristics to include (or exclude) individuals as belonging (or not) to desirable (or undesirable) group categories.

Chapter 15
Linking Practices and Standards to Strategies that Enhance Writing for Students With Disabilities

Destini N. Braxton
St. Mary's College of Maryland, USA

Karen L. Terrell
https://orcid.org/0000-0003-1094-5940
Loyola University Maryland, USA

ABSTRACT

Because mathematics and language are interconnected, often students are asked to "explain" their answers or "justify" their choices. It is critical that students are able to verbally express their thoughts, as this is considered an "authentic practice of the discipline." The Standards for Mathematical Practice require that all students be able to make sense of problems, and also communicate both in writing and orally; skills which the communication standard has promoted for over two decades. Moreover, in the case of SWDs, those who are dually identified often show more confidence when asked to write their logic, as opposed to speaking. This chapter, therefore, explores strategies that will assist and allow students to exhibit their acquired knowledge through their writing skills in mathematics and enable the building of their confidence when explaining their logic and reasoning – especially those with diverse learning needs.

INTRODUCTION

In a reflective editorial, Cai and his colleagues (2020) declare, "Maximizing *learning* for every student has long been and continues to be a critical and challenging goal for teachers and researchers" (p. 12). Today's mathematics classroom is riddled with guidelines and principles from various sources such as *Principles and Standards for School Mathematics* (NCTM, 2000), *Common Core State Standards*

DOI: 10.4018/978-1-6684-6538-7.ch015

for Mathematics (NGACBP/CCSSO, 2010), and *Standards for Preparing Teachers of Mathematics* (AMTE, 2017). When we consider our students with disabilities (SWDs), we also incorporate and lean heavily upon the tenets of Universal Design for Learning (UDL; CAST, 2018) for methods of reducing academic barriers that diverse students face in inclusive classroom settings (Root et al., 2022). UDL "does not separate students with disabilities and assume that their fundamental needs and/or learning processes are different from those of students without disabilities. Instead, UDL is focused on learner variability…" (Lambert, 2020, p. 4), and thus, offers a support system not only to SWDs, but also multilingual learners (MLs).

According to UDL (CAST, 2018), there are three networks within the brain that function together and contribute most to learning. The recognition network comprises multiple means of representation that provide options for perception, language and symbols, and comprehension; while the strategic network allows for action, expression and communication, and executive function as means of action and expression. The process of writing – the focus of this chapter – combines these two networks, requiring both the knowledge of words and phrases that comprise language. This method of communication also involves the ability to connect and express terms in meaningful ways that are appropriate for a given context. In addition, because writing also "presents an opportunity for the student to go back and examine his or her mathematical thinking" (Bixby, 2018, p.144), this communicative mode embraces the third UDL network, affective, by providing an option for a student to self-monitor and regulate their progress as a mean of action (CAST, 2018).

Because mathematics and language are interconnected (Lim & Pugalee, 2004), often students are asked to "explain" their answers or "justify" their choices. This, of course, can be done symbolically and visually; however, it is critical that students also are able to verbally express their thoughts via oracy or literacy, as this is considered an "authentic practice of the discipline" (p.2). NCTM (2020) purports, "When students are challenged to think and reason about mathematics and to communicate the results of their thinking to others orally or in writing, they learn to be clear and convincing" (p.60). The Standards for Mathematical Practice require that all students be able to reason abstractly and quantitatively, construct viable arguments, attend to precision, and express regularity (NGACBP/CCSSO, 2010). All of these functions involve students' abilities to communicate both in writing and verbally, skills which the Communication Standard (NCTM, 2000) has promoted for over two decades.

In the case of SWDs, those who are dually identified – classified as both English language learners and students receiving special education services (Lei et al., 2020) – often show more confidence when asked to write their logic, as opposed to speaking. This can be due to students placing higher priority on the linguistic and interactive scaffolds and supports during instruction, rather than the mathematics itself. Lei and colleagues (2020) defined *linguistic scaffolds* as "effective and responsive support for students' language output performance, which requires teachers to use language that is comprehensible to students when providing them with new and more sophisticated knowledge" (p. 127) and *interactive scaffolds* as " a strategic back-and-forth between teachers and students or among students to facilitate comprehension of content and related language use." Thus, it is critical that emphasis be placed on writing in mathematics not just for general education students, but those with diverse needs. After all, "Students who have opportunities, encouragement, and support for speaking, writing, reading, and listening in mathematics classes reap dual benefits: they communicate to learn mathematics, and they learn to communicate mathematically (NCTM, 2000, p.60).

WRITING IN MATHEMATICS CLASSROOMS

The construction of learning opportunities should be thoughtfully designed to connect the teaching of mathematical concepts to the exhibition of students' learning (Cai et al., 2020). This chapter, therefore, explores strategies that will assist and allow students to exhibit their acquired knowledge through their writing skills in mathematics and enable the building of their confidence when explaining their logic and reasoning – especially those with diverse learning needs. We specifically stray away from the typical behaviorist approaches to researching the learning of SWDs in mathematics, and embrace a more constructivist qualitative methodology, focused upon a more narrative design to relay the impact of writing strategies in inclusive mathematical settings (Lambert, 2020). More specifically, we will provide narrative and empirical evidence from the implementation of the "Word Problem Roulette" strategy (Barton & Heidema, 2009) in a sixth-grade inclusive mathematics classroom.

Flores and Brittain (2003) contend, "Writing can help students become active participants in their own learning… [and] allows students to engage in an interaction with the subject or content area" (p.112). The Roulette strategy assists this interaction by requiring participants to not only work collaboratively to solve a problem, but also to relay the steps involved to other classmates so that they can reach the same solution as well (Barton & Heidema, 2009). This process coincides with what we know high-stakes assessments evaluate in students' writing – that they can "extend their written communication skills through both writing about their solutions and critiquing one another's work in class discussions" (Pugalee, 2001, p.298). Further, the Roulette strategy allows students to be formatively assessed and receive peer and teacher feedback prior to any formal summative evaluation (Lim & Pugalee, 2004). Lambert (2020) adds, "When such central investigations are well designed, they offer opportunities for *every child* [emphasis added] in a class to be part of a collective inquiry" (p. 8).

In this chapter, we present artifacts and reflections from the execution of the Roulette strategy, which is set in a context that can cause some mathematics teachers to be more hesitant to attempt such an activity. However, all students, despite learning challenges, deserve engagement with rigorous mathematics given the appropriate accommodations and modifications to enable them to acquire the content and experience success (NGACBP/CCSSO, 2010; NCTM, 2000). In addition, we will clearly articulate the connections of such approaches in mathematics to UDL (CAST, 2018), the Standards of Mathematical Practice (NGACBP/CCSSO, 2010), and even the functional language approach based in systemic functional linguistics (Halliday, 1985) and now encouraged in the most recent language standards (WIDA, 2020). As Pugalee (2001) concludes, "Providing such experiences is pivotal in developing communication processes that promote mathematical literacy for all students" – our ultimate aspiration.

MIDDLE-SCHOOL IMPLEMENTATION

The context in which this activity was implemented was an inclusive ability mathematics classroom, consisting of 50% students with disabilities and 25% English language learners. The racial/ethnic demographic of the course was 75% African American students and 25% Latinx students. The grade level was seventh grade (average age range of 13-14), where students actually learn more eighth-grade mathematics standards (i.e., pre-algebra) in a statewide initiative to have all eighth-grade students (average age range of 14-15) enrolled in Algebra I or Geometry. It is important to note that the teacher is a special education teacher and that this was her first time implementing the Roulette strategy in her mathematics class and used various scaffolds (e.g., sentence strips, timers) and grouping methods (e.g., homogeneous ability groups, heterogeneous ability groups).

Sentence strips were an important scaffold implemented during the Word Problem Roulette routine because it demonstrated to students how detailed and specific the steps should be when written out. It also emphasized that students should be able to read the steps and accurately evaluate the problem, should another individual be asked to solve the given problem and steps. However, the end goal is to promote academic writing in the math classroom and by eventually eliminating the use of sentence strips will allow for the promotion of independent writing and math skill development. The course curriculum for the year had been completed, and the teacher was preparing students for the state's end-of-the-year standardized assessment. Mathematical writing had been a concern and interest for the course throughout the school year; however, the Roulette exercise was the first writing strategy that the teacher was able to employ in the class.

Word Problem Roulette Routine

The Word Problem Roulette routine seems to have surfaced in the 1990s, being applied to secondary mathematics in 1994 by Davis and Gerber as a means of embedding mathematics with reading and writing strategies to elicit students' thinking and assess their understanding (Heidema, 2009). According to the researchers,

- Students are divided into collaborative groups.
- Each group is given a word problem and told to try to solve it verbally.
- After the discussion, students write steps to the solutions in words rather than math symbols.
- Each student should write one sentence and pass the solution sheet to the next student to write the next sentence.
- After the group has written the steps which solve the problem, a group member should read the solution while another group member transcribes the symbolic representation on the chalkboard. (Davis & Gerber, 1994, p.56)

Barton and Heidema (2002) revamp the final step to have a group member serve as "reporter" and read the steps from the solution sheet to another class member. The latter person performs the steps read aloud and obtains a solution, which should be identical to the solution at which the group arrived. According to Heidema (2009), "Students benefit from communicating their own thinking and from hearing how other students think about a problem. They have a chance to try out different ideas and to come to an agreement on a suitable method to solve a problem" (p.6). It is because of both the speaking and writing communicative modes embedded in this routine and its direct connection to the *Sustaining Effort & Persistence* and *Expression & Communication* UDL options (CAST, 2018) that the teacher decided to attempt its use with her dually identified students.

First Implementation

In the initial implementation of the activity, the teacher pre-selected students' work groups based on their learning tiers, making the groups homogeneous. These tiers were as follows:

a) Tier 1: Independent learning for students who have fully mastered and grasped the mathematical concepts and content for the learning standard.
b) Tier 2: Students who have some understanding of the mathematical concepts and content but still need teacher support to reach full mastery of the learning standard.
c) Tier 3: Intensive student support is still needed from the teacher in order to fully grasp and understand the mathematical concepts and content for the learning standard.

Out of a class of 22 students, the majority of the class fell into the Tier 3; whereas there was only one student considered Tier 1 and 2-3 students in the Tier 2 category.

To prepare students for the Roulette task, the teacher briefed the class the day before about the nature of the collaborative work they would be doing in preparation for the state assessment. The goal of this task was not only to assess students' content knowledge in collaborative form, but also familiarize them with the language of the exam. On the day of implementation the teacher gave students explicit directions and expectations for the task. First, students were to sit in predetermined groups, by finding their placements written on the board as they arrived. They were then provided blank sheets of paper (1 per group) and Desmos online calculators on Chromebooks. As the teacher projected a problem on the screen, students were to work collaboratively to generate a written explanation of how to solve the problem. The three problems the class was given were:

1) What is the value of ?

A 64 B 160 C 256 D 1,024

2) If x = 2 and t = 4, what is the value of ?

A 12 B 4 C -72 D -144

3) Dora bought a total of 48 cupcakes. Each cupcake costs $0.55, including tax. Of the cupcakes she bought, 3/8 were vanilla cupcakes. What was the total cost of only the vanilla cupcakes?

A $26.40 B $16.50 C $9.90 D $6.60

The order of the problems was determined by the level of complexity and verbiage – lowest to highest.

During the first problem of the activity, the teacher observed that students were disengaged and refused to participate. Some of the reasons that they reported were, "I do not feel like writing," "I don't know how to do this," "This is too much writing," and "I don't even know what the problem is asking. How will I explain it in this group?" The students seemed unsure of their abilities to solve the problems, although they had previously learned the material and had been working towards mastery of multi-step problems. Students were also very uneasy about communicating their thoughts about mathematics. They used little mathematics vocabulary in their discussions – a practice that the class had been developing over the course of the school year. Thus, during the administration of the first problem, the task switched from collaborative group work to whole-class instruction to reengage students and reiterate the purpose of the process. After completing this example with the class, students were told that the activity will continue next class, i.e., they would work on the second and third problems in groups again. Upon reflection, the teacher determined from this process that more definitive directions needed to be provided in order to promote more self-regulation among the students (CAST, 2018). Also, as the tiered grouping did not assist in fostering collaboration and community as the UDL guidelines promote, she determined that heterogeneous placements may result in more fruitful interactions going forward.

Second Implementation

For the second attempt, students again found their placements on the board and each group was given a blank sheet of paper and the Desmos calculator. Also, the teacher set a time of 15 minutes for each problem (#2 and #3 from day 1) and projected a timer on the SmartBoard to help the class keep track of time as well. According to Cohen and Lotan (2014), "Making a realistic time schedule for each phase (and sticking to it) is an indispensable management tool" (p.84). Further, instruction was given for each group to designate a *scriber*, who would be responsible for writing the steps as the group brainstormed together but was not required to contribute to the brainstorming for the sake of eliminating confusion for the student. The other group members were responsible for actively participating in the group's discussion, explaining the steps and logical reasoning behind each step written down by the scriber. "Roles… contribute to the smooth functioning of groups, thereby allowing the teacher to observe, provide feedback, and push the students' thinking by posing challenging questions" (p.115).

During this implementation of the Word Problem Roulette strategy, the teacher observed more peer teaching and active engagement during the small-group activity. She noticed much more group discussion and that the scribers were often tier 3 students. This produced an unforeseen and undesirable collaboration outcome because it limited the voice of the students who were identified as a tier 3 student. While circulating and assisting the students, the teacher heard an appropriate amount of mathematics vocabulary used orally during the discussion (e.g., *coefficient, variable, multiply, inverse operation*). However, to the teacher's surprise, very little terminology was written down. Most scribers chose to only write down the algebraic expressions instead of the verbiage needed to explain each step. During this iteration, each group read each problem aloud in its entirety – a change from the prior class experience – and students overall seemed more willing to attempt each of the problems. There was no time for the groups to share their results with the class, due to an unexpected fire drill.

Third Implementation

A week and a half later, the teacher implemented a third attempt at incorporating the Roulette strategy. In the previous attempts, she noticed that students often had difficulty expressing their thoughts in writing, and thus, the teacher decided to revise the activity by 1) giving the groups copies of the problems and 2) incorporating an additional scaffold of providing sentence strips. Sentence strips are an increasingly utilized scaffolding tool, interchangeable with sentence frames or starters, integrated in classroom activities to facilitate student learning of text creation and writing standards (e.g., Cioe-Peña, 2022; Echevarría & Graves, 2014; Sunseri & Sunseri, 2019). The strips used in this activity consisted of the verbal expressions of the steps required to evaluate the mathematics problems (see Appendix A) correctly. Students were required to cut and paste the sentence strips in the correct order and talk with their groups about why they chose the order that they did. This accommodation assisted with the mathematical conversation and provided more time for the groups to explain each step to the whole class.

When students saw that their steps matched the presenting group's, they were excited and their confidence increased as they demonstrated their understanding and mathematical reasoning skills. Thus,

Figure 1. Example of students' work for the Roulette writing task

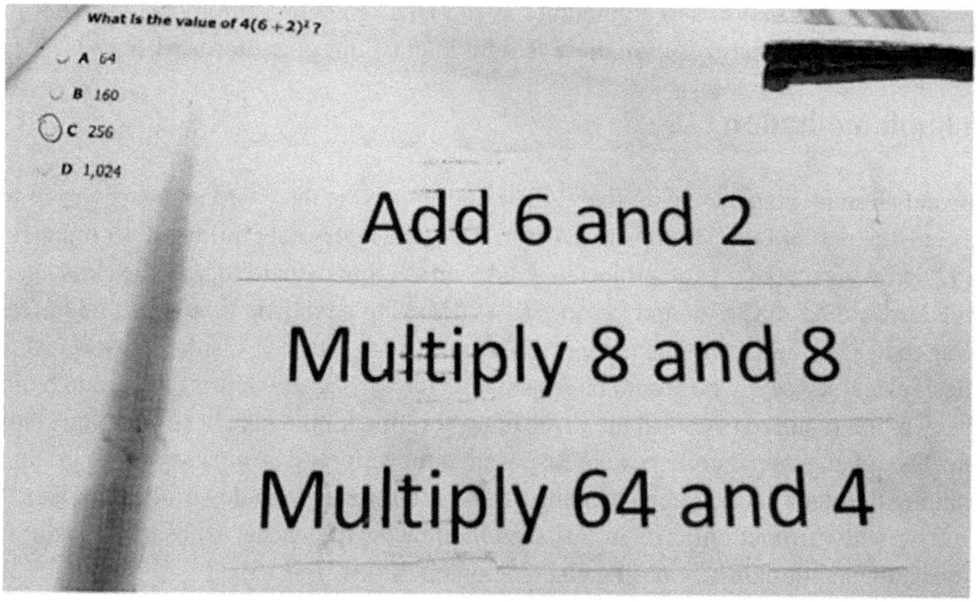

Figure 2. Example of students' work for the Roulette writing task
Note. Students' names were marked out to eliminate student identification.

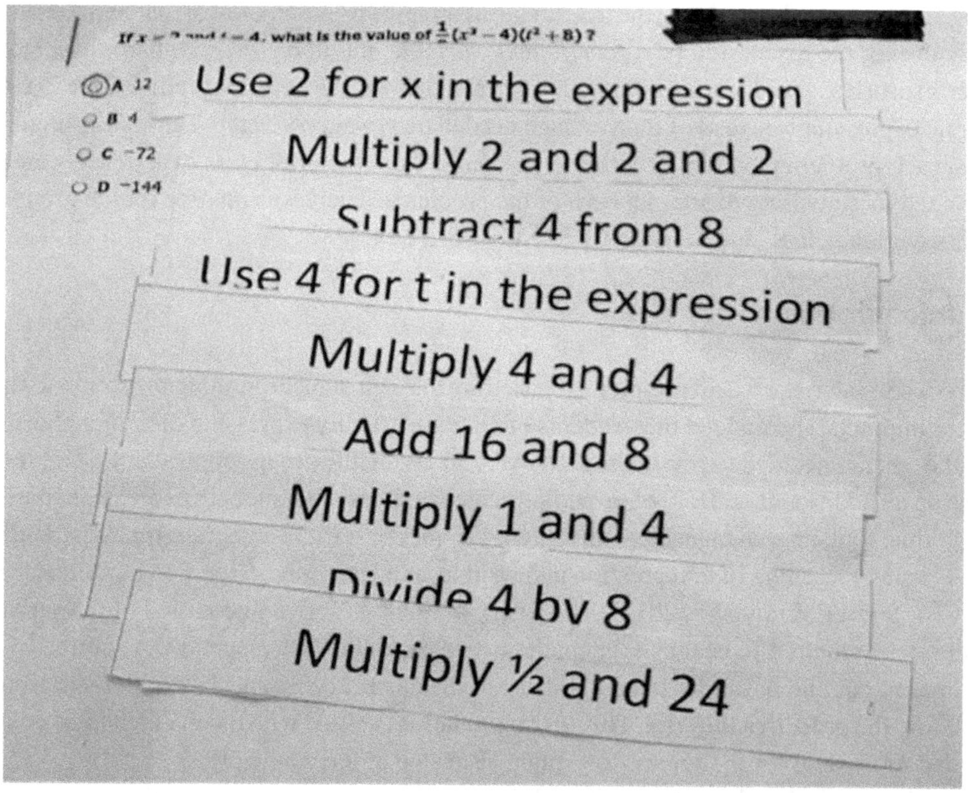

incorporating the sentence strips in this inclusive class setting not only increased students' use of language and writing in math class but also affected their confidence regarding their math ability. Figure 1 displays examples of completed student work during the third implementation of the Roulette task. While the Roulette step of having a reporter read the steps to a student outside of the presenting group did not occur, multiple means of engagement (CAST, 2018) were renewed with the scaffolding of the activity. Students were not as distracted, they worked more collaboratively with one another, and their motivation increased as their choices were confirmed by other groups.

CONNECTIONS TO CURRENT MATHEMATICS METHODS

According to Meyer et al. (2014), there are four attributes for effective goal setting in a UDL classroom: separating the means from the ends, considering all three learning networks, challenging all learners, and actively involving learners. With the primary goals of this task being to relate writing to mathematics, review the school year's course work, and familiarize students with the language of the state exam, employing the Word Problem Roulette strategy offered a means to achieve these goals. Students in this inclusive classroom found the content and the process of writing challenging, but then with appropriate support (e.g. sentence strips), the task actively involved all the learners of the classroom. The collaborative task offered a means of engagement (i.e., the effective network), provided options for comprehension and for language use (i.e., the recognition network), and allowed for communication in all four interpretive and expressive modes (reading, writing, speaking, and listening), as well as kinesthetic or physical action (i.e., the strategic network).

Koestler et al. (2013) purport, "The inclusion of the eight Standards of Mathematical Practice in the *Common Core State Standards for Mathematics* (CCSSM; CCSSI 2010) helps fully represent what it means to *do mathematics*" (p.121). In this activity, students were required to go beyond mere computations and consider the processes involved in completing the problems. Each child had to make sense of each problem and persevere in solving them (SMP #1). This took a few attempts, but with the right scaffolding, students were able to arrive at solutions both in computation and in writing. With the assistance of the Desmos calculator and the sentence strips, the class needed to utilize their tools appropriately (SMP #5). Finally, the Roulette routine called for attention to precision both numerically and linguistically (SMP #6). Engaging students in this task allowed students to not only demonstrate their procedural fluency, but also enhanced their productive disposition (NRC, 2001).

WIDA (2020) states that one of the goals of the modern classroom has to be enabling our students "to become increasingly aware and strategic in their use of language to negotiate meaning and achieve their purposes in various contexts" (p.20) – i.e., the functional approach to language. The Roulette task involved the explanation function (or *genre*), requiring students to clearly state how or why something occurs. This typically invites a sequence or order of events, thus, lending a prime opportunity to incorporate sentence strips with this activity. Visuals can be an integral part of demonstrating explanations, and so having these "pieces" that must be ordered in some way to express how the solutions may be achieved offers a kinesthetic medium through which students can come to understand (and produce) mathematical explanations.

LESSONS LEARNED

Based on the three implementations, the teacher reported that using heterogeneous groups, simple, direct instructions, and scaffolding (e.g., sentence strips, timers) were effective in engaging students and facilitating their small group collaboration, whole group interactions, and learning. The use of sentence strips should initially set the tone for the Word Problem Roulette. As students get more comfortable with the routine, the teacher will begin to gradually remove this scaffold, using sentence frames and other strategies to build students' independence with writing. Many challenges of the activity emerged from a lack of explicit directions. The teacher noted that she would incorporate the Roulette strategy again in the future, making the following alterations to the implementation:

- Setting a time for question completion. The set time allowed for more structure and allowed appropriate allocation of time to promote student engagement and focus. This also requires the teacher to differentiate effectively based on each group's needs.
- Assigning handouts with the questions printed instead of displaying each on the board. This allowed students to have something tangible to refer back to with the problem already provided fully, and allowed groups to move on to the next question instead of at the teacher's discretion.
- Producing a class example before the small group activity. Providing a fully executed example to the class before expecting the class to collaborate with their peers will demonstrate to students the importance of the writing activity, display how the teacher desires for the activity to be completed, and set the tone for the class activity without the teacher's assistance.
- Distributing sentence strips to groups, but requiring students to write the steps out instead of cutting and pasting. This will fully enforce the concept of writing in math and eliminate the time designated for cutting and pasting. Also, only implementing the sentence strip scaffold on the first and second implementation or according to the teacher's best judgment. This is due to the fact that the goal of the activity is to witness students progressing in writing out their math steps when solving various problems and to promote academic independence.
- Including problems with various kinds of answer responses (e.g., multiple choice, short answer). This will allow students to write out their logical reasoning and processing when answer choices are and are not provided to them in the math problem.
- Implementing the Roulette strategy throughout the school year and not only during preparation for the state assessment. This will ensure that the students complete the activity with fidelity and confidence. It will also reemphasize to students the importance of writing in mathematics during various mathematical standards.

Overall, the Roulette writing task is one of many strategies and activities that can better assist mathematics teachers as they begin incorporating writing into the content area. The hope is that this chapter will provide a realistic glance into the challenges and successes of beginning this endeavor. We believe the incorporation of writing strategies in mathematics is crucial as we, as a field, continue on our quest to provide high-quality education to all students.

REFERENCES

Association of Mathematics Teacher Educators. (2017). *Standards for Preparing Teachers of Mathematics*. AMTE. https://amte.net/sites/default/files/SPTM.pdf

Barton, M. L., & Heidema, C. (2002). *Teaching reading in mathematics* (2nd ed.). Mid-continent Research for Education and Learning.

Bixby, M. M. (2018). Effective and efficient use of mathematics writing tasks. *Mathematics Teacher*, *112*(2), 143–146. doi:10.5951/mathteacher.112.2.0143

Cai, J., Morris, A., Hohensee, C., Hwang, S., Robison, V., Cirillo, M., Kramer, S. L., Hiebert, J., & Bakker, A. (2020). Maximizing the quality of learning opportunities for every student. *Journal for Research in Mathematics Education*, *51*(1), 12–25. doi:10.5951/jresematheduc.2019.0005

CAST. (2018). *Universal design for learning guidelines version 2.2*. CAST.

Cioe-Peña, M. (2022, June). TrUDL, a path to full inclusion: The intersectional possibilities of translanguaging and Universal Design for Learning. *TESOL Quarterly*, *56*(2), 799–812. doi:10.1002/tesq.3074

Cohen, E. G., & Lotan, R. A. (2014). *Designing groupwork: Strategies for the heterogeneous classroom* (3rd ed.). Teachers College Press.

Davis, S. J., & Gerber, R. (1994, September). Open to suggestion: Content area strategies in secondary mathematics classrooms. *Journal of Reading*, *38*(1), 55–57.

Echevarría, J. L., & Graves, A. W. (2014). *Sheltered content instruction: Teaching English learners with diverse abilities* (5th ed.). Pearson.

Flores, A., & Brittain, C. (2003, October). Writing to reflect in a mathematics methods course. *Teaching Children Mathematics*, *10*(2), 112–118. doi:10.5951/TCM.10.2.0112

Halliday, M. K. (1985). *An introduction to functional grammar*. Arnold.

Heidema, C. (2009). Reading and writing to learn in Mathematics: Strategies to improve problem solving. *Adolescent Literacy In Perspective*, 2-9.

Lambert, R. (2020, February). *Increasing Access to Universally Designed Mathematics Classrooms*. Policy Analysis for California Education.

Lei, Q., Xin, Y. P., Morita-Mullaney, T., & Tzur, R. (2020). Instructional Scaffolds in Mathematics Instruction for English Learners With Learning Disabilities: An Exploratory Case Study. *Learning Disabilities (Weston, Mass.)*, *18*(1), 123–144.

Lim, L. and Pugalee, D.K. (2004). Using journal writing to explore “They communicate to learn mathematics and they learn to communicate mathematically.” *Ontario Action Researcher*, *7*(2), Article 2.

Meyer, A., Rose, D. H., & Gordon, D. (2014). *Universal Design for Learning: Theory and practice*. CAST, Inc.

National Council of Teachers of Mathematics. (2000). *Principles and standards for school mathematics*. NCTM.

National Governors Association Center for Best Practices & Council of Chief State School Officers. (2010). *Common Core State Standards for Mathematics*. NCTM.

National Research Council. (2001). *Adding it up: Helping children learn mathematics*. National Academy Press.

Pugalee, D. K. (2001, January). Using communication to develop students' mathematical literacy. *Mathematics Teaching in the Middle School*, *6*(5), 296–299. doi:10.5951/MTMS.6.5.0296

Root, J. R., Jimenez, B., & Saunder, A. (2022, February). Leveraging the UDL framework to plan grade-aligned mathematics in inclusive settings. *In Practice*, *1*(1), 13–22.

Sunseri, A. B., & Sunseri, M. A. (2019). The write aid for ELLs: The strategies bilingual student teachers use to help their ELL students write effectively. *The CATESOL Journal*, *31*(1), 117–131.

WIDA. (2020). *WIDA English language development standards framework, 2020 edition: Kindergarten–grade 12*. Board of Regents of the University of Wisconsin System.

Table 1. Sentence strips strips for problem one

Add 6 and 2
Multiply 8 and 8
Multiply 64 and 4

Table 2. Problem two

Use 2 for x in the expression
Multiply 2 and 2 and 2
Subtract 4 from 8
Use 4 for t in the expression
Multiply 4 and 4
Add 16 and 8
Multiply 1 and 4
Divide 4 by 8
Multiply ½ and 24

APPENDIX A

Table 3. Problem three

Multiply 48 and 3
Divide 144 by 8
Multiply 18 and .55

Chapter 16
Constructing Written Arguments for Developing Mathematical Reasoning for Students With Learning Difficulties in the Elementary Grades

Sharlene A. Kiuhara
https://orcid.org/0000-0003-3541-5495
University of Utah, USA

ABSTRACT

Using argument writing to enhance students' mathematics learning has been shown to increase students' content knowledge. Investigating argument writing has also been found to increase the content knowledge and quality of mathematical reasoning of students with learning challenges. This chapter examines the instructional challenges for implementing mathematics writing to students who experience learning difficulties, evidence-based practices to address these students' math and writing needs, and a closer look at how constructing written arguments can enhance fraction knowledge and genre knowledge for students with learning difficulties.

INTRODUCTION

Mathematical reasoning and communication involve a combination of mathematical practices and language ability for successful learning (Ball & Cohen, 1990; Boaler, 2016; National Council of Teachers of Mathematics [NCTM], 2000). To engage students in developing their mathematical reasoning, scholars recommend constructing written arguments and critiquing peers' reasoning to facilitate students' learning (Casa et al., 2016). Meta-analyses of studies conducted in the United States and other countries

DOI: 10.4018/978-1-6684-6538-7.ch016

have strong evidence to suggest that supporting students' mathematics learning by constructing written arguments in academic contexts enhances learning for students who are 6 to 18 years old in primary, secondary, and post-secondary classrooms (Bangert-Drowns et al., 2004; Graham et al., 2020). Although embedding writing activities has shown to be an empirically effective instructional practice to enhance mathematics learning, few empirical studies have investigated the effects of using written arguments for developing mathematical knowledge and reasoning of students with learning difficulties (Hacker et al., 2019; Kiuhara et al., 2020; Kiuhara, Levin, Tolbert, Erickson et al., 2023; Kiuhara, Levin, Tolbert, O'Keeffe et al., 2023).

This chapter will begin with a brief overview of the language and mathematics difficulties students with learning challenges may encounter when engaging in activities that use writing to learn mathematics. Next, a brief overview of the evidence supporting the genre of argumentative writing to develop students' mathematical reasoning will be presented. Finally, to illustrate how argumentative writing can be used to develop students mathematical reasoning, especially for students with learning challenges, an instructional model will be presented that embeds the explicit teaching of both argumentative writing and mathematical reasoning targeted for students with learning challenges who are 9 to 12 years old and represent Grades 4 to 6 in the United States (Hacker et al., 2019; Kiuhara et al., 2020; Kiuhara, Levin, Tolbert, Erickson et al., 2023; Kiuhara, Levin, Tolbert, O'Keeffe et al., 2023). The mathematics content focuses on learning part-whole and partitioning fractions and foundational concepts, such as comparing the magnitude of two fractions between 0 and 1, equivalence, and measure (Charalambous & Pitta-Pantazi, 2007; Wu, 2011), which aligns with the Common Core State Standards for Mathematics (2010) adopted by many states across the United States. Moreover, combining mathematical practices, such as making sense of problems and preserving when solving them, within an evidence-based instructional framework engages students in the processes of constructing written arguments when learning foundational fraction concepts, supports the learning and attention needs of students with learning challenges, develops their strategic competence and productive dispositions toward learning mathematics, and shows promise for developing the quality of their mathematical reasoning and their fraction knowledge (Hacker et al., 2019; Kiuhara et al., 2020; Kiuhara, Levin, Tolbert, Erickson et al., 2023; Kiuhara, Levin, Tolbert, O'Keeffe et al., 2023).

CHARACTERISITCS OF LANGUAGE AND MATHEMATICS DIFFICULTIES

One in 5 children in U. S. classrooms experience learning and attention challenges, and 6% of students receive specialized instruction under their state's rules and regulations for a specific learning disability (Horowitz et al., 2017). However, the majority of 9 to 10-year-old students across the U. S. are failing to meet proficiency standards in mathematics and writing, including students who qualify for special services and accommodations (National Assessment of Educational Progress [NAEP] for Mathematics, 2013, 2015; NAEP for Writing, 2011). Recent mathematics assessment data collected after the pandemic showed a significant learning loss from 2020 to 2022, which places these students even further behind in mathematics (NAEP, 2022). To effectively teach mathematics writing to students with learning challenges, and because writing and mathematics are cognitively demanding tasks for all students, it is essential to address areas that individual students find challenging when learning and to understand the learning supports students need to engage them in mathematics writing.

KEY CHARACTERISTICS OF LEARNING CHALLENGES

Examining key learning characteristics of students who struggle with acquiring the skills and knowledge needed to compose argumentative text may help identify improved ways to enrich students' learning and productive dispositions in mathematics and writing. Children who experience learning challenges often experience one or a combination of cognitive or behavioral challenges that may seem insurmountable when learning mathematics (Geary, 2004; Hallet et al., 2010; Mazzocco et al., 2013). When implementing mathematics writing activities, supporting students' use of language becomes essential. Children use language to make sense and meaning of problems, discuss patterns and misconceptions to justify their solutions, and collaborate and share their solutions with peers. However, students with learning difficulties often exhibit language and vocabulary development delays, which may impede learning (Desoete, 2015; Forsyth & Powel, 2017; Kastberg et al., 2016; Krowka & Fuchs, 2017).

Fraction Learning

Many children find fractions difficult to learn and fail to learn foundational concepts through secondary schooling (Mazzocco et al., 2013), which creates further roadblocks for them when learning advanced mathematics (Cirino et al., 2016; Siegler et al., 2012). Children who lack foundational fraction knowledge exhibit problems understanding that fractions are used to represent two quantities and often make computational errors (Mazzocco et al., 2013). For example, students often apply whole-number logic when adding or subtracting fractions (e.g., $\frac{3}{8}-\frac{2}{4}=\frac{1}{4}$) (Namkung & Fuchs, 2015). Unlike whole numbers, a fraction is represented by two numbers (e.g., $\frac{2}{3}$) that conveys a single numerical value (Bailey et al., 2015; Hallet et al., 2010; Jordan et al., 2013). Understanding fractions requires students to think abstractly by viewing magnitude, as determined by the numerator and denominator, holistically rather than separately (Fuchs et al., 2013; Newton et al., 2014; Namkung & Fuchs, 2015). In addition to understanding these concepts, students must demonstrate procedural knowledge and computational accuracy when performing fraction arithmetic and solving fraction problems involving multiple steps and word problems involving fractions (Bryant, 2008; Geary et al., 2008; NMAP, 2008).

Research-based writing practices are also essential when addressing the language learning needs of students with learning challenges. These include providing systematic instruction (i.e., systematic instruction is intentional instruction that begins with what students know and builds on their current knowledge) and supporting students' use of mathematics language to communicate their understanding of mathematical concepts, such as vocabulary and the difference between using colloquial and mathematical terms (e.g., the top number of a fraction vs. numerator) (Fuchs et al., 2021). Furthermore, using representations to illustrate mathematical concepts and procedures (e.g., fraction blocks and number lines) may help identify relevant vocabulary and address other challenges students may encounter. Other issues include working memory capacity that affects students' ability to parse tasks into meaningful steps, such as solving multistep problems (e.g., 3/6+ 1/3 =4/9; Swanson & Fung, 2016); misinterpreting mathematical concepts and numerical representations (e.g., applying whole number logic when adding or subtracting fractions, Numkung & Fuchs, 2015), and misusing symbolic notation to represent mathematical concepts (e.g., <, =, >; Geary, 2004; Griffin & Jitendra, 2009; Mazzocco et al., 2014; Swanson & Fung, 2016).

Writing Learning

Students with learning challenges may also exhibit difficulties managing the processes involved in writing (e.g., mindful planning of what to write, generating and organizing ideas, and drafting and revising text: Graham & Harris, 2018; Troia, 2006), which may require extended and extensive instructional support (Troia, 2006). Students who exhibit learning challenges rarely engage in advanced planning, demonstrate difficulties with writing mechanics that interfere with the writing process (e.g., spelling, handwriting), and prioritize writing mechanics over reconstructing their ideas after receiving feedback or discussing patterns and inconsistencies in their solutions (Graham, 2018; Harris & Graham, 2018). Students may also exhibit limited background knowledge, genre knowledge, and difficulties processing information, which may impede their ability to understand the mathematics writing activity or task.

Writing research evidence suggests that providing students with systematic and evidence-based instruction improves their writing quality, writing dispositions, and strategic planning and composing processes (Graham & Harris, 2018; Gillespie & Graham, 2012; Graham & Perin, 2007). Graham and colleagues (2018) present four research-based practices to address the writing needs of children in elementary grades: (1) provide a minimum of one hour per day for students to write (i.e., a minimum of 30 min focused on teaching a variety of writing strategies and skills and providing time for students to practice and applying the skills they learned, such as integrating writing tasks in mathematics lessons); (2) implement explicit instruction for each component of the writing process (e.g., generating ideas, planning, drafting, sharing, evaluating, editing, and publishing) for the purpose of persuading and analyzing; and (3) allow class time for students to master basic writing skills, such as handwriting, spelling, sentence construction, typing and word processing); and (4) create a supportive classroom environment in which writing is valued (e.g., plan for regular and structured opportunities for students to collaborate and problem solve during writing and provide opportunities for students to engage with each other about their writing, such as giving and receiving feedback from peers and making judicious decisions to restructure or revise their writing based on the feedback they receive).

Mathematics Learning Using Argument Writing

In a meta-analysis of empirical studies (k = 56), Graham et al. (2020) found that writing enhances students' learning in mathematics (Effect size [ES] = .30). When using writing to learn content material, the authors also identified writing argumentative text with the highest effect size (ES = .42, k = 7) compared to informative text (ES = .34, k = 18), journal writing (ES = .33, k = 18), and using graphic representations (ES = .19, k = 8). Because writing involves genre knowledge, students with learning challenges may have difficulty meeting the rhetorical goals for constructing argumentative text (e.g., identifying the text structure, audience, and purpose for writing) and attending to whole-text organization (Graham & Harris, 2003, 2018).

Using argument writing to facilitate mathematics learning supports the view that the rhetorical structure of argumentative text may deepen students' understanding of the concepts being taught (Klein, 1999; Klein et al., 2017). Constructing written arguments may provide opportunities for students to reason through a mathematics problem by addressing the rhetorical structure of a written argument. When constructing a written argument, students are prompted to present their conceptual understanding of and solution to a problem (i.e., state a claim) and address a counter solution to the problem (i.e., state a counterclaim). A counter solution encourages students to reflect and elaborate on why a different answer may be inaccurate or different from their claim (De La Paz, 2005; Nussbaum & Kardash, 2005; Wiley & Voss, 1999) and provides an opportunity for students with learning challenges to address common misconceptions about the mathematical concepts being taught (Hacker et al., 2019; Kiuhara et al., 2020). The text structure for constructing a written argument justifying an answer to a mathematics problem shows promise for increasing students' understanding of the mathematics concept being taught and increasing the quality of their mathematical reasoning (Cross, 2009; Hacker et al., 2019; Kiuhara et al., 2020; Kiuhara, Levin, Tolbert, Erickson et al., 2023; Kiuhara, Levin, Tolbert, O'Keeffe et al., 2023).

For example, presenting (a) a statement that represents the problem to be solved (e.g., When comparing the magnitude of two fractions, $\frac{1}{4}$ and $\frac{2}{3}$; (b) a stated claim or answer to the mathematics problem (e.g., 2/3 is greater than 1/4; (c) provide evidence to support their conceptual understanding of the problem (e.g., because when I represent these fractions on a number line, 2/3 is closest to one whole); and (d) present a common misconception as a counterclaim with reason to support (e.g., others may argue that 1/4 is greater than 2/3 because 4 is larger than 3) (Kiuhara, Levin, Tolbert, Erickson, et al., 2023). When asked to justify their solution to a mathematics problem, the less proficient writers may write an answer to a problem with reasons to support their answer without addressing other functional elements of an argumentative text, such as presenting a counter-solution or misconception with supporting evidence (Hacker et al., 2019; Kiuhara et al., 2020).

Finally, writing is a self-directed endeavor, which compounds the learning challenges students encounter when learning to write. Students often face challenges with their attitudes toward writing, writing self-efficacy, motivation to see through the writing process, setting self-reflective goals, and self-monitoring their writing progress and affective needs (Graham & Harris, 2018; Graham et al., 2017). For example, once students understand the rhetorical elements for constructing an argument, they can begin self-monitoring their writing to justify their solutions and understand the mathematics concepts taught. The instructional model presented below will illustrate how the rhetorical elements for constructing a written argument can be used for learning fractions.

ARGUMENT WRITING FOR LEARNING FRACTIONS: AN INSTRUCTIONAL MODEL

In response to the poor national achievement outcomes in mathematics and writing for children in the United States, Kiuhara and colleagues developed an argumentative writing strategy called FACT + R^2C^2 (herein referred to as FACT), which uses argument writing to learn foundational concepts of fractions (see Hacker et al., 2019; Kiuhara et al., 2020; Kiuhara, Levin, Tolbert, Erickson et al., 2023; Kiuhara, Levin, Tolbert, O'Keeffe et al., 2023). The FACT strategies were designed using the Self-Regulated Strategy Development (SRSD) instructional framework (see SRSD for FACT described below; for a complete description of the theoretical constructs and application of SRSD, see Harris & Graham, 1996; Harris et al., 2014; Harris et al., 2008).

Each letter of FACT represents steps to use argument writing as a tool to develop students' reasoning and communication of their problem-solving process: **F** = Figure out a plan (e.g., What is my task? Do I understand the problem? What do I need to know? What tools do I need to solve the problem?); **A** = Act on it (What reasons, evidence and support will I use to support my proposed solution? What mathematics and writing words will I choose?); and **C** = Compare my reasoning with a peer's (How is my claim similar or different? What are my reasons? Does it make sense? Can we make improvements?). The **T** = Tie it up in a written argument prompts students to construct a written argument based on their findings: **R** = How should I restate the task?; **R** = What reasons, evidence, and support should I include?; **C** = How do I approach addressing an answer differently from mine?; and **C** = What concluding statement could I provide?

The FACT strategies use the SRSD instructional framework to teach mathematics writing to students with learning difficulties. SRSD was developed is an evidence-based practice with over 100 SRSD efficacy studies establishing its effectiveness across all student abilities for developing students' genre knowledge, language and writing skills, and competencies to become engaged and successful writers (Graham & Harris, 2018). SRSD consists of multiple components incorporating evidence-based writing practices within its instructional framework. The evidence-based practices incorporated within the SRSD framework are also effective in their own right for addressing individual students' learning and attention needs (e.g., giving and receiving feedback from teachers and peers, collaborative problem-solving and writing, self-monitoring, setting specific learning and writing goals, inquiry activities, analyzing good models of writing, and the using process writing approach: Gillespie & Graham, 2014; Graham et al., 2012; Graham et al, 2016; Perin & Graham, 2007). Students' writing development is scaffolded within the writing process through explicit strategies instruction (Harris & Graham, 1996; Harris et al., 2014; Harris et al., 2008). The gradual release of learning responsibility to students allows them to take ownership of their writing and mathematics learning.

SRSD for FACT instruction embeds six stages of SRSD for learning the FACT strategies: (1) develop students' *background and language knowledge and genre knowledge* for using the strategies; (2) collectively *discuss* the FACT strategies to be learned, why the strategies might be helpful for students, and address the learning challenges students may experience when constructing written arguments and learning fractions; (3) *model* how to apply the FACT strategies through a problem solving think aloud process; (4) have students take time to *paraphrase and memorize* the steps of the FACT strategies so the sequence of the strategies make sense for them, (5) continue to *support* students' mathematics writing until (6) students can take full ownership for using the FACT and affective strategies without the support of the teacher.

Although SRSD for FACT instruction provides students with explicit instruction, it is also student-centered. That is, as students learn the skills and strategies of FACT, the teacher's role is to facilitate students' learning while students engage in multiple opportunities to develop and use language through speaking and writing activities (Harris & Graham, 2009; Kiuhara et al., 2020). Throughout instruction, students discuss their learning challenges and learning goals, share and reflect on their learning, and participate in writing activities with the teacher and peers until they are ready to construct written arguments independently (Gillespie & Kiuhara, 2017; McKeown et al., 2016). Teachers monitor students' learning before moving on to the next lesson. This may require teachers to reteach a concept, review content from a prior lesson, or provide multiple opportunities with focused activities for students to practice applying the skills learned.

EVIDENCE SUPPORTING ARGUMENTATIVE WRITING TO TEACH FRACTIONS

Kiuhara and colleagues have conducted a series of efficacy trials to test the effects of FACT on students' fraction learning involving 9- to 12-year-old students with learning challenges (Hacker et al., 2019;

Kiuhara et al., 2020; Kiuhara, Levin, Tolbert, Erickson et al., 2023; Kiuhara, Levin, Tolbert, O'Keeffe et al., 2023). Although the design and results of these studies are reported in detail elsewhere, evidence from the preliminary study by Kiuhara et al., 2020 will be briefly summarized here.

Context and Participants

The initial efficacy cluster-based randomized controlled trial occurred in a suburban school district in the intermountain region of the United States. The first author worked with key district personnel to identify 10 teachers and their students across seven primary grade schools to participate in the study. The teachers provided mathematics instruction to 9- to 12- year-old students with learning difficulties. The students received either supplemental instruction in mathematics or specialized instruction to support their learning due to a learning disability in mathematics. The teachers' classrooms were randomly assigned to the FACT condition (n = 5; 28 students) and the control condition (n = 5; 31 students). Initial tests of equivalence were conducted, and no differences were found at the student level between the treatment and control conditions for mathematics calculation, written expression, grade level, or students receiving specialized instruction in mathematics.

Data Collection

Students in the treatment and control conditions were administered a fraction test and a writing test before FACT instruction began and at the end of the study. The fraction test consisted of 25 multiple-choice questions that assessed the students' understanding of partitioning, equivalence, comparison of fractions for magnitude, and order fractions from least to greatest. For the writing test, students were asked to look at the relationship of two fractions, decide how best to represent the equivalence or magnitude (e.g., draw a number line, use an area model, or jot down notes), and then construct a written argument justifying their answer. Two research assistants typed the students' papers verbatim into a word processing program, and the typed responses were checked for reliability.

Five FACT Lessons

The FACT scope and sequence are presented in Table 1. FACT consisted of five lessons aligned with the teaching standards adopted by several states across the United States for fractions and writing argumentative text (CCSS-Mathematics, 2010; CCSS-Language Arts, 2010).

Lesson One

The first lesson aligns with the SRSD stages of developing students' background knowledge for learning fractions, modeling using mathematics vocabulary, and facilitating discussion to develop students' mathematical reasoning about solving fraction problems. The teacher emphasizes mathematical practices for using multiple representations when solving a problem (e.g., when would a number line, area model. or fraction blocks be helpful?); the importance of attending to precision (e.g., what symbols are important to solve this problem or what mathematical language can you use?); and make sense of problems and persevere when tackling a difficult problem (e.g., how would you describe the problem in your own words or how might you show your thinking or problem-solving?) (Kiuhara, Levin, Tolbert, Erickson, et al., 2023).

The teacher discusses common misconceptions with students when learning fractions and models teaching fraction vocabulary. Teachers begin posing content questions (e.g., Which tool do you prefer using to represent a fraction between 0 and 1 and why did you choose this tool?) and incorporate using multiple representations to illustrate the part-whole relationship (e.g., fraction cubes, area models, and number lines). Students use a notebook to keep track of useful information, such as mathematics vocabulary and intentional goals for learning and writing, and a folder to track their progress and new mathematics vocabulary.

Lesson Two

The second lesson aligns with the discuss, memorize, and modeling stages of SRSD and focuses on introducing FACT's two mathematics writing strategies. The teacher models using the strategies when comparing the magnitude of two fractions and introduces new vocabulary for students to incorporate as they problem-solve. The students analyze model papers for the rhetorical parts of an argument (e.g., stated claims and supporting evidence, counterclaims and supporting evidence, and a concluding statement). The teacher also introduces the concepts of self-monitoring, setting learning goals, and how to address affective behaviors when working through a challenging problem.

Examples of students' statements for learning are shown in Figure 1. These self-statements help students manage the mathematics writing process. The teacher can ask the students to write motivational statements of things they can say to help them persevere when solving difficult problems. For example, students can create their statements to get them started (e.g., "I'm excited to work together with Jose to figure out this fraction problem"); when they feel frustrated or tired (e.g., "I'm hungry, but I'm going to persevere until I arrive at an answer"); when they come across a problem or task that is difficult (e.g., "I don't know how to solve this so I will ask Jessica if she can help me get started"); when they check their written arguments (e.g., "Did I meet my goal for including all the parts of an argument?"). They can also write things to say when working with a peer (e.g., "Does my solution make sense?") and when providing feedback to peers about their writing (e.g., "You did a good job including the correct answer, but can you provide a more precise counterclaim and reason to support your answer?").

When students begin applying the FACT strategies, they self-monitor their progress by tracking argumentative elements and the number of mathematics and writing vocabulary they use in their writing. Figure 2 presents examples of students' self-monitoring for argumentative writing. After constructing a written argument, students analyze their writing for parts of an argument. They underline words and phrases in their writing that represent mathematics vocabulary and transition words or phrases for writing. They begin graphing their progress each time they use FACT. The squares on the rocket represent each step for constructing a written argument so they can evaluate their written argument. The students use their chart to set learning goals for the next time they use FACT. The students in this study collectively decided to monitor their use of mathematics vocabulary and transition words and phrases in their written paragraphs by representing each with a star. The students also drew a planet to note and celebrate when they reached their writing goal. The teacher provided time at the end of each class session for students to reflect on their writing and to set new writing goals (e.g., "Next time I justify my answer, I will include a counterclaim with supporting evidence," or "I will be more precise in communicating my answer by referring to the denominator instead of using 'bottom number."

Lesson Three

In this lesson, teachers continue to discuss and model solving fraction problems using FACT as students build on their learning of magnitude and equivalence and use multiple representations to compare two fractions. Independently or together with peers, the students continue to solve fraction problems using FACT and construct written arguments to justify their answers.

Lesson Four

In Lesson 4, the teacher gradually releases responsibility for using the FACT strategies to students. Depending on the learning needs of the students, they begin using the FACT strategies collaboratively with peers or individually during classroom instruction. Students may continue to discuss their proposed solutions with peers, individually compose their written arguments, and then provide feedback to peers on their written arguments.

Lesson Five

In this lesson, the teacher continues to build on students' fraction knowledge as students begin solving word problems. By this time, students are actively engaged in argumentative writing by applying the FACT strategies, sharing their learning and writing processes, and consulting with peers to critique their written arguments. The teacher monitors students' discussions, guides learning by posing questions about their writing process, and determines which students would benefit from further instructional support.

Findings

At the end of the study, students receiving FACT instruction increased their fraction knowledge on a fraction test (g = .60), their writing quality (g = 1.82), and total words written (g = 1.92). Students' genre knowledge was measured by the number of argumentative elements included in their writing (g = 3.20). Although the students in the FACT conditions outperformed students in the control condition, it is important to note that the students who received FACT instruction and specialized instruction in mathematics (n = 16) outscored students who also had learning difficulties but were not receiving special education services (g = 1.04).

A sixth-grade student's reasoning before and after learning the FACT strategies is presented in Figure 3. After learning the FACT strategies, the student's writing was captured during Lesson 5 without teacher support. This student did not qualify for special education services but was provided supplemental mathematics instruction. The student's pre-FACT writing reveals that she had some knowledge that a fraction represented part of a whole by segmenting a circle into three parts, but her model showed that the segments were not equivalent. In her later writing, the student validated her solution by using three approaches for solving the story problem, including a linear model to demonstrate the concept of equivalent segments. Another example of the pre-and post-test writing of a student is presented in Figure 4. Sarah is an 8-year-old student receiving speech and language services. As noted in the figure on Sarah's pretest, she understood that she needed to provide a reason for her proposed solution. Her response, "I write this answer because the others are wrong. A was right," indicated that Sarah would benefit from instructional support for fraction learning and writing. In her post-test writing, she was able to justify her answer with evidence; however, her concluding statement contradicted her answer. It is difficult to know what her reasoning was for her concluding statement. Because writing is a complex and cognitively challenging endeavor (Troia, 2006), Sarah may be experiencing fatigue or would benefit from rechecking her response.

Discussion

FACT using SRSD is a language-focused mathematics writing intervention with activities for developing the language and text structure of an argument. It provides opportunities for students to give and provide feedback on their reasoning and proposed solutions. FACT using SRSD also provides multiple opportunities for students to self-monitor their learning intentions and reflect on their attitudes and beliefs about writing and mathematics and time to practice writing (Case et al., 1992; Graham et al., 2018; Geary et al., 2008; Woodward et al., 2012). Incorporating evidence-based practices and intentional instructional support is a way to address the learning needs of students who find mathematics and constructing written arguments.

IMPLICATIONS FOR TEACHING AND RESEARCH

Using argumentative writing to support students' mathematics learning allows students to develop their mathematical reasoning and understanding of mathematics. However, children with learning difficulties may need additional instructional academic support in mathematics and writing. Effective mathematics writing instruction for students with learning challenges requires teachers to identify the unique learning needs of students and incorporate evidence-based practices to support students' language development. Although we have empirical evidence for implementing evidence-based practices for mathematics and writing instruction, it is essential to understand the learning and attention needs of students and build on what students know rather than what they are doing incorrectly (Lewis, 2014; Lewis & Fisher, 2016; Lambert & Tan, 2017). For practitioners, mathematics writing is a way to engage students in constructing written arguments collaboratively with peers and extends classroom activities that go beyond procedural tasks, such as memorizing rules and formulas to calculate answers (Bryant & Pedrotty Bryant, 2008; Jonsson et al., 2014; Lewis & Fisher, 2016). Using argumentative writing to support students' mathematics learning provides a way for students to develop their mathematical reasoning and understanding of mathematics. Having students construct written arguments to support their learning provides teachers with further insight into students' current understandings.

Finally, for researchers, developing and investigating ways to use argumentative writing to enhance students' learning and communication about mathematics, especially for students with learning challenges, may allow students to access mathematics learning and participate and feel included in a classroom or school culture. We have ample evidence to support instructional practices that support the mathematics and writing of students with learning challenges (Fuchs et al., 2021; Graham et al., 2018; Woodward et al., 2012). However, we have limited knowledge about the lived experiences of these children participating in academic school learning and the teachers supporting them. In sum, implementing mathematics writing for students with learning challenges, such as FACT, shows promise for enhancing their mathematical knowledge and ability to process information critically and construct new understandings.

REFERENCES

Bailey, D. H., Zhou, X., Zhang, Y., Cui, J., Fuchs, L. S., Jordan, N. C., Gersten, R., & Siegler, R. S. (2015). Development of fraction concepts and procedures in U.S. and Chinese children. *Journal of Experimental Child Psychology, 129*, 68–83. doi:10.1016/j.jecp.2014.08.006 PMID:25268902

Ball, D. L., & Cohen, D. K. (1999). Developing practice, developing practitioners: Toward a practice-based professional education. In G. Sykes & L. Darling-Hammond (Eds.), *Teaching as the learning profession: Handbook of policy and practice* (pp. 2–21). Jossey-Bass.

Bangert-Drowns, R. L., Hurley, M. M., & Wilkinson, B. (2004). The effects of school-based writing-to-learn interventions on academic achievement: A meta-analysis. *Review of Educational Research, 74*(1), 29–58. doi:10.3102/00346543074001029

Boaler, J. (2016). *Mathematical mindsets: Unleashing students' potential through creative math, inspiring messages, and innovative teaching*. Jossey-Bass.

Bryant, B. R., & Pedrotty Bryant, D. (2008). Introduction to the Special Series: Mathematics and Learning Disabilities. *Learning Disability Quarterly, 31*(1), 3–8. doi:10.2307/30035521

Casa, T. M., Firmender, J. M., Cahill, J., Cardetti, F., Choppin, J. M., Cohen, J., . . . Zawodniak, R. (2016). *Types of and purposes for elementary mathematics writing: Task force recommendations.* Retrieved from https://mathwriting.education.uconn.edu/

Case, L. P., Harris, K. R., & Graham, S. (1992). Improving the mathematical problem-solving skills of students with learning disabilities: Self-regulated strategy development. *The Journal of Special Education, 26*(1), 1–19. doi:10.1177/002246699202600101

Charalambous, Y., & Pitta-Pantazi, D. (2007). Drawing on a theoretical model to study students' understanding of fractions. *Educational Studies in Mathematics, 64*(3), 293–316. doi:10.100710649-006-9036-2

Common Core State Standards Initiative. (2010). *Common core state standards for mathematics*. CCSSI. https://learning.ccsso.org/ common-core-state-standards-initiative.

Common Core State Standards Initiative. (2010). *Common core state standards for language arts*. CCSSI. https://learning.ccsso.org/ common-core-state-standards-initiative.

De La Paz, S. (2005). Effects of historical reasoning instruction and writing strategy mastery in culturally and academically diverse middle school classrooms. *Journal of Educational Psychology, 97*(2), 139–156. doi:10.1037/0022-0663.97.2.139

Desoete, A. (2015). Language and math. In P. Aunio, R. Mononen, & A. Laine, Mathematical learning difficulties—snapshots of current European research. LUMAT, 5, 647–674.

Forsyth, S. R., & Powell, S. R. (2017). Differences in the mathematics vocabulary knowledge of fifth-grade students with and without learning difficulties. *Learning Disabilities Research & Practice, 32*(4), 231–245. doi:10.1111/ldrp.12144

Fuchs, L. S., Newman-Gonchar, R., Schumacher, R., Dougherty, B., Bucka, N., Karp, K. S., Woodward, J., & Morgan, S. (2021). *Assisting Students Struggling with Mathematics: Intervention in the Elementary Grades (WWC 2021006)*. Washington, DC: National Center for Education Evaluation and Regional Assistance (NCEE), Institute of Education Sciences, U.S. Department of Education. https://whatworks.ed.gov/

Geary, D. C. (2004). Mathematics and learning disabilities. *Journal of Learning Disabilities*, *37*(1), 4–15. doi:10.1177/00222194040370010201 PMID:15493463

Geary, D. C., Boykin, A. W., Embretson, S., Reyna, V., Siegler, R., Berch, D. B., & Graban, J. (2008). Chapter 4: Report of the task group on learning processes. In U.S. Department of Education (Ed.), *The final report of the National Mathematics Advisory Panel* (pp. 4-xi– 4-221). Washington, DC: U.S. Department of Education. http://www.ed.gov/ about/bdscomm/list/mathpanel/report/learning-processes.pdf

Gillespie Rouse, A., & Kiuhara, S. A. (2017). SRSD in writing and professional development for teachers: Practice and promise for elementary and middle school students with learning disabilities. *Learning Disabilities Research & Practice*, *32*(3), 180–188. doi:10.1111/ldrp.12140

Graham, S. (2018). A writer(s) within community model of writing. In C. Bazerman, V. Berninger, D. Brandt, S. Graham, J. Langer, S. Murphy, P. Matsuda, D. Rowe, & M. Schleppegrell (Eds.), *The lifespan development of writing* (pp. 272–325). National Council of English. doi:10.1080/00461520.2018.1481406

Graham, S. & Bollinger, A. (2018). *Teaching elementary school students to be effective writers: A practice guide*. Institute of Education Sciences, U.S. Department of Education. https://ies.ed.gov/ncee/wwc/Docs/PracticeGuide/WWC_Elem_Writing_PG_Dec182018.pdf

Graham, S., & Harris, K. (2003). Students with learning disabilities and the process of writing: A meta-analysis of SRSD studies. In H. L. Swanson, K. R. Harris, & S. Graham (Eds.), *Handbook of learning disabilities* (pp. 323–344). The Guilford Press.

Graham, S., & Harris, K. R. (2018). An examination of the design principles underlying a self-regulated strategy development study. *Journal of Writing Research*, *10*(2), 139–187. doi:10.17239/jowr-2018.10.02.02

Graham, S., Kiuhara, S. A., Harris, K. R., & Fishman, E. (2017). The relationship among strategic writing behavior, writing motivation, and writing performance with young developing writers. *The Elementary School Journal*, *118*(1), 82–104. doi:10.1086/693009

Graham, S., Kiuhara, S. A., & MacKay, M. (2020). The Effects of Writing on Learning in Science, Social Studies, and Mathematics: A Meta-Analysis. *Review of Educational Research*, *90*(2), 179–226. doi:10.3102/0034654320914744

Graham, S., McKeown, D., Kiuhara, S. A., & Harris, K. R. (2012). A meta-analysis of writing instruction for students in the elementary grades. *Journal of Educational Psychology*, *104*(4), 879–896. doi:10.1037/a0029185

Graham, S., & Perin, D. (2007). *Writing next: Effective strategies to improve writing of adolescents in middle and high schools*. A Carnegie Corporation Report. Alliance for Excellent Education.

Griffin, C. C., & Jitendra, A. K. (2009). Word problem-solving instruction in inclusive third-grade mathematics classrooms. *The Journal of Educational Research*, *102*(3), 187–202. doi:10.3200/JOER.102.3.187-202

Hacker, D. J., Kiuhara, S. A., & Levin, J. R. (2019). A metacognitive intervention for teaching fractions to students with or at-risk for learning disabilities in mathematics. Special issue on metacognition in mathematics education. *ZDM Mathematics Education, 51*(4), 601–612. doi:10.100711858-019-01040-0

Hallet, D., Nunes, T., & Bryant, P. (2010). Individual differences in conceptual and procedural knowledge when learning fractions. *Journal of Educational Psychology, 102*(2), 395–406. doi:10.1037/a0017486

Harris, K. R., & Graham, S. (1996). *Making the writing process work: Strategies for composition and self-regulation* (2nd ed.). Brookline Books.

Harris, K. R., & Graham, S. (2009). Self-regulated strategy development in writing: Premises, evolution, and the future. *Teaching and Learning Writing, 6*, 113–135. doi:10.1348/978185409X422542

Harris, K. R., & Graham, S. (2016). Self-regulated strategy development in writing: Policy implications of an evidence-based practice. *Policy Insights from the Behavioral and Brain Sciences, 3*(1), 77-84. 4. doi:10.1177/2372732215624216

Harris, K. R., & Graham, S. (2018). Self-regulated strategy development: Theoretical bases, critical instructional elements, and future research. In R. Fidalgo, K. R. Harris, & M. Braaksma (Eds.), *Design principles for teaching effective writing: Theoretical and empirical grounded principles* (pp. 119–151). Brill.

Harris, K. R., Graham, S., & Mason, L. H. (2003). Self-regulated strategy development in the classroom: Part of a balanced approach to writing instruction for students with disabilities. *Focus on Exceptional Children, 35*, 1–16.

Harris, K. R., Graham, S., Mason, L. H., & Friedlander, B. (2008). *Powerful writing strategies for all students*. Brooks Publishing.

Harris, K. R., Lane, K. L., Graham, S., Driscoll, S. A., Sandmel, K., Brindle, M., & Schatschneider, C. (2012). Practice-based professional development for self-regulated strategies development in writing: A randomized controlled study. *Journal of Teacher Education, 63*(2), 103–119. doi:10.1177/0022487111429005

Horowitz, S. H., Rawe, J., & Whittaker, M. C. (2017). *The State of Learning Disabilities: Understanding the 1 in 5*. National Center for Learning Disabilities.

Jonsson, B., Norqvist, M., Liljekvist, Y., & Lithner, J. (2014). Learning mathematics through algorithmic and creative reasoning. *The Journal of Mathematical Behavior, 36*, 20–32. doi:10.1016/j.jmathb.2014.08.003

Kastberg, D., Chan, J. Y., & Murray, G. (2016). *Performance of U.S. 15-year-old students in science, reading, and mathematical literacy in an international context: First look at PISA 2015* (NCES 2017–048). Washington, DC: National Center for Education Statistics, U.S. Department of Education. https://nces.ed.gov/pubsearch

Kiuhara, S. A., Gillespie Rouse, A., Dai, T., Witzel, B., Morphy, P., & Unker, B. (2020). Constructing Written Arguments to Develop Fraction Knowledge. *Journal of Educational Psychology, 112*(3), 584–607. doi:10.1037/edu0000391

Kiuhara, S. A., Levin, J. R., Tolbert, M., Erickson, M., & Kruse, K. (2023). Can Argumentative Writing Improve Math Knowledge for Elementary Students with a Mathematics Learning Disability? A Single-Case Classroom Intervention Investigation. In X. Liu, M. Hebert, & R. A. Alves (Eds.), *The Hitchhiker's Guide to Writing Research*. Springer. doi:10.1007/978-3-031-36472-3_11

Kiuhara, S. A., Levin, J. R., Tolbert, M., O'Keeffe, B. V., O'Neill, R. E., & Jameson, M. J. (2023). Teaching argument writing in math class: Challenges and solutions to improve the performance of 4th and 5th graders with disabilities. Special issue on teaching writing. *Reading and Writing*. doi:10.100711145-023-10459-7 PMID:37359030

Klein, P. D. (1999). Reopening inquiry into cognitive processes in writing-to-learn. *Educational Psychology Review*, *11*(3), 203–270. doi:10.1023/A:1021913217147

Krowka, S. K., & Fuchs, L. S. (2017). Cognitive profiles associated with responsiveness to fraction intervention. *Learning Disabilities Research & Practice*, *32*(4), 216–230. doi:10.1111/ldrp.12146 PMID:29123333

Lambert, R., & Tan, P. (2017). Conceptualizations of students with and without disabilities as mathematical problem solvers in educational research: A critical review. *Education Sciences*, *7*(2), 51. doi:10.3390/educsci7020051

Lewis, K. E. (2014). Difference not deficit: Reconceptualizing mathematical learning disabilities. *Journal for Research in Mathematics Education*, *45*(3), 351–396. doi:10.5951/jresematheduc.45.3.0351

Lewis, K. E., & Fisher, M. B. (2016). Taking stock of 40 years of research on mathematical learning disability: Methodological issues and future directions. *Journal for Research in Mathematics Education*, *47*(4), 338–371. doi:10.5951/jresematheduc.47.4.0338

Mazzocco, M. M. M., Myers, G. F., Lewis, K. E., Hanich, L. B., & Murphy, M. M. (2013). Limited knowledge of fraction representations differentiates middle school students with mathematics learning disability (dyscalculia) versus low mathematics achievement. *Journal of Experimental Child Psychology*, *115*(2), 371–387. doi:10.1016/j.jecp.2013.01.005 PMID:23587941

McKeown, D., Brindle, M., Harris, K. R., Graham, S., Collins, A. A., & Brown, M. (2016). Illuminating growth and struggles using mixed methods: Practice-based professional development and coaching for differentiating SRSD instruction in writing. *Reading and Writing*, *29*(6), 1105–1140. doi:10.100711145-016-9627-y

National Assessment of Education Progress. (2022). *Reading and mathematics scores for 9-year-olds decline during pandemic: Results from 2022 NAEP long-term trend assessment*. NAEP. www.nationsreportcard.gov/highlights/ltt/2022/supporting_files/ltt-2022-age9-infographic.pdf

National Center for Education Statistics, Institute of Education Sciences, National Assessment of Educational Progress for Mathematics. (2013). The nation's report card. NCES.

National Center for Education Statistics, Institute of Education Sciences, National Assessment of Educational Progress for Mathematics. (2015). The nation's report card. NCES. https:// nces.ed.gov/nationsreportcard/mathematics/

National Center for Education Statistics, Institute of Education Sciences, National Assessment of Educational Progress for Writing. (2011). *National assessment of educational progress at grades 8 and 12*. NCES. https://nces.ed.gov/nationsreportcard/pdf/main2011/2012470.pdf

National Council of Teachers of Mathematics. (2000). *Principles and standards for school mathematics*. NCTM.

National Mathematics Advisory Panel. (2008). *Foundations for success: The final report of the National Mathematics Advisory Panel*. Washington, DC: U.S. Department of Education. www2.ed.gov/about/bdscomm/list/mathpanel/report/final-report.pdf

Swanson, H. L., & Fung, W. (2016). Working memory components and problem-solving accuracy: Are there multiple pathways? *Journal of Educational Psychology*, *108*(8), 1153–1177. doi:10.1037/edu0000116

Troia, G. A. (2006). Writing instruction for students with learning disabilities. In C. A. MacArthur, S. Graham, & J. Fitzgerald (Eds.), *Handbook of writing research* (pp. 324–336). Guilford Press.

Woodward, J., Beckmann, S., Dirscoll, M., Franke, M., Herzig, P., Jitendra, A., & Ogbuehi, P. (2012). *Improving mathematical problem solving in grades 4 through 8: A practice guide* (NCEE, 2012–4055). Washington, DC: National Center for Education Evaluation and Regional Assistance, Institute of Education Sciences, U.S. Department of Education. http://ies.ed.gov/ncee/wwc/publications_ reviews.aspx#pubsearch/

Compilation of References

Abidin, Z. (2018). Mathematical communication characteristics of pre-service primary school teacher in explaining the area of trapezoid reviewed from school origin. *Journal of Research and Advances in Mathematics Education, 3*(2), 118–126. https://eric.ed.gov/?id=EJ1283918. doi:10.23917/jramathedu.v3i2.6784

Aguirre, J. M., & Bunch, G. C. (2012). What's language got to do with it?: Identifying language demands in mathematics instruction for English language learners. In S. Celedón-Pattichis & N. G. Ramirez's (Eds.) Beyond good teaching: Advancing mathematics education for ELLs (pp. 183-194). The National Council of Teachers of Mathematics.

Aguirre, J. M., & Bunch, G. C. (2012). What's language got to do with it?: Identifying language demands in mathematics instruction for English language learners. In S. Celedón-Pattichis & N. Ramirez (Eds.), *Beyond good teaching: Advancing mathematics education for ELLs* (pp. 183–194). National Council of Teachers of Mathematics.

Albano, G., Swidan, O., & Pierri, A. (2023). A model for analyzing the explanatory writing of undergraduate students when solving mathematical tasks. *International Journal of Mathematical Education in Science and Technology, 54*(2), 180–194. doi:10.1080/0020739X.2021.1949757

Alcock, L., Gilmore, C., & Inglis, M. (2013). Experimental methods in mathematics education research. *Research in Mathematics Education, 15*(2), 97–99. doi:10.1080/14794802.2013.797731

Alexander, R. (2008). *Towards dialogic teaching: Rethinking classroom talk* (4th ed.). Dialogos.

American Educational Research Association (2006). Do the math: Cognitive demand makes a difference. *Research Points, 4*(2).

Andersson, C., & Palm, T. (2018). Reasons for teachers' successful development of a formative assessment practice through professional development–a motivation perspective. *Assessment in Education: Principles, Policy & Practice, 25*(6), 576–597. doi:10.1080/0969594X.2018.1430685

An, H., Shin, S., & Lim, K. (2009). The effects of different instructor facilitation approaches on students' interactions during asynchronous online discussions. *Computers & Education, 53*(3), 749–760. doi:10.1016/j.compedu.2009.04.015

An, S., Tinajero, J., Tillman, D., & Kim, S. J. (2019). Preservice teachers' development of literacy-themed mathematics instruction for early childhood classrooms. *International Journal of Early Childhood, 51*(1), 41–57. doi:10.100713158-019-00232-9

Applebee, A. N., & Langer, J. A. (2011). "EJ" Extra: A Snapshot of Writing Instruction in Middle Schools and High Schools. *English Journal, 100*(6), 14–27.

Arsenault, T. L., Powell, S. R., Hebert, M. A., King, S. G., Lin, X., & Lang, D. (2023). Mathematics-writing profiles for students with mathematics difficulty. *Reading and Writing, 36*(8), 2025–2052. doi:10.100711145-022-10375-2

Association of Mathematics Teacher Educators in Connecticut. (2022). *Equity in mathematics education: A joint position statement for Connecticut.* Associated Teachers of Mathematics in Connecticut, & Connecticut Council of Leaders of Mathematics. https://portal.ct.gov/-/media/SDE/Board/BoardMaterials120722/Position_Statement_on_Joint_Equity_in_Mathematics_Education.pdf

Association of Mathematics Teacher Educators. (2017). *Standards for Preparing Teachers of Mathematics.* AMTE. https://amte.net/sites/default/files/SPTM.pdf

Atasoy, E., & Adnan, B. A. K. İ. (2020). Investigation of students' cognitive learning in mathematics lessons supported with writing activities. *Turkish Journal of Computer and Mathematics Education, 11*(2), 528–583. doi:10.16949/turkbilmat.703648

Australian Curriculum, Assessment and Reporting Authority. (2022a). General Capabilities: Critical and Creative Thinking version 9.0 – About the general capability.

Australian Curriculum, Assessment and Reporting Authority. (2022b). Mathematics F-10 Version 9.0: About the learning area. Retrieved from https://v9.australiancurriculum.edu.au/downloads/learning-areas

Baek, J. M. (2017). Numerical reasoning: Making sense of numbers and operations through multiplication in grades 3-5. In M. T. Battista (Ed.), *Reasoning and sense making in the mathematics classroom: Grades 3-5* (pp. 23–42). National Council of Teachers of Mathematics.

Bailey, D. H., Zhou, X., Zhang, Y., Cui, J., Fuchs, L. S., Jordan, N. C., Gersten, R., & Siegler, R. S. (2015). Development of fraction concepts and procedures in U.S. and Chinese children. *Journal of Experimental Child Psychology, 129*, 68–83. doi:10.1016/j.jecp.2014.08.006 PMID:25268902

Balka, D. S. (1974). Creative ability in mathematics. *The Arithmetic Teacher, 21*(7), 633–636. doi:10.5951/AT.21.7.0633

Ball, D. L. (2000). Bridging practices: Intertwining content and pedagogy in teaching and learning to teach. *Journal of Teacher Education, 51*(3), 241–247. doi:10.1177/0022487100051003013

Ball, D. L., & Cohen, D. K. (1999). Developing practice, developing practitioners: Toward a practice-based professional education. In G. Sykes & L. Darling-Hammond (Eds.), *Teaching as the learning profession: Handbook of policy and practice* (pp. 2–21). Jossey-Bass.

Ball, D. L., Hill, H. C., & Bass, H. (2005). Knowing mathematics for teaching: Who knows mathematics well enough to teach third grade, and how can we decide? *American Educator*, (Fall), 14–46. http://hdl.handle.net/2027.42/65072

Ball, D. L., Thames, M. H., & Phelps, G. (2008). Content knowledge for teaching: What makes it special? *Journal of Teacher Education, 59*(5), 389–407. doi:10.1177/0022487108324554

Bangert-Drowns, R. L., Hurley, M. M., & Wilkinson, B. (2004). The effects of school-based writing-to-learn interventions on academic achievement: A meta-analysis. *Review of Educational Research, 74*(1), 29–58. doi:10.3102/00346543074001029

Barnes, D. (1992). *From communication to curriculum.* Boynton/Cook.

Bartholomew, S. R., & Jones, M. D. (2022). A systematized review of research with Adaptive Comparative Judgment (ACJ) in higher education. *International Journal of Technology and Design Education, 32*(2), 1159–1190. doi:10.100710798-020-09642-6

Barton, M. L., & Heidema, C. (2002). *Teaching reading in mathematics* (2nd ed.). Mid-continent Research for Education and Learning.

Barwell, R. (2005). Ambiguity in the mathematics classroom. *Language and Education, 19*(2), 117–125. doi:10.1080/09500780508668667

Barwell, R. (2018). Writing in mathematics classrooms. In A. L. Bailey, C. A. Maher, & L. C. Wilkinson (Eds.), *Language, literacy and learning in the STEM disciplines: How language counts for English learners* (pp. 101–114). Routledge. doi:10.4324/9781315269610-6

Barwell, R., Leung, C., Morgan, C., & Street, B. (2005). Applied linguistics and mathematics education: More than words and numbers. *Language and Education, 19*(2), 141–146. doi:10.1080/09500780508668670

Barwell, R., Moschkovich, J., & Setati Phakeng, M. (2017). *Language diversity and mathematics: Second language, bilingual and multilingual learners. Compendium for research in mathematics education.* National Council of Teachers of Mathematics.

Baxter, J. A., Woodward, J., & Olson, D. (2005). Writing in mathematics: An alternative form of communication for academically low-achieving students. *Learning Disabilities Research & Practice, 20*(2), 119–135. doi:10.1111/j.1540-5826.2005.00127.x

Baxter, J. A., Woodward, J., Olson, D., & Robyns, J. (2002). Blueprint for writing middle school mathematics. *Mathematics Teaching in the Middle School, 8*(1), 52–56. doi:10.5951/MTMS.8.1.0052

Baxter, J., Woodward, J., & Olson, D. (2001). Effects of reform-based mathematics instruction in five third-grade classrooms. *The Elementary School Journal, 101*(5), 529–547. doi:10.1086/499686

Benjamin, A. (2013). *Math in plain English: Literacy strategies for the mathematics classroom.* Routledge. doi:10.4324/9781315855127

Berkenkotter, C. (1982). Writing and problem solving. In T. Fulwiler & A. Young (Eds.) *Language connection: Writing and reading across the curriculum.* (pp. 33—44). National Council of Teachers of English. https://wac.colostate.edu/books/language_connections/chapter3.pdf

Berube, C. T. (2004). Are standards preventing good teaching? *The Clearing House: A Journal of Educational Strategies, Issues and Ideas, 77*(6), 264–267. doi:10.3200/TCHS.77.6.264-268

Bicer, A., Capraro, R. M., & Capraro, M. M. (2013). Integrating writing into mathematics classroom to increase students' problem solving skills. *International Online Journal of Educational Sciences, 5*(2), 361–369.

Bicer, A., Lee, Y., Perihan, C., Capraro, M. M., & Capraro, R. M. (2020). Considering mathematical creative self-efficacy with problem posing as a measure of mathematical creativity. *Educational Studies in Mathematics, 105*(3), 457–485. doi:10.100710649-020-09995-8

Bicer, A., Perihan, C., & Lee, Y. (2018). The impact of writing practices on students' mathematical attainment. *International Electronic Journal of Mathematics Education, 13*(3), 305–313. doi:10.12973/iejme/3922

Bicknell, B. (1999). The writing of explanation and justification in mathematics: Differences and dilemmas. In J. Truran & K. M. Truran (Eds.), *Proceedings of the 22 Annual Conference of the Mathematics Education Research Group of Australasia* (pp. 75—83). Mathematics Education Research Group of Australasia.

Bieda, K. N., & Staples, M. (2020). Justification as an equity practice. *Mathematics Teacher: Learning and Teaching PK-12, 113*(2), 102-108.

Binary Arts Corporation. (1995). *Visual brainstorms: The smart thinking game.* BAC.

Bixby, M. M. (2018). Effective and efficient use of math writing tasks. *Mathematics Teacher, 112*(2), 143–146. doi:10.5951/mathteacher.112.2.0143

Blachowicz, C., & Fisher, P. (2000). Vocabulary instruction. In M. L. Kamil, P. B. Mosenthal, P. D. Pearson, & R. Barr (Eds.), *Handbook of Reading Research* (Vol. 3, pp. 545–561). Erlbaum.

Blanton, M. L., Brizuela, B. M., Stephens, A., Knuth, E., Isler, I., Gardiner, A. M., & Stylianou, D. (2018). Implementing a framework for early algebra. In C. Kieran, C. Kieran, & Ohmer's (Eds.) Teaching and learning algebraic thinking with 5-to 12-year-olds: The global evolution of an emerging field of research and practice (pp.27-49). Springer. doi:10.1007/978-3-319-68351-5_2

Blanton, M. L., Stylianou, D. A., & David, M. (2009). Understanding instructional scaffolding in classroom discourse on proof. In D.A. Stylianou, M.L. Blanton, & E.J. Knuth's (Eds.) Teaching and learning proof across the grades: A k-16 perspective (pp.290-306). Routledge.

Boaler, J. (2002). The development of disciplinary relationships: Knowledge, practice and identity in mathematics classrooms. *For the Learning of Mathematics, 22*(1), 42–47.

Boaler, J. (2016). *Mathematical mindsets.* Wiley.

Boaler, J. (2016). *Mathematical mindsets: Unleashing students' potential through creative math, inspiring messages, and innovative teaching*. Jossey-Bass.

Bossé, M. J., & Faulconer, J. (2008). Learning and assessing mathematics through reading and writing. *School Science and Mathematics, 108*(1), 8–9. doi:10.1111/j.1949-8594.2008.tb17935.x

Bransford, J., Darling-Hammond, L., & LePage, P. (2005). Introduction. In L. Darling-Hammond, J. & Bransford (Eds.), Preparing teachers for a changing world: What teachers should learn and be able to do (pp. 35–117). Wiley.

Bransford, J. D., Brown, A. L., & Cocking, R. R. (2000). *How people learn.* National Academy Press.

Brenner, M. E. (2006). Interviewing in Educational Research. In J. Green, G. Camilli, & P. Elmore (Eds.), *Complementary Methods for Research in Education* (pp. 357–370). American Educational Research Association/Erlbaum.

Brenner, M. E., Mayer, R. E., Moseley, B., Brar, T., Duran, R., Reed, B. S., & Webb, D. (1997). Learning by understanding: The role of multiple representations in learning algebra. *American Educational Research Journal, 34*(4), 663–689. doi:10.3102/00028312034004663

Brown, A. L., & Palincsar, A. (1982). Inducing strategic learning from texts by means of informed, self-control training. *Technical Report No.* 262.

Brown, T. A. (2015). *Confirmatory factor analysis for applied research* (2nd ed.). The Guilford Press. doi:10.4324/9781315757421

Brozo, W. G., & Crain, S. (2018). Writing in math: A disciplinary literacy approach. *The Clearing House: A Journal of Educational Strategies, Issues and Ideas, 91*(1), 7–13. doi:10.1080/00098655.2017.1342435

Bruner, J. S., & Kenney, H. J. (1965). Representation and mathematics learning. *Monographs of the Society for Research in Child Development, 30*(1), 50–59. doi:10.2307/1165708

Bruun, F., Diaz, J. M., & Dykes, V. J. (2015). The language of mathematics. *Teaching Children Mathematics, 21*(9), 530–536. doi:10.5951/teacchilmath.21.9.0530

Bryant, B. R., & Pedrotty Bryant, D. (2008). Introduction to the Special Series: Mathematics and Learning Disabilities. *Learning Disability Quarterly, 31*(1), 3–8. doi:10.2307/30035521

Bunch, G. C. (2014). The language of ideas and the language of display: Reconceptualizing “academic language” in linguistically diverse classrooms. *International Multilingual Research Journal*, *8*(1), 70–86. doi:10.1080/19313152.2014.852431

Bunch, G. C., & Willett, K. (2013). Writing to mean in middle school: Understanding how second language writers negotiate textually-rich content-area instruction. *Journal of Second Language Writing*, *22*(2), 141–160. doi:10.1016/j.jslw.2013.03.007

Bundock, K., Hawken, L. S., Kiuhara, S. A., O’Keeffe, B. V., O’Neill, R. E., & Cummings, M. B. (2021). Teaching rate of change and problem solving to high school students with high incidence disabilities at tier 3. *Learning Disability Quarterly*, *44*(1), 35–49. doi:10.1177/0731948719887341

Burns, M. (2012). *Math reasoning inventory. Overview: About the assessments.*

Burns, M. (1995). *Writing in math class: A resource for grades 2-8*. Math Solutions.

Burns, M. (2004). Writing in math. *Educational Leadership*, 30–33.

Burton, L., & Morgan, C. (2000). Mathematicians writing. *Journal for Research in Mathematics Education*, *31*(4), 429–453. doi:10.2307/749652

Bywater, J. P., Chiu, J. L., Hong, J., & Sankaranarayanan, V. (2019). The teacher responding tool: Scaffolding the teacher practice of responding to student ideas in mathematics classrooms. *Computers & Education*, *139*, 16–30. doi:10.1016/j.compedu.2019.05.004

Bywater, J. P., Lilly, S., & Chiu, J. L. (2022). Examining technology-supported teacher responding and students’ written mathematical explanations. *Journal of Mathematics Teacher Education*, 1–23.

Cai, J., Morris, A., Hohensee, C., Hwang, S., Robison, V., Cirillo, M., Kramer, S. L., Hiebert, J., & Bakker, A. (2020). Maximizing the quality of learning opportunities for every student. *Journal for Research in Mathematics Education*, *51*(1), 12–25. doi:10.5951/jresematheduc.2019.0005

Caldwell, J. H., Karp, K., & Bay-Williams, J. M. (2011). *Developing essential understanding of addition and subtraction for teaching mathematics in prekindergarten-grade 2*. National Council of Teachers of Mathematics.

Calkins, L. (1983). *Lessons from a child: On the teaching and learning of writing*. Heinemann.

Campbell, K. J., Collis, K. F., & Watson, J. M. (1995). Visual processing during mathematical problem solving. *Educational Studies in Mathematics*, *28*(2), 177–194. doi:10.1007/BF01295792

Campbell, T. G., King, S., & Zelkowski, J. (2021). Comparing middle grade students’ oral and written arguments. *Research in Mathematics Education*, *23*(1), 21–38. doi:10.1080/14794802.2020.1722960

Cantrell, S. C., Burns, L. D., & Callaway, P. (2008). Middle-and high-school content area teachers’ perceptions about literacy teaching and learning. *Literacy Research and Instruction*, *48*(1), 76–94. doi:10.1080/19388070802434899

Carpenter, T. P., Fennema, E., & Franke, M. L. (1996). Cognitively guided instruction: A knowledge base for reform in primary mathematics instruction. *The Elementary School Journal*, *97*(1), 3–20. doi:10.1086/461846

Carpenter, T. P., Franke, M. L., & Levi, L. (2003). *Thinking mathematically*. Heinemann.

Carpenter, T. P., Franke, M. L., & Levi, L. (2003). *Thinking mathematically: Integrating arithmetic and algebra in elementary school*. Heinemann.

Carpenter, T. P., & Moser, J. M. (1984). The acquisition of addition and subtraction concepts in grades one through three. *Journal for Research in Mathematics Education, 15*(3), 179–202. doi:10.2307/748348

Casa, T. M. (2022). *Making things write in maths class*. Presentation to the 2022 South Australian Department for Education's Literacy Summit (online). https://rise.articulate.com/share/L98Q9teAf10nsBwvtTYurzxKiidxvyfn

Casa, T. M., Choppin, J. M., & Moschkovich, J. N. (2020). *Crafting a research agenda for writing to reason mathematically.* Research Gate. https://www.researchgate.net/publication/343319379_Crafting_a_Research_Agenda_for_Writing_to_Reason_Mathematically

Casa, T. M., Evans, K., Firmender, J. M., & Colonnese, M. W. (2017). Why should students write in math class? *Educational Leadership, 74*(5). ASCD. http://www.ascd.org/publications/educational-leadership/feb17/vol74/num05/Why-Should-Students-Write-in-Math-Class¢.aspx

Casa, T. M., Firmender, J. M., Cahill, J., Cardetti, F., Choppin, J. M., Cohen, J., & Zawodniak, R. (2016). *Types of and purposes for elementary mathematical writing: Task force recommendations.* Math Writing. http://mathwriting.eduation.uconn.edu

Casa, T. M., Firmender, J. M., Cahill, J., Cardetti, F., Choppin, J. M., Cohen, J., & Zawodniak, R. (2016). *Types of and purposes for elementary mathematical writing: Task force recommendations.* Math Writing. https://mathwriting.education.uconn.edu/

Casa, T. M., Firmender, J. M., Cahill, J., Cardetti, F., Choppin, J. M., Cohen, J., & Zawodniak, R. (2016). *Types of and purposes for elementary mathematical writing: Task force recommendations.* Mathwriting. http://mathwriting.education.uconn.edu

Casa, T. M., Firmender, J. M., Cahill, J., Cardetti, F., Choppin, J. M., Cohen, J., & Zawodniak, R. (2016). Types of and purposes for elementary mathematical writing: Task force recommendations. University of Connecticut. http://mathwriting.education.uconn.edu

Casa, T. M., Firmender, J. M., Cahill, J., Cardetti, F., Choppin, J. M., Cohen, J., & Zawodniak, R. (2016). *Types of and purposes for primary mathematical writing: Task force recommendations.* Math Writing. http://mathwriting.education.uconn.edu

Casa, T. M., Firmender, J. M., Cahill, J., Cardetti, F., Choppin, J. M., Cohen, J., . . . Zawodniak, R. (2016). *Types of and purposes for elementary mathematics writing: Task force recommendations.* Retrieved from https://mathwriting.education.uconn.edu/

Casa, T. M., Firmender, J. M., Cahill, J., Cardetti, F., Choppin, J. M., Cohen, J., Cole, S., Colonnese, M. W., Copley, J., DiCicco, M., Dieckmann, J., Dori, J., Gavin, M. K., Hebert, M. A., Karp, K. S., LaBella, E., Moschkovich, J. N., Moylan, K., Olinghouse, N. G., & Zawodniak, R. (2016). *Types of and purposes for elementary mathematical writing: Task force recommendations.* https://mathwriting.education.uconn.edu.

Casa, T. M., Firmender, J. M., Cahill, J., Cardetti, F., Choppin, J. M., Cohen, J., Cole, S., Colonnese, M. W., Copley, J., DiCicco, M., Dieckmann, J., Dorl, J., Gavin, M. K., Hebert, M. A., Karp, K. S., LaBella, E., & Moschkovich, J. N., Moylan, K. Olinghouse, N. G., & Zawodniak, R. (2016). *Types of and purposes for elementary mathematical writing: Task force recommendations.* Elementary Mathematical Writing Task Force. http://mathwriting.education.uconn.edu

Casa, T. M., Firmender, J. M., Cahill, J., Cardetti, F., Choppin, J. M., Cohen, J., Cole, S., Colonnese, M. W., Copley, J., DiCicco, M., Dieckmann, J., Dorl, J., Gavin, M. K., Hebert, M. A., Karp, K. S., LaBella, E., Moschkovich, J. N., Moylan, K., Olinghouse, N. G., & Zawodniak, R. (2016). *Types of and purposes for elementary mathematical writing: Task force recommendations.* Math Writing. https://mathwriting.education.uconn.edu/wpcontent/uploads/sites/1454/2016/04/Types_of_and_Purposes_for_Elementary_Mathematical_Writing_for_Web-2.pdf

Casa, T. M., Firmender, J. M., Cahill, J., Cardetti, F., Choppin, J. M., Cohen, J., Cole, S., Colonnesse, M., Copley, J., DiCicco, M., & Dieckmann, J. (2016). *Types of and purposes for elementary mathematical writing: Task force recommendations.* published by the University of Connecticut. http://mathwriting.education.uconn.edu

Casa, T. M., Gilson, C., Bruce-Davis, M. N., Gubbins, E. J., Hayden, S. M., & Canavan, E. J. (2022). Maximizing the potential of mathematical writing prompts. *Mathematics Teacher: Learning and Teaching Prek-12, 115*(8), 538–550.

Casa, T., Gilson, C. M., Bruce-Davis, M. N., Gubbins, E. J., Hayden, S. M., & Canavan, E. J. (2022). Maximizing the potential of mathematical writing prompts. *Mathematics Teacher: Learning & Teaching PK-12, 115*(8), 538–550.

Casa, T. M. (2013). Capturing thinking on the talk frame. *Teaching Children Mathematics, 19*(8), 516–523. doi:10.5951/teacchilmath.19.8.0516

Casa, T. M., Arsenault, T. L., Powell, S. R., & Hebert, M. A. (2023). Assessing and scoring elementary mathematical writing: Research and practice considerations. In K. L. Wright & T. S. Hodges (Eds.), *Assessing disciplinary writing in both research and practice* (pp. 151–187). IGI Global. doi:10.4018/978-1-6684-8262-9.ch006

CasaT. M.CardettiF.ColonneseM. W.FirmenderJ. M. (in preparation)

Casa, T. M., Firmender, J. M., Gavin, M. K., & Carroll, S. R. (2017). The influence of challenging geometry and measurement units on the mathematics achievement of kindergarteners. *Gifted Child Quarterly, 61*(1), 52–72. doi:10.1177/0016986216671806

Casa, T. M., MacSwan, J. R., LaMonica, K. E., Colonnese, M. W., & Firmender, J. M. (2019). An analysis of the amount and characteristics of writing prompts in grade 3 mathematics student books. *School Science and Mathematics, 119*(4), 176–189. doi:10.1111sm.12333

Case, L. P., Harris, K. R., & Graham, S. (1992). Improving the mathematical problem-solving skills of students with learning disabilities: Self-regulated strategy development. *The Journal of Special Education, 26*(1), 1–19. doi:10.1177/002246699202600101

CAST. (2018). *Universal design for learning guidelines version 2.2.* CAST.

Caygill, R., & Zhao, B. (2021). *Progress and achievement and the context of mathematics and statistics learning in New Zealand (English-medium education).* Ministry of Education. https://assets.education.govt.nz/public/Literacy-and-Maths-strategy-development-in-2021/Progress-and-achievement-and-the-context-of-mathematics-and-statistics-learning-in-New-Zealand.pdf

Chan, W. W. L., & Kwan, J. L. Y. (2021). Pathways to word problem solving: The mediating roles of schema construction and mathematical vocabulary. *Contemporary Educational Psychology, 65*, 1–12. doi:10.1016/j.cedpsych.2021.101963

Chapin, S. H., O'Connor, C., & Anderson, N. C. (2009). *Classroom discussions: Using math talk to help students learn.* Math Solutions.

Charalambous, Y., & Pitta-Pantazi, D. (2007). Drawing on a theoretical model to study students' understanding of fractions. *Educational Studies in Mathematics, 64*(3), 293–316. doi:10.100710649-006-9036-2

Charalombous, Y. C., Hill, H. C., & Ball, D. (2011). Prospective teachers' learning to provide instructional explanations: How does it look and what it might take? *Journal of Mathematics Teacher Education, 14*(6), 441–463. doi:10.100710857-011-9182-z

Chi, M. T. H., de Leeuw, N., Chiu, M., & LaVancher, C. (1994). Eliciting self-explanations improves understanding. *Cognitive Science, 18*(3), 439–477. doi:10.1016/0364-0213(94)90016-7

Chiu, J. L., McElhaney, K., Zhang, N., Biswas, G., Fried, R., Basu, S., & Alozie, N. (2019). *A principled approach to NGSS-aligned curriculum development: A pilot study* [Paper presentation]. *NARST Annual International Conference*, Baltimore, MD.

Chow, J. C., & Ekholm, E. (2019). Language domains differentially predict mathematics performance in young children. *Early Childhood Research Quarterly, 46*, 179–186. doi:10.1016/j.ecresq.2018.02.011

Christie, F. (2002). Classroom discourse analysis: A functional perspective. *Continuum.*

Cioe-Peña, M. (2022, June). TrUDL, a path to full inclusion: The intersectional possibilities of translanguaging and Universal Design for Learning. *TESOL Quarterly, 56*(2), 799–812. doi:10.1002/tesq.3074

Civil, M., & Hunter, R. (2015). Participation of non-dominant students in argumentation in the mathematics classroom. *Intercultural Education, 26*(4), 296–312. doi:10.1080/14675986.2015.1071755

Clarke, D. J., Waywood, A., & Stephens, M. (1993). Probing the structure of mathematical writing. *Educational Studies in Mathematics, 25*(3), 235–250. doi:10.1007/BF01273863

Clements, D. H., & Sarama, J. (2004). Learning trajectories in mathematics education. *Mathematical Thinking and Learning, 6*(2), 81–89. doi:10.120715327833mtl0602_1

Cobb, P., Boufi, A., McClain, K., & Whitenack, J. (1997). Reflective discourse and collective reflection. *Journal for Research in Mathematics Education, 28*(3), 258–277. doi:10.2307/749781

Cobb, P., & Hodge, L. L. (2002). A relational perspective on issues of cultural diversity and equity as they play out in the mathematics classroom. *Mathematical Thinking and Learning, 4*(2-3), 249–284. doi:10.1207/S15327833MTL04023_7

Coffey, J. E., Hammer, D., Levin, D. M., & Grant, T. (2011). The missing disciplinary substance of formative assessment. *Journal of Research in Science Teaching, 48*(10), 1109–1136. doi:10.1002/tea.20440

Cohen, E. G., & Lotan, R. A. (2014). *Designing groupwork: Strategies for the heterogeneous classroom* (3rd ed.). Teachers College Press.

Cohen, J. A., Casa, T. M., Miller, H. C., & Firmender, J. M. (2015). Characteristics of second graders' mathematical writing. *School Science and Mathematics, 115*(7), 344–355. doi:10.1111sm.12138

Colonesse, M. W., Amspaugh, C. M., LeMay, S., Evans, K., & Field, K. (2018). Writing in the disciplines: How math fits into the equation. *The Reading Teacher, 72*(3), 379–387. doi:10.1002/trtr.1733

Colonnese, M. W. (2020). The development of instructional guidelines for elementary mathematical writing. *School Science and Mathematics, 120*(3), 129–143. doi:10.1111sm.12391

Colonnese, M. W. (2022). Exploratory writing to support mathematical sense making. *Mathematics Teacher PK-12. Learning and Teaching, 115*(4), 265–271.

Colonnese, M. W., Cardetti, F., & Casa, T. M. (2023). *School Science and Mathematics.*

Colonnese, M. W., & Casto, A. R. (2023). Analysis of pre-service teachers' written mathematical reasoning. *Mathematics Teacher Education and Development*, *25*(1), 23–42.

Colonnese, M. W., Reinke, L. T., & Polly, D. (2022). An analysis of the questions elementary education teacher candidates pose to elicit mathematical thinking. *Action in Teacher Education*, *44*(3), 196–211. doi:10.1080/01626620.2021.2020696

Common Core State Standards Initiative. (2010). *Common core state standards for language arts*. CCSSI. https://learning.ccsso.org/ common-core-state-standards-initiative.

Common Core State Standards Initiative. (2010). *Common core state standards for mathematics*. CCSSI. https://learning.ccsso.org/ common-core-state-standards-initiative.

Common Core State Standards Initiative. (2010). *Common Core State Standards for Mathematics*. Core Standards Initiative. http://www.corestandards.org/assets/CCSSI_Math%20Standards.pdf

Common Core State Standards. (2010). *Common Core State Standards for mathematical practice*. CCSS. http://www.corestandards.org/Math/Practice

Conference Board of the Mathematical Sciences (2012). *The mathematical education of teachers II*. American Mathematical Society.

Confrey, J., Gianopulos, G., McGowan, W., Shah, M., & Belcher, M. (2017). Scaffolding learner-centered curricular coherence using learning maps and diagnostic assessments designed around mathematics learning trajectories. *ZDM Mathematics Education*, *49*(5), 717–734. doi:10.100711858-017-0869-1

ConneC.T. (n.d.). *Connecticut Schools on the Web*. State of Connecticut. https://www.dir.ct.gov/kids/Schools_Libraries/ctschools.asp?ctportalPNavCtr=|27298|#27301

Connolly, P. (1989). Writing and the ecology of learning. In P. Connolly & T. Vilardi (Eds.), *Writing to learn mathematics and science* (pp. 1–14). Teachers College Press.

Correnti, R., Matsumura, L. C., Hamilton, L., & Wang, E. (2013). Assessing students' skills at writing analytically in response to texts. *The Elementary School Journal*, *114*(2), 142–177. doi:10.1086/671936

Craig, T. S. (2016). The role of expository writing in mathematical problem solving, *African Journal of Research in Mathematics. Science and Technology Education*, *20*(1), 57–66. doi:10.1080/10288457.2016.1147803

Cramer, K., & Henry, A. (2002). Using manipulative models to build number sense for addition and fractions. In B. Litwiller & G. Bright (Eds.), *Making sense of fractions, ratios, and proportions* (pp. 41–48). National Council of Teachers of Mathematics.

Cramer, K., & Whitney, S. (2010). Learning rational number concepts and skills in elementary school classrooms. In D. V. Lambdin & F. K. Lester (Eds.), *Teaching and learning mathematics: Translating research for elementary school teachers* (pp. 15–22). National Council of Teachers of Mathematics.

Cramer, K., Wyberg, T., & Leavitt, S. (2008). The role of representations in fraction addition and subtraction. *Mathematics Teaching in the Middle School*, *13*(8), 490–496. doi:10.5951/MTMS.13.8.0490

Crawford, L., Quebec-Fuentes, S., Huscroft-D'Angelo, J., & Lambert, M. (2020, April). *Using latent class analysis to validate a classroom based numerical reasoning tool* [Panel presentation]. American Educational Research Association (AERA), San Francisco, CA, United States.

Crawford, L., Quebec-Fuentes, S., Huscroft-D'Angelo, J., & Higgins, K. N. (2018). Evaluating quantitative reasoning strategies for comparing fractions: A tool for teachers. *Intervention in School and Clinic*, *54*(1), 225–234.

Crespo, S. (2000). Seeing more than right and wrong answers: Prospective teachers' interpretations of students' mathematical work. *Journal of Mathematics Teacher Education, 3*(2), 155–181. doi:10.1023/A:1009999016764

Creswell, J. W., & Clark, V. L. P. (2017). *Designing and conducting mixed methods research.* Sage publications.

Cross, D. I. (2009). Alignment, cohesion, and change: Examining mathematics teachers' belief structures and their influence on instructional practices. *Journal of Mathematics Teacher Education, 12*(5), 324–346. doi:10.100710857-009-9120-5

Cross, D. I. (2009). Creating optimal mathematics learning environments: Combining argumentation and writing to enhance achievement. *International Journal of Science and Mathematics Education, 7*(5), 905–930. doi:10.100710763-008-9144-9

Dacey, L. & Polly, D. (2012). The *Common Core State Standards for Mathematics*: The big picture. Invited article in *Teaching Children Mathematics, 18*(6), 378-383.

Dacey, L. S., Salemi, R. E., & O'Connell, K. H. (2018). *Why write in math class? K-5.* Stenhouse Publishers.

Danielson, C. (2010). Writing papers in math class: A tool for encouraging mathematical exploration by preservice primary teachers. *School Science and Mathematics, 110*(8), 374–381. doi:10.1111/j.1949-8594.2010.00049.x

Davies, B., Alcock, L., & Jones, I. (2020). Comparative judgement, proof summaries and proof comprehension. *Educational Studies in Mathematics, 105*(2), 181–197. doi:10.100710649-020-09984-x

Davis, E. A., Zembal-Saul, C., & Kademian, S. M. (2019). Working towards a vision of sensemaking in elementary science. In E. A. Davis, C. Zembal-Saul, & S. M. Kademian (Eds.), Sensemaking in elementary science: Supporting teacher learning (pp. 1–11). Routledge.

Davis, B. (1997). Listening for differences: An evolving conception of mathematics teaching. *Journal for Research in Mathematics Education, 28*(3), 355–376. doi:10.2307/749785

Davis, S. J., & Gerber, R. (1994, September). Open to suggestion: Content area strategies in secondary mathematics classrooms. *Journal of Reading, 38*(1), 55–57.

de Araujo, Z., Roberts, S. A., Willey, C., & Zahner, W. (2018). English learners in K–12 mathematics education: A review of the literature. *Review of Educational Research, 88*(6), 879–919. doi:10.3102/0034654318798093

de Araujo, Z., & Smith, E. (2022). Examining English language learners' learning needs through the lens of algebra curriculum materials. *Educational Studies in Mathematics, 109*(1), 65–87. doi:10.100710649-021-10081-w PMID:34934250

De La Paz, S. (2005). Effects of historical reasoning instruction and writing strategy mastery in culturally and academically diverse middle school classrooms. *Journal of Educational Psychology, 97*(2), 139–156. doi:10.1037/0022-0663.97.2.139

Deatline-Buchman, A., & Jitendra, A. K. (2006). Enhancing argumentative essay writing of fourth-grade students with learning disabilities. *Learning Disability Quarterly, 29*(1), 39–54. doi:10.2307/30035531

Department for Education South Australia. (2022a). *Episode 4: Learning to write, writing to learn.* Department of Education South Australia. https://www.education.sa.gov.au/schools-and-educators/curriculum-and-teaching/teach-podcast/podcasts/episode-4-learning-to-write-writing-to-learn

Derewianka, B., & Jones, P. (2016). *Teaching language in context* (2nd ed.). Oxford University Press.

Desoete, A. (2015). Language and math. In P. Aunio, R. Mononen, & A. Laine, Mathematical learning difficulties—snapshots of current European research. LUMAT, 5, 647–674.

DeWolfe, M., & Vosniadou, S. (2015). The representation of fraction magnitudes and the whole number bias reconsidered. *Learning and Instruction, 37*, 39–49. doi:10.1016/j.learninstruc.2014.07.002

Dietiker, L., Baldinger, E., Kassarjian, M. and Shreve, B. (2013). *Core Connections 3*. V(5). Elk Grove, CA: CPM Educational Program.

Dougherty, B. J. (1996). The write way: A look at journal writing in first-year algebra. *Mathematics Teacher, 89*(7), 556–560. doi:10.5951/MT.89.7.0556

Duke, N. K. (2000). For the rich it's richer: Print experiences and environments offered to children in very low- and very high-socioeconomic first-grade classrooms. *American Educational Research Journal, 37*(2), 441–478. doi:10.3102/00028312037002441

Dyer, E. B., & Sherin, M. G. (2016). Instructional reasoning about interpretations of student thinking that supports responsive teaching in secondary mathematics. *ZDM Mathematics Education, 48*(1-2), 69–82. doi:10.100711858-015-0740-1

Echevarría, J. L., & Graves, A. W. (2014). *Sheltered content instruction: Teaching English learners with diverse abilities* (5th ed.). Pearson.

Ediger, M. (2006). Writing in the mathematics curriculum. *Journal of Instructional Psychology, 33*(2), 120–123.

Education Committee of the European Mathematical Society (ECEMS). (2011). Do theorems admit exceptions? Solid findings in mathematics education on empirical proof schemes. *EMS Newsletter, 82*, 50–53.

Ellis, A. B. (2007). Connections between generalizing and justifying: Students' reasoning with linear relationships. *Journal for Research in Mathematics Education, 38*(3), 194–229.

Ellis, A. E., Lockwood, E., Williams, C. C. W., Dogan, M. F., & Knuth, E. (2012). Middle school students' example use in conjecture exploration and justification. In L. R. Van Zoest, J. J. Lo, & J. L. Kratky (Eds.), *Proceedings of the 34th Annual Meeting of the North American Chapter of the Psychology of Mathematics Education* (pp. 135–142). Western Michigan University.

Engelhard, G. Jr. (1992). The measurement of writing ability with a Many-Faceted Rasch Model. *Applied Measurement in Education, 5*(3), 171–191. doi:10.120715324818ame0503_1

Erbilgin, E. (2019). Two mathematics teacher educators' efforts to improve teaching and learning processes: An action research study. *Teaching and Teacher Education, 78*, 28–38. doi:10.1016/j.tate.2018.11.005

Erickson, F. (1986). *Teachers' practical ways of seeing and making sense: A final report*. Office of Educational Research and Improvement.

Espinas, D. R., & Fuchs, L. S. (2022). The effects of language instruction on math development. *Child Development Perspectives, 16*(2), 69–75. doi:10.1111/cdep.12444 PMID:35493789

Evans, S., & Swan, M. (2014). Developing students' strategies for problem solving in mathematics: The role of pre-designed "Sample Student Work". *Educational Designer, 2*(7), 1–31. https://www.educationaldesigner.org/ed/volume2/issue7/article25/

Evens, H., & Houssart, J. (2004). Categorizing pupils' written answers to a mathematics test question: 'I know but I can't explain.'. *Educational Research, 46*(3), 269–282. doi:10.1080/0013188042000277331

Evi-Colombo, A., Cattaneo, A., & Bétrancourt, M. (2020). Technical and pedagogical affordances of video annotation: A literature review. *Journal of Educational Multimedia and Hypermedia, 29*(3), 193–226.

Evidence for Learning. (2020). *Improving mathematics in upper primary and lower secondary*. Evidence for Learning.

Falchikov, N., & Goldfinch, J. (2000). Student peer assessment in higher education: A meta-analysis comparing peer and teacher marks. *Review of Educational Research, 70*(3), 287–322. doi:10.3102/00346543070003287

Faltis, C. J., & Valdés, G. (2016). Preparing teachers for teaching in and advocating for linguistically diverse classrooms: A vade mecum for teacher educators. In D. Gitomer & C. Bell's (Eds.) Handbook of Research on Teaching (pp. 549-592). American Educational Research Association. doi:10.3102/978-0-935302-48-6_8

Fang, Z. (2012a). Language correlates of disciplinary literacy. *Topics in Language Disorders*, *32*(1), 19–34. doi:10.1097/TLD.0b013e31824501de

Fang, Z. (2012b). The challenges of reading disciplinary texts. In T. L. Jetton & C. Shanahan (Eds.), *Adolescent literacy in the academic disciplines: General principles and practical strategies* (pp. 34–68). Guilford Publications.

Fennell, F., & Karp, K. (2017). Fraction sense: Foundational understandings. *Journal of Learning Disabilities*, *50*(6), 648–650. doi:10.1177/0022219416662030 PMID:27506552

Fernández, C., Llinares, S., & Valls, J. (2012). Learning to notice students' mathematical thinking through on-line discussions. *ZDM Mathematics Education*, *44*(6), 747–759. doi:10.100711858-012-0425-y

Firmender, J. M., Casa, T. M., & Colonnese, M. W. (2017). Write on: Reasoning through mathematical writing. *Teaching Children Mathematics*, *24*(2), 84–92. doi:10.5951/teacchilmath.24.2.0084

Firmender, J. N., Dilley, A., Amspaugh, C., Field, K., LeMay, S., & Casa, T. M. (2017). Beyond *doing* mathematics: Engaging talented students in mathematically creative writing. *Gifted Child Today*, *40*(4), 205–211. doi:10.1177/1076217517722180

Fitzgerald, L., Hunter, J., & Hunter, R. (2021). Shifting teacher practices in relation to grouping: Gap gazing or strengths focused approaches. *Mathematics Teacher Education and Development*, *23*(3), 97-110. https://eric.ed.gov/?id=EJ1321057

Flores, A., & Brittain, C. (2003, October). Writing to reflect in a mathematics methods course. *Teaching Children Mathematics*, *10*(2), 112–118. doi:10.5951/TCM.10.2.0112

Flores, N., & Rosa, J. (2015). Undoing appropriateness: Raciolinguistic ideologies and language diversity in education. *Harvard Educational Review*, *85*(2), 149–171. doi:10.17763/0017-8055.85.2.149

Forman, E. (1996). Learning mathematics as participation in classroom practice: Implications of sociocultural theory for educational reform. In L. Steffe, P. Nesher, P. Cobb, G. Goldin, & B. Greer (Eds.), *Theories of mathematical learning* (pp. 115–130). Lawrence Erlbaum Associates.

Forsyth, S. R., & Powell, S. R. (2017). Differences in the mathematics-vocabulary knowledge of fifth-grade students with and without learning difficulties. *Learning Disabilities Research & Practice*, *32*(4), 231–245. doi:10.1111/ldrp.12144

Freeman, B., Higgins, K. N., & Horney, M. (2016). How students communicate mathematical ideas: An examination of multimodal writing using digital technologies. *Contemporary Educational Technology*, *7*(4), 281–313. doi:10.30935/cedtech/6178

Fuchs, L. S., Newman-Gonchar, R., Schumacher, R., Dougherty, B., Bucka, N., Karp, K. S., Woodward, J., & Morgan, S. (2021). *Assisting Students Struggling with Mathematics: Intervention in the Elementary Grades (WWC 2021006).* Washington, DC: National Center for Education Evaluation and Regional Assistance (NCEE), Institute of Education Sciences, U.S. Department of Education. https://whatworks.ed.gov/

Fuchs, L. S., Fuchs, D., Compton, D. L., Powell, S. R., Seethaler, P. M., Capizzi, A. M., Schatschneider, C., & Fletcher, J. M. (2006). The cognitive correlates of third-grade skill in arithmetic, algorithmic computation, and arithmetic word problems. *Journal of Educational Psychology*, *98*(1), 29–43. doi:10.1037/0022-0663.98.1.29

Fuchs, L. S., Seethaler, P. M., Sterba, S. K., Craddock, C., Fuchs, D., Compton, D. L., Geary, D. C., & Changas, P. (2021). Closing the word-problem achievement gap in first grade: Schema-based word-problem intervention with embedded language comprehension instruction. *Journal of Educational Psychology, 113*(1), 86–103. doi:10.1037/edu0000467 PMID:33776137

Fukawa-Connelly, T., & Buck, S. (2010). Using portfolio assessments to assess students' mathematical thinking. *Mathematics Teacher, 103*(9), 649–654. doi:10.5951/MT.103.9.0649

Fuson, K. C. (1988). *Children's counting and concepts of number.* Springer-Verlag. doi:10.1007/978-1-4612-3754-9

Gaudin, C., & Chaliès, S. (2015). Video viewing in teacher education and professional development: A literature review. *Educational Research Review, 16*, 41–67. doi:10.1016/j.edurev.2015.06.001

Gavin, M. K., Casa, T. M., Adelson, J. L., Carroll, S. R., & Sheffield, L. J. (2009). The impact of advanced curriculum on the achievement of mathematically promising elementary students. *Gifted Child Quarterly, 53*(3), 188–202. doi:10.1177/0016986209334964

Gavin, M. K., Casa, T. M., Adelson, J. L., & Firmender, J. M. (2013). The impact of advanced geometry and measurement units on the achievement of grade 2 students. *Journal for Research in Mathematics Education, 3*(44), 478–509. doi:10.5951/jresematheduc.44.3.0478

Gavin, M. K., Casa, T. M., Firmender, J. M., & Carroll, S. R. (2013). The impact of advanced geometry and measurement curriculum units on the mathematics achievement of first-grade students. *Gifted Child Quarterly, 57*(2), 71–84. doi:10.1177/0016986213479564

Geary, D. C., Boykin, A. W., Embretson, S., Reyna, V., Siegler, R., Berch, D. B., & Graban, J. (2008). Chapter 4: Report of the task group on learning processes. In U.S. Department of Education (Ed.), *The final report of the National Mathematics Advisory Panel* (pp. 4-xi– 4-221). Washington, DC: U.S. Department of Education. http://www.ed.gov/ about/ bdscomm/list/mathpanel/report/learning-processes.pdf

Geary, D. C. (2004). Mathematics and learning disabilities. *Journal of Learning Disabilities, 37*(1), 4–15. doi:10.1177 /00222194040370010201 PMID:15493463

Gebhard, M. (2019). *Teaching and researching ELLs' disciplinary literacies: Systemic functional linguistics in action in the context of U.S. school reform.* Routledge. doi:10.4324/9781315108391

Geller, E. H., Son, J. Y., & Stigler, J. W. (2017). Conceptual explanations and understanding fraction comparisons. *Learning and Instruction, 52*, 122–129. doi:10.1016/j.learninstruc.2017.05.006

Gerard, L. F., Spitulnik, M., & Linn, M. C. (2010). Teacher use of evidence to customize inquiry science instruction. *Journal of Research in Science Teaching, 47*(9), 1037–1063. doi:10.1002/tea.20367

Gerard, L., Bradford, A., & Linn, M. C. (2022). Supporting teachers to customize curriculum for self-directed learning. *Journal of Science Education and Technology, 31*(5), 660–679. doi:10.100710956-022-09985-w

Gillespie Rouse, A., & Kiuhara, S. A. (2017). SRSD in writing and professional development for teachers: Practice and promise for elementary and middle school students with learning disabilities. *Learning Disabilities Research & Practice, 32*(3), 180–188. doi:10.1111/ldrp.12140

Goffney, I., Gutiérrez, R., & Boston, M. (2018). *Rehumanizing mathematics for Black, Indigenous, and Latinx students.* National Council of Teachers of Mathematics.

Goldsby, D. S., & Cozza, B. (2002). Writing samples to understand mathematical thinking. *Mathematics Teaching in the Middle School, 7*(9), 517–520. doi:10.5951/MTMS.7.9.0517

Goldsmith, L. T., & Seago, N. (2011). Using classroom artifacts to focus teachers' noticing. In M. G. Sherin, V. R. Jacobs, & R. A. Philipp (Eds.), *Mathematics teacher noticing: Seeing through teachers' eyes* (pp. 169–187). Routledge.

Goodwin, C. (1994). Professional vision. *American Anthropologist, 96*(3), 606–633. doi:10.1525/aa.1994.96.3.02a00100

Graham, S. & Bollinger, A. (2018). *Teaching elementary school students to be effective writers: A practice guide.* Institute of Education Sciences, U.S. Department of Education. https://ies.ed.gov/ncee/wwc/Docs/PracticeGuide/WWC_Elem_Writing_PG_Dec182018.pdf

Graham, S., & Perin, D. (2007). *Writing next: Effective strategies to improve writing of adolescent middle and high school—A report to Carnegie Corporation of New York. Alliance for Excellent Education.* Carnegie Corporation of New York. https://production-carnegie.s3.amazonaws.com/filer_public/3c/f5/3cf58727-34f4-4140-a014-723a00ac56f7/ccny_report_2007_writing.pdf

Graham, S., & Perin, D. (2007). *Writing next: Effective strategies to improve writing of adolescents in middle and high schools.* A Carnegie Corporation Report. Alliance for Excellent Education.

Graham, S., Bollinger, A., Booth Olson, C., D'Aoust, C., MacArthur, C., McCutchen, D., & Olinghouse, N. (2012). *Teaching elementary school students to be effective writers: A practice guide* (NCEE 2012-4058). National Center for Education Evaluation and Regional Assistance, Institute of Education Sciences, U.S. Department of Education. http://ies.ed.gov/ncee/wwc/publications_reviews.aspx#pubsearch

Graham, S. (2018). A revised writer(s)-within-community model. *Educational Psychologist, 53*(4), 258–279. doi:10.1080/00461520.2018.1481406

Graham, S. (2019). Changing how writing is taught. *Review of Research in Education, 43*(1), 277–303. doi:10.3102/0091732X18821125

Graham, S., Gillespie, A., & McKeown, D. (2013). Writing: Importance, development, and instruction. *Reading and Writing, 26*(1), 1–15. doi:10.100711145-012-9395-2

Graham, S., & Harris, K. (2003). Students with learning disabilities and the process of writing: A meta-analysis of SRSD studies. In H. L. Swanson, K. R. Harris, & S. Graham (Eds.), *Handbook of learning disabilities* (pp. 323–344). The Guilford Press.

Graham, S., & Harris, K. R. (2018). An examination of the design principles underlying a self-regulated strategy development study. *Journal of Writing Research, 10*(2), 139–187. doi:10.17239/jowr-2018.10.02.02

Graham, S., Hebert, M., Paige Sandbank, M., & Harris, K. R. (2016). Assessing the writing achievement of young struggling writers: Application of generalizability theory. *Learning Disability Quarterly, 39*(2), 72–82. doi:10.1177/0731948714555019

Graham, S., Kiuhara, S. A., Harris, K. R., & Fishman, E. (2017). The relationship among strategic writing behavior, writing motivation, and writing performance with young developing writers. *The Elementary School Journal, 118*(1), 82–104. doi:10.1086/693009

Graham, S., Kiuhara, S. A., & MacKay, M. (2020). The effects of writing on learning in science, social studies, and mathematics: A meta-analysis. *Review of Educational Research, 90*(2), 179–226. doi:10.3102/0034654320914744

Graham, S., McKeown, D., Kiuhara, S. A., & Harris, K. R. (2012). A meta-analysis of writing instruction for students in the elementary grades. *Journal of Educational Psychology, 104*(4), 879–896. doi:10.1037/a0029185

Graves, D. H. (1983). *Writing: Teachers and children at work.* Heineman.

Griffin, C. C., & Jitendra, A. K. (2009). Word problem-solving instruction in inclusive third-grade mathematics classrooms. *The Journal of Educational Research, 102*(3), 187–202. doi:10.3200/JOER.102.3.187-202

Guilford, J. P. (1957). Creative abilities in the arts. *Psychological Review, 64*(2), 110–118. doi:10.1037/h0048280 PMID:13420286

Gutiérrez, R. (2009). Framing equity: Helping students "play the game" and "change the game. *Teaching for Excellence and Equity in Mathematics, 1*(1), 4-8.0.

Gutierrez, K., & Rogoff, B. (2003). Cultural ways of learning: Individual traits or repertoires of practice. *Educational Researcher, 32*(5), 19–25. doi:10.3102/0013189X032005019

Gutiérrez, R. (2012). Context matters: How should we conceptualize equity in mathematics education? In B. Herbel-Eisenmann, J. Chopin, D. Wagner, & D. Pimm (Eds.), *Equity in Discourse for Mathematics Education* (pp. 17–33). Springer. doi:10.1007/978-94-007-2813-4_2

Hackenberg, A. J. (2013). The fractional knowledge and algebraic reasoning of students with the first multiplicative concept. *The Journal of Mathematical Behavior, 32*(3), 538–563. doi:10.1016/j.jmathb.2013.06.007

Hacker, D. J., Kiuhara, S. A., & Levin, J. R. (2019). A metacognitive intervention for teaching fractions to students with or at-risk for learning disabilities in mathematics. Special issue on metacognition in mathematics education. *ZDM Mathematics Education, 51*(4), 601–612. doi:10.100711858-019-01040-0

Hair, J., Anderson, R. E., Tatham, R. L., & Black, W. C. (1995). *Multivariate data analysis* (4th ed.). Prentice-Hall Inc.

Håland, A. (2016). Disciplinary literacy in elementary school: How a struggling student positions herself as a writer. *The Reading Teacher, 70*(4), 457–468. doi:10.1002/trtr.1541

Hallet, D., Nunes, T., & Bryant, P. (2010). Individual differences in conceptual and procedural knowledge when learning fractions. *Journal of Educational Psychology, 102*(2), 395–406. doi:10.1037/a0017486

Halliday, M. A. K. (1985). *An introduction to functional grammar*. Hodder Arnold.

Hamburger, L., & Chu, H. (2019). Making slope a less slippery concept for English Learners. In A. Walqui & G. Bunch's (Eds.) Amplifying the curriculum: Designing quality learning opportunities for English Learners (115-137), Teachers College Press.

Hammer, D., Goldberg, F., & Fargason, S. (2012). Responsive teaching and the beginnings of energy in a third grade classroom. *Review of Science. Mathematics and ICT Education, 6*(1), 51–72.

Harel, G., & Sowder, L. (1998). Students' proof schemes: Results from exploratory studies. American Mathematical Society, 7, 234-283.

Harris, K. R., & Graham, S. (2016). Self-regulated strategy development in writing: Policy implications of an evidence-based practice. *Policy Insights from the Behavioral and Brain Sciences, 3*(1), 77-84. 4. doi:10.1177/2372732215624216

Harris, K. R., & Graham, S. (1996). *Making the writing process work: Strategies for composition and self-regulation* (2nd ed.). Brookline Books.

Harris, K. R., & Graham, S. (2009). Self-regulated strategy development in writing: Premises, evolution, and the future. *Teaching and Learning Writing, 6*, 113–135. doi:10.1348/978185409X422542

Harris, K. R., & Graham, S. (2018). Self-regulated strategy development: Theoretical bases, critical instructional elements, and future research. In R. Fidalgo, K. R. Harris, & M. Braaksma (Eds.), *Design principles for teaching effective writing: Theoretical and empirical grounded principles* (pp. 119–151). Brill.

Harris, K. R., Graham, S., & Mason, L. H. (2003). Self-regulated strategy development in the classroom: Part of a balanced approach to writing instruction for students with disabilities. *Focus on Exceptional Children*, *35*, 1–16.

Harris, K. R., Graham, S., Mason, L. H., & Friedlander, B. (2008). *Powerful writing strategies for all students*. Brooks Publishing.

Harris, K. R., Lane, K. L., Graham, S., Driscoll, S. A., Sandmel, K., Brindle, M., & Schatschneider, C. (2012). Practice-based professional development for self-regulated strategies development in writing: A randomized controlled study. *Journal of Teacher Education*, *63*(2), 103–119. doi:10.1177/0022487111429005

Hattie, J. (2009). *Visible learning: A synthesis of over 800 meta-analyses relating to achievement*. Routledge.

Hattie, J., Fisher, D., Frey, N., Gojak, L. M., Moore, S. D., & Mellman, W. (2016). *Visible learning for mathematics, Grades K-12: What works best to optimize student learning (Corwin Mathematics Series)*. Corwin.

Hattie, J., Fisher, D., Frey, N., Gojak, L. M., Moore, S. D., & Mellman, W. (2016). *Visible learning for mathematics, Grades K-12: What works best to optimize student learning*. Corwin.

Hattie, J., & Timperley, H. (2007). The power of feedback. *Review of Educational Research*, *77*(1), 81–112. doi:10.3102/003465430298487

Hauk, S., & Isom, M. A. (2009). Fostering college students' autonomy in written mathematical justification. *Investigations in Mathematics Learning*, *2*(1), 49–78. doi:10.1080/24727466.2009.11790290

Healy, L., & Hoyles, C. (2000). A study of proof conceptions in algebra. *Journal for Research in Mathematics Education*, *31*(4), 396–428. doi:10.2307/749651

Hebert, M. A., & Powell, S. R. (2016). Examining fourth-grade mathematics writing: Features of organization, mathematics vocabulary, and mathematical representations. *Reading and Writing*, *29*(7), 1511–1537. doi:10.100711145-016-9649-5

Heidema, C. (2009). Reading and writing to learn in Mathematics: Strategies to improve problem solving. *Adolescent Literacy In Perspective*, 2-9.

Heinze, J. A., Gruessing, M., & Lipowsky, F. (2018). Instructional approaches to foster third graders' adaptive use of strategies: An experimental study on the effects of two learning environments on multi-digit addition and subtraction. *Instructional Science*, *46*(6), 869–891. doi:10.100711251-018-9457-1

Helstad, K., Solbrekke, T. D., & Wittek, A. L. (2017). Exploring teaching academic literacy in mathematics in teacher education. *Education Inquiry*, *8*(4), 318–336. doi:10.1080/20004508.2017.1389225

Henningsen, M., & Stein, M. K. (1997). Mathematical tasks and student cognition: Classroom-based factors that support and inhibit high-level mathematical thinking and reasoning. *Journal for Research in Mathematics Education*, *28*(5), 524–549. doi:10.2307/749690

Herbel-Eisenmann, B. A., Steele, M. D., & Cirillo, M. (2013). (Developing) teacher discourse moves: A framework for professional development. *Mathematics Teacher Educator*, *1*(2), 181–196. doi:10.5951/mathteaceduc.1.2.0181

Herbst, P., Aaron, W., & Chieu, V. M. (2013). LessonSketch: An environment for teachers to examine mathematical practice and learn about its standards. In D. Polly (Ed.), *Common core mathematics standards and implementing digital technologies* (pp. 281–294). IGI Global. doi:10.4018/978-1-4666-4086-3.ch019

Herringer, N. (2013). *Making sense of mathematics through reasoning*. 3planning. https://www.3plearning.com/blog/making-sense-mathematics-reasoning-2/

Heuer, L. (2005). Graphic representation in the mathematics classroom. In J. M. Kenney (Ed.), *Literacy strategies for improving mathematics instruction* (pp. 51–71). Association for Supervision and Curriculum Development.

Heyes, J. R. (2012). Modeling and remodeling writing. *Written Communication*, *29*(3), 369–388. doi:10.1177/0741088312451260

Hiebert, J. (1990). The role of routine procedures in the development of mathematical competence. In T. Cooney & C. Hirsch (Eds.), *Teaching and learning mathematics in the 1990's* (pp. 31–40). National Council of Teachers of Mathematics.

Hiebert, J. (1997). *Making sense: Teaching and learning mathematics with understanding*. Heinemann.

Hiebert, J., & Carpenter, T. P. (1992). Learning and teaching with understanding. In D. A. Grouws (Ed.), *Handbook of research on mathematics teaching and learning: A project of the National Council of Teachers of Mathematics* (pp. 65–97). Macmillan Publishing.

Hiebert, J., & Grouws, D. (2007). The effects of classroom mathematics teaching on students' learning. In F. Lester (Ed.), *Second handbook of research on mathematics teaching and learning* (pp. 371–404). National Council of Teachers of Mathematics.

Hill, H. C. (2010). The nature and predictors of elementary teachers' mathematical knowledge for teaching. *Journal for Research in Mathematics Education*, *41*(5), 513–545. doi:10.5951/jresematheduc.41.5.0513

Hillman, A. M. (2014). A literature review on disciplinary literacy: How do secondary teachers apprentice students into mathematical literacy? *Journal of Adolescent & Adult Literacy*, *57*(5), 397–406. doi:10.1002/jaal.256

Hillocks, G. (2006). Middle and high school composition. In P. Smagorinsky's (Ed.). Research on composition: Multiple perspectives on two decades of change (pp.48-77). Teachers College Press.

Hillocks, G. (2005). At last: The focus on form vs. content in teaching writing. *Research in the Teaching of English*, *40*, 238–248.

Hodds, M., Alcock, L., & Inglis, M. (2014). Self-explanation training improves proof comprehension. *Journal for Research in Mathematics Education*, *45*(1), 62–101. doi:10.5951/jresematheduc.45.1.0062

Horowitz, S. H., Rawe, J., & Whittaker, M. C. (2017). *The State of Learning Disabilities: Understanding the 1 in 5*. National Center for Learning Disabilities.

Howard, C. D., Barrett, A. F., & Frick, T. W. (2010). Anonymity to promote peer feedback: Pre-service teachers' comments in asynchronous computer-mediated communication. *Journal of Educational Computing Research*, *43*(1), 89–112. doi:10.2190/EC.43.1.f

Howden, H. (1989). Teaching number sense. *The Arithmetic Teacher*, *36*(6), 6–11. doi:10.5951/AT.36.6.0006

Hughes, E. M., & Lee, J.-Y. (2019). Effects of a mathematical writing intervention on middle school students' performance. *Reading & Writing Quarterly*, *36*(2), 1–17. doi:10.1080/10573569.2019.1677537

Hughes, E. M., Marketz, A. M., & Cozad, L. E. (2019). Evaluating various undergraduate perspectives of elementary-level mathematical writing. *International Journal of Science and Mathematics Education*, *17*(5), 1031–1048. doi:10.100710763-018-9903-1

Hughes, E. M., Powell, S. R., & Stevens, E. A. (2016). Supporting clear and concise mathematics language: Say this, not that. *Teaching Exceptional Children*, *49*(1), 7–17. doi:10.1177/0040059916654901

Hughes, E. M., Riccomini, P. J., & Lee, J.-Y. (2020). Investigating written expressions of mathematical reasoning for students with learning disabilities. *The Journal of Mathematical Behavior*, *58*, 100775. doi:10.1016/j.jmathb.2020.100775

Huitzilopochtli, S. (2022). *Multilingual Students' Arguments in Early Algebra: Writing, Refining, and Revising through Conferencing*. University of California.

Huixilopotchli, S. (2022). *Multilingual students' arguments in early algebra: Writing, refining, and revising through conferencing* [Unpublished doctoral dissertation, University of California, Santa Cruz].

Hu, L., & Bentler, P. M. (1999). Cutoff criteria for fit indexes incovariance structure analysis: Conventional criteria versus new alternatives. *Structural Equation Modeling, 6*(1), 1–55. doi:10.1080/10705519909540118

Humphrey, S., & Macnaught, L. (2016). Developing teachers' professional knowledge for discipline literacy instruction. In H. de Silva Joyce (Ed.), *Language at work in social contexts: Analysing language use in work, educational, medical and museum contexts* (pp. 68–87). Cambridge Scholars Publishing.

Hunter, J., & Jones, I. (2018). Free-response tasks in primary mathematics: A window on students' thinking. In J. Hunter, P. Perger, L. Darragh (Eds.), *Proceedings of the 41st Annual Conference of the Mathematics Education Research Group of Australasia, 41*, 400–407. Auckland, New Zealand.

Ivanič, R. (1998). *Writing and identity*. John Benjamins. doi:10.1075wll.5

Jacobs, V. R., Empson, S. B., Jessup, N. A., Dunning, A., Pynes, D. A., Krause, G., & Franke, T. M. (2022). Profiles of teachers' expertise in professional noticing of children's mathematical thinking. *Journal of Mathematics Teacher Education*, 1–30. doi:10.100710857-022-09558-z

Jacobs, V. R., Lamb, L. L., & Philipp, R. A. (2010). Professional noticing of children's mathematical thinking. *Journal for Research in Mathematics Education, 41*(2), 169–202. doi:10.5951/jresematheduc.41.2.0169

Jacobs, V. R., & Spangler, D. A. (2017). Research on core practices in K-12 mathematics teaching. In J. Cai (Ed.), *Compendium for research in mathematics education* (pp. 766–792). National Council of Teachers of Mathematics.

Jansen, A. (2020). *Rough draft math: Revising to learn*. Stenhouse Publishers.

Jeffery, J. V. (2009). Constructs of writing proficiency in US state and national writing assessments: Exploring variability. *Assessing Writing, 14*(1), 3–24. doi:10.1016/j.asw.2008.12.002

Jitendra, A. K., Nelson, G., Pulles, S. M., Kiss, A. J., & Houseworth, J. (2016). Is mathematical representation of problems an evidence-based strategy for students with mathematics difficulties? *Exceptional Children, 83*(1), 8–25. doi:10.1177/0014402915625062

Johanning, D. I. (2011). Estimation's role in calculations with fractions. *Mathematics Teaching in the Middle School, 17*(2), 96–102. doi:10.5951/mathteacmiddscho.17.2.0096

Johnson, H., Watson, P. A., Delahunty, T., McSwiggen, P., & Smith, T. (2011). What is it they do? Differentiating knowledge and literacy practices across content disciplines. *Journal of Adolescent & Adult Literacy, 55*(2), 100–109. doi:10.1002/JAAL.00013

Jones, I. (2021). Improving conceptual understanding through comparing and contrasting students' mathematical writing. *5th Northeastern Research in Undergraduate Mathematics Education Conference*, Rutgers, New Jersey (online event). https://sites.google.com/view/pcrg/events/nerume/5th-ne-rume-conference?authuser=0

Jones, I., & Sirl, D. (2017). Peer assessment of mathematical understanding using comparative judgement. *Nordic Studies in Mathematics Education, 22*(4), 147–164. https://hdl.handle.net/2134/28439

Jones, I., & Alcock, L. (2014). Peer assessment without assessment criteria. *Studies in Higher Education, 39*(10), 1774–1787. doi:10.1080/03075079.2013.821974

Jones, I., & Davies, B. (2023). Comparative judgement in education research. *International Journal of Research & Method in Education*, 1–12. doi:10.1080/1743727X.2023.2242273

Jones, I., & Inglis, M. (2015). The problem of assessing problem solving: Can comparative judgement help? *Educational Studies in Mathematics*, *89*(3), 337–355. doi:10.100710649-015-9607-1

Jones, I., & Wheadon, C. (2015). Peer assessment using comparative and absolute judgement. *Studies in Educational Evaluation*, *47*, 93–101. doi:10.1016/j.stueduc.2015.09.004

Jonsson, B., Norqvist, M., Liljekvist, Y., & Lithner, J. (2014). Learning mathematics through algorithmic and creative reasoning. *The Journal of Mathematical Behavior*, *36*, 20–32. doi:10.1016/j.jmathb.2014.08.003

Jordan, N. C., Kaplan, D., Locuniak, M. N., & Ramineni, C. (2007). Predicting first grade math achievement from developmental number sense trajectories. *Learning Disabilities Research & Practice*, *22*(1), 36–46. doi:10.1111/j.1540-5826.2007.00229.x

Jordan, N. C., Kaplan, D., Ramineni, C., & Locuniak, M. N. (2009). Early math matters: Kindergarten number competence and later mathematics outcomes. *Developmental Psychology*, *45*(3), 850–867. doi:10.1037/a0014939 PMID:19413436

Jurdak, M., & Rihab, A. Z. (1998). The effect of journal writing on achievement in and attitudes toward mathematics. *School Science and Mathematics*, *98*(8), 412–419. doi:10.1111/j.1949-8594.1998.tb17433.x

Juzwik, M. M., Curcic, S., Wolbers, K., Moxley, K. D., Dimling, L. M., & Shankland, R. K. (2006). Writing into the 21st century: An overview of research on writing, 1999 to 2004. *Written Communication*, *23*(4), 451–476. doi:10.1177/0741088306291619

Kaiser, G., Busse, A., Hoth, J., König, J., & Blömeke, S. (2015). About the complexities of video-based assessments: Theoretical and methodological approaches to overcoming shortcomings of research on teachers' competence. *International Journal of Science and Mathematics Education*, *13*(2), 369–387. doi:10.100710763-015-9616-7

Kalir, R. H., & Garcia, A. (2021). *Annotation*. MIT Press. doi:10.7551/mitpress/12444.001.0001

Kaput, J. J. (2008). What is algebra? What is algebraic reasoning. In J.J. Kaput, D.W. Carraher, & M.L. Blanton's (Eds.) Algebra in the early grades (pp. 5-17), Routledge.

Karp, K. S., Bush, S. B., & Dougherty, B. J. (2022). 13 Rules That Expire. *Mathematics Teacher: Learning and Teaching PK-12*, *115*(2), 154-163.

Karp, K. S., Bush, S. B., & Dougherty, B. J. (2015). 12 math rules that expire in the middle grades. *Mathematics Teaching in the Middle School*, *21*(4), 208–215. doi:10.5951/mathteacmiddscho.21.4.0208

Kastberg, D., Chan, J. Y., & Murray, G. (2016). *Performance of U.S. 15-year-old students in science, reading, and mathematical literacy in an international context: First look at PISA 2015* (NCES 2017– 048). Washington, DC: National Center for Education Statistics, U.S. Department of Education. https://nces.ed.gov/pubsearch

Kauffman, D., Johnson, S. M., Kardos, S. M., Liu, E., & Peske, H. G. (2002). "Lost at sea": New teachers' experiences with curriculum and assessment. *Teachers College Record*, *104*(2), 273–300. doi:10.1111/1467-9620.00163

Kaufman, J. C., & Beghetto, R. A. (2009). Beyond big and little: The Four C model of creativity. *Review of General Psychology*, *13*(1), 1–12. doi:10.1037/a0013688

Kazemi, E., & Stipek, D. (2001). Promoting conceptual thinking in four upper-elementary mathematics classrooms. *The Elementary School Journal*, *102*(1), 59–80. doi:10.1086/499693

Kenney, R. H., & Montan, N. (2015). Using writing to encourage PSMTS' reflections on ambiguity in mathematical language. *North American Chapter of the International Group for the Psychology of Mathematics Education.* https://files.eric.ed.gov/fulltext/ED584314.pdf

Kenney, J. M. (2005). *Literacy strategies for improving mathematics instruction.* Association for Supervision and Curriculum Development.

Kenney, R., Shoffner, M., & Norris, D. (2013). Reflecting to learn mathematics: Supporting pre-service teachers' pedagogical content knowledge with reflection on writing prompts in mathematics education. *Reflective Practice, 14*(6), 787–800. doi:10.1080/14623943.2013.836082

Kenney, R., Shoffner, M., & Norris, D. (2014). Reflecting on the use of writing to promote mathematical learning: An examination of preservice mathematics teachers' perspectives. *Teacher Educator, 49*(1), 28–43. doi:10.1080/08878730.2013.848002

Kersaint, G. (2015). *Orchestrating mathematical discourse to enhance student learning.* Ready Curriculum Associates. https://ttaconline.org/Document/zxbIhX_YCJNP0qvIYsAjT0x-qdzE3V1X/WP-Curriculum_Associates%20Orchestrating_Mathematical_Discourse.pdf0.pdf

Ketterlin-Geller, L. R., & Perry, L. (2018). Measuring early mathematical reasoning skills: Developing tests of numeric relational reasoning and spatial reasoning. *Poster presented at the 2018 National Science Foundation Principal Investigators Meeting.* CADREK. https://cadrek12.org/posters/measuring-early-mathematical-reasoning-skills-developing-tests-numeric-relational-reasoning

Kieran, C. (2007). Learning and teaching algebra at the middle school through college levels: Building meaning for symbols and their manipulation. In F.K. Lester's (Ed.) Second handbook of research on mathematics teaching and learning (Vol. 2, pp.707-762). NCTM.

Kilpatrick, J., Swafford, J., & Findell, B. (2001). Adding it up: Helping children learn mathematics. J. Kilpatrick, J. Swafford, & B. Findell (Eds.). National Research Council. Washington, DC: National Academy Press.

Kim, H., & Oh, Y. (2014). The effect of essay writing-centered mathematics teaching on problem solving and mathematical disposition. *Communications of Mathematical Education, 28*(1), 131–154. doi:10.7468/jksmee.2014.28.1.131

Kingsdorf, S., & Krawec, J. (2014). Error analysis of mathematical word problem solving across students with and without learning disabilities. *Learning Disabilities Research & Practice, 29*(2), 66–74. doi:10.1111/ldrp.12029

Kiuhara, S. A., Levin, J. R., Tolbert, M., Erickson, M., & Kruse, K. (2023). Can Argumentative Writing Improve Math Knowledge for Elementary Students with a Mathematics Learning Disability? A Single-Case Classroom Intervention Investigation. In X. Liu, M. Hebert, & R. A. Alves (Eds.), *The Hitchhiker's Guide to Writing Research.* Springer. doi:10.1007/978-3-031-36472-3_11

Kiuhara, S. A., Levin, J. R., Tolbert, M., O'Keeffe, B. V., O'Neill, R. E., & Jameson, M. J. (2023). Teaching argument writing in math class: Challenges and solutions to improve the performance of 4th and 5th graders with disabilities. Special issue on teaching writing. *Reading and Writing.* doi:10.100711145-023-10459-7 PMID:37359030

Kiuhara, S. A., Rouse, A. G., Dai, T., Witzel, B. S., Morphy, P., & Unker, B. (2019). Constructing written arguments to develop fraction knowledge. *Journal of Educational Psychology, 112*(3), 584–607. doi:10.1037/edu0000391

Klein, P. D. (1999). Reopening inquiry into cognitive processes in writing-to-learn. *Educational Psychology Review, 11*(3), 203–270. doi:10.1023/A:1021913217147

Kleve, B., & Penne, S. (2016). Learning subjects in school – being outsiders or insiders in the disciplinary discourses on mathematics and Language 1. *International Journal of Educational Research*, *78*, 41–49. doi:10.1016/j.ijer.2016.05.014

Kline, S. L., & Ishii, D. K. (2008). Procedural explanations in mathematics writing: A framework for understanding college students' effective communication practices. *Written Communication*, *25*(4), 441–461. doi:10.1177/0741088308322343

Kling, G., & Bay-Williams, J. B. (2014). Assessing basic fact fluency. *Teaching Children Mathematics*, *20*(8), 488–497. doi:10.5951/teacchilmath.20.8.0488

Koestler, C., Felton, M., Bieda, K., & Otten, S. (2013). *Connecting the NCTM process standards and the CCSSM practices*. National Council of Teachers of Mathematics.

König, J., Santagata, R., Scheiner, T., Adleff, A. K., Yang, X., & Kaiser, G. (2022). Teacher noticing: A systematic literature review of conceptualizations, research designs, and findings on learning to notice. *Educational Research Review*, *36*, 100453. doi:10.1016/j.edurev.2022.100453

Kosko, K. (2016). Writing in mathematics. A survey of K-12 teachers' reported frequency in the classroom. *School Science and Mathematics*, *116*(5), 276–285. doi:10.1111sm.12177

Kosko, K. W., & Zimmerman, B. S. (2019). Emergence of argument in children's mathematical writing. *Journal of Early Childhood Literacy*, *19*(1), 82–106. doi:10.1177/1468798417712065

Kostos, K., & Shin, E. K. (2010). Using math journals to enhance second graders' communication of mathematical thinking. *Early Childhood Education Journal*, *38*(3), 223–231. doi:10.100710643-010-0390-4

Krainer, K. (2006). Action research and mathematics teacher education. *Journal of Mathematics Teacher Education*, *9*, 213–219. doi:10.100710857-006-9009-5

Kramarski, B., & Mevarech, Z. R. (2003). Enhancing mathematical reasoning in the classroom: The effects of cooperative learning and metacognitive training. *American Educational Research Journal*, *40*(1), 281–310. doi:10.3102/00028312040001281

Krebs, A. S. (2005). Analyzing student work as a professional development activity. *School Science and Mathematics*, *105*(8), 402–411. doi:10.1111/j.1949-8594.2005.tb18060.x

Krowka, S. K., & Fuchs, L. S. (2017). Cognitive profiles associated with responsiveness to fraction intervention. *Learning Disabilities Research & Practice*, *32*(4), 216–230. doi:10.1111/ldrp.12146 PMID:29123333

Kuzle, A. (2013). Promoting writing in mathematics: Prospective teachers' experiences and perspectives on the process of writing when doing mathematics as problem solving. *CEPS Journal*, *3*(4), 41–59. doi:10.26529/cepsj.222

Lakatos, I. (1976). *Proofs and refutations: the logic of mathematical discovery*. Cambridge University Press. doi:10.1017/CBO9781139171472

Lambert, R. (2020, February). *Increasing Access to Universally Designed Mathematics Classrooms*. Policy Analysis for California Education.

Lambert, R., & Tan, P. (2017). Conceptualizations of students with and without disabilities as mathematical problem solvers in educational research: A critical review. *Education Sciences*, *7*(2), 51. doi:10.3390/educsci7020051

Lampert, M. (1990). When the problem is not the question and the solution is not the answer: Mathematical knowing and teaching. *American Educational Research Journal*, *27*(1), 29–63. doi:10.3102/00028312027001029

Lappan, G., Fey, J. T., Fitzgerald, W. M., Friel, S. N., & Phillips, E. D. (1998). *Connected Mathematics*. Dale Seymour Publications.

Lee, E., & Oh, S. Y. (2017). *Computer-mediated communication.* Oxford Bibleographies. https://www.oxfordbibliographies.com/display/document/obo-9780199756841/obo-9780199756841-0160.xml

Lei, Q., Xin, Y. P., Morita-Mullaney, T., & Tzur, R. (2020). Instructional Scaffolds in Mathematics Instruction for English Learners With Learning Disabilities: An Exploratory Case Study. *Learning Disabilities (Weston, Mass.), 18*(1), 123–144.

Lepak, J. R., Wernet, J. L., & Ayieko, R. A. (2018). Capturing and characterizing students' strategic algebraic reasoning through cognitively demanding tasks with focus on representations. *The Journal of Mathematical Behavior, 50*, 57–73. doi:10.1016/j.jmathb.2018.01.003

Lepak, J., & Going, T. (2019). Designing scaffolds for students' written arguments. *Mathematics Teaching in the Middle School, 24*(5), 300–303. doi:10.5951/mathteacmiddscho.24.5.0300

Lewis, K. E. (2014). Difference not deficit: Reconceptualizing mathematical learning disabilities. *Journal for Research in Mathematics Education, 45*(3), 351–396. doi:10.5951/jresematheduc.45.3.0351

Lewis, K. E., & Fisher, M. B. (2016). Taking stock of 40 years of research on mathematical learning disability: Methodological issues and future directions. *Journal for Research in Mathematics Education, 47*(4), 338–371. doi:10.5951/jresematheduc.47.4.0338

Liljedahl, P. (2023). *Building Thinking Classrooms, Grades K-12: Fourteen Teaching Practices for Enhancing Learning*. Corwin.

Lilly, S., Fick, S. J., Chiu, J. L., & McElhaney, K. W. (2020). Supporting elementary students to develop mathematical models within design-based integrated science and mathematics projects. In *Proceedings of the 14th international conference of the learning sciences.* International Society of the Learning Sciences.

Lilly, S., McAlister, A. M., Fick, S. J., & Chiu, J. L. (2023). A comparison of elementary teachers' verbal supports for students in inclusive and general classroom contexts during an NGSS-aligned science, engineering, and computer science unit. *Science Education, 107*(4), 853–883. doi:10.1002ce.21788

Lim, L. and Pugalee, D.K. (2004). Using journal writing to explore "They communicate to learn mathematics and they learn to communicate mathematically." *Ontario Action Researcher, 7*(2), Article 2.

Lim, W., Son, J. W., & Kang, S. H. (2021). How reducing discomfort impacts peer assessments of preservice teachers. *Sustainability (Basel), 13*(11), 6435. doi:10.3390u13116435

Lin, X., Peng, P., & Zeng, J. (2021). Understanding the relation between mathematics vocabulary and mathematics performance: A meta-analysis. *The Elementary School Journal, 121*(3), 357–540. doi:10.1086/712504

Lithner, J. (2000). Mathematical reasoning in task solving. *Educational Studies in Mathematics, 41*(2), 165–190. https://www.jstor.org/stable/3483188

Locuniak, M. N. (2010). *Early predictors of calculation fluency in second grade* (Publication No. 3397013) [Doctoral dissertation, University of Delaware]. ProQuest Dissertations Publishing.

Lyle, J. (2003). Stimulated recall: A report on its use in naturalistic research. *British Educational Research Journal, 29*(6), 861–878. doi:10.1080/0141192032000137349

Magiera, M. T., & Zambak, V. C. (2020). Exploring prospective teachers' ability to generate and analyze evidence-based explanatory arguments. *International Journal of Education and Science, (IJRES), 6*(2), 327—346.

Magiera, M. T., & Zambak, V. S. (2020). Exploring prospective teachers' ability to generate and analyze evidence-based explanatory arguments. *International Journal of Research in Education and Science, 6*(2), 327–346. https://eric.ed.gov/?id=EJ1250671. doi:10.46328/ijres.v6i2.765

Magiera, M. T., & Zambak, V. S. (2021). Prospective K-8 teachers' noticing of student justifications and generalizations in the context of analyzing written artifacts and video-records. *International Journal of STEM Education, 8*(1), 1–21. doi:10.118640594-020-00263-y

Mainali, B. (2021). Representation in teaching and learning mathematics. [IJEMST]. *International Journal of Education in Mathematics, Science, and Technology, 9*(1), 1–21. doi:10.46328/ijemst.1111

Malone, A. S., & Fuchs, L. S. (2017). Error patterns in ordering fractions among at-risk fourth-grade students. *Journal of Learning Disabilities, 50*(3), 337–352. doi:10.1177/0022219416629647 PMID:26966153

Marks, G., & Mousley, J. (1990). Mathematics education and genre: Dare we make the process writing mistake again? *Language and Education, 4*(2), 117–135. doi:10.1080/09500789009541278

Marshall, N., Shaw, K., Hunter, J., & Jones, I. (2020). Assessment by comparative judgement: An application to secondary statistics and English in New Zealand. *New Zealand Journal of Educational Studies, 55*(1), 49–71. doi:10.100740841-020-00163-3

Martin, C. L. (2015). Writing as a tool to demonstrate mathematical understanding. *School Science and Mathematics, 115*(6), 302–313. doi:10.1111sm.12131

Martin, C. S., & Polly, D. (2019). Examining the use of multiple writing and discourse tasks in 5th grade mathematics. *The Journal of Educational Research, 112*(6), 663–675. doi:10.1080/00220671.2019.1678106

Martínez Hinestroza, J. (2022, June). (Re)Learning what it means to participate: Bringing student and teacher perspectives into dialogue. *The Elementary School Journal, 122*(4), 616–641. doi:10.1086/719465

Martin, J. R. (1995). Text and clause: Fractal resonance. *Text & Talk, 15*(1), 5–42.

Martin, J. R., & Rose, D. (2008). *Genre relations: Mapping culture.* Equinox.

Mason, J. (2002). *Researching your own practice: The discipline of noticing.* Routledge. doi:10.4324/9780203471876

Mathematics Assessment Resource Service (MARS). (2015) *Evaluating Statements: Consecutive Sums.* Mathematics Assessment Project Classroom Challenges: A formative assessment lesson. http://map.mathshell.org

Mathematics Assessment Resource Service. (2015). *Mathematics Assessment Project classroom challenges: Formative assessment lessons.* MARS. http://map.mathshell.org

Mayer, R. E. (1985). Implications of cognitive psychology for instruction in mathematical problem solving. In E. A. Silver (Ed.), *Teaching and learning mathematical problem solving: Multiple research perspectives* (pp. 123–138). Lawrence Erlbaum Associates.

Mazzocco, M. M. M., Myers, G. F., Lewis, K. E., Hanich, L. B., & Murphy, M. M. (2013). Limited knowledge of fraction representations differentiates middle school students with mathematics learning disability (dyscalculia) versus low mathematics achievement. *Journal of Experimental Child Psychology, 115*(2), 371–387. doi:10.1016/j.jecp.2013.01.005 PMID:23587941

McCarthy, D. S. (2008). Communication in mathematics: Preparing preservice teachers to include writing in mathematics teaching and learning. *School Science and Mathematics, 108*(7), 334–340. doi:10.1111/j.1949-8594.2008.tb17846.x

McCarthy, L. A., & Faulkner, M. (2019). Integrating writing and mathematics: Journaling to increase learning and enjoyment while reducing anxiety. *Teaching Mathematics and its Applications: An International. The Journal of IMA / Islamic Medical Association of North America, 39*, 145–159.

McCloskey, M., Caramazza, A., & Basili, A. (1985). Cognitive mechanisms in number processing and calculation: Evidence form dyscalculia. *Brain and Cognition, 4*(2), 171–196. doi:10.1016/0278-2626(85)90069-7 PMID:2409994

McCormick, K. (2010). Experiencing the power of learning mathematics through writing. *Issues in the Undergraduate Mathematics Preparation of School Teachers: The Journal, 4*, 1–8.

McIntosh, M. E., & Draper, R. J. (2001). Using learning logs in mathematics: Writing to learn. *Mathematics Teacher, 94*(7), 554–557. doi:10.5951/MT.94.7.0554

McKeown, D., Brindle, M., Harris, K. R., Graham, S., Collins, A. A., & Brown, M. (2016). Illuminating growth and struggles using mixed methods: Practice-based professional development and coaching for differentiating SRSD instruction in writing. *Reading and Writing, 29*(6), 1105–1140. doi:10.100711145-016-9627-y

Meel, D. E. (1999). Email dialogue journals in a college calculus classroom: A look at the implementation and benefits. *Journal of Computers in Mathematics and Science Teaching, 18*(4), 387–413.

Meyer, A., Rose, D. H., & Gordon, D. (2014). *Universal Design for Learning: Theory and practice.* CAST, Inc.

Ministry of Education. (2007). *The New Zealand curriculum.* Learning Media.

Ministry of Education. (2020). *The Ontario curriculum, grades 1–8: Mathematics.* King's Printer for Ontario. https://assets-us-01.kc-usercontent.com/fbd574c4-da36-0066-a0c5-849ffb2de96e/90439c6e-f40c-4b58-840c-557ed88a9345/The%20 Ontario%20Curriculum%20Grades%201%E2%80%938%20-%20Mathematics,%202020%20(January%202021).pdf

Minnameier, G. (2004). Peirce-suit of truth–why inference to the best explanation and abduction ought not to be confused. *Erkenntnis, 60*(1), 75–105. doi:10.1023/B:ERKE.0000005162.52052.7f

Monoyiou, A., Xistouri, X., & Philippou, G. (2006). Primary students' reasoning in problem solving and teachers' evaluation of their arguments. *International Group for the Psychology of Mathematics Education, 4*, 177.

Monson, D., Krupa, E., Lesseig, K., & Casey, S. (2020). Developing secondary prospective teachers' ability to respond to student work. *Journal of Mathematics Teacher Education, 23*(2), 209–232. doi:10.100710857-018-9420-8

Moran, A. S., Swanson, H. L., Gerber, M. M., & Fung, W. (2014). The effects of paraphrasing interventions on problem-solving accuracy for children at risk for math disabilities. *Learning Disabilities Research & Practice, 29*(3), 97–105. doi:10.1111/ldrp.12035

Morgan, C. (1998). *Writing mathematically: The discourse of investigation.* Falmer Press.

Morgan, C. (2001). Mathematics and human activity: Representation in mathematical writing. *Research in Mathematics Education, 3*(1), 169–182. doi:10.1080/14794800008520091

Morgan, C. (2004). Word, definitions and concepts in discourses of mathematics, teaching and learning. *Language and Education, 18*, 1–15.

Moschkovich, J. (2018). Considering silence and precision: Communication practices that support equitable mathematical discussions for marginalized students. In R. Hunter's (Ed.) Mathematical discourse that breaks barriers and creates space for marginalized learners (pp. 61-83). Brill.

Moschkovich, J. (2013). Principles and guidelines for equitable mathematics teaching practices and materials for English language learners. *Journal of Urban Mathematics Education, 6*(1), 45–57. doi:10.21423/jume-v6i1a204

Moschkovich, J. N. (1996). Moving up and getting steeper: Negotiating shared descriptions of linear graphs. *Journal of the Learning Sciences*, *5*(3), 239–277. doi:10.120715327809jls0503_4

Moschkovich, J. N. (2002). A situated and sociocultural perspective on bilingual mathematics learners. *Mathematical Thinking and Learning*, *4*(2 & 3), 189–212. doi:10.1207/S15327833MTL04023_5

Moschkovich, J. N. (2003). What counts as mathematical discourse? *International Group for the Psychology of Mathematics Education*, *3*, 325–332.

Moschkovich, J. N. (2007). Examining mathematical discourse practices. *For the Learning of Mathematics*, *27*(1), 24–30.

Moschkovich, J. N. (2008). "I went by twos, he went by one:" Multiple interpretations of inscriptions as resources for mathematical discussions. *Journal of the Learning Sciences*, *17*(4), 551–587. doi:10.1080/10508400802395077

Moschkovich, J. N. (2010). Language(s) and learning mathematics: Resources, challenges, and issues for research. In J. N. Moschkovich (Ed.), *Language and mathematics education: Multiple perspectives and directions for research* (pp. 1–28). Information Age Publishing.

Moschkovich, J. N. (2010). Recommendations for research on language and mathematics education. In J. N. Moschkovich (Ed.), *Language and mathematics education: Multiple perspectives and directions for research* (pp. 151–170). Information Age Publishing, Inc.

Moschkovich, J. N. (2013a). Equitable practices in mathematics classrooms: Research based recommendations. *Teaching for Excellence and Equity in Mathematics*, *5*(1), 26–34.

Moschkovich, J. N. (2013b). Issues regarding the concept of mathematical practices. In Y. Li & J. N. Moschkovich (Eds.), *Proficiency and Beliefs in Learning and Teaching Mathematics: Learning from Alan Schoenfeld and Günter Toerner* (pp. 257–275). Sense Publishers. doi:10.1007/978-94-6209-299-0_15

Moschkovich, J. N. (2015). Academic literacy in mathematics for English learners. *The Journal of Mathematical Behavior*, *40*, 43–62. doi:10.1016/j.jmathb.2015.01.005

Moschkovich, J. N. (2015b). A sociocultural approach to academic literacy in mathematics for adolescent English Learners: Integrating mathematical proficiency, practices, and discourse. In D. Molle, E. Sato, T. Boals, & C. Hedgspeth (Eds.), *Multilingual Learners and Academic Literacies: Sociocultural Contexts of Literacy Development in Adolescents* (pp. 75–104). Routledge.

Mo, Y., & Troia, G. A. (2017). Similarities and differences in constructs represented by U. S. States' middle school writing tests and the 2007 National Assessment of Educational Progress writing assessment. *Assessing Writing*, *33*, 48–67. doi:10.1016/j.asw.2017.06.001

Mudaly, V., & De Villiers, M. (2000). Learners' needs for conviction and explanation within the context of dynamic geometry. *Pythagoras*, *52*(8), 20–23.

Munter, C., Stein, M., & Smith, M. (2015). Dialogic and direct instruction: Two distinct models of mathematics instruction and the debate(s) surrounding them. *Teachers College Record*, *117*(11), 1–32. doi:10.1177/016146811511701102

Murphy, P., & Hall, K. (Eds.). (2008). *Learning and practice: Agency and identities*. Sage.

Murray, D. (1968). *A writer teaches writing: A practical method of teaching composition*. Houghton Mifflin.

Muthén, B., du Toit, S. H. C., & Spisic, D. (1997). Robust inference using weighted least squares and quadratic estimating equations in latent variable modelling with categorical and continuous outcomes (Technical Report). University of California, Los Angeles.

Nachowitz, M. (2018). Intent and engagement: Writing in mathematics for conceptual understanding. *Investigations in Mathematics Learning, 11*(4), 245–257. doi:10.1080/19477503.2018.1461051

Namkung, J. M., & Bricko, N. (2021). The effects of algebraic equation solving intervention for students with mathematics learning difficulties. *Journal of Learning Disabilities, 54*(2), 111–123. doi:10.1177/0022219420930814 PMID:32508218

Namkung, J. M., Hebert, M., Powell, S. R., Hoins, M., Bricko, N., & Torchia, M. (2020). Comparing and validating four methods for scoring mathematics writing. *Reading & Writing Quarterly, 36*(2), 157–175. doi:10.1080/10573569.2019.1700858

Narciss, S., & Huth, K. (2004). How to design informative tutoring feedback for multi-media learning. In H. M. Niegemann, D. Leutner, & R. Brunken (Eds.), *Instructional design for multimedia learning* (pp. 181–195). Waxmann.

Nathan, M. J., & Koedinger, K. R. (2000). An investigation of teachers' beliefs of students' algebra development. *Cognition and Instruction, 18*(2), 209–237. doi:10.1207/S1532690XCI1802_03

National Academies of Sciences, Engineering, and Medicine (2017). *Promoting the educational success of children and youth learning English: Promising futures.* National Academies Press.

National Academies of Sciences, Engineering, and Medicine (2018). *English Learners in STEM Subjects: Transforming Classrooms, Schools, and Lives.* The National Academies Press. . doi:10.17226/25182

National Assessment of Education Progress. (2022). *Reading and mathematics scores for 9-year-olds decline during pandemic: Results from 2022 NAEP long-term trend assessment.* NAEP. www.nationsreportcard.gov/highlights/ltt/2022/supporting_files/ltt-2022-age9-infographic.pdf

National Center for Education Statistics, Institute of Education Sciences, National Assessment of Educational Progress for Mathematics. (2013). The nation's report card. NCES.

National Center for Education Statistics, Institute of Education Sciences, National Assessment of Educational Progress for Writing. (2011). *National assessment of educational progress at grades 8 and 12.* NCES. https://nces.ed.gov/nationsreportcard/pdf/main2011/2012470.pdf

National Center for Education Statistics. (2023). Mathematics Performance. *Condition of Education.* U.S. Department of Education, Institute of Education Sciences. https://nces.ed.gov/programs/coe/indicator/cnc

National Council of Supervisors of Mathematics. TODOS: Mathematics for all (2016). *Mathematics Education Through the Lens of Social Justice: Acknowledgment, Actions, and Accountability.* NCSM. https://www.todos-math.org/socialjustice

National Council of Teachers of Mathematics. (1989). *Curriculum and evaluation standards for school mathematics.* NCTM.

National Council of Teachers of Mathematics. (1989). *Curriculum and evaluation standards.* NCTM.

National Council of Teachers of Mathematics. (1991). *Professional standards for teaching mathematics.* NCTA.

National Council of Teachers of Mathematics. (2000). *Principles and standards for school mathematics* (Vol. 1). NCTM.

National Council of Teachers of Mathematics. (2000). *Principles and standards for school mathematics.* NCTM.

National Council of Teachers of Mathematics. (2014). *Principles to actions: Ensuring mathematical success for all.* NCTM.

National Council of Teachers of Mathematics. (2020). *Standards for the preparation of middle level mathematics teachers.* NCTM. https://www.nctm.org/uploadedFiles/Standards_and_Positions/NCTM_Middle_School_2020_Final.pdf

National Council of Teachers of Mathematics. (NCTM). (2000). *Principles and standards for school mathematics.* NCTM. https://www.nctm.org/Standards-and-Positions/Principles-and-Standards/

National Council of Teachers of Mathematics. (NCTM). (2014). *Principles to action: Executive summary.* NCTM. https://www.nctm.org/uploadedFiles/Standards_and_Positions/PtAExecutiveSummary.pdf

National Governors Association & Council of Chief State School Officers (NGA & CCSSO). (2010a). *Common Core State Standards for mathematics.* CCSSO. https://learning.ccsso.org/wp-content/uploads/2022/11/Math_Standards1.pdf

National Governors Association & Council of Chief State School Officers (NGA & CCSSO). (2010b). *Common Core State Standards for English language arts & literacy in history/social studies, science, and technical subjects.* CCSSO. https://learning.ccsso.org/wp-content/uploads/2022/11/ELA_Standards1.pdf

National Governors Association Center for Best Practices & Council of Chief State School Officers. (2010). *Common Core State Standards for Mathematics.* NCTM.

National Governors Association Center for Best Practices & Council of Chief State School Officers. (2010). *Common Core State Standards for Mathematics.* NCTM. http://www.corestandards.org/wp-content/uploads/Math_Standards1.pdf

National Governors Association Center for Best Practices & Council of Chief State School Officers. (2010). *Common Core State Standards mathematics.* SSO.

National Governors Association Center for Best Practices and Council of Chief State School Officers. (2010). *Common core state standards for mathematics.* National Governors Association Center for Best Practices and Council of Chief State School Officers. http://www.corestandards.org/Math/

National Governors Association Center for Best Practices and the Council of Chief State School Officers. (2010a). *Common core state standards for English language arts & literacy in history/social studies, science, and technical subjects.* CCSSO. https://learning.ccsso.org/wp-content/uploads/2022/11/ELA_Standards1.pdf

National Governors Association Center for Best Practices and the Council of Chief State School Officers. (2010b). *Common core state standards for mathematics.* CCSSO. https://learning.ccsso.org/wp-content/uploads/2022/11/ADA-Compliant-Math-Standards.pdf

National Governors Association Center for Best Practices, & the Council of Chief State School Officers. (2010). *Common Core State Standards for mathematics.* NCTM. http://www.corestandards.org/assets/CCSSI_Math%20Standards.pdf

National Governors Association Centre for Best Practices, Council of Chief State School Officers. (2010). *Common core state standards.* NCTM.

National Mathematics Advisory Panel. (2008). *Foundations for success: The final report of the National Mathematics Advisory Panel.* Washington, DC: U.S. Department of Education. www2.ed.gov/about/bdscomm/list/mathpanel/report/final-report.pdf

National Research Council. (2001). *Adding it up: Helping children learn mathematics.* National Academy Press.

National Research Council. (2001). *Adding It Up: Helping Children Learn Mathematics.* The National Academies Press.

NCTM. (2014). *Ensuring Mathematical Success for All.* National Council of Teachers of Mathematics.

Nemirovsky, R., Barros, A., Noble, T., Schnepp, M., & Solomon, J. (2005). Learning Mathematics in High School: Symbolic Places and Family Resemblances. In T.A. Romberg, T.P. Carpenter, & F. Dremock's (Eds) Understanding Mathematics and Science Matters (pp.185-206). Routledge.

Ntenza, S. (2006). Investigating forms of children's writing in grade 7 mathematics classrooms. *Educational Studies in Mathematics*, *61*(3), 321–345. doi:10.100710649-006-5891-0

O'Connor, C., Michaels, S., Chapin, S., & Harbaugh, A. G. (2016). The silent and the vocal: Participation and learning in whole-class discussion. *Learning and Instruction*.

O'Halloran, K. (2000). Classroom discourse in mathematics: A multisemiotic analysis. *Linguistics and Education*, *10*(3), 359–388. doi:10.1016/S0898-5898(99)00013-3

O'Halloran, K. (2005). Mathematical discourse: Language, symbolism and visual images. *Continuum.*

O'Halloran, K. L. (2008). *Mathematical Discourse: Language, Symbolism and Visual Images*. A&C Black.

O'Halloran, K. L. (2015). The language of learning mathematics: A multimodal perspective. *The Journal of Mathematical Behavior*, *40*, 63–74. doi:10.1016/j.jmathb.2014.09.002

Odden, T. O. B., & Russ, R. S. (2019). Defining sensemaking: Bringing clarity to a fragmented theoretical construct. *Science Education*, *103*(1), 187–205. doi:10.1002ce.21452

Orsmond, P., Merry, S., & Reiling, K. (1996). The importance of marking criteria in the use of peer assessment. *Assessment & Evaluation in Higher Education*, *21*(3), 239–250. doi:10.1080/0260293960210304

Ortmeier-Hooper, C. (2013). *The ELL writer: Moving beyond basics in the secondary classroom.* Teachers College Press.

Otto, A. D., Caldwell, J. H., Lubinski, C. A., & Hancock, S. W. (2011). *Developing essential understanding of multiplication and division for teaching mathematics in grades 3-5.* National Council of Teachers of Mathematics.

Our Statewide School District. (n.d.). Hawai'i Department of Education. https://www.teachinhawaii.org/district-profile/#:~:text=On%20the%20island%20of%20O,ll%20find%20Kaua'i%20District

Parrish, S. D. (2011). Number talks build numerical reasoning. *Teaching Children Mathematics*, *18*(3), 198–206. doi:10.5951/teacchilmath.18.3.0198

Paxton, M. (2007). Tensions between textbook pedagogy and the literacy practices of the disciplinary community: A study of writing in first year economics. *Journal of English for Academic Purposes*, *6*(2), 109–125. doi:10.1016/j.jeap.2007.04.003

Pearson, P., & Gallagher, M. (1983). The instruction of reading comprehension. *Contemporary Educational Psychology*, *8*(3), 317–344. doi:10.1016/0361-476X(83)90019-X

Peng, P., & Lin, X. (2019). The relation between mathematics vocabulary and mathematics performance among fourth graders. *Learning and Individual Differences*, *69*, 11–21. doi:10.1016/j.lindif.2018.11.006

Peng, P., Lin, X., Ünal, Z. E., Lee, K., Namkung, J., Chow, J., & Sales, A. (2020). Examining the mutual relations between language and mathematics: A meta-analysis. *Psychological Bulletin*, *146*(7), 595–634. doi:10.1037/bul0000231 PMID:32297751

Pennsylvania Department of Education. (2023). *The Pennsylvania System of School Assessment Grade 3 Mathematics Items and Scoring Sampler*. PA DoE. https://www.education.pa.gov/Documents/K-12/Assessment%20and%20Accountability/PSSA/Item%20and%20Scoring%20Samples/2023%20PSSA%20ISS%20Math%20Grade%203.pdf

Perry, B., & Dockett, S. (2002). Young children's access to powerful mathematical ideas. In L. D. English (Ed.), *Handbook of international research in mathematics education: Directions for the 21st century* (pp. 81–111). Lawrence Erlbaum.

Plucker, J. A., Beghetto, R. A., & Dow, G. T. (2004). Why isn't creativity more important to educational psychologists? Potentials, pitfalls, and future directions in creativity research. *Educational Psychologist, 39*(2), 83–96. doi:10.120715326985ep3902_1

Pollitt, A. (2012). The method of Adaptive Comparative Judgement. *Assessment in Education: Principles, Policy & Practice, 19*(3), 281–300. doi:10.1080/0969594X.2012.665354

Polly, D. (2021). Advancing equity-based mathematics teaching in the primary grades: The case of two clinical practice experiences. *International Journal of Teacher Education and Professional Development, 4*(1), 68–88. doi:10.4018/IJTEPD.2021010105

Polly, D., & Martin, C. S. (in press). Considering the design and use of differentiated activities and fluency games to advance equity-based mathematics practices. *Educational Designer.*

Pólya, G. (1945). *How to solve it.* Princeton University Press. doi:10.1515/9781400828678

Polya, G. (1957). *How to solve it: A new aspect of mathematical method* (2nd ed.). Doubleday.

Porter, M. K., & Masingila, J. O. (2000). Examining the effect of writing on conceptual and procedural knowledge in calculus. *Educational Studies in Mathematics, 42*(2), 165–177. doi:10.1023/A:1004166811047

Powell, S. R., & Driver, M. K. (2015). The influence of mathematics vocabulary instruction embedded within addition tutoring for first-grade students with mathematics difficulty. *Learning Disability Quarterly, 38*(4), 221–233. doi:10.1177/0731948714564574

Powell, S. R., & Hebert, M. A. (2016). Influence of writing ability and computation skill on mathematics writing. *The Elementary School Journal, 117*(2), 310–335. doi:10.1086/688887

Powell, S. R., & Hebert, M. A. (2022). Thinking beyond symbols: Writing and reading in mathematics. In Z. A. Philippakos & S. Graham (Eds.), *Writing and reading connections: Bridging research and classroom practice.*

Powell, S. R., Hebert, M. A., Cohen, J. A., Casa, T. M., & Firmender, J. M. (2017). A synthesis of mathematics writing: Assessments, interventions, and surveys. *Journal of Writing Research, 8*(3), 493–526. doi:10.17239/jowr-2017.08.03.04

Powell, S. R., Hebert, M. A., & Hughes, E. M. (2021). How educators use mathematics writing in the classroom: A national survey of mathematics educators. *Reading and Writing, 34*(2), 417–447. doi:10.100711145-020-10076-8

Prediger, S., & Erath, K. (2014). Content, interaction, or both? Synthesizing two German traditions in a video study on learning to explain the mathematics classroom microcultures. *Eurasia Journal of Mathematics, Science and Technology Education, 10*(4), 313–327. doi:10.12973/eurasia.2014.1085a

Pressley, M. (2000). What should comprehension instruction be the instruction of? In M. L. Kamil, P. B. Mosenthal, P. D. Pearson, & R. Barr (Eds.), *Handbook of reading research* (Vol. 3, pp. 545–561). Erlbaum.

Price, J. N. (2001). Action research, pedagogy and change: The transformative potential of action research in pre-service teacher education. *Journal of Curriculum Studies, 33*(1), 43–74. doi:10.1080/00220270118039

Pugalee, D. (2005). *Writing to develop mathematical understanding*. Christopher-Gordon.

Pugalee, D. K. (2001). Writing, mathematics, and metacognition: Looking for connections through students' work in mathematical problem solving. *School Science and Mathematics, 101*(5), 236–245. doi:10.1111/j.1949-8594.2001.tb18026.x

Pugalee, D. K. (2001, January). Using communication to develop students' mathematical literacy. *Mathematics Teaching in the Middle School, 6*(5), 296–299. doi:10.5951/MTMS.6.5.0296

Pugalee, D. K. (2004). A comparison of verbal and written descriptions of students' problem solving processes. *Educational Studies in Mathematics, 55*(1), 27–47. doi:10.1023/B:EDUC.0000017666.11367.c7

Pugalee, D. K., DiBiase, W., & Wood, K. (1999). Writing and the development of problem solving in mathematics and science. *Middle School Journal, 30*(5), 48–52.

Purpura, D. J., Logan, J. A. R., Hassinger-Das, B., & Napoli, A. R. (2017). Why do early mathematics skills predict later reading? The role of mathematical language. *Developmental Psychology, 53*(9), 1633–1642. doi:10.1037/dev0000375 PMID:28758785

Remillard, J. T., & Bryans, M. B. (2004). Teachers' orientations toward mathematics curriculum materials: Implications for teacher learning. *Journal for Research in Mathematics Education, 35*(5), 352–388. doi:10.2307/30034820

Renzulli, J. S., Leppien, J. H., & Hays, T. S. (2000). *The multiple menu model: A practical guide for developing differentiated curriculum.* Creative Learning Press.

Reys, B. (1994). Promoting number sense in the middle grades. *Mathematics Teaching in the Middle School, 1*(3), 114–120. doi:10.5951/MTMS.1.2.0114

Riccomini, P. J., Smith, G. W., Hughes, E. M., & Fries, K. M. (2015). The language of mathematics: The importance of teaching and learning mathematical vocabulary. *Reading & Writing Quarterly, 31*(3), 235–252. doi:10.1080/10573569.2015.1030995

Riccomini, P. J., Witzel, B. S., & Deshpande, D. S. (2022). Combining visual representations and a powerful retention strategy with peer-mediated strategies to improve mathematical outcomes for students with EBD. *Beyond Behavior, 31*(1), 42–52. doi:10.1177/10742956211072555

Ritchey, K. D., & Coker, D. L. Jr. (2013). An investigation of the validity and utility of two curriculum-based measurement writing tasks. *Reading & Writing Quarterly, 29*(1), 89–119. doi:10.1080/10573569.2013.741957

Rittle-Johnson, B., Schneider, M., & Star, J. R. (2015). Not a one-way street: Bidirectional relations between procedural and conceptual knowledge of mathematics. *Educational Psychology Review, 27*(4), 587–597. doi:10.100710648-015-9302-x

Rittle-Johnson, B., & Star, J. R. (2007). Does comparing solution methods facilitate conceptual and procedural knowledge? An experimental study on learning to solve equations. *Journal of Educational Psychology, 99*(3), 561–574. doi:10.1037/0022-0663.99.3.561

Rivera, F. D., & Becker, J. R. (2007). Abduction–induction (generalization) processes of elementary majors on figural patterns in algebra. *The Journal of Mathematical Behavior, 26*(2), 140–155. doi:10.1016/j.jmathb.2007.05.001

Robertson, A. D., Atkins, L. J., Levin, D. M., & Richards, J. (2016). What is responsive teaching? In A. D. Robertson, R. E. Scherr, & D. Hammer (Eds.), *Responsive teaching in science and mathematics* (pp. 1–34). Routledge.

Rodriguez, A. J., & Zozakiewicz, C. (2010). Facilitating the integration of multiple literacies through science education and learning technologies. In A. J. Rodriguez (Ed.), *Science education as a pathway to teaching language literacy* (pp. 23–45). Sense Publishers. doi:10.1163/9789460911316_005

Rogoff, B. (2014). Learning by observing and pitching in to family and community endeavors: An orientation. *Human Development, 57*(2-3), 69–81. doi:10.1159/000356757

Rogoff, B., Paradise, R., Mejía Arauz, R., Correa-Chávez, M., & Angelillo, C. (2003). Firsthand learning through intent participation. *Annual Review of Psychology, 54*(1), 175–203. doi:10.1146/annurev.psych.54.101601.145118 PMID:12499516

Root, J. R., Jimenez, B., & Saunder, A. (2022, February). Leveraging the UDL framework to plan grade-aligned mathematics in inclusive settings. *In Practice*, *1*(1), 13–22.

Rose, D., & Martin, J. R. (2012). Knowledge about language. In D. Rose & J. R. Martin (Eds.), *Learning to write, reading to learn: Genre, knowledge and pedagogy in the Sydney school* (pp. 206–264). Equinox.

Rossbridge, J., & Rushton, K. (2014). The critical conversation about text: Joint construction. *Primary English Teachers Association Australia Paper 196*. PETAA. https://petaa.edu.au/w/w/Store/Item_Detail.aspx?iProductCode=PP196&Category=PEN

Rothery, J. (1994). *Write it right: Exploring literacy in school English.* Disadvantaged Schools Program, Metropolitan East Region, NSW Department of School Education.

Rothery, J. (1996). Making changes: Developing an educational linguistics. In R. Hasan & G. Williams (Eds.), *Literacy in society*. Addison Wesley Longman.

Rowland, T. (1999). *The pragmatics of mathematics education: Vagueness in mathematical discourse*. Routledge, Taylor & Francis Group.

SACE Board of South Australia. (2015). *Mathematical methods subject outline*. SACE. https://www.sace.sa.edu.au/web/mathematical-methods/stage-2/planning-to-teach/subject-outline

Sadler, D. R. (1989). Formative assessment and the design of instructional systems. *Instructional Science*, *18*(2), 119–144. doi:10.1007/BF00117714

Saldaña, J. (2013). The coding manual for qualitative researchers. *Sage (Atlanta, Ga.)*.

Saldaña, J. (2016). *The coding manual for qualitative researchers.* Sage Publications Inc.

Salinas, T. M. (2004). Effects of reflective notebooks on perceptions of learning and mathematics anxiety. *PRIMUS (Terre Haute, Ind.)*, *14*(4), 315–327. doi:10.1080/10511970408984096

Sanders, C. V. (2009). Exploring and writing geometry. *Mathematics Teacher*, *102*(6), 432–439. doi:10.5951/MT.102.6.0432

Santagata, R., König, J., Scheiner, T., Nguyen, H., Adleff, A. K., Yang, X., & Kaiser, G. (2021). Mathematics teacher learning to notice: A systematic review of studies of video-based programs. *ZDM Mathematics Education*, *53*(1), 119–134. doi:10.100711858-020-01216-z

Santos, L., & Semana, S. (2015). Developing mathematics written communication through expository writing supported by assessment strategies. *Educational Studies in Mathematics*, *88*(1), 65–87. doi:10.100710649-014-9557-z

Savignon, S. (1991). Communicative language teaching. *TESOL Quarterly*, *25*(2), 261–277. doi:10.2307/3587463

Scheiner, T. (2016). Teacher noticing: Enlightening or blinding? *ZDM Mathematics Education*, *48*(1-2), 227–238. doi:10.100711858-016-0771-2

Scheiner, T. (2021). Towards a more comprehensive model of teacher noticing. *ZDM Mathematics Education*, *53*(1), 85–94. doi:10.100711858-020-01202-5

Schoenfeld, A. (2009). The soul of mathematics (Series editor's forward). In D.A. Stylianou, M.L. Blanton, & E.J. Knuth's (Eds.) Teaching and learning proof across the grades: A K-16 perspective (pp. xii-xvi). Routledge.

Schoenfeld, A. H. (1987). *What's all the fuss about metacognition.* Cognitive Science and Mathematics Education.

Selmer, S., Lindstrom, D., & Lampen, E. (2022). A Case Study of Prospective Teachers Engaged in Professional Noticing of their Students' Mathematical Thinking. *Education Sciences*, *12*(10), 656. doi:10.3390/educsci12100656

Seo, B. (2016). Written discourse analysis: Looking at adolescent mathematical writing. *SAGE Research Methods Cases*, (Part 2), 1–21.

Seto, M., & Meel, E. D. (2006). Writing in mathematics: Making it work. *PRIMUS: Problems, Resources, and Issues in Mathematics Undergraduate Studies, 16*(3), 204–232. doi:10.1080/10511970608984147

Sezer, R. (2010). Effects of reflective writing in mathematics methods courses on pre-service teachers' perceptions. *Eurasian Journal of Educational Research, 40*(1), 188–206.

Sfard, A. (2007). When the rules of discourse change, but nobody tells you: Making sense of mathematics learning from a commognitive standpoint. *Journal of the Learning Sciences, 16*(4), 565–613. doi:10.1080/10508400701525253

Shanahan, C., & Shanahan, T. (2018). Disciplinary literacy. In D. Lapp & D. Fisher (Eds.), *Handbook of research on teaching the English language arts* (pp. 281–308). Routledge.

Shanahan, T., & Shanahan, C. (2008). Teaching disciplinary literacy to adolescents: Rethinking content-area literacy. *Harvard Educational Review, 78*(1), 40–59. doi:10.17763/haer.78.1.v62444321p602101

Shanahan, T., & Shanahan, C. (2012). What is disciplinary literacy and why does it matter? *Topics in Language Disorders, 32*(1), 7–18. doi:10.1097/TLD.0b013e318244557a

Sherin, M. G. (2007). The development of teachers' professional vision in video clubs. In R. Goldman, R. Pea, B. Barron, & S. Derry (Eds.), *Video research in the learning sciences* (pp. 383–395). Routledge.

Sherin, M. G., Jacobs, V. R., & Philipp, R. A. (2011). Situation awareness in teaching: What educators can learn from video-based research in other fields. In M. G. Sherin, V. R. Jacobs, & R. A. Philipp (Eds.), *Mathematics teacher noticing: Seeing through teachers' eyes* (pp. 81–95). Routledge. doi:10.4324/9780203832714

Sherin, M. G., Linsenmeier, K. A., & van Es, E. A. (2009). Selecting video clips to promote mathematics teachers' discussion of student thinking. *Journal of Teacher Education, 60*(3), 213–230. doi:10.1177/0022487109336967

Sherin, M. G., & van Es, E. A. (2009). Effects of video club participation on teachers' professional vision. *Journal of Teacher Education, 60*(1), 20–37. doi:10.1177/0022487108328155

Shield, M., & Galbraith, P. (1998). The analysis of student expository writing in mathematics. *Educational Studies in Mathematics, 36*(1), 29–52. doi:10.1023/A:1003109819256

Shute, V. J. (2008). Focus on formative feedback. *Review of Educational Research, 78*(1), 153–189. doi:10.3102/0034654307313795

Siebert, D., & Draper, R. J. (2008). Why content-area literacy messages do not speak to mathematics teachers: A critical content analysis. *Literacy Research and Instruction, 47*(4), 229–245. doi:10.1080/19388070802300314

Siegler, R. S., Carpenter, T., Fennel, F., Geary, D., Lewis, J., Okamoto, Y., & Wray, J. (2010). *IES practice guide: Developing effective fractions instruction for kindergarten through 8th grade* (NCEE Report 2010-4039). National Center for Education Evaluation.

Siegler, R. S., Thompson, C. A., & Schneider, M. (2011). An integrated theory of whole number and fractions development. *Cognitive Psychology, 62*(4), 273–296. doi:10.1016/j.cogpsych.2011.03.001 PMID:21569877

Siemon, D., Horne, M., Clements, D., Confrey, J., Maloney, A., Samara, J., & Watson, A. (2017). Researching and using learning progressions (trajectories) in mathematics education. In *Proceedings of the 41st Conference of the International Group for the Psychology of Mathematics Education* (Vol. 1, pp. 109–136).

Simpson, A., & Haltiwanger, L. (2017). "This is the first time I've done this": Exploring secondary prospective mathematics teachers' noticing of students' mathematical thinking. *Journal of Mathematics Teacher Education*, *20*(4), 335–355. doi:10.100710857-016-9352-0

Singer, N. R., & Zeni, J. (2004). Building bridges: Creating an online conversation community for preservice teachers. *English Education*, *37*(1), 30–49.

Sleep, L., & Boerst, T. A. (2012). Preparing beginning teachers to elicit and interpret students' mathematical thinking. *Teaching and Teacher Education*, *28*(7), 1038–1048. doi:10.1016/j.tate.2012.04.005

Smith, E. (2008). Representational thinking as a framework for introducing functions in the elementary curriculum (133-160). In J.J. Kaput, D.W. Carraher, & M.L. Blanton's (Eds.) Algebra in the early grades. Routledge.

Smith, E. P., Kelly, J., Sappington, S., Warren, K., & Jansen, A. (2023). Using judicious telling to strengthen literacy in mathematics. *Mathematics Teacher: Learning and Teaching Pre-K–12, 116*(2), 115-123.

Smith, E., Smith, R. A., Powell, S., & Hebert, M. (2021). Examining the value, importance, self-efficacy and prior experiences of mathematical writing for preservice teachers. In D. Olanoff, K. Johnson, & S. M. Spitzer (Eds.) P*roceedings of the forty-third annual meeting of the North American Chapter of the International Group for the Psychology of Mathematics*. PMENA.

Smith, E., Smith, R. A., Price, M., & Powell, S. (2023). Assessing future teachers' mathematical writing: An assessment tool and insights. In C. Cobbs and D. Kombe (Eds.) *Proceedings of the 50th Annual Conference of the Research Council on Mathematics Learning*. https://www.rcml-math.org/assets/Proceedings/RCML%202023_Final_Proceedings.pdf

Smith, E., Smith, R. A., Price, M., Powell, S., & Hebert, M. (2023). *An examination of preservice teachers' experiences with, beliefs about, and instructional practices for mathematical writing.*

Smith, E. P. (2021). *Semantically linked syntactic literacy affordances in secondary mathematics.* In D. Olanoff, K. Johnson, & S. Spitzer (Eds.), *Proceedings of the forty-third annual meeting of the North American Chapter of the International Group for the Psychology of Mathematics Education* (pp. 1469-1478). Philadelphia, PA.

Smith, E., & Smith, R. A. (2024). *Elementary preservice teachers' assessment of and instructional decisions for mathematical writing* [Report presentation]. *Annual Conference of the Association of Mathematics Teacher Educators*, Orlando, FL, United States.

Smith, J. P. (1996). Efficacy and teaching mathematics by telling: A challenge for reform. *Journal for Research in Mathematics Education*, *27*(4), 387–402. doi:10.5951/jresematheduc.27.4.0387

Smith, M. S., & Stein, M. K. (1998). Reflections on practice: Selecting and creating mathematical tasks: From research to practice. *Mathematics Teaching in the Middle School*, *3*(5), 344–350. doi:10.5951/MTMS.3.5.0344

Smith, M. S., & Stein, M. K. (2011). *5 practices for orchestrating productive mathematics discussions.* National Council of Teachers of Mathematics.

Snow, C. E. (1987). The development of definitional skill. *Journal of Child Language*, *17*(3), 697–710. doi:10.1017/S0305000900010953 PMID:2269704

Solomon, Y., & O'Neill, J. (1998). Mathematics and narrative. *Language and Education*, *12*(3), 210–221. doi:10.1080/09500789808666749

Soto-Johnson, H., & Fuller, E. (2012). Assessing proofs via oral interviews. *Investigations in Mathematics Learning*, *4*(3), 1–14. doi:10.1080/24727466.2012.11790313

Steele, D. (2007). Understanding students' problem-solving knowledge through their writing. *Mathematics Teaching in the Middle School, 13*(2), 102–109. doi:10.5951/MTMS.13.2.0102

Steen, L. A., Turner, R., & Burkhardt, H. (2007). Developing mathematical literacy. In *Modelling and applications in mathematics education* (pp. 285–294). Springer. doi:10.1007/978-0-387-29822-1_30

Steffe, L. P., Olive, J. (2010). Articulation of the reorganization hypothesis. *Children's fractional knowledge*, 49-74.

Steffe, L. P. (2012). On the construction of learning trajectories of children: The case of commensurate fractions. In *Hypothetical Learning Trajectories* (pp. 129–162). Routledge.

Steffe, L. P., Cobb, P., & von Glasersfeld, E. (1988). *Construction of arithmetical meanings and strategies*. Springer-Verlag. doi:10.1007/978-1-4612-3844-7

Stein, M. K., Engle, R. A., Smith, M. S., & Hughes, E. K. (2008). Orchestrating productive mathematical discussions: Five practices for helping teachers move beyond show and tell. *Mathematical Thinking and Learning, 10*(4), 313–340. doi:10.1080/10986060802229675

Stein, M. K., Grover, B. W., & Henningsen, M. (1996). Building student capacity for mathematical thinking and reasoning: An analysis of mathematical tasks used in reform classrooms. *American Educational Research Journal, 33*(2), 455–488. doi:10.3102/00028312033002455

Stephens, A. C., Ellis, A. B., Blanton, M., & Brizuela, B. M. (2017). Algebraic thinking in the elementary and middle grades. In J. Cai's (Ed.) Compendium for research in mathematics education. National Council of Teachers of Mathematics.

Stockero, S. L., Leatham, K. R., Van Zoest, L. R., & Peterson, B. E. (2017). Noticing distinctions among and within instances of student mathematical thinking. In E. O. Schack, M. H. Fisher, & J. A. Wilhelm (Eds.), *Teacher noticing: Bridging and broadening perspectives, contexts, and frameworks* (pp. 467–480). Springer. doi:10.1007/978-3-319-46753-5_27

Stonewater, J. K. (2002). The mathematics writer's checklist: The development of a preliminary assessment tool for writing in mathematics. *School Science and Mathematics, 102*(7), 324–334. doi:10.1111/j.1949-8594.2002.tb18216.x

Strauss, A., & Corbin, J. (1997). *Grounded theory in practice*. Sage Publications.

Stylianides, A. J. (2019). Secondary students' proof constructions in mathematics: The role of written versus oral mode of argument representation. *Review of Education, 7*(1), 156–182. doi:10.1002/rev3.3157

Stylianides, A. J., & Stylianides, G. J. (2006, July). Content knowledge for mathematics teaching: the case of reasoning and proving. In *Proceedings 30th conference of the international group for the psychology of mathematics education* (Vol. 5, pp. 201-208). IEEE.

Stylianides, G. J., & Stylianides, A. J. (2009). Facilitating the transition from empirical arguments to proof. *Journal for Research in Mathematics Education, 40*(3), 314–352. doi:10.5951/jresematheduc.40.3.0314

Stylianou, D. A., & Silver, E. A. (2004). The role of visual representations in advanced mathematical problem solving: An examination of expert-novice similarities and differences. *Mathematical Thinking and Learning, 6*(4), 353–387. doi:10.120715327833mtl0604_1

Stylianou, D., & Blanton, M. (2018). *Teaching with mathematical argument: Strategies for supporting everyday instruction*. Heinemann.

Sunseri, A. B., & Sunseri, M. A. (2019). The write aid for ELLs: The strategies bilingual student teachers use to help their ELL students write effectively. *The CATESOL Journal, 31*(1), 117–131.

Sutherland, L., Howard, S., & Markauskaite, L. (2010). Professional identity creation: Examining the development of beginning preservice teachers' understanding of their work as teachers. *Teaching and Teacher Education*, *26*(3), 455–465. doi:10.1016/j.tate.2009.06.006

Swafford, J., & Bryan, J. (2000). Instructional strategies for promoting conceptual change: Supporting middle school students. *Reading & Writing Quarterly*, *16*(2), 139–161. doi:10.1080/105735600278006

Swain, M. (2001). Integrating language and content teaching through collaborative tasks. *Canadian Modern Language Review*, *58*(1), 44–63. doi:10.3138/cmlr.58.1.44

Swain, M. (2009). Languaging, agency and collaboration in advanced second language proficiency. In H. Byrnes' (Ed.), *Advanced language learning: The contribution of Halliday and Vygotsky* (pp. 95–108). Continuum.

Swanson, H. L., & Fung, W. (2016). Working memory components and problem-solving accuracy: Are there multiple pathways? *Journal of Educational Psychology*, *108*(8), 1153–1177. doi:10.1037/edu0000116

Swanson, H. L., Kong, J. E., Moran, A. S., & Orosco, M. J. (2019). Paraphrasing interventions and problem-solving accuracy: Do generative procedures help English language learners with math difficulties? *Learning Disabilities Research & Practice*, *34*(2), 68–84. doi:10.1111/ldrp.12194

Tan, T., & Garces-Bacsal, R. M. (2013). The effect of journal writing on mathematics achievement among high-ability students in Singapore. *Gifted and Talented International*, *28*(1), 173–184. doi:10.1080/15332276.2013.11678412

Taylor, J. A., & McDonald, C. (2007). Writing in groups as a tool for non-routine problem-solving in first year university mathematics. *International Journal of Mathematical Education in Science and Technology*, *38*(5), 639–655. doi:10.1080/00207390701359396

Teledahl, A. (2017). How young students communicate their mathematical understanding. *International Journal of Mathematical Education in Science and Technology*, *48*(4), 555–572. doi:10.1080/0020739X.2016.1256447

Temple, C., & Doerr, H. M. (2018). How do teachers develop and enact a disciplinary view of literacy in mathematics? *Journal of Adolescent & Adult Literacy*, *61*(5), 483–488. doi:10.1002/jaal.664

Teuscher, D., Kulinna, P. H., & Crooker, C. (2015). Writing to learn mathematics: An update. *The Mathematics Educator*, *24*(2).

Thomas, D. P. (2019). Rapid decline and gender disparities in the NAPLAN writing data. *Australian Educational Researcher*, *47*(5), 777–796. doi:10.100713384-019-00366-8

Tomlinson, C. A., Kaplan, S. N., Renzulli, J. S., Purcell, J., Leppien, J., & Burns, D. (2002). *The parallel curriculum: A design to develop high potential and challenge high-ability learners*. Corwin Press.

Topping, K. J. (2009). Peer Assessment. *Theory into Practice*, *48*(1), 20–27. doi:10.1080/00405840802577569

Totten, S. (2005). Writing to learn for preservice teachers. *The Quarterly, 27*(2), 17-20, 28.

Toulmin, S. E. (2003). The uses of argument (updated ed.). New York, NY: Cambridge University Press.

Toulmin, S. (2003). *The uses of argument*. Cambridge University Press. doi:10.1017/CBO9780511840005

Troia, G. A. (2006). Writing instruction for students with learning disabilities. In C. A. MacArthur, S. Graham, & J. Fitzgerald (Eds.), *Handbook of writing research* (pp. 324–336). Guilford Press.

Tuttle, C. L. (2005). Writing in the mathematics classroom. In J. M. Kenney (Ed.), *Literacy strategies for improving mathematics instruction* (pp. 24–50). Association for Supervision and Curriculum Development.

United Kingdom Department for Education. (2014). *Mathematics programmes of study: Key stages 1 and 2: National curriculum in England.* UKDoE. https://www.gov.uk/government/publications/nationalcurriculum-in-england-mathematics-programmes-of-studyn

Valdés, G. (2017). Entry visa denied: The construction of symbolic language borders in educational settings. In O. García, N. Flores, & M. Spotti's. (Eds.) The Oxford handbook of language and society (pp. 321-348). Oxford University Press.

Van de Walle, J. A., Karp, K. S., & Bay-Williams, J. M. (2019). *Elementary and middle school mathematics: Teaching developmentally* (10th ed.). Pearson Education.

van den Berg, I., Admiraal, W., & Pilot, A. (2006). Design principles and outcomes of peer assessment in higher education. *Studies in Higher Education, 31*(3), 341–356. doi:10.1080/03075070600680836

van Es, E. A., & Sherin, M. G. (2002). Learning to notice: Scaffolding new teachers' interpretations of classroom interactions. *Journal of Technology and Teacher Education, 10*(4), 571–596.

van Es, E. A., & Sherin, M. G. (2010). The influence of video clubs on teachers' thinking and practice. *Journal of Mathematics Teacher Education, 13*(2), 155–176. doi:10.100710857-009-9130-3

van Es, E. A., & Sherin, M. G. (2021). Expanding on prior conceptualizations of teacher noticing. *ZDM Mathematics Education, 53*(1), 17–27. doi:10.100711858-020-01211-4

Van Garderen, D., & Montague, M. (2003). Visual-spatial representation, mathematical problem solving, and students of varying abilities. *Learning Disabilities Research & Practice, 18*(4), 246–254. doi:10.1111/1540-5826.00079

Verhavert, S., Bouwer, R., Donche, V., & Maeyer, S. D. (2019). A meta-analysis on the reliability of comparative judgement. *Assessment in Education: Principles, Policy & Practice, 26*(5), 541–562. doi:10.1080/0969594X.2019.1602027

Villeneuve, E. F., Hajovsky, D. B., Mason, B. A., & Lewno, B. M. (2019). Cognitive ability and math computation developmental relations with math problem solving: An integrated, multigroup approach. *School Psychology, 34*(1), 96–108. doi:10.1037pq0000267 PMID:29985022

Vygotsky, L. S. (1978). *Mind in society: The development of higher psychological processes.* Harvard University Press.

Wagner, D., Herbel-Eisenmann, B., & Choppin, J. M. (2012). Inherent connections between discourse and equity in mathematics classrooms. In B. Herbel-Eisenmann, J. M. Choppin, D. Wagner, & D. Pimm (Eds.), *Equity in Discourse for Mathematics Education: Theories, Practices and Policies* (pp. 1–16). Springer. doi:10.1007/978-94-007-2813-4_1

Wall, J., Selmer, S., & Bingham Brown, A. (2016). Assessing elementary prospective teachers' mathematical explanations after engagement in online mentoring modules. *Contemporary Issues in Technology & Teacher Education, 16*(3), 373–414.

Walqui, A., & Bunch, G. C. (Eds.). (2019). *Amplifying the curriculum: Designing quality learning opportunities for English learners.* Teachers College Press.

Wang, S. H., Lang, N., Bunch, G. C., Basch, S., McHugh, S. R., Huitzilopochtli, S., & Callanan, M. (July 2021). Dismantling Persistent Deficit Narratives About the Language and Literacy of Culturally and Linguistically Minoritized Children and Youth: Counter-Possibilities. *Frontiers in Education* (*Vol. 6*). Frontiers Media SA. doi:10.3389/feduc.2021.641796

Ward, R. A. (2005). Using children's literature to inspire K–8 preservice teachers' future mathematics pedagogy. *The Reading Teacher, 59*(2), 132–143. doi:10.1598/RT.59.2.3

Waywood, A. (1992). Journal writing and learning mathematics. *For the Learning of Mathematics, 12*(2), 34–43.

Webb, N. M., Franke, M. L., Ing, M., Wong, J., Fernandez, C. H., Shin, N., & Turrou, A. C. (2014). Engaging with others' mathematical ideas: Interrelationships among student participation, teachers' instructional practices, and learning. *International Journal of Educational Research, 63*, 79–93. doi:10.1016/j.ijer.2013.02.001

Weber, E., & Lockwood, E. (2014). The duality between ways of thinking and ways of understanding: Implications for learning trajectories in mathematics education. *The Journal of Mathematical Behavior, 35*, 44–57. doi:10.1016/j.jmathb.2014.05.002

Weyers, J., König, J., Santagata, R., Scheiner, T., & Kaiser, G. (2023). Measuring teacher noticing: A scoping review of standardized instruments. *Teaching and Teacher Education, 122*, 103970. doi:10.1016/j.tate.2022.103970

WIDA. (2020). *WIDA English language development standards framework, 2020 edition: Kindergarten–grade 12*. Board of Regents of the University of Wisconsin System.

Widjaja, W., Vale, C., Groves, S., & Doig, B. (2017). Teachers' professional growth through engagement with lesson study. *Journal of Mathematics Teacher Education, 20*(4), 357–383. doi:10.100710857-015-9341-8

Wilcox, B., & Monroe, E. E. (2011). Integrating writing and mathematics. *The Reading Teacher, 64*(7), 521–529. doi:10.1598/RT.64.7.6

Wilkinson, G. S., & Robertson, G. J. (2006). *Wide Range Achievement Test 4 professional manual*. Psychological Assessment Resources.

Williams, B., Onsmans, A., & Brown, T. (2010). Exploratory factor analysis: A five-step guide for novices. *Australasian Journal of Paramedicine, 8*(3), 1–13. doi:10.33151/ajp.8.3.93

Wilson, P. H., Sztajn, P., Edgington, C., & Confrey, J. (2014). Teachers' use of their mathematical knowledge for teaching in learning a mathematics learning trajectory. *Journal of Mathematics Teacher Education, 17*(1), 149–175. doi:10.100710857-013-9256-1

Witzel, B. S. (2005). Using CRA to teach algebra to students with math difficulties in inclusive settings. *Learning Disabilities (Weston, Mass.), 3*(2), 49–60.

Witzel, B. S., Ferguson, C. J., & Mink, D. V. (2012). Number sense: Strategies for helping preschool through grade 3 children develop math skills. *Young Children, 67*(3), 89–94.

Wood, T., Williams, G., & McNeal, B. (2006). Children's mathematical thinking in different classroom cultures. *Journal for Research in Mathematics Education, 37*(3), 222–252. doi:10.2307/30035059

Woodward, J., Beckmann, S., Dirscoll, M., Franke, M., Herzig, P., Jitendra, A., & Ogbuehi, P. (2012). *Improving mathematical problem solving in grades 4 through 8: A practice guide* (NCEE, 2012–4055). Washington, DC: National Center for Education Evaluation and Regional Assistance, Institute of Education Sciences, U.S. Department of Education. http://ies.ed.gov/ncee/wwc/publications_ reviews.aspx#pubsearch/

Yig, K. G. (2022). Research trends in mathematics education: A quantitative content analysis of major journals 2017-2021. *Journal of Pedagogical Research, 6*(3), 137–153.

Yin, R. K. (2018). *Case study research and applications: Design and methods*. SAGE Books.

Zahner, D., & Corter, J. E. (2010). The process of probability problem solving: Use of external visual representations. *Mathematical Thinking and Learning, 12*(2), 177–204. doi:10.1080/10986061003654240

Zarestky, J., & Bigler, M. (2021). Reflective journaling in mathematics: Insights into the development of future teachers. *New Horizons in Adult Education and Human Resource Development, 33*(3), 52–65. doi:10.1002/nha3.20334

Zhang, N., Biswas, G., McElhaney, K., Basu, S., McBride, E., & Chiu, J. L. (2020). Studying the interactions between science, engineering, and computational thinking in a learning-by-modeling environment. In I. I. Bittencourt, M. Cukurova, K. Muldner, R. Luckin, & E. Millán (Eds.), *Proceedings of the international conference on artificial intelligence in education* (pp. 598-609). Springer. 10.1007/978-3-030-52237-7_48

Zwiers, J., Dieckmann, J., Rutherford-Quach, S., Daro, V., Skarin, R., Weiss, S., & Malamut, J. (2017). *Principles for the design of mathematics curricula: Promoting language and content development.* Stanford University, UL/SCALE. https://ell.stanford.edu/content/mathematics-resources-additional-resources

About the Contributors

Madelyn W. Colonnese is an assistant professor in the Reading and Elementary Education Department at the University of North Carolina at Charlotte. She teaches undergraduate elementary education courses and graduate courses in mathematics education. Her research focuses on mathematical writing. Students are often encouraged to write in mathematics class however, this writing does not always engage students in writing to advance their mathematical understanding. As a member of the Elementary Mathematical Writing Task Force, she helped to clarify the purpose of mathematical writing, as writing to reason and to communicate mathematically. She is currently investigating ways to support teachers with this instructional practice and to engage students in mathematical writing.

Tutita M. Casa is an Associate Professor in the Neag School of Education at the University of Connecticut with expertise in elementary mathematics education. Her scholarship is focused on discourse, and she is a leading scholar in mathematical writing. She directed the Elementary Mathematical Writing Task Force, with funding from the National Science Foundation, to concretely establish this aspect of discourse in the scholarly literature. Dr. Casa is dedicated to creating and testing the impact of materials and models to advance elementary students' abilities to write, particularly to ensure those typically marginalized in mathematics classrooms have more opportunities to have a voice to reason mathematically. She also has extensive experience developing and researching the efficacy of advanced mathematics units for elementary students through Project M2: Mentoring Young Mathematicians (Grades K-2) and Project M3: Mentoring Mathematical Minds (Grades 3-5). These two federally funded endeavors produced units fostering a discourse-rich mathematics learning community, eight of which were awarded the Curriculum Studies Award by the National Association for Gifted Children. Dr. Casa also has extensively published for research and practitioner audiences, including in Educational Leadership, Gifted Child Quarterly, Journal for Research in Mathematics, Journal of Writing Research, Mathematics Teacher: Learning and Teaching PreK-12, School Science and Mathematics, Teaching Children Mathematics, and Young Children.

Fabiana Cardetti is Professor of Mathematics at the University of Connecticut. Her research focuses on ways to improve the mathematics education of students and teachers. In particular, she works on examining learning experiences that provide opportunities for each learner to authentically engage in doing mathematics. Fabiana's research work includes the design of such experiences as well as the study of their implementation, uptake, and subsequent revisions or refinement. In addition, she works collaboratively in interdisciplinary groups capitalizing on the synergy across different disciplines to investigate the enhancement of STEM education from Kindergarten to Graduate School. She currently leads or co-leads several projects supported by external funding agencies, including the National Science Foundation (NSF).

***

Mohammad S. Al-younes is a Postdoctoral Research Associate in Mathematics Education in the Department of Mathematical and Statistical Sciences at Marquette University. His research focuses on prospective teachers' mathematical reasoning in the context of problem-solving, facilitation of mathematical discourse, and game-based learning.

Tessa Arsenault is a Doctoral Candidate in Special Education at UT-Austin.

Destini Braxton is a Visiting Assistant Professor at St. Mary's College of Maryland. She served five years as a special education mathematics teacher for Richmond Public Schools. She received her Ph.D. in Educational Psychology at Virginia Commonwealth University. During her time as a special education teacher, she worked closely with students who were dually identified as English Language Learners and students with a disability. During her doctoral journey, she was a member of the Multilingual Learners in Schools lab, led by Dr. Luciana de Oliveira, and Motivation in Context Lab, led by Dr. Sharon Zumbrunn. Her research examines how motivational beliefs, student emotions, and trust impact teacher-student relationships and engagement for underrepresented students in STEM while positively addressing the overrepresentation of Black and Latinx students with learning disabilities.

James Bywater is an Assistant Professor of Learning Design and Technology at James Madison University. He holds a bachelor's degree in physics from the University of Oxford and a Ph.D. in instructional technology from the University of Virginia. His research focuses on the design and implementation of interactive technologies that support teaching and learning, with a focus on mathematics education and the use of natural language processing technologies to support equitable teaching practices. Prior to his doctoral work he was a high school mathematics teacher for ten years and taught in public schools in Chicago, Maine, and Virginia.

Jennifer Chiu is an Associate Professor of STEM education at the University of Virginia. She holds a bachelor's degree in engineering from Stanford University and a master's and Ph.D. from the University of California, Berkeley. Her research explores the design of educational technologies to support student and teacher learning within science, mathematics, engineering, and computer science domains. Prior to her doctoral work she worked at Hewlett-Packard in their mobile computing division and as a high school science and mathematics teacher. Chiu is currently a PI and Co-PI on multiple federally- and foundation-funded projects and on the editorial board of multiple journals.

Lindy Crawford, Ph.D., is the Executive Director of Lee Pesky Learning Center where she is honored to lead a caring and committed team of professionals that helps young people turn obstacles into opportunities. A native of the northwest with deep roots in Idaho, Lindy earned a Ph.D. in Special Education at the University of Oregon and most recently was the endowed chair of special education at TCU in Fort Worth, TX. Lindy began her career as a preschool teacher and director of an early childhood center in Washington state. Dr. Crawford has served the field of education for over 30 years as a teacher, leader, researcher, author, and advocate for children who learn differently. Some of her publications can be found in The Journal of Special Education, the Journal of Learning Disabilities, Remedial and Special Education, and the Bilingual Research Journal.

Michael Hebert is an Associate Professor at the University of California, Irvine and the Director of the UCI Writing Project. His research and teaching focus on writing and literacy for students with disabilities. Some of his current projects involve pandemic recovery for reading in early elementary grades, mathematical writing, and writing for students with visual impairments.

Jodie Hunter is a Professor in Mathematics Education at Massey University, New Zealand.

Jacqueline Huscroft-D'Angelo is currently a Research Scientist at the Oregon Research Institute. She also serves as adjunct faculty at the University of Nebraska-Lincoln and Omaha. Her research focuses on developing and evaluating interventions to address the educational needs of students with or at-risk of emotional/ behavioral disorders and those in residential or foster care settings. Specifically focusing on improving these students' educational outcomes and health literacy, parents' involvement in school and understanding of special education, and professional development for educators working with these students, as well as mathematical reasoning of students with learning disabilities.

Joo-Young Lee, Ph.D., is a Postdoctoral Research Associate in the College of Education and Human Development at the University of Maine. Her research primarily focuses on designing and implementing effective mathematics interventions and assessments to support students with mathematics difficulties in inclusive classrooms.

Sarah Lilly is a post-doctoral research associate in the Department of Curriculum, Instruction and Special Education at the University of Virginia. She holds a B.S. in Mathematics and English and an M.A.Ed. in Secondary Education from The College of William and Mary. Sarah completed her Ph.D. in Curriculum and Instruction at the University of Virginia in 2022. She defended her dissertation on March 14, 2022, in celebration of Pi Day. Her research centers on STEM+CS education, particularly using qualitative methods to understand the integration of mathematics and science concepts with computational modeling and engineering design practices in technology-enhanced learning environments. Prior to beginning doctoral work, she taught secondary mathematics for four years as well as created and implemented an interdisciplinary, project-based mathematics, science, and principles-of-technology curriculum for freshmen and sophomore high school students in Albemarle County, Virginia.

Marta T. Magiera is a Professor of Mathematics Education in the Department of Mathematical and Statistical Sciences at Marquette University. She is interested in the mathematical cognition of doing mathematics and conducts research in both theoretical and applied aspects of mathematics education and mathematics teacher education. Her primary work focuses on the nature of mathematical thinking, teaching, and learning and addresses practicing and pre-service mathematics teachers' development in the area of mathematical reasoning and proof, problem-solving, and mathematical modeling.

Christie Martin is an Associate professor in the Instruction and Teacher Education Department of the University of South Carolina, Columbia. Dr. Martin teaches undergraduate and graduate courses in elementary education and research. Her research focuses on content area literacy in mathematics and science.

Judit Moschkovich is Professor of Mathematics Education in the Education Department at the University of California, Santa Cruz. Her research uses socio-cultural approaches to study mathematical thinking and learning, and in particular algebraic thinking, mathematical discourse, and mathematics learners who are bilingual, learning English, and/or Latine. Her publications have examined mathematics learning, especially for bilingual and/or Latine adolescent learners, and documented the assets and strengths these students use during classroom discussions of mathematics problems. She uses these examples to contradict deficit views of these students as mathematics learners. She completed her PhD at UC Berkeley and received a Spencer Dissertation Fellowship and a NAEd Postdoctoral Fellowship. She is a Fellow of the American Educational Research Association (AERA, 2018) and received the 2019 Distinguished Scholar Award, Special Interest Group for Research in Mathematics Education (SIG-RME) in AERA.

Christopher Nazelli is an associate professor of teaching at Wayne State University and the director of curriculum and instruction for its Center for Excellence and Equity in Mathematics. His research interests include mathematics teacher learning and lesson study.

S. Asli Özgün-Koca is a professor of mathematics education at Wayne State University. She teaches secondary mathematics methods and mathematics education graduate courses in the College of Education. Her research interests include the use of technology in mathematics education, the education of mathematics teachers, and using lesson study as a tool for teacher education.

Drew Polly is a professor in the Department of Reading and Elementary Education at the University of North Carolina at Charlotte. His research agenda focuses on examining how to support the implementation of technology and standards-based pedagogies. More information can be found at: .

Sarah Powell is an associate professor of Special Education at UT-Austin.

Madeline Price is a doctoral student in Teacher and Learning at UNLV.

Sarah Quebec Fuentes is currently a Professor in mathematics education at Texas Christian University. Prior to receiving her doctoral degree, she was a middle and high school mathematics teacher for 10 years. Her research focuses on classroom discourse, instructional leadership, preservice teacher education, teacher knowledge, educative curriculum materials, teacher self-efficacy, collaboration, and developing fraction sense.

Ashley Rowe is a math interventionist at Lee Pesky Learning Center and a certified Math Coach. Her focus has been advocating for conceptual learning for all math learners, understanding learning trajectories in mathematics, equity and inclusion, and utilizing these understandings to empower all students.

Erin Smith is an assistant professor of Mathematics Education at UNLV.

Ethan P. Smith is an assistant professor of mathematics education at Washington State University Tri-Cities. His research focuses on the relationships between mathematical writing and spoken discourse in high school mathematics, as well as the aligned or divergent perspectives between teachers and their instructional coaches regarding the implementation of ambitious mathematics curriculum materials. He has previously worked as a middle school mathematics teacher and mathematics teacher educator in Mississippi. He has served as a founding member of the Mississippi Mathematics Specialists Network and worked as the Social Media Coordinator for the Mississippi Council of Teachers of Mathematics.

Alex Smith has research interests that include writing and using assessment to individualize instruction for students receiving special education services. He received his PhD in special education from the University of Missouri-Columbia and is currently an assistant professor of special education at the University of Nevada, Las Vegas.

Karen Terrell has a mission of providing high-quality education for all, and this has guided Dr. Terrell's teaching practice, service, and research, as evidenced by the varied experiences throughout her career since 2000. She has taught high-school mathematics, middle-school mathematics, and both undergraduate and graduate courses. She also has served as a Turnaround School Mathematics Coach for Boston Public Schools, and she later consulted for both charter and private schools in Massachusetts and Rhode Island as well as the Massachusetts Department of Youth Services. In all of these positions, Dr. Terrell was responsible for staying abreast of and employing best practices and innovative strategies in education, coaching teachers to ensure that students of diverse backgrounds were provided equitable access to curricula, and building relationships with other stakeholders both at the site, district, and/or program levels. She received her Doctor of Philosophy degree from Boston College in 2012 and is now Assistant Professor of Mathematics Education at Loyola University Maryland. Her specialties include mathematics education, content and language integration, and assessment. Her research interests include these topics as well as genre pedagogy and technology integration for diverse learners.

Deborah Zopf is a recently retired mathematics instructor from Henry Ford College. She teaches the elementary education mathematics content, developmental, and STEM mathematics courses. Her research interests include strategies for improving student success, use of technology in mathematics education, and the education of mathematics teachers.

Index

A

Abductive Reasoning 302, 315
Algorithmic Reasoning Strategy 216, 237
argumentative math writing 16
Argumentative MW 18, 21, 26-29, 34, 243
Argumentative Writing 5, 14, 95, 116, 177, 181, 264, 269, 272, 287, 295-296, 298, 300, 304, 309, 329, 332-333, 335-337, 341
artifacts 97, 102, 110-111, 115-116, 119-120, 126-127, 130-131, 157, 318
assessment 2, 19-24, 29, 32-34, 38, 42-44, 52, 59, 99, 101, 113, 129, 136-139, 142-147, 149-155, 157, 159, 167-168, 175, 177, 181, 186, 188-189, 194-195, 206, 208-209, 228, 231, 236, 244, 246, 260-262, 285, 295, 304, 312, 319-320, 324, 329, 341-342

B

Bilingual 285, 292, 302, 310, 326

C

Casa 1-6, 10-11, 13, 17, 23, 30, 32, 34, 36, 57, 60, 94-97, 109, 111-113, 116, 129, 132, 153, 155, 158-159, 161, 163-165, 169, 171-172, 177, 190, 194, 196-197, 206, 208, 212, 232, 239, 242-243, 254, 256, 258, 261, 269, 284, 292, 295, 310, 313, 328, 338
Classroom Practice 32, 108, 135, 149-151, 155, 233, 308
Collaboration 66, 78, 87, 89-90, 111, 156, 168, 197, 231, 308, 314, 321, 324
College 1, 20, 22, 29, 31, 57, 59-63, 67, 71-72, 79, 82, 113, 132, 191, 193, 311-314, 316, 325
Common Core 12, 18, 32, 63, 91, 93, 113, 130, 155, 161, 170, 172, 176-177, 190-191, 207, 213, 235, 237, 239, 261, 272, 284, 287-288, 292-293, 312, 316, 323, 326, 329, 338
Communication 1, 12, 17, 21-22, 24, 30, 32, 34-35, 37, 39, 42-44, 52, 55-56, 59-60, 62-66, 77, 80, 85, 87, 89, 91, 93-94, 97, 99, 109-110, 131, 189, 191, 194, 207, 210, 213, 226, 238-242, 256, 258-259, 262, 265-266, 271-272, 284, 295, 298, 312, 316-319, 323, 326, 328, 332, 337
Comparative Judgment 135-139, 141-145, 147-152
Computer-Mediated 62, 66, 91
Conceptual Understanding 12, 21-22, 99, 117, 143, 146-147, 150, 153, 161, 176, 188, 204, 214-215, 222, 226, 229, 231, 237, 243, 255, 266-268, 270-271, 275, 283, 287, 331-332
Confirmatory Factor Analysis 238, 248-249, 258
Constructs 238, 240-241, 244, 246, 254, 256-257, 259-260, 332
Contextual Problem Solving 35
Contextual Problems 35-36, 39-40, 55-56, 60
creative math writing 16
Creative MW 18, 29, 34

D

Deductive Reasoning 315
Department for Education South Australia 155, 169
Dialogic Instruction 93-96, 111
disabilities 30-31, 33-34, 233-235, 256, 258-262, 316-318, 325, 338-342
Disciplinary Literacy 57, 95, 114, 171, 191-192, 210-211, 213, 216, 232-234, 236-237
Disciplinary Writing 209-211, 213, 215, 226, 228, 231-232, 237
Disciplinary Writing in Mathematics 209, 211, 213, 226, 228, 232, 237
Discourse 1-3, 6, 9-10, 12-13, 34-35, 37, 60, 93-97, 99-100, 102, 108-109, 111-113, 130, 155, 157, 159, 169-170, 174, 189-191, 193-194, 196, 235, 256, 265-266, 268-269, 273, 282, 284-287, 310, 312

E

Education 1-2, 9-13, 17, 19-20, 22-23, 29-33, 35-36, 39, 57-60, 62, 66, 68, 79, 90-91, 93-94, 112-114, 116, 128-134, 136, 139, 150-172, 190-193, 205-208, 210, 216, 226, 232-233, 235-236, 239, 244, 246, 248, 258-261, 265, 284-286, 293, 295, 309-314, 317-318, 324-325, 336, 338-342
Elementary 1-4, 9-12, 14, 23, 25, 30, 32-36, 39, 55-58, 62-63, 67-68, 89-90, 95, 97, 112, 121, 129-131, 135, 137-139, 142, 150, 152, 156, 159-162, 169, 177, 190, 192-196, 205-206, 208-210, 214, 228, 232-233, 236, 256, 258, 260-261, 284, 293, 310, 312-314, 328, 331, 338-339, 341
Elementary Mathematics 1, 23, 95, 142, 150, 338
Elementary Preservice Teachers 32, 35
Elementary School 10-12, 14, 32, 34-36, 55-56, 58, 62-63, 68, 89, 129, 135, 138, 150, 152, 196, 208, 210, 233, 256, 258, 260-261, 312, 339
Elementary Students 1-2, 4, 11, 23, 33, 89, 131, 137, 161-162, 193, 195, 205, 341
Elizabeth van Es 93
engaged teacher noticing 115, 118-120
equity 13, 30, 66, 158, 168, 174-175, 178, 184-186, 188-190, 266-268, 271, 285, 297, 310
explanatory math writing 16
Explanatory Mathematical Writing 35, 39, 43, 56, 115-120, 123, 126, 128, 204, 238
Explanatory MW 18-21, 26, 28, 34, 239, 243, 254
Explanatory Writing 1, 14, 35-43, 55-57, 60, 95, 110, 116, 120-121, 126-127, 161, 193, 199, 236, 239, 243, 256-257, 269, 272, 274, 287, 292, 295, 298, 301, 303, 309
Explanatory Writing in Mathematics 35, 60, 236, 239, 256-257
Exploratory Factor Analysis 238, 248, 262
exploratory math writing 16
Exploratory MW 17, 22, 29, 34
Exploratory Writing 1, 5, 14, 17, 57, 107-108, 116, 287, 309

F

Formative Assessment 42, 129, 159, 175, 181, 186, 188, 195, 206, 208-209, 228, 285, 304, 312
Fraction Magnitude 209, 215-217, 219
Fraction Sense 215-216, 222, 226, 228, 231, 233, 237
Fractions 75, 86, 136-137, 193, 195, 197-202, 204, 208, 214-216, 222, 226, 231-237, 277, 329-330, 332-336, 338, 340
Framework 38-41, 60, 111, 130, 166, 168, 172, 174-179, 181-182, 184-186, 188-189, 191, 209-210, 215-218, 226, 228, 231-232, 237, 264-265, 268-270, 273-275, 282-284, 295, 308, 310, 313, 326, 329, 332-333

G

Genre 17, 32, 155-168, 170-172, 235, 323, 328-329, 331, 333, 336
Genre map 156-168, 172
Gradual Release of Responsibility 155, 166

H

higher education 150, 152-154

I

inclusive 128, 131, 158, 262, 316-318, 323, 326, 339
Inductive Reasoning 315
Informative Writing 4, 14, 178
Intervention 18, 22, 31, 33-34, 193-197, 199-205, 233, 259, 261, 337, 339-341

J

Justifying 44, 54, 98, 105, 109, 139, 164-165, 213, 264, 268, 270, 272, 282, 287, 291, 293, 295, 297-298, 304, 308, 311, 315, 332, 334

L

Language 7-8, 13, 22, 24, 27, 30-33, 39, 52, 55, 57, 60, 63, 89, 94-95, 98, 100, 112-113, 120, 128, 132, 143, 155-158, 161, 163-165, 167-172, 176, 180, 184, 189, 191, 211-213, 216, 231, 233-239, 241-242, 244, 248, 255, 258-259, 261-262, 264, 266-278, 281-287, 291-295, 298, 300-301, 303-304, 308-318, 320, 323, 326, 328-330, 333-334, 336-338
Learning 2-4, 6-13, 17, 19-20, 22-24, 29-39, 55-59, 63, 65-66, 68-69, 73, 76, 78-79, 83, 87-91, 94-102, 109, 111-119, 123-124, 127-133, 135-138, 141-143, 145, 148-159, 166-172, 174-177, 180-181, 184, 189-191, 193-197, 199-200, 206-210, 213-214, 216, 228, 231, 233-236, 239, 258-262, 264-268, 270-276, 282-287, 290-292, 294-295, 298, 300, 304, 308-314, 316-321, 323-325, 328-342

M

Macrogenre 155, 159-160, 164, 172
Math Practices (MPs) in "Common Core" 287
Math Teacher Educator 34
Math Writing 16, 30, 34, 129, 206, 258
Mathematical Argument 18, 164-165, 168, 196, 198, 204-205, 291, 295-298, 309, 313
Mathematical Conference 291
Mathematical Discourse 9, 12, 35, 37, 60, 113, 170, 194, 235, 256, 268-269, 273, 282, 285-287, 312
mathematical literacy 174, 189, 192, 234, 318, 326, 340
Mathematical Practice 21, 63, 130, 163, 177, 213, 272-274, 276, 278, 282, 284, 287-288, 315-318, 323, 328
Mathematical Practices 21, 149-150, 158, 176, 264-276, 281-283, 285, 287, 294-295, 298, 328-329, 334
Mathematical Problem Solving and Reasoning 238, 240-242, 255
Mathematical Proficiency 147, 266, 268-273, 282, 285, 287
Mathematical Writing 1-12, 14, 17, 24, 30-39, 41, 43, 55-60, 62-63, 65, 69, 71, 78-79, 82, 85-90, 93-97, 100-102, 108-112, 115-129, 135-143, 145-151, 153, 155-159, 163, 167-169, 172, 174-177, 189-191, 193-199, 201, 204-206, 208, 210-212, 232, 235, 238-240, 244-245, 258-260, 264-268, 270, 272-274, 276, 278, 282-284, 287-288, 291-292, 294-296, 298, 300, 302, 308-310, 319
Mathematically Creative Writing 5, 11, 14, 116, 287
Mathematics 1-7, 9-14, 16-17, 19-25, 27-40, 55-60, 62-65, 67-70, 72-75, 83, 85-91, 93-102, 105, 109-116, 118, 120-121, 123-124, 128-139, 142-143, 146-172, 174-181, 184, 186, 188-199, 205-214, 217, 226, 228, 231-246, 248-262, 264-276, 278, 282-287, 291-294, 296-298, 301, 308-315, 317-321, 323-326, 328-342
mathematics teacher educators 16, 32, 35, 37, 56, 60, 91, 158, 168, 195, 286, 325
Metacognitive 3, 34, 37, 99, 117, 172, 193-194, 207, 340
Metalanguage 172
Miriam Sherin 93

N

Number and Operations 149, 209, 214-215, 228, 231-232, 237
Numerical Reasoning 209-210, 214-216, 232-233, 235, 237

P

Practices 2-3, 7, 10, 12-13, 17-22, 25, 27, 29, 32-33, 39-40, 60, 63, 68, 87, 90-91, 95-96, 102, 109, 111, 114, 116, 124, 128-129, 131, 143, 149-150, 152, 154-159, 170, 174, 176-177, 184, 191, 195, 207, 210, 228, 234, 239, 256, 258, 261, 264-276, 282-283, 285-287, 293-295, 297-298, 303, 308, 312, 315-316, 326, 328-331, 333-334, 337
Pre-Service Teacher 34, 91, 115, 120
Problem Explanation 53, 60
Professional Development 9, 34, 120, 130, 135-138, 142, 148, 150-151, 153, 191, 205-206, 284, 339-341
Proof Schemes 291, 295-297, 304, 311

R

Reasoning 4-5, 8-9, 11, 13-14, 18, 21, 27-28, 31, 34-44, 46-57, 60, 64, 83, 93-94, 97, 99-101, 105-106, 109, 111, 113, 116-117, 119, 124, 130, 143, 155-157, 159-160, 163-165, 167, 169, 174-175, 194-195, 197, 199, 201-205, 207-235, 237-245, 248-257, 260, 264-272, 278, 281-283, 286-287, 292-295, 298, 302, 304, 308-309, 311, 314-316, 318, 321, 324, 328-329, 332, 334, 336-338, 340
Reliability 137, 142, 154, 216-217, 248, 334
Representation 12-13, 72-73, 77, 79, 81, 84, 86, 96, 148-149, 171-172, 197, 202-204, 212-213, 220, 222-223, 225-226, 233-234, 238-241, 245, 247-253, 255, 257-258, 260, 262, 270, 294, 302, 317, 319
responding 19, 56, 64, 77, 79, 107, 110, 115, 119-121, 123-124, 126-129, 206
Revising 22, 28, 39, 113, 178, 264-265, 267-274, 276-278, 282-286, 288, 304, 311, 331
Revision 19, 27, 79, 101, 159-160, 176, 184, 265, 277, 282, 307

S

Scaffolding 88, 100, 114, 129, 133, 152, 167, 172, 188, 303, 310, 321, 323-324
Self-regulated strategy development 328, 332, 338-340
Shaping 93, 95, 102, 105, 108-110, 112
Strategic Reasoning 35-37, 43-44, 46-47, 50, 56, 60
strategies 4, 8, 14, 22, 28-29, 38-40, 42-44, 46, 52, 54-55, 57, 59, 64, 67, 88, 94, 96, 109, 117-118, 128, 131, 138, 144-145, 152, 157, 164-166, 174-175, 177-178, 181, 186-189, 210, 214-215, 218, 223, 232-234, 236, 243, 255, 262, 268, 291,

293, 304, 313, 316, 318-319, 324-326, 331-333, 335-336, 339-340
Symbolic Language Border 309, 315

T

Teacher Learning 113, 130, 133, 136-138, 149
Teacher Noticing 93, 95, 110-111, 114-121, 123-124, 127-128, 130-131, 133-134, 314
Teaching 1-2, 9-13, 16, 19-22, 25, 28-32, 36, 56-59, 62-64, 66-68, 70, 76-77, 87-91, 94, 96-98, 100, 102, 109-112, 114-116, 118-120, 127-134, 138-139, 148-149, 152, 154-159, 165-172, 174-178, 181-182, 184, 188-192, 196, 206-208, 210, 213, 228, 231-236, 256, 259-260, 262, 266-267, 271, 273, 284-286, 293, 304, 309-314, 318, 321, 325-326, 329, 331, 334-335, 337-341
Teaching and Learning Cycle 166-167, 172
text 3, 18-19, 66, 115-116, 120-121, 123, 126-128, 159, 170-172, 177, 242, 244, 247, 267, 270, 272, 275-278, 280, 283, 296, 308, 321, 330-332, 334, 337

U

UDL 316-319, 321, 323, 326
University 1, 13, 16, 23, 35, 58-59, 62-63, 67, 71-72, 79, 83, 85, 93, 115, 135, 137, 155-156, 169, 172, 174, 190, 193, 208-209, 234, 238, 261, 264, 284, 286, 291, 311-312, 314, 316, 326, 328

V

video 23-25, 27, 58, 66, 94-95, 97, 102, 104, 110, 114-116, 118-120, 127-128, 130, 133-135

W

WIDA 316, 318, 323, 326
Writing 2-14, 16-27, 29-34, 36-43, 45, 47, 49, 52, 54-60, 62-65, 68-75, 77-79, 82-90, 93-102, 105, 107-129, 131-133, 135-143, 145-151, 153, 155-172, 174-182, 184, 186-191, 193-202, 204-216, 222, 226, 228, 231-233, 235-237, 239-262, 264-278, 282-284, 287-288, 291-292, 294-296, 298, 300-304, 308-311, 313-314, 316-325, 328-342
Writing Tasks 40, 110, 129, 139, 147, 159-164, 262, 325, 331
Writing-to-learn 3, 10, 36, 94, 109, 111, 176, 189, 195, 206, 232, 260, 310, 328, 338, 341

www.ingramcontent.com/pod-product-compliance
Lightning Source LLC
LaVergne TN
LVHW062117300926
849161LV00010B/2476
9781668465387